A Systems Biology Approach to Advancing Adverse Outcome Pathways for Risk Assessment

Natàlia Garcia-Reyero • Cheryl A. Murphy
Editors

A Systems Biology Approach to Advancing Adverse Outcome Pathways for Risk Assessment

Springer

Editors
Natàlia Garcia-Reyero
Environmental Laboratory
US Army Engineer Research &
Development Center
Vicksburg, MS, USA

Cheryl A. Murphy
Department of Fisheries and Wildlife
Michigan State University
East Lansing, MI, USA

ISBN 978-3-030-09758-5 ISBN 978-3-319-66084-4 (eBook)
DOI 10.1007/978-3-319-66084-4

Printed on acid-free paper

This Springer imprint is published by Springer Nature
The registered company is Springer International Publishing AG
The registered company address is: Gewerbestrasse 11, 6330 Cham, Switzerland

We would like to thank the authors of these chapters – without their expertise, vision and patience, this book would not have been possible.

Preface

In 2007, the National Research Council (NRC) released a document titled "Toxicity Testing in the twenty-first Century: A Vision and a Strategy", that called for a paradigm shift in toxicology testing. The NRC report advocated for a testing platform to be based on *in vitro* methods instead of whole animal testing, and that takes a pathway approach by studying perturbations of biological systems and key toxicity pathways. This approach would ideally use a combination of computational biology and a comprehensive array of high-throughput *in vitro* tests, preferably with cells and tissues. The adverse outcome pathway (AOP) framework was born out of this NRC's call for action. The concepts underlying the AOP framework are not necessarily new. Risk assessors and researchers had already adopted mode-of-action based approaches to determine mechanisms underlying adverse toxic effects, and biologists and ecologists had espoused translating stress responses across levels of biological organization for decades. However, what was new was the organizing framework and structure, the common terminology and a convergence of new tools (omics, computational, crowd-sourcing, global connectivity) that helped solidify the framework and propel it forward. Now, almost a decade after its conception, we have made great progress and the momentum is on the side of further development and advances. Currently, there is a worldwide community of scientists that contribute to the online knowledgebase, and there are regularly scheduled workshops and meetings that continue to move the science and framework forward, bringing in an increasingly broader range of expertise. Those that work on AOPs are no longer just biologists, but also include computer scientists, mathematicians, modelers, and social scientists. The framework started as an approach to collect and organize biological information with the original purpose to determine how toxic chemicals can perturb the biological pathways and affect apical endpoints relevant to individual and population risk assessment. However, because the AOP framework is chemically agnostic, it can eventually be used to determine the impacts of any stressor, and as such can potentially unite biologists that work at every level of biological organization. The goal of this book was to explore the current state of the science and regulatory aspects, but also to think a little outside the box and bring in authors that could discuss areas of research that have not been addressed fully but would be required to move the AOP framework forward. While the title of this book implies the use of systems biology approaches to advance the AOP framework, we also wanted to include chapters focusing on novel technologies or approaches to advance the understanding of potential molecular initiating events, key events or different levels of biological organization. We asked authors to discuss topics such as epigenetics, omics, genetic engineering, cell free assays, life history and adaptation, behavior and social acceptance. We also asked authors to discuss non-model species, invertebrates, plants and the potential of the zebrafish embryo. We wanted to describe novel quantitative and weight of evidence approaches that have the potential to overcome some barriers to prediction and we also wanted to reach scientists that have not been very active in this field yet. We hope that by including these topics and authors in this collection that this helps to advance the AOP framework by connecting to a broader range of scientific expertise and by embracing new areas of research.

Vicksburg, MS, USA Natàlia Garcia-Reyero
East Lansing, MI, USA Cheryl A. Murphy

Contents

List of Contributors

Philipp Antczak Institute for Integrative Biology, University of Liverpool, Liverpool, UK

Adeline Arini Faculty of Agricultural and Environmental Sciences, McGill University, Montreal, QC, Canada

Brandon M. Armstrong Department of Fisheries and Wildlife, Michigan State University, East Lansing, MI, USA

Niladri Basu Faculty of Agricultural and Environmental Sciences, McGill University, Montreal, QC, Canada

Department of Environmental Health Sciences, University of Michigan School of Public Health, Ann Arbor, MI, USA

Shannon Bell National Institute of Environmental Health Sciences, Research Triangle Park, NC, USA

Erica Brockmeier University of Liverpool, Institute of Integrative Biology, Liverpool, UK

Patience Browne US Environmental Protection Agency, Office of Science Coordination and Policy, Washington, DC, USA

Emma Butler Safety & Environmental Assurance Centre, Unilever, Colworth Science Park, Sharnbrook, Bedfordshire, UK

Michael J. Carvan III School of Freshwater Sciences, University of Wisconsin – Milwaukee, Milwaukee, WI, USA

Warren M. Casey National Toxicology Program Interagency Center for the Evaluation of Alternative Toxicological Methods, Research Triangle Park, NC, USA

John Colbourne School of Biosciences, University of Birmingham, Birmingham, UK

Diana Coman Schmid Eawag, Swiss Federal Institute for Aquatic Science and Technology, Department of Environmental Toxicology, Dübendorf, Switzerland

David J. Dix US Environmental Protection Agency, Office of Science Coordination and Policy, Washington, DC, USA

Stephen Edwards National Health and Environmental Effects Research Laboratory, United States Environmental Protection Agency, Research Triangle Park, NC, USA

Kevin C. Elliott Lyman Briggs College, Department of Fisheries and Wildlife, Department of Philosophy, Michigan State University, East Lansing, MI, USA

Suzanne Fitzpatrick Center for Food Safety and Applied Nutrition, Food and Drug Administration, College Park, MD, USA

Christy M. Foran Environmental Laboratory, U.S. Army Engineer Research and Development Center, Concord, MA, USA

Natàlia Garcia-Reyero Environmental Laboratory, US Army Engineer Research & Development Center, Vicksburg, MS, USA

Andre Gergs gaiac — Research Institute for Ecosystem Analysis and Assessment, Aachen, Germany

Steve Gutsell Safety & Environmental Assurance Centre, Unilever, Colworth Science Park, Sharnbrook, Bedfordshire, UK

Markus Hecker School of the Environment & Sustainability and Toxicology Centre, University of Saskatchewan, Saskatoon, SK, Canada

Geoff Hodges Safety & Environmental Assurance Centre, Unilever, Colworth Science Park, Sharnbrook, Bedfordshire, UK

Dries Knapen Zebrafishlab, Veterinary Physiology and Biochemistry, Department of Veterinary Sciences, University of Antwerp, Wilrijk, Belgium

Jeremy K. Larson School of Freshwater Sciences, University of Wisconsin – Milwaukee, Milwaukee, WI, USA

Jeremy Leonard Oak Ridge Institute for Science and Education, Oak Ridge, TN, USA

Konstadia Lika Department of Biology, University of Crete, Voutes University Campus, Heraklion, Greece

Igor Linkov Environmental Laboratory, U.S. Army Engineer Research and Development Center, Concord, MA, USA

Teresa Mathews Environmental Sciences Division, Oak Ridge National Laboratory, Oak Ridge, TN, USA

Olivia Massey Contractor to the Environmental Laboratory, U.S. Army Engineer Research and Development Center, Concord, MA, USA

Mary T. McBride Global Market Regulations and Standards Strategy Agilent Technologies, Inc., Santa Clara, CA, USA

Bette Meek McLaughlin Centre for Population Health Risk Assessment, University of Ottawa, Ottawa, ON, Canada

Krittika Mittal Faculty of Agricultural and Environmental Sciences, McGill University, Montreal, QC, Canada

Francisco X. Mora-Zamorano School of Freshwater Sciences, University of Wisconsin – Milwaukee, Milwaukee, WI, USA

Erik B. Muller Marine Science Institute, University of California, Santa Barbara, CA, USA

Cheryl A. Murphy Department of Fisheries and Wildlife, Michigan State University, East Lansing, MI, USA

Diane Nacci US Environmental Protection Agency, Office of Research and Development, Narragansett, RI, USA

Mark Nelms Oak Ridge Institute for Science and Education, Oak Ridge, TN, USA

Roger M. Nisbet Department of Ecology, Evolution and Marine Biology, University of California, Santa Barbara, CA, USA

Noffisat Oki Oak Ridge Institute for Science and Education, Oak Ridge, TN, USA

Angela Peace Department of Mathematics and Statistics, Texas Tech University, Lubbock, TX, USA

Edward Perkins US Army Engineer Research and Development Center, Environmental Laboratory, Vicksburg, MS, USA

Smitha Pillai Department of Environmental Toxicology, Eawag, Swiss Federal Institute for Aquatic Science and Technology, Dübendorf, Switzerland

Christopher H. Remien Department of Mathematics, University of Idaho, Moscow, ID, United States

Cecilie Rendal Safety & Environmental Assurance Centre, Unilever, Colworth Science Park, Sharnbrook, Bedfordshire, UK

Taylor Rycroft Contractor to the Environmental Laboratory, U.S. Army Engineer Research and Development Center, Concord, MA, USA

Kristin Schirmer Department of Environmental Toxicology, Eawag, Swiss Federal Institute for Aquatic Science and Technology, Dübendorf, Switzerland

Irvin R. Schultz National Marine Fisheries Service, Lynker Tech & NW Fisheries Science Center, NOAA, Seattle, Washington, United States

Yu-Mei Tan National Exposure Research Laboratory, United States Environmental Protection Agency, Research Triangle Park, NC, USA

Nadine Taylor School of Biosciences, University of Birmingham, Birmingham, UK

Steven Van Cruchten Applied Veterinary Morphology, Department of Veterinary Sciences, University of Antwerp, Wilrijk, Belgium

Lucia Vergauwen Zebrafishlab, Veterinary Physiology and Biochemistry, Department of Veterinary Sciences, University of Antwerp, Wilrijk, Belgium

Christopher Warner US Army Engineer Research and Development Center, Environmental Laboratory, Vicksburg, MS, USA

Karen H. Watanabe School of Mathematical and Natural Sciences, Arizona State University, Tempe, AZ, USA

Carl Westmoreland Safety & Environmental Assurance Centre, Unilever, Colworth Science Park, Sharnbrook, Bedfordshire, UK

Catherine Willett The Humane Society of the United States, Washington, DC, USA

Kristine L. Willett University of Mississippi, University, MS, USA

Clemens Wittwehr European Commission Joint Research Centre, Ispra, Italy

Anze Zupanic Department of Environmental Toxicology, Eawag, Swiss Federal Institute for Aquatic Science and Technology, Dübendorf, Switzerland

Chapter 1
Advancing Adverse Outcome Pathways for Risk Assessment

Natàlia Garcia-Reyero and Cheryl A. Murphy

Abstract The Adverse Outcome Pathway (AOP) framework was first proposed by Ankley and colleagues back in 2010 (Ankley et al. Environ Toxicol Chem 29:730–741, 2010). AOPs organize information across biological levels of organization, with common terminology and concepts and with the goal of informing human and ecological risk assessment. Not only was the framework rapidly embraced, it also spearheaded an unprecedented amount of research both nationally and internationally dedicated to understanding, developing, and accepting AOPs. Although developing AOPs has made an impressive start, there are still areas of research that need to be focused on. Many uncertainties remain in the use and acceptance of AOPs for regulatory purposes and this book explores the advancement of AOPs for risk assessment by focusing on different aspects of AOP development such as incorporating behavior, non-model species, invertebrates, plants, synthetic biology and epigenetics. Novel methods for developing predictive tools via quantitative methods are explored, as well as social considerations of barriers to AOP acceptance.

1.1 Background

Risk assessment has long relied on mechanistic information for hazard prediction. Some of the earlier endeavors include dose-response modeling efforts (Clewell et al. 1995), and mode-of-action efforts such as the ones developed by the International Program on Chemical Safety (IPCS) to determine modes-of-action of pesticides and industrial chemicals of human relevance (Willett et al. 2014). Conceivably, one of the first main efforts for pathway-based approaches is the Mode of Action (MoA) framework for human health risk assessment. MoA is a series of key events (KE) along a biological pathway from the initial chemical

N. Garcia-Reyero (✉)
Environmental Laboratory, US Army Engineer Research & Development Center, Vicksburg, MS, USA
e-mail: natalia@icnanotox.org

C.A. Murphy
Department of Fisheries and Wildlife, Michigan State University, East Lansing, MI, USA

N. Garcia-Reyero, C.A. Murphy (eds.), *A Systems Biology Approach to Advancing Adverse Outcome Pathways for Risk Assessment*,
DOI 10.1007/978-3-319-66084-4_1

interaction to the toxicological outcome, with KE being defined as measurable and necessary precursors events to the adverse outcome (see Chap. 17 for more information). The National Research Council further developed this concept by envisioning a network of pathways leading to a predictive, hypothesis-driven toxicity assessment (NRC 2007). This toxicity pathway was defined as a cellular response pathway that, when sufficiently perturbed, is expected to result in adverse health effects. More recently, this concept was further characterized for both human health and ecological risk assessment as the adverse outcome pathway (AOP) (Ankley et al. 2010). An AOP was defined as a conceptual construct that portrays existing knowledge concerning the linkage between a direct molecular initiating event and an adverse outcome that is relevant to risk assessment. AOPs are modular and composed of reusable elements, key events (KEs) and key event relationships (KERs). They are considered living documents that will evolve over time as new information is available (Villeneuve et al. 2014). From the initial dose-response modeling efforts to the MoA or AOP frameworks, it is clear that these pathway-based approaches to understand and organize mechanistic information are the base of the remarkable changes in the way risk assessment is performed (reviewed in (Willett et al. 2014)). Delineating and understanding mechanisms and the physiological differences between test species and target species, are the only path forward for cross-species extrapolations, particularly for sensitive populations that are at risk of extinction. Further, understanding mechanisms allows for the development of quantitative models to aid prediction, which in turn can be used to understand multiple stressor scenarios.

1.2 AOP Development

Many challenges remain in the advancement of informative and predictive AOPs. Particularly, there is a need to establish credible links between responses at the molecular or cellular level and adverse outcomes measured at higher levels of biological organization. Therefore, computational tools and models that quantify KERs within an AOP are of special interest and large efforts are being made to develop them. There is also a need to understand how pathways differ by conditions and states such as life stages, sex, exposure, and species. In this chapter, we explore some of the main efforts being developed as well as some new potential areas of interest to AOP development (see Fig. 1.1).

1.2.1 Alternative Methods and Non-model Species for AOP Development

A very exciting aspect of AOPs is their potential to aid in the development of alternative methods and in vitro/in silico models that could lead to reducing and eventually eliminating animal testing (Garcia-Reyero 2015). Many ongoing international

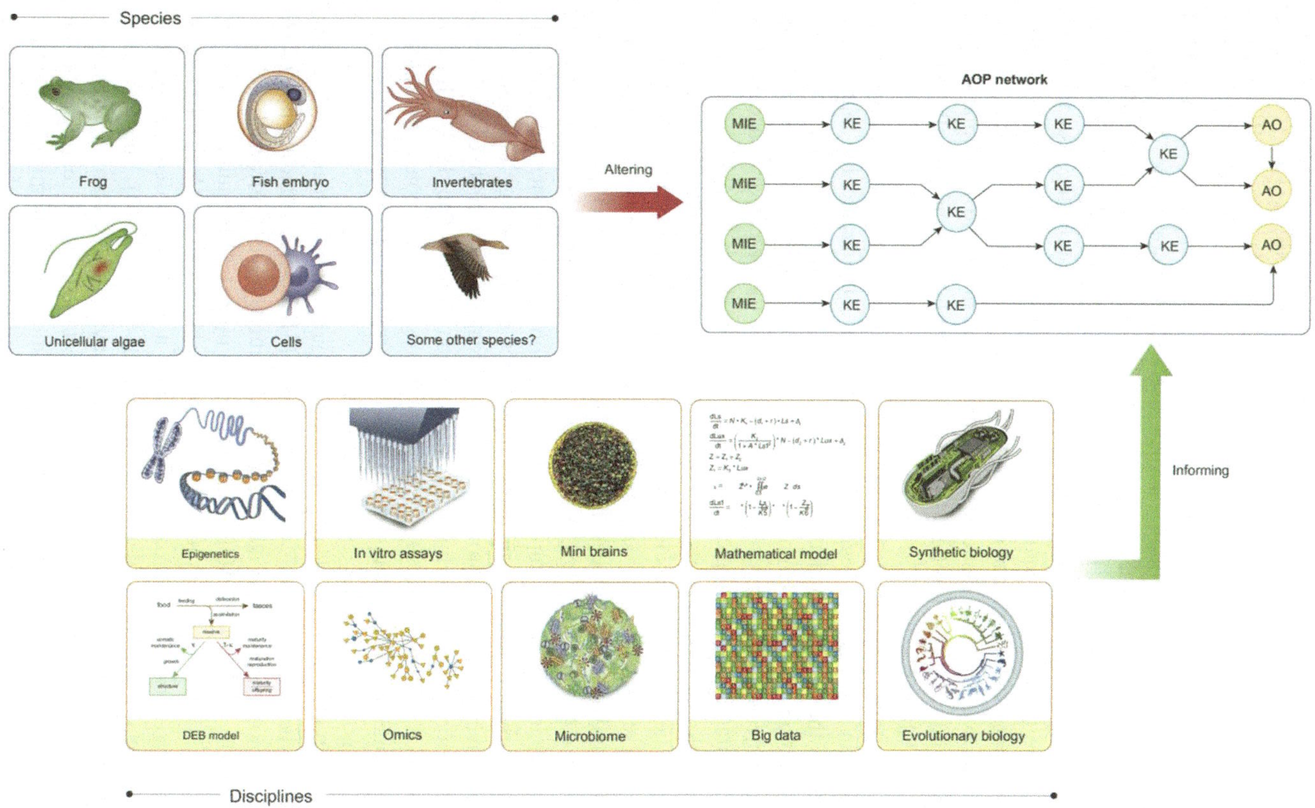

Fig. 1.1 The AOP development presents many challenges, but also many opportunities. The large amount of species with many different strategies and sensitivities calls for state-of-the-art species comparison tools and for a better understanding of the systems. Many systems, technologies, and disciplines can not only affect AOPs but also allow for a better and more quantitative description

efforts are focused on developing more predictive in vitro/in vivo methods to reduce animal testing. For instance, the National Toxicology Program Interagency Center for the Evaluation of Alternative Toxicological Methods (NICEATM) is an office within the US National Institute of Environmental health Sciences (NIEHS) that supports the development and evaluation of new, revised, and alternative methods to identify potential hazards to human health and the environment, with a focus on replacing, reducing, or refining animal use. Furthermore, the Interagency Coordinating Committee on the Validation of Alterative Methods (ICCVAM), a permanent committee of the NIEHS under NICEATM, is composed of representatives from fifteen US Federal regulatory and research agencies that require, use, generate, or disseminate toxicological and safety testing information. This committee also maintains a page listing alternative testing methods accepted by US and international regulatory authorities that can reduce animal use and improve animal welfare (https://ntp.niehs.nih.gov/pubhealth/evalatm/iccvam/acceptance-of-alternative-methods/index.html).

There are many other efforts focused on what is known as 3Rs (reduce, refine, and replace) in research and regulation with the goal of guaranteeing that animal welfare meets the highest standards and that the minimum use of animal studies are performed. For instance, the Human Toxicology Project consortium (https://human-toxicologyproject.org) is a group of stakeholders with the objective of accelerating the implementation of a biological pathway-based approach to toxicology, which will help develop better predictive tools and hasten the replacement of animal use in toxicology. The American Society for Cellular and Computational Toxicology (ASCCT) is a scientific society dedicated to the promotion of toxicology testing and research that reduces and replaces the use of animals. The John Hopkins Center for Alternatives to Animal Testing (CAAT) is part of the John Hopkins University and promotes humane science by supporting the creation, development, validation and use of alternatives to animals in research, product safety testing, and education. They even have an official journal, ALTEX, dedicated to Alternatives to Animal Experimentation, (http://altweb.jhsph.edu/altex/index.html). The PETA International Science Consortium http://www.piscltd.org.uk/) promotes non-animal research methods and coordinates the scientific and regulatory expertise of its members with the goal of replacing tests on animals.

These methods can help identify potential toxicity of chemicals or mixtures, particularly when the molecular initiating events (MIE) or KEs leading to adverse outcomes they measure have already been identified. Several efforts have been made to link in vitro tests to AOPs. For instance, Vinken and Blaauboer developed an in vitro basal cytotoxicity testing strategy for new chemicals that lack information on potential toxicity. This approach was based on a newly proposed generic AOP linking chemical insult to cell death (Vinken and Blaauboer 2017). The skin sensitization AOP is another example where in vitro assays can provide an accurate prediction of an adverse outcome. Three non-animal test methods addressing either the MIE, KE2 or KE3 are accepted as OECD (Organisation for Economic Co-operation and Development) test guidelines, therefore accelerating the development of integrated approaches for testing and assessment (reviewed in (Ezendam et al. 2016)).

Another example of high-throughput in vitro screening to detect MIEs and KEs is the US EPA Endocrine Disruptor Screening Program (EDSP, see Chap. 2). The EDSP is a regulatory program designed to screen and test chemicals for potential endocrine bioactivity and the risk of endocrine disruption in humans and wildlife. Other US federal programs such as the EPA's Toxcast program (http://www2.epa.gov/chemical-research/toxicity-forecasting) or the Tox21 collaboration (http://www.ncats.nih.gov/tox21) also use high throughput assays to screen thousands of chemicals for hundreds of molecular targets as potential MIEs and KEs.

It is worth noting that the majority of these 3Rs efforts are focused on human health-related AOPs. Nevertheless, there is increasing interest on efforts to develop them for environmental-related AOPs. Chapter 3 explores the use of cell-free assays as species agnostic, in vitro toxicity-testing tools of potential relevance to ecological risk assessment. Similarly, Schroeder and colleagues advocate the use of high throughput toxicity testing coupled with AOP knowledge for environmental monitoring and risk assessment (Schroeder et al. 2016). Arguably, the knowledge, techniques and expertise acquired from the human health arena will be also applicable to the development of environmental toxicology related AOPs.

1.2.1.1 Model and Non-model Species

Toxicity testing of chemicals is extremely costly in money, time, and animal lives. This provides limitations to fully understand the hazard potential of many compounds. While high throughput in vitro assays can rapidly provide accurate information about the mechanisms of action or MIE of thousands of chemicals (Knudsen et al. 2011; Kleinstreuer et al. 2014), they often fail to capture the potential adverse effects at the organism level due to the lack of a complete system. The fish embryo, and particularly the zebrafish (*Danio rerio*) embryo, has been proposed as a model to address these limitations (reviewed in (Planchart 2016)). While fish embryo models are of interest because of their low maintenance and husbandry costs, they also had reduced animal welfare concerns during the embryonic stages. The National Institutes of Health Office of Laboratory Animal Welfare (NIH OLAW) considers fish as live animals after hatching, which is now described to be at 72 h post fertilization (hpf) for zebrafish. It also states that zebrafish larvae under 8 days of age do not feel pain or distress. Nevertheless, new developments in the field are likely to affect the standards and IACUC policies applied to zebrafish embryo research (Moulder 2016; Bartlett and Silk 2016). (See Chap. 4 for more information on the fish embryo for AOP development).

There is also increasing interest in using invertebrate model species for the development of AOPs. Invertebrates provide many advantages over the use of vertebrate species such as generally shorter life cycles that allows for faster chronic and full cycle toxicity tests (see Chap. 5).

Current testing strategies for defining toxicity reference values in ecological risk assessment rely on extensive animal testing with selected model species. Results are then extrapolated to other species of interest. Nevertheless, this could lead to great uncertainty due to unknown species sensitivity differences. Toxicity pathway-based, in vitro, in silico, and read-across approaches have been proposed to decrease uncertainty in cross-species extrapolation for risk assessment or toxicity prediction on non-model species (see Chap. 6).

1.2.2 Novel Approaches for AOP Development

1.2.2.1 Systems Approaches

There are many different approaches being used to advance AOPs. For instance, omics technologies can provide mechanistic information on the effects of chemicals and can therefore help elucidate mechanisms of toxicity (see Chap. 9). In recent times, efforts have been focused on developing measurable linkages between KEs in order to establish quantitative AOPs (qAOPs). Different systems and modeling techniques are being considered and applied to develop measurable KERs such as flux balance analysis, reverse toxicokinetic models, or physiologically-based models (see Chaps. 13 and 14). In particular, the linkages between qAOPs and dynamic energy budgets (Chap. 14) could improve risk assessment by tapping into 30 years of established metabolic theory and to constrain qAOPs within realistic energetic demands of organismal function. Physiologically-based qAOPs that incorporate cell-free assays can, in principle, be used to interpret the impact of multiple contaminants on ecologically-relevant endpoints such as egg production (Chap. 16). Leonard and colleagues advocate the use of a tiered approach to incorporate AOPs into risk assessment, both in poor and rich data scenarios, and explore the use of systems approaches to develop AOPs (see Chap. 12). Systems approaches can also lead to the development of computationally predicted AOPs (cpAOPs). These cpAOPs can serve as scaffolds to accelerate the expert curation of AOPs and provide guidance on testing strategies, such as identifying pathway targets that lack genomic markers or high-throughput screening tests (Oki et al. 2016; Bell et al. 2016; Oki and Edwards 2016).

Other efforts involving systems approaches include the use of machine learning models to predict adverse outcomes from in vitro assays. Strickland and colleagues combined data from *in chemico* and in vitro assays as well as physicochemical properties and in silico read-across prediction of skin sensitization hazards into groups. The groups were then evaluated using two machine learning approaches, logistic regression and support vector machine. The models performed better at prediction than any of the alternative methods alone or test batteries combining data from the individual methods (Strickland et al. 2016). Models were also built to predict potency categories using four machine-learning approaches. A two-tiered strategy modeling sensitizer/non-sensitizer responses and then classifying the sensitizers as strong or weak provided the best performance (Zang et al. 2017). These results suggest that computational models using non-animal methods may provide valuable information to predict adverse outcomes.

Computational models of biological systems at different scales can therefore provide means and platforms to integrate biological understanding to facilitate inference and extrapolation. Furthermore, the systematic organization of knowledge into AOP frameworks can inform and direct design and development of predictive models to enhance the use of mechanistic and in silico data for hazard assessment (Wittwehr et al. 2016). In particular, models that can integrate suborganismal processes to predict outcomes at higher levels of biological organization, such as population or community level responses, are needed. Integration with dynamic energy budgets and individual-based models is one such approach (Chap. 14) but there are also many other ways to approach these problems. In order to advance the development of qAOPs for ecological risk assessment Wittwehr and colleagues suggest encouraging the engagement of the modeling community through crowd-sourcing challenges. An example of a successful crowd-sourcing effort is the Dialogue on Reverse Engineering Assessment and Methods (DREAM, (Stolovitzky et al. 2007)). The DREAM challenge has revolutionized the use of systems biology approaches and has pioneered the development of many of the algorithms that are now used. Furthermore, the challenge not only brings researchers together to work towards a common goal but also produces robust performance evaluation criteria (Wittwehr et al. 2016). Thus, a similar approach could be used for the advancement of qAOPs.

1.2.2.2 Behavior

Behavioral assays are widely used in toxicology research and can be powerful indicators of dysfunction because behavior integrates molecular, physiological, and environmental stimuli. However, such assays are challenging to incorporate into the AOP framework because of the difficulties in anchoring a behavioral change to molecular response (Chap. 8) and then to inform human and ecological risk assessments (Murphy et al. 2008). Recently, there has been a focus on understanding the molecular processes involved in behavioral change (e.g., Raferty and Volz 2015; Jin et al. 2016), but this area of research is in its infancy. Rather than assuming significance to any behavioral perturbation, behavioral endpoints must be categorized and validated as relevant for risk assessment for human or ecological health (Chap. 8), because then mechanistic linkages to higher levels of biological organization are possible.

1.2.2.3 Synthetic Biology and Genetic Engineering

The revolution in the field of synthetic biology and genetic modification has led to developments and advancements hard to imagine just a few years ago (see Chap. 10). Within the last 10 years, numerous tools have been developed for the genetic modification of many different species (Baltimore et al. 2015). These recent advancements include a powerful gene-editing technology known as CRISPR that has been described as *the biggest game changer to hit biology since PCR* (Ledford 2015).

While these methods hold great promise in becoming standard techniques to understand gene function in both model and non-model organisms, many are worried that this fast developing field pace leaves little time for addressing the ethical and safety issues that can raise from these types of experiments (Ledford 2015). For instance, a recent study developed a gene drive system targeting female reproduction in the malaria mosquito vector that could expedite the process to suppress mosquito populations to levels that do not support malaria transmission (Hammond et al. 2016). These gene drive experiments that could manipulate wild populations should be considered and evaluated carefully in order to assess context-dependent risks (Champer et al. 2016).

Genetic and synthetic biology approaches can also be used to elucidate MIEs, including protein binding and function. For instance, using amino acid substitutions can help understand specificity, and binding sites and could be useful for species extrapolation. Targeted knockouts can help elucidate specific pathways and KEs, and genetic devices can be used to elucidate both MIEs and KEs (see Chap. 10).

1.2.2.4 Epigenetics

The term epigenetics refers to both heritable processes independent of the DNA sequences, and transcriptional regulatory processes that influence many cellular properties (see Chap. 11). While it is now believed that an epigenetic change can be either a molecular initiating event or a key event leading to adverse outcomes, epigenetic events have hardly been considered as part of an AOP. This is not only due to the uncertainty related to how to incorporate them but also to the lack of understanding of the basic mechanisms underlying epigenetic regulation. Nevertheless, the field is rapidly advancing and there is no doubt that epigenetics will be an important part of heritable adverse effects understanding in the near future.

1.2.2.5 Metagenomics and the Microbiome

The term microbiome refers to the full collection of genes of all the microbes in a community, even though it is often used to refer to the full collection of microbes in such community, also known as microbiota. The importance of the microbiome has been gaining recognition in the last years, even being described as the "*last organ under active research*" (Baquero and Nombela 2012) or "*microbial organ*" (Spor et al. 2011). Many researchers now have shown the close relationship between the microbiome, resistance, and susceptibility to stressors and diseases.

Claus and colleagues evaluated the relationship between (human) gut bacteria and environmental pollutants in order to understand the relevance of the bacteria-toxicant relationship for the host (Claus et al. 2016). Many factors can affect the composition of the microbiome, including environmental and other stochastic factors as well as the host genetics (Spor et al. 2011; Claus et al. 2016). This is relevant because the microbiome influences many critical roles in essential host processes,

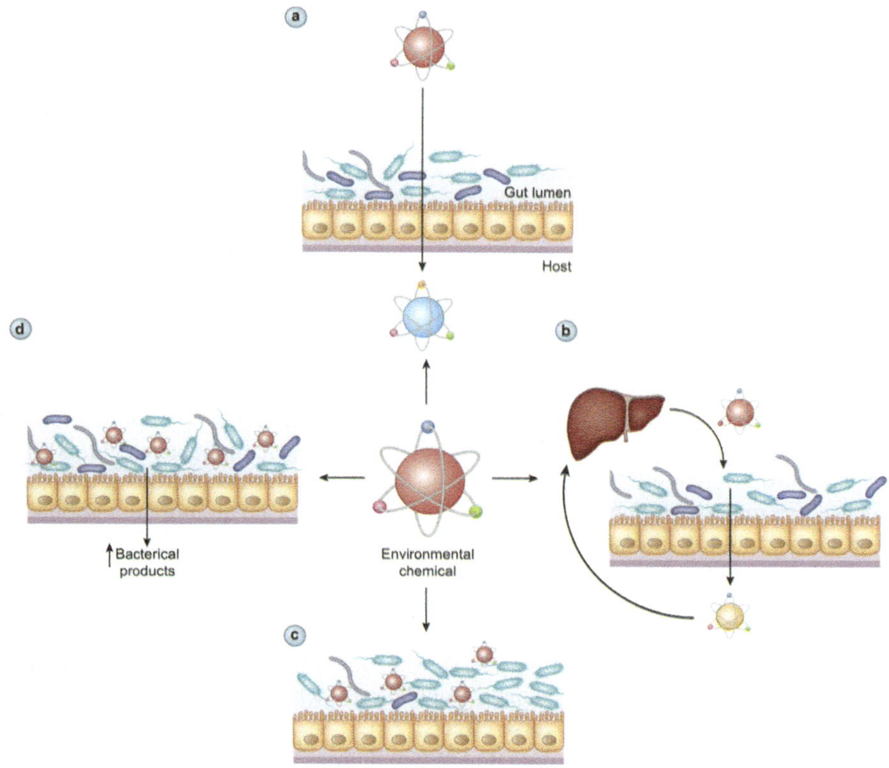

Fig. 1.2 Environmental chemicals and the gut microbiota can interact via multiple mechanisms. (**a**) Environmental chemicals may be directly metabolized by the gut microbiota. (**b**) Xenobiotics can be readily absorbed from the GI tract, then transported by the portal blood to the liver for detoxification. The liver tends to oxidize xenobiotics, forming conjugates with glucuronic acid, sulfate, or glutathione that can be excreted in the bile and enter the intestine where microbiota metabolism can take place. The GI microbiota generally deconjugates and reduces the hepatic xenobiotic metabolites, resulting in the formation of non-polar molecules of lower molecular weight, which are readily reabsorbed. Microbiota-mediated deconjugation of metabolites previously conjugated by the liver may regenerate the original xenobiotic or form new toxic metabolites. (**c**) Environmental chemicals can interfere with the composition of microbiota. (**d**) Pollutants can also change the metabolic activity of the microbiota (Adapted from Claus et al. (2016))

such as digestion, immunity, epithelial development, or disease outbreak in humans and other vertebrates including fish (Nayak 2010; Giatsis et al. 2015). Human gut microbiomes have the ability to metabolize chemicals and can be classified broadly within five different core enzymatic families (azoreductases, nitroreductases, β-glucuronidases, sulfatases and β-lyases) which are involved in the metabolism of many environmental pollutants (Claus et al. 2016). It is clear that bacterial metabolism of pollutants can affect their toxicity for the host. At the same time, pollutants can alter the composition of the microbiome, which could also contribute to their toxicity (Fig. 1.2).

It is clear that the microbiome can play a role in the relative toxicity of a compound and could be considered as a potential influence on KERs and even AOP networks. While a better understanding of the microbiome influence on adverse outcomes will need more intensive research, it should certainly be considered to fully understand the toxicity of chemicals and/or their metabolites.

1.2.2.6 Genomics, Evolution and Adaptation

Ecotoxicology and the AOP framework are involved in understanding how chemicals or stressors affect individuals, populations, and ecosystems. However, concerns have often been raised by the scientific community about the oversimplification of real ecological conditions (Calow and Forbes 2003; De Schamphelaere et al. 2011). One of those oversimplifications relate to the fact that conventional AOPs are mostly focused on understanding the adverse effects of a stressor on an individual/population without taking into account genetic variability and adaptability, often using a single genotype (De Schamphelaere et al. 2011). This increases robustness and predictability of the adverse outcomes but might fail in predicting effects on evolving and adapting populations (Barata et al. 1998; Messiaen et al. 2010). Natural selection during stressor exposure might therefore be favoring more resistant genotypes that could eventually lead to adapted populations, which could have significant implications when assessing adverse effects.

Several studies illustrate the potential of populations to adapt to stressors. One of the best-known examples involves the Elizabeth River system in southeastern Virginia and its Atlantic killifish (*Fundulus heteroclitus*) populations. This aquatic system is heavily contaminated with polycyclic aromatic hydrocarbons (PAHs). While in some areas the populations were clearly impacted, some subpopulations displayed a remarkable resistance to the PAHs toxic effects on embryonic development (Di Giulio and Clark 2016). There is also evidence of an evolved tolerance to PAHs due to changes in enzymes related to oxidative phosphorylation metabolism in killifish hepatocytes (Du et al. 2015), as well as genetic differentiation at specific nucleotides in the aryl hydrocarbon receptors AHR1 and AHR2, and specific AHR2 single nucleotide polymorphisms (SNPs) associated with a PCB-resistant killifish population (Reitzel et al. 2014). Nacci and colleagues also provided genetic evidence for killifish adaptation to pollutants, therefore providing an example of contemporary evolution driven by human-mediated selection on natural populations (Nacci et al. 2016). Furthermore, a follow up study identified the AhR-based signaling pathway as a target of selection for the killifish evolutionary adaptation, also suggesting that killifish high nucleotide diversity has likely been crucial for rapid adaptation (Reid et al. 2016).

While genetic variability and adaptation of populations might be extremely hard to understand, quantify, and incorporate into the AOP framework, they certainly warrant further study, particularly when the AOP framework is considered for environmental monitoring, or susceptible and vulnerable populations and species. Mechanistic understanding underlying evolutionary theory, such as energetic tradeoffs may help formalize this endeavor (Groh et al. 2015). For example, the AOP link to dynamic energy budgets theory may provide a way to incorporate life history traits into AOPs which may facilitate cross-species extrapolations (Chap. 14).

1.3 Current International Efforts and Challenges

International efforts are ongoing to further develop the AOP framework, including a large project effort coordinated by the OECD known as the AOP knowledge base (AOP-KB; http://aopkb.org) that provides a single point of access to several modules used for AOP development, exploration and description as well as AOP repository (Fig. 1.3, Chap. 18). The AOP-KB is organized in a systematic, searchable, and transparent manner according to an established set of guidelines and principles that

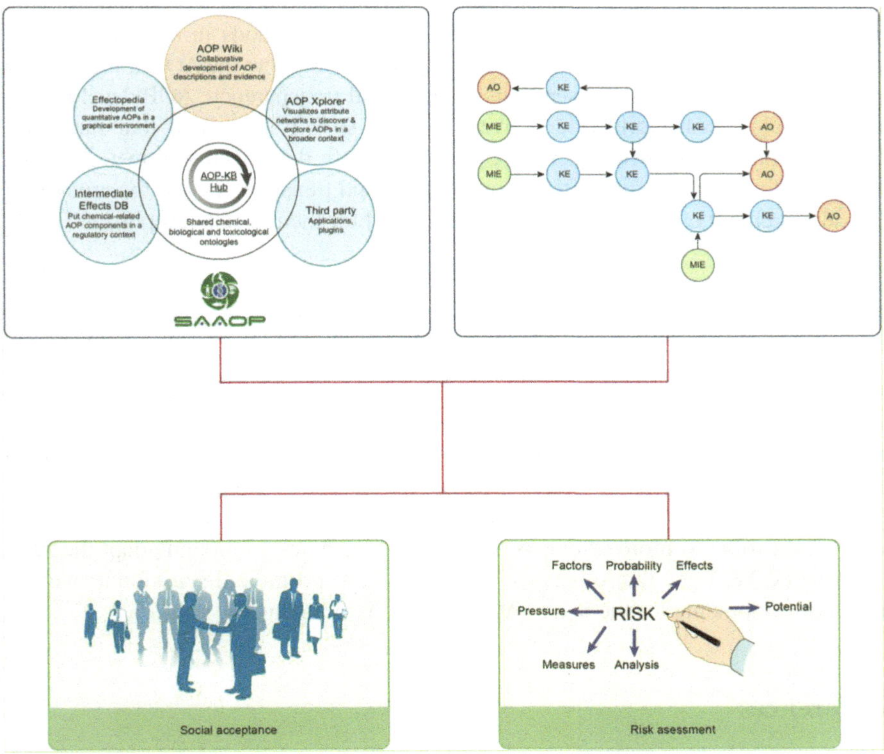

Fig. 1.3 The AOP-KB is an international effort to aid in the development and acceptance of AOPs and eventually AOP networks for both social acceptance and risk assessment

facilitates evaluation of the suitability for various regulatory applications (Villeneuve et al. 2014). The AOP Wiki (https://aopwiki.org/) is a collaborative international effort and represents a central repository for AOPs. The AOP-Xplorer module is a computational tool that enables the automated graphical representation of AOPs and AOP networks among them. The Effectopedia module is a modeling platform designed for collaborative development and utilization of AOPs. The Intermediate Effects database will host chemical-related data derived from non-apical endpoint methods and inform how individual compounds trigger MIEs and KEs.

The Society for the Advancement of AOPs (SAAOP) was created in 2014. The purpose of SAAOP is to promote and advance scientific research that fosters the development and use of adverse outcome pathways. The SAAOP maintains the AOP-Wiki under the guidance of the OECD Expert Advisory Group on Molecular Screening and Toxicogenomics (EAGMST).

In these times of social and political instability and overload of contradicting information, it is important to ensure that novel approaches to risk assessment and policy-making are transparent in order to avoid conflict and mistrust. AOPs are no exception, particularly during the developmental stage when a clear quantitative correlation between KE has not yet been established and assessment can be perceived as biased. Elliot and colleagues (see Chap. 19) recommend that AOPs be employed in "win-win" situations such as the assessment of alternative methods in order to improve acceptance, while stressing the two principles that will allow the AOP framework to move further with social consent: engagement and transparency. AOP development exponential growth worldwide is overwhelming so it is important that standards, quality controls, and strict peer-review processes are developed and met. As mentioned earlier, collaborative international efforts and transparency will be crucial for the advancement of AOPs for risk assessment and for their social acceptance.

1.4 Conclusions and Future Considerations

Regardless of the many challenges, we believe that AOPs will continue revolutionizing the (eco)toxicology and risk assessment world and will hopefully be key in the development of novel, robust, and truly predictive alternative methods for animal testing. AOPs unite biologists that work across all levels of biological organization and because of a common framework and language, we expect AOPs to continue to grow and evolve as more scientists and funding agencies adopt and adapt the AOP framework. We hope that this book will inspire and promote discussion as well as novel developments for the use of AOPs in risk assessment.

References

Ankley GT, Bennett RS, Erickson RJ et al (2010) Adverse outcome pathways: a conceptual framework to support ecotoxicology research and risk assessment. Environ Toxicol Chem 29:730–741

Baltimore D, Berg P, Botchan M et al (2015) Biotechnology. A prudent path forward for genomic engineering and germline gene modification. Science 348:36–38. doi:10.1126/science.aab1028
Baquero F, Nombela C (2012) The microbiome as a human organ. Clin Microbiol Infect 18:2–4. doi:10.1111/j.1469-0691.2012.03916.x
Barata C, Baird DJ, Markich SJ (1998) Influence of genetic and environmental factors on the tolerance of Daphnia magna Straus to essential and non-essential metals. Aquat Toxicol 42:115–137. doi:10.1016/S0166-445X(98)00039-3
Bartlett DH, Silk SB (2016) Office of Laboratory Animal Welfare Comments. Zebrafish 13:563–564. doi:10.1089/zeb.2016.1344
Bell SM, Angrish MM, Wood CE, Edwards SW (2016) Integrating publicly available data to generate computationally predicted adverse outcome pathways for fatty liver. Toxicol Sci 150:510–520. doi:10.1093/toxsci/kfw017
Calow P, Forbes VE (2003) Does ecotoxicology inform ecological risk assessment? Environ Sci Technol 37:146A–151A
Champer J, Buchman A, Akbari OS (2016) Cheating evolution: engineering gene drives to manipulate the fate of wild populations. Nat Rev Genet 17:146–159. doi:10.1038/nrg.2015.34
Claus SP, Guillou H, Ellero-Simatos S (2016) The gut microbiota: a major player in the toxicity of environmental pollutants? NPJ Biofilms Microbiomes 2:16003. doi:10.1038/npjbiofilms.2016.3
Clewell HJ, Gentry PR, Gearhart JM et al (1995) Considering pharmacokinetic and mechanistic information in cancer risk assessments for environmental contaminants: examples with vinyl chloride and trichloroethylene. Chemosphere 31:2561–2578. doi:10.1016/0045-6535(95)00124-Q
De Schamphelaere KAC, Glaholt S, Asselman J et al (2011) Will genetic adaptation of natural populations to chemical pollution result in lower or higher tolerance to future climate change? Integr Environ Assess Manag 7:141–143. doi:10.1002/ieam.149
Di Giulio RT, Clark BW (2016) Uteb-18-259. J Toxicol Environ Health, Part B 18:259–298. doi:10.1080/15320383.2015.1074841
Du X, Crawford DL, Oleksiak MF (2015) Effects of anthropogenic pollution on the oxidative phosphorylation pathway of hepatocytes from natural populations of Fundulus heteroclitus. Aquat Toxicol 165:231–240. doi:10.1016/j.aquatox.2015.06.009
Ezendam J, Braakhuis HM, Vandebriel RJ (2016) State of the art in non-animal approaches for skin sensitization testing: from individual test methods towards testing strategies. Arch Toxicol 90:2861–2883. doi:10.1007/s00204-016-1842-4
Garcia-Reyero N (2015) Are adverse outcome pathways here to stay? Environ Sci Technol 49:3–9. doi:10.1021/es504976d
Giatsis C, Sipkema D, Smidt H et al (2015) The impact of rearing environment on the development of gut microbiota in tilapia larvae. Sci Rep 5:18206. doi:10.1038/srep18206
Groh KJ, Carvalho RN, Chipman JK et al (2015) Development and application of the adverse outcome pathway framework for understanding and predicting chronic toxicity: I. Challenges and research needs in ecotoxicology. Chemosphere 120:778–792. doi:10.1016/j.chemosphere.2014.09.068
Hammond A, Galizi R, Kyrou K et al (2016) A CRISPR-Cas9 gene drive system targeting female reproduction in the malaria mosquito vector Anopheles gambiae. Nat Biotechnol 34:78–83. doi:10.1038/nbt.3439
Jin Y, Zhu Z, Wang Y, Yang E, Feng X, Fu Z (2016) The fungicide imazalil induces developmental abnormalities and alters locomotor activity during early developmental stages in zebrafish. Chemosphere 153:455–461. doi:10.1016/j.chemosphere.2016.03.085
Kleinstreuer NC, Yang J, Berg EL et al (2014) Phenotypic screening of the ToxCast chemical library to classify toxic and therapeutic mechanisms. Nat Biotechnol 32:583–591. doi:10.1038/nbt.2914
Knudsen TB, Houck KA, Sipes NS et al (2011) Activity profiles of 309 ToxCastâ„¢ chemicals evaluated across 292 biochemical targets. Toxicology 282:1–15. doi: 10.1016/j.tox.2010.12.010
Ledford H (2015) CRISPR, the disruptor. Nature 522:20–24

Messiaen M, De Schamphelaere KAC, Muyssen BTA, Janssen CR (2010) The micro-evolutionary potential of Daphnia magna population exposed to temperature and cadmium stress. Ecotoxicol Environ Saf 73:1114–1122. doi:10.1016/j.ecoenv.2010.05.006

Moulder GL (2016) Hatching fish—when should animal tracking begin? Zebrafish 13:S–150–S–152. doi:10.1089/zeb.2016.1243

Murphy CA, Rose KA, Alvarez MC, Fuiman LA (2008) Modeling larval fish behavior: scaling the sublethal effects of methylmercury to population relevant endpoints. Aquat Toxicol 86:470–484. doi:10.1016/j.aquatox.2007.12.009

Nacci D, Proestou D, Champlin D et al (2016) Genetic basis for rapidly evolved tolerance in the wild: adaptation to toxic pollutants by an estuarine fish species. Mol Ecol 25:5467–5482. doi:10.1111/mec.13848

Nayak SK (2010) Role of gastrointestinal microbiota in fish. Aquac Res 41:1553–1573. doi:10.1111/j.1365-2109.2010.02546.x

NRC (2007) Toxicity testing in the 21st century: a vision and a strategy. The National Academies Press, Washington, DC

Oki NO, Edwards SW (2016) An integrative data mining approach to identifying adverse outcome pathway signatures. Toxicology 350–352:49–61. doi:10.1016/j.tox.2016.04.004

Oki NO, Nelms MD, Bell SM et al (2016) Accelerating adverse outcome pathway development using publicly available data sources. Curr Envir Health Rep 3:53–63. doi:10.1007/s40572-016-0079-y

Planchart A (2016) Advancing toxicology research using in vivo high throughput toxicology with small fish models. ALTEX. doi:10.14573/altex.1601281

Raftery TD, Volz DC (2015) Abamectin induces rapid and reversible hypoactivity within early zebrafish embryos. Neurotoxicol Teratol 49:10–18. doi:10.1016/j.ntt.2015.02.006

Reid NM, Proestou DA, Clark BW et al (2016) The genomic landscape of rapid repeated evolutionary adaptation to toxic pollution in wild fish. Science 354:1305–1308. doi:10.1126/science.aah4993

Reitzel AM, Karchner SI, Franks DG et al (2014) Genetic variation at aryl hydrocarbon receptor (AHR) loci in populations of Atlantic killifish (Fundulus heteroclitus) inhabiting polluted and reference habitats. BMC Evol Biol 14:6. doi:10.1186/1471-2148-14-6

Schroeder AL, Ankley GT, Houck KA, Villeneuve DL (2016) Environmental surveillance and monitoring-the next frontiers for high-throughput toxicology. Environ Toxicol Chem 35:513–525. doi:10.1002/etc.3309

Spor A, Koren O, Ley R (2011) Unravelling the effects of the environment and host genotype on the gut microbiome. Nat Rev Micro 9:279–290. doi:10.1038/nrmicro2540

Stolovitzky G, Monroe D, Califano A (2007) Dialogue on reverse-engineering assessment and methods: the DREAM of high-throughput pathway inference. Ann N Y Acad Sci 1115:1–22. doi:10.1196/annals.1407.021

Strickland J, Zang Q, Paris M et al (2016) Multivariate models for prediction of human skin sensitization hazard. J Appl Toxicol 37:347–360. doi:10.1002/jat.3366

Villeneuve DL, Crump D, Garcia-Reyero N et al (2014) Adverse outcome pathway (AOP) development I: strategies and principles. Toxicol Sci 142:312–320. doi:10.1093/toxsci/kfu199

Vinken M, Blaauboer BJ (2017) In vitro testing of basal cytotoxicity: establishment of an adverse outcome pathway from chemical insult to cell death. Toxicol In Vitro 39:104–110. doi:10.1016/j.tiv.2016.12.004

Willett C, Caverly Rae J, Goyak KO et al (2014) Pathway-based toxicity: history, current approaches and liver fibrosis and steatosis as prototypes. ALTEX. doi:10.14573/altex.1401283

Wittwehr C, Aladjov H, Ankley G et al (2016) How adverse outcome pathways can aid the development and use of computational prediction models for regulatory toxicology. Toxicol Sci. doi:10.1093/toxsci/kfw207

Zang Q, Paris M, Lehmann DM et al (2017) Prediction of skin sensitization potency using machine learning approaches. J Appl Toxicol. doi:10.1002/jat.3424

Chapter 2
Use of High-Throughput and Computational Approaches for Endocrine Pathway Screening

Patience Browne, Warren M. Casey, and David J. Dix

Abstract The Endocrine Disruptor Screening Program (EDSP) screens and tests environmental chemicals for potential effects in the estrogen, androgen, and thyroid hormone pathways, and is one of the only regulatory programs designed around a mode of action framework. A variety of biological systems affect apical endpoints used in regulatory risk assessments and without mechanistic data, endocrine disruption cannot be determined. When the EDSP was developed in 1998, computational and high throughput approaches were intended to be part of the screening process, however, methods at that time were limited in availability and performance. Recently, the revolution in automated in vitro testing and computational toxicology has generated excellent tools that can be used for endocrine screening. Toxicity pathway and Adverse Outcome Pathway frameworks facilitate integrating diverse data for screening chemicals for potential endocrine activity. In addition, pathway frameworks can be used to evaluate performance of computational approaches as alternatives for low throughput and animal-based assays. Similarly, pathway frameworks may be used to evaluate the predictive performance of one or more computational models to predict downstream key events. Computational approaches such as these may provide an alternative to the EDSP Tier 1 battery and used for weight of evidence screening of a chemical's potential endocrine activity.

Disclaimer The views expressed in this chapter are those of the authors and do not necessarily reflect the views or policies of the U.S. EPA or NIH

P. Browne (✉) • D.J. Dix
US Environmental Protection Agency, Office of Science Coordination and Policy, Washington, DC, USA
e-mail: Patience.Browne@oecd.org

W.M. Casey
National Toxicology Program, Interagency Center for the Evaluation of Alternative Toxicological Methods, Research Triangle Park, NC, USA

N. Garcia-Reyero, C.A. Murphy (eds.), *A Systems Biology Approach to Advancing Adverse Outcome Pathways for Risk Assessment*,
DOI 10.1007/978-3-319-66084-4_2

2.1 The Endocrine Disruptor Screening Program

The US EPA's Endocrine Disruptor Screening Program (EDSP) is a regulatory program designed to screen and test chemicals for potential endocrine bioactivity, and the risk of endocrine disruption in humans and wildlife. The EDSP was established in 1998 in response to amendments of the Food Quality Protection Act (FQPA) and Federal Food, Drug and Cosmetic Act (FFDCA) compelling EPA screen chemicals for potential estrogenic effects in humans (FQPA 1996; SDWA 1996). In response, EPA convened the Endocrine Disruption Screening and Testing Advisory Committee comprised of regulatory, industry, and academic experts to make recommendations to the agency on development and implementation of an endocrine disruptor screening program. The committee recommended expanding the scope to include effects of chemicals on the androgen and thyroid pathways in wildlife and humans, and to do so employing a two-tiered screening and testing strategy (EDSTAC 1998). Tier 1 was developed to screen chemicals for their potential to interfere with estrogen, androgen, and thyroid signaling pathways in both sexes of several vertebrate taxa. The Tier 1 screening battery includes five in vitro assays that provide mechanistic data and six short term, in vivo assays include bioassays measuring changes in organ weights, as well as more complicated assays conducted in organisms with functional neuroendocrine axes (Fig. 2.1). The resulting battery of 11 complementary assays, when considered collectively in a weight of evidence evaluation, was expected to maximize sensitivity for identifying chemicals potential with endocrine activity while reducing the limitations of individual assays. Tier 2 was developed to characterize dose-response relationships and test for adverse effects of chemical exposures. Also developed were four longer term, definitive Tier 2 assays that test for endocrine disruption in mammals, fish, amphibians and birds, that include apical endpoints necessary for risk assessment (Fig. 2.1).

Evaluating results from multiple screening and testing assays conducted at various levels of biological organization can present a challenge for interpretation. In order to rigorously screen chemicals in the EDSP Tier 1 data were conceptually organized in "estrogenic", "anti-estrogenic", "androgenic", "anti-androgenic", and "thyroid-active" endocrine pathways (EDSTAC 1998, US EPA 2011; Fig. 2.1). The apical endpoints of Tier 2 testing assays used in the EDSP and risk assessment relate to changes in growth, development and reproduction that are regulated by endocrine and non-endocrine biological pathways. Linking upstream events and mechanistic data from EDSP Tier 1 to adverse effects in Tier 2 requires confidence in the causality of an endocrine-specific mechanism. The EDSP screening and testing strategy links mechanistic data to apical endpoints and is a unique regulatory program designed around a toxicological mode of action framework (Fig. 2.2).

The biological and chemical domains of the EDSP are determined by the FQPA and FFDCA statues under which the program was established. The EDSP is responsible for evaluating potential endocrine effects of all pesticide active and inert ingredients, and chemicals found in drinking water sources which conceivably could include almost any chemicals in commerce (US EPA 2012). The universe of

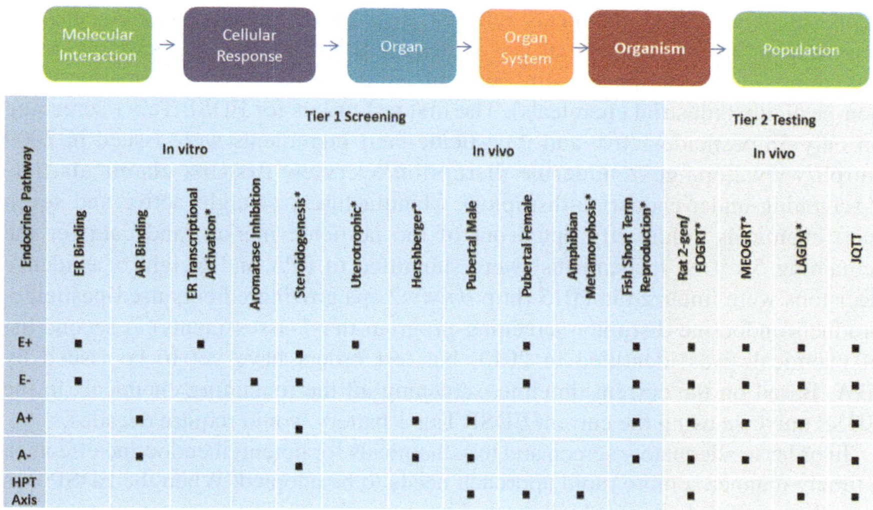

Fig. 2.1 The U.S. EPA's Endocrine Disruptor Screening Program (EDSP) screening battery of 11 Tier 1 assays and definitive Tier 2 tests to identify dose-response relationships and adverse effects. Screening and testing data are interpreted by endocrine pathway. Though overly simplistic because whole-animal in vivo studies include multiple endpoints that measure effects at different levels of biological organization, a generic AOP (top) can be overlaid on the Tier 1 screening and Tier 2 testing assays. E+ = estrogenic, E− = Anti-estrogenic, A+ = androgenic, A− = anti-androgenic, HPG axis = hypothalamic pituitary gonadal axis, HPT axis = hypothalamic pituitary thyroid axis. (*EPA guidelines harmonized with OECD. *EOGRT* extended one generation reproductive toxicity, *MEOGRT* Medaka extended one generation reproductive toxicity, *LAGDA* larval amphibian growth and development assay, *JQTT* Japanese quail toxicity test)

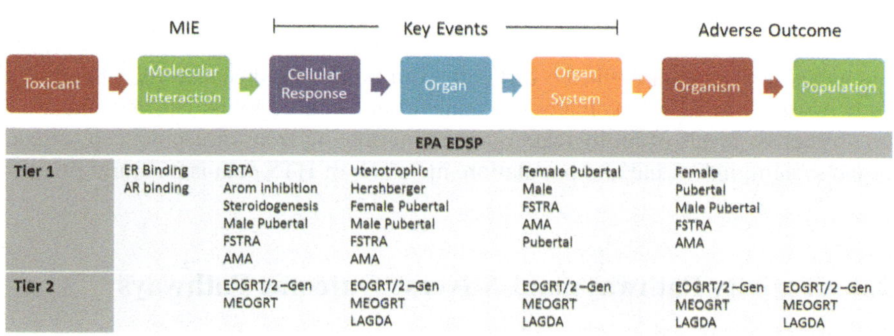

Fig. 2.2 The EPA EDSP Tier 1 and Tier 2 assays and endocrine screening and testing assays that are part of the OECD Conceptual Framework with endpoints mapped to a generic Adverse Outcome Pathway. *MIE* = Molecular Initiating Event. (*MIE* molecular initiating event, *ER* estrogen receptor, *AR* androgen receptor, *ERTA* estrogen receptor transactivation assay, *FSTRA* fish short term reproduction assay, *AMA* amphibian metamorphosis assay, *EOGRT* extended one generation reproductive toxicity, *MEOGRT* Medaka extended one generation reproductive toxicity, *LAGDA* larval amphibian growth and development assay)

approximately 10,000 chemicals relevant to the EDSP includes both data-rich chemicals subject to substantial in vivo testing prior to use (e.g., pesticide active ingredients), and data-poor chemicals with limited data or use information (e.g., non-pesticide industrial chemicals). The first test orders for EDSP Tier 1 screening on only 58 pesticide-active and 9 pesticide-inert ingredients were issued in 2009 (http://www2.epa.gov/endocrine-disruption/overview-first-list-chemicals-tier-1-screening-under-endocrine-disruptor). Manufacturers of eight active and seven inert chemicals voluntarily opted out of the pesticide market, and data for the remaining 52 'List 1 chemicals' were submitted to EPA and weight of evidence decisions were finalized in 2015 (http://www2.epa.gov/ingredients-used-pesticide-products/endocrine-disruptor-screening-program-tier-1-assessments). A second list of chemicals was identified in 2013, but test orders have yet to be issued by EPA. Based on the current timeline, screening all the remaining chemicals in the EDSP universe using the current EDSP Tier 1 battery would require decades.

In order to adequately screen and test chemicals for potential endocrine effects in a timely manner, a more rapid approach needs to be adopted. When the EDSP was initially conceived, in vitro high throughput screening (HTS) assays were proposed as an initial step to provide mechanistic data and prioritize chemicals for further in vivo screening. However, at the time, the availability and reliability of commercially available assays were limited. In subsequent years, the technological revolution in biology has produced a number of reliable and readily available HTS tools available for toxicity testing. US Federal programs such as the Tox21 collaboration (http://www.ncats.nih.gov/tox21), and EPA's ToxCast program (http://www2.epa.gov/chemical-research/toxicity-forecasting) are now using HTS assays to screen thousands of chemicals for hundreds of molecular targets, and ToxCast and Tox21 include many HTS assays relevant to estrogen, androgen, and thyroid pathways. These HTS tools have obvious application to the EDSP program and can increase the rate of chemical screening, identifying chemicals likely to pose the greatest risk to wildlife and human health. Integrating high throughput and traditional animal-based toxicology data could be difficult to interpret, but because the underlying framework of the EDSP evaluates mechanistic and whole animal data and considers effects across levels of biological organization ranging from molecule, cell, organ, organ system, individual and population, inclusion of HTS data is a natural fit.

2.2 Toxicity Pathways and Adverse Outcome Pathways

Toxicity pathways, described in the National Resource Council report on *Toxicity testing in the twenty-first Century* (NRC 2007), are cellular response pathways that when sufficiently perturbed results in adverse health effects, but do not necessarily include a molecular initiating event (MIE) or an adverse outcome. The Adverse Outcome Pathway (AOP) framework was derived from the toxicity pathway concept and is a framework for organizing biological and toxicological knowledge (Ankley et al. 2010). There is substantial diversity in definitions of and components included in toxicity pathways (Whelan and Anderson 2013). Recent efforts have attempted to avoid similar confusion by developing of precise vocabulary and defining criteria for evaluating candidate AOPs (Villeneuve et al. 2014a). AOPs begin with a molecular initiating event and terminate with an adverse outcome, linked by a series of biologically plausible and measurable intermediate key events at increasingly complex levels of biological organization from cell to tissue, organ, and organism or population. Relationships between levels of biological organization may be causal, inferential, or putative and may be based on in vitro, in vivo or computational data. Originally developed for ecotoxicology, population-level effects were considered to be an adverse outcome (Ankley et al. 2010; Kramer et al. 2011). As the framework has been adopted for human health assessment, adversity is generally considered a detrimental effect observed in an organism (Patlewicz et al. 2015). For the purposes of this discussion, a toxicity pathway may be considered a part of a (putative) AOP (Fig. 2.3). While both toxicity pathways and AOPs represent a simplification of complex biological processes, they provide organizing frameworks to link mechanistic information to data collected over different biological scales and evaluate underlying biology knowledge (or gaps therein).

To support AOP development and foster collaboration and coordination among an international community, an AOP Wiki was developed by the US EPA, US Army Corps of Engineers, EU Joint Research Centre and other partners (https://aopkb.org/aopwiki/index.php/Main_Page). In addition to its function as an open repository of AOP information, this resource is also expected to promote collective participation of a broader scientific and regulatory community in AOP development, evaluation, exploration and application. Once an AOP is described, the supporting weight-of-evidence and strength of predictive relationships between key events and adverse outcomes can be evaluated using modified Bradford-Hill criteria to assess the strength of experimental methods and biological relevance of the observed responses (Becker et al. 2015; Vinken 2013; http://www.oecd.org/chemicalsafety/testing/adverse-outcome-pathways-molecular-screening-and-toxicogenomics.htm).

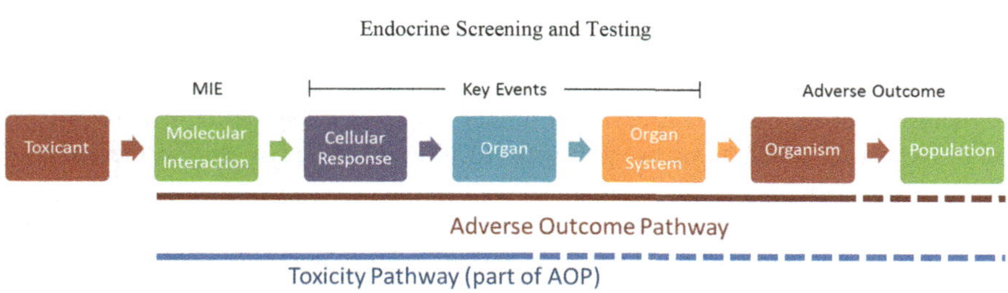

Fig. 2.3 A generic Adverse Outcome Pathway including a molecular initiating event (MIE), several key events, and terminating in an adverse outcome which is at the level of the organism in human health assessment and at the level of population for ecotoxicology. For the purposes of this discussion, a Toxicity Pathway can be considered part of an AOP that may not include an adverse outcome. EDSP Tier 1 screening includes potential molecular initiating events, but not adverse outcomes. Similarly, EDSP Tier 2 testing assays provide organismal and population level apical effects, but lack mechanistic data

The AOP concept was intended to provide information on apical endpoints considered in risk assessment and regulatory decisions, although the initial development of AOPs has focused on highly specific biological pathways, using the framework primarily to identify gaps in biological pathways and generate research hypotheses. A single MIE (e.g., ligand binding to the estrogen receptor) may be associated with many separate AOPs, and similarly, an adverse outcome (e.g., reduced fecundity) may result from the perturbation in any one of several separate pathways. Development of detailed individual AOPs may provide valuable insights into underlying toxicological and physiological processes, but such fine scale consideration of biological pathways is not generally applicable to regulatory science. Alternatively, linking multiple AOPs in an AOP network that integrates several MIEs leading to common key events and terminating in the same apical response (Knapen et al. 2015; Villeneuve et al. 2014b) has clear utility as a framework for organizing and identifying points of biological convergence common to more than one MIE. For endocrine screening, portions of a multitude of putative AOPs are assessed in the course of identifying bioactivity in relevant toxicity pathways.

2.3 Screening and Testing for Endocrine Bioactivity and Potential Risk for Disruption

Toxicity pathway and AOP concepts are a natural fit in the EDSP evaluation of a chemical's potential endocrine effects. The AOP conceptual framework relies on defined relationships between the MIE and downstream key events, relationships that have been well established for the estrogen, androgen, and thyroid pathways and inherent in the EDSP screening and testing approach. The EDSP screening and testing integrates data collected at different levels of biological complexity and was designed around a mode of action framework (EDSTAC 1998; US EPA 2011). Endocrine perturbation, if sufficiently strong, may impact apical endpoints, but may be initially expressed as more subtle changes at cellular, organ, and organismal levels. These subtle effects resulting from chemical exposure may be overlooked in traditional acute and chronic toxicity studies if more fine-scale biological endpoints are not observed or apical responses may be incorrectly attributed to some other toxicity pathway in the absence of endocrine-specific mechanistic data.

The EDSP screening and testing approach assumes underlying biological links between endpoints measured in different assays and at different scales. While overly simplistic because some in vivo EDSP assays measure cellular, organ, and organismal endpoints, the putative biological relationships between endpoints in each endocrine pathway can mapped to a generic AOP (Fig. 2.1). Tier 1 screening assays represent a toxicity pathway rather than a complete AOP (Figs. 2.1 and 2.3). The five in vitro screening assays are potential MIE or key events based on molecular or cellular responses. Two of the six in vivo assays (Uterotrophic and Hershberger) provide organ responses, and the four intact animal models (Male and Female

Pubertal, Fish Short Term Reproduction Assay or FSTRA, and Amphibian Metamorphosis Assay or AMA) provide data at the level of the organ system or organism, but do not include endpoints considered adverse outcomes (Figs. 2.1 and 2.3). Tier 2 assays include apical endpoints that may be altered through a variety of biological pathways such as impaired growth or reproduction, but do not include information regarding a specific mechanistic of action (Figs. 2.1 and 2.3). Together, Tier 1 and Tier 2 data can be integrated as a full AOP including both the molecular initiating event and the adverse outcome (Figs. 2.1 and 2.3). Given the mode of action framework inherent in the EDSP, inclusion of assays measuring different levels of biological complexity and pathway-based organization for interpreting data, the EDSP is excellent example of how application of AOP concepts can strengthen science for regulatory decisions.

The EDSP is now incorporating HTS data in the endocrine screening and testing framework (US EPA 2015; Browne et al. 2015). As mentioned previously, endocrine screening was always meant to include HTS data, and the recent availability of hundreds of diverse HTS assays in programs such as ToxCast and Tox21 can elucidate MIEs and the sequence of early key events for thousands of chemicals structures. In addition to providing a framework for interpreting diverse biological data, toxicity pathways or AOPs provide a context for incorporating additional data (e.g., HTS) with Tier 1 screening battery and Tier 2 assay data in order to evaluate the endocrine activity of environmental chemicals. Moreover, toxicity pathways or AOPs can provide a context for comparing and evaluating the performance of alternative methods (e.g., HTS assays).

To increase available information and reduce the number of animals used to evaluate the safety of chemicals, there is widespread interest in using computational and high throughput screening alternatives to traditional toxicological methods. When initially proposed, Ankley et al. (2010) recognized adverse outcome pathways as potential frameworks for integrating mechanistic data with conventional animal-based studies and for building predictive models. The toxicity pathway or AOP framework can be used to evaluate the performance of HTS alternatives to traditional, lower throughput in vitro assays that measure MIEs and key events, and can also help characterize the ability of in vitro HTS methods to predict effects downstream in the pathway, including in vivo responses (Fig. 2.4).

Adoption of new scientific methods requires the new method to be appropriately interrogated to establish the soundness of the data produced (i.e. validation). High throughput and ultra-high throughput assays are usually conducted in the few, suitably equipped laboratories capable of rapidly screening thousands of chemicals. Traditional inter-laboratory validation studies may take years to complete and rely on relatively few chemicals tested in multiple labs, and are both not appropriate for high throughput methods and fail to exploit the advantages of HTS. In contrast, implementing a performance-based approach allows for single lab validations by examining the performance of high throughput methods against large sets of structurally diverse reference chemicals that are active (or inactive) over a wide range of potencies. For each molecular target, candidate reference chemicals can be identified from traditional toxicological methods and may be independent of the specific

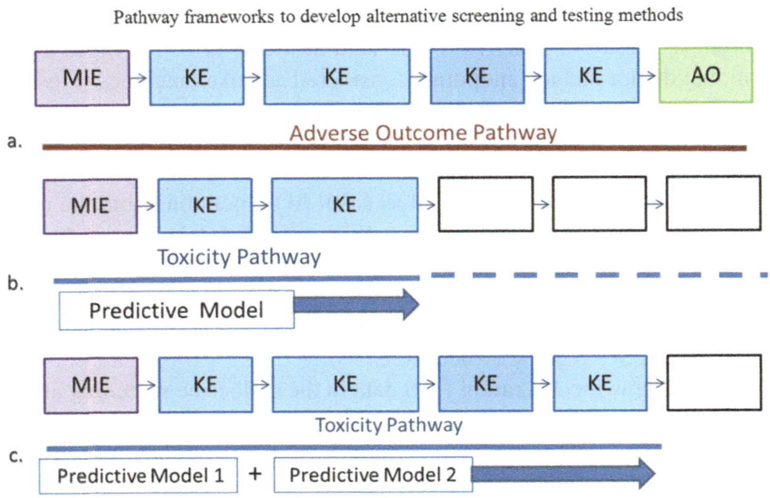

Fig. 2.4 A generic Adverse Outcome Pathway (AOP;) is shown (**a**) including a Molecular Initiating Event (MIE) is indicated in purple, Key Events (KE) indicated in blue, and Adverse Outcomes (AO) indicated in green. The pathway framework may be used to develop alternative methods and determine predictive performance. A Toxicity Pathway (**b**) may be part of an AOP and be used as an organizing frame work to determine how well predictive models predict downstream key events. Several models (**c**) may be combined to predict more complex biological outcomes and ultimately may predict the adverse outcome

assay method used to identify the chemical activity. For example, chemicals that are active and inactive in estrogen receptor (ER) signaling may be identified from ER binding, ER transactivation, cell proliferation, or ER cofactor recruitment assays. Reference chemicals active in more than one type of assay reduces inclusion of chemicals with erroneous activities due to their interaction with a particular assay technology (e.g., chemophores, cytotoxic chemicals, etc.). Extending this logic, reference chemicals identified using this approach are likely to be active across multiple levels of a toxicity pathway. A case study using the estrogen receptor agonist toxicity pathway is given below.

2.3.1 Estrogen Receptor Model

The EDSP is including HTS assay results to identify estrogen receptor agonist activity and provide mechanistic data for inclusion in an AOP/toxicity pathway context (US EPA 2015; Browne et al. 2015). Eighteen HTS assays that measure multiple points in the ER signaling pathway using a variety of technologies include high throughput analogues of Tier 1 in vitro ER assays (e.g., ER binding and ER transactivation assays). Concentration-response data from these 18 ER assays were integrated into an ER model, the output of which provides a model score of the potential agonist and antagonist activity, chemical potency, and a measure of assay-specific false positive activity of each chemical run in ToxCast (Judson et al. 2015). The redundancy of the 18 assays and inclusion of a variety of assays technologies represents a substantial benefit compare to the low throughput, animal-dependent Tier 1 EDSP in vitro assays.

The performance of the ER model was evaluated against a relatively large set of structurally diverse reference chemicals. In vitro ER reference chemicals were identified by the Interagency Coordinating Committee on the Validation of Alternative Test Methods (ICCVAM; http://ntp.niehs.nih.gov/pubhealth/evalatm/iccvam/test-method-evaluations/endocrine-disruptors/in-vitro-assay-review/brd/index.html) and OECD (2012) for the express purpose of validating novel in vitro assays. Forty ER agonist reference chemicals with reproducible in vitro assay results included 28 agonists of differing potencies indicated by a range in AC_{50} (Activity Concentration at 50% of maximum)and 12 inactive chemicals (Judson et al. 2015). The consensus list of reference chemicals were positive or negative in multiple assay types and for this reason, the results obtained were likely biologically relevant rather than artifacts of a single assay technology. The ER model predicted the activity of in vitro reference chemicals with an overall accuracy of 93% and a false negative rate of 7% (Browne et al. 2015).

In addition to evaluating the ER model as a one-for-one data alternative to the low throughput ER binding and ER transactivation in vitro assays in the existing EDSP Tier 1 battery, the ER model performance was evaluated against the results of the rodent Uterotrophic bioassay measuring in vivo ER activation driving changes in rodent uterine weight. A systematic review of Uterotrophic studies published in scientific journals was undertaken to identify studies that were methodologically consistent with the EDSP Tier 1 guideline (Kleinsteuer et al. 2015). "Guideline-like" studies were identified for 103 chemicals and study details including chemical, dosing, and uterine weight were extracted into a database (Kleinsteuer et al. 2015; http://ntp.niehs.nih.gov/pubhealth/evalatm/tox21-support/endocrine-disruptors/edhts.html). Of the 103 chemicals with guideline-like Uterotrophic studies, 43 chemicals had consistent ER agonist activities which was indicated by change in uterine weight (or lack thereof) in two or more independent guideline-like studies and were considered in vivo reference chemicals (Kleinsteuer et al. 2015). The in vivo reference chemicals were then used to evaluate the ER model predictions of the in vivo response. Again, the ER model performance was excellent against in vivo reference chemicals with an accuracy of 86% with a false negative rate of 3% (Browne et al. 2015).

Based on the performance of the ER model against the 40 in vitro reference chemicals and 43 of in vivo ER agonist reference chemicals (65 unique chemicals), EPA published a Federal Register Notice stating the intention of the agency to accept computational tools and predictive models as alternative data for the current EDSP Tier 1 ER binding, ERTA, and rodent uterotrophic screening assays (US EPA 2015). The performance-based validation approach used to evaluate the ER model predictions against both in vitro and in vivo assays relies on presumptive relationships between the MIE (i.e. ER binding), and changes at the level of the protein, cell, and organ (i.e. change in uterine weight) consistent with the organization and interpretation of the EDSP Tier 1 screening battery data. Computational methods can be examined as one-to-one alternatives for the current low throughput EDSP Tier 1 in vitro analogs, but as in the case of the ER model, when evaluated in the context of an AOP/toxicity pathway framework, the alternative method accurately predicts downstream key events in the estrogen agonist pathway (e.g., Fig. 2.4b).

2.3.2 Pathway Frameworks for Evaluating Computational Methods for EDSP Tier 1 Assays

Currently, the EDSP Tier 1 battery includes low throughput in vitro assays for androgen receptor (AR) binding, aromatase inhibition, and alteration of steroidogenesis (Fig. 2.1). The ToxCast and Tox21 programs incorporate HTS alternatives for these assays, and similarly to the ER model, the biological signaling pathway is more extensively covered by multiple HTS assays that rely on different assay technologies. The EPA intends to adopt a similar pathway-based approach for validating the one-to-one HTS alternatives for existing low throughput EDSP Tier 1 in vitro assays, and to use an AOP organizing framework to investigate the performance of the HTS assays and computational models to predict downstream in vivo endpoints, following the endocrine pathway approach for interpreting data outlined in the weight of evidence guidance (US EPA 2012). Though performance-based validation requires identifying a large, robust set of reference chemicals for each key event in the AOP currently included in the EDSP screening battery, the need to generate novel in vivo animal data for these purposes may be reduced by leveraging data in the scientific literature following the example of Kleinsteuer et al. (2015), and can further be used to populate endocrine AOPs.

In addition to examining computational alternatives for existing endpoints in the Tier 1 screening battery, the ToxCast and Tox21 programs, along with other emerging toxicological methods, provide mechanistic data that are not included in current EDSP screening. For example, several in vitro assays for potential thyroid hormone pathway MIEs are now available (https://www.ncbi.nlm.nih.gov/pcassay?term=thyroid). In addition to HTS assays for thyroid hormone receptor interactions, which are not expected to be a common mechanism of action for thyroid active environmental chemicals, other MIEs such as thyroid peroxidase (TPO) inhibition (Paul-Friedman et al. 2016), thyroid releasing hormone receptor binding, thyroid stimulating hormone receptor, and alteration of the sodium/iodide symporter (NIS; Lacotte et al. 2013) are now available. These and other in vitro assays that provide mechanistic information helpful for interpreting the endocrine toxicity of environmental chemicals were not initially available for inclusion in EDSP screening but can be easily incorporated into a weight of evidence evaluation using an AOP or toxicity pathway framework.

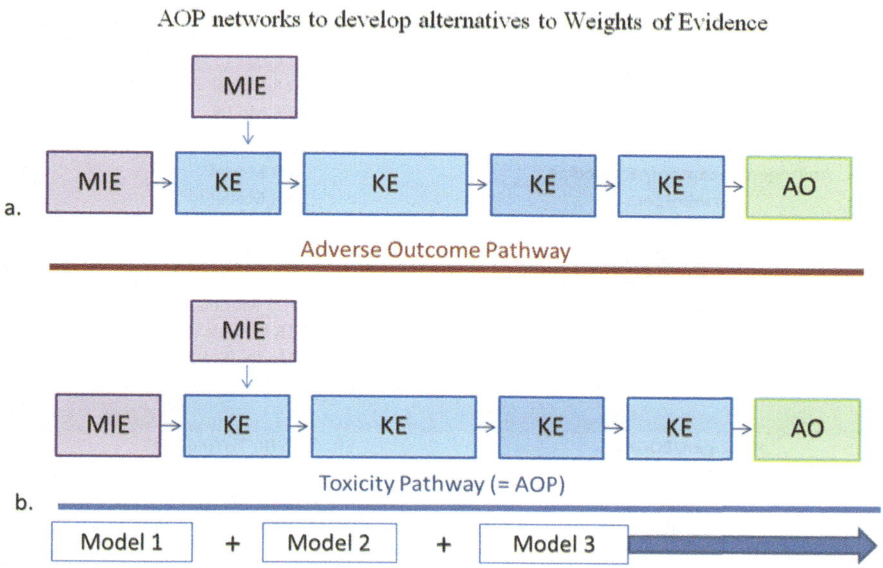

Fig. 2.5 A generic Network Adverse Outcome Pathway (AOP;) is shown (**a**) including several Molecular Initiating Events (MIE) indicated in purple, Key Events (KE) indicated in *blue*, and an Adverse Outcomes (AO) indicated in *green*. The AOP network may be used to develop alternative methods for a weight of evidence to determine a chemical's potential for endocrine bioactivity (**b**) and may be considered in lieu of EDSP Tier 1 screening data

Initial attempts to include alternative toxicological approaches in the EDSP have been limited to examining the performance of new technologies to predict an MIE and early key events immediately downstream from the MIE (Fig. 2.4b). Evaluations to date have been limited to the organ weight changes and have not included whole-animal responses. Alternative methods are not likely to replace in vivo methods on a like-for-like basis. For most chemicals, predicting whole-animal responses measured in organisms with intact neuroendocrine axes (e.g., pubertal, fish, and amphibian assays) will be more complex and may require integration of several predictive models (Fig. 2.4c), multiple MIEs or Key Events (Fig. 2.5), and multiple endocrine pathways (Fig. 2.6). Predictive models may be validated separately using the appropriate endocrine toxicity pathway, and then models can be integrated to predict complicated biological responses (Fig. 2.6). AOP concepts can continue to provide a valuable framework for organizing data and interpreting the biological basis for integrated approaches to testing and assessment used for making regulatory decisions, or identify additional data needs (Burden et al. 2015, Allen et al. 2014; Patlewicz et al. 2015; Fig. 2.6). Ultimately, in vitro tests may be implemented into integrated testing strategies and provide alternatives to conventional in vivo endocrine toxicity testing (Vinken 2013).

EDSP Tier 1 Battery of Assays	HTS Assays & Predictive Model Alternatives
Estrogen Receptor (ER) Binding	ER Model
Estrogen Receptor Transactivation (ERTA)	ER Model
Uterotrophic	ER Model
Androgen Receptor (AR) Binding	AR Model
Hershberger	AR Model
Aromatase	STR
Steroidogenesis (STR)	STR
Female Rat Pubertal	ER, STR , THY Models (Future)
Male Rat Pubertal	AR, STR , THY Models (Future)
Fish Short Term Reproduction	ER, AR, STR Models (Future)
Amphibian Metamorphosis	THY Model (Future)
EDSP Tier 2 Tests	**HTS Assays & Predictive Model Alternatives**
Rat 2-gen/EOGRT	ER, AR, STR, THY(Future)
MEOGRT	ER, AR, STR (Future)
LAGDA	THY (Future)
Quail	ER, AR, STR, THY (Future)

Fig. 2.6 EDSP Tier 1 screening battery assays and Tier 2 testing assays and the high throughput screening (HTS) assays and predictive model alternatives. In the case of whole-animal in vivo assays (e.g., Female Rat Pubertal assay), several predictive models may be needed to predict the outcome in an integrated approach to testing and assessment (IATA). *ER* estrogen receptor, *AR* androgen receptor, *STR* steroidogenesis, *THY* thyroid

2.3.3 *Pathway Frameworks for Evaluating Computational Methods for Weight of Evidence Determinations of Endocrine Activity*

Interpretation of the potential endocrine activity of a chemical screened in the EDSP Tier 1 battery is made by a weight of evidence determination, and a pathway-based interpretation of Tier 1 screening battery results (Fig. 2.1). Because the more complicated, whole-animal assays contain more than one endpoint measured at different biological levels, mapping all Tier 1 endpoints in each endocrine pathway to a toxicity pathway or AOP may improve the interpretation of Tier 1 battery results and add to the underlying biological plausibility. The weight of evidence guidance describes key lines of inquiry including agreement of outcomes within an individual assay (i.e. "complementarity") and among the different assays in the battery (i.e. "redundancy"). Using the AOP/toxicity pathway framework to organize and evaluate EDSP data, the consideration of redundancy in cellular and organ responses measured by different assays becomes easier to identify. Ultimately, predictive models developed and integrated as alternatives for individual Tier 1 assays may be used, alone or in combinations, as alternatives to the current screening data required for the EDSP Tier 1 weight of evidence decisions on a chemicals potential endocrine activity. Further, consideration of any other relevant scientific data can also be integrated in pathway frameworks, and in combination with predictive models, may be adequate to determine if a chemical is a candidate for further testing and which tests are appropriate.

Currently, pathway concepts are being used as an organizational framework for validating alternative methods and integrating data for Tier 1 endocrine screening, and as mentioned previously, Tier 1 assays do not include apical endpoints required for risk assessment. Individual and population level responses that define adverse outcomes for human health and ecotoxicology, respectively, are captured in the longer, animal-intensive EDSP Tier 2 tests. As AOPs continue to be developed, populated, and validated, the hope is that eventually, quantitative AOPs may be used to model organism or population responses from upstream events in lieu of Tier 2 tests currently needed for risk assessment (Groh et al. 2015), but may take years to develop and validate. Alternatively, working backward from an adverse outcome to identify an upstream, measurable key event in an AOP or a point of convergence of several AOPs may be interpreted as a qualitative biomarker or "tipping point" toward the adverse outcome and may replace the need to demonstrate adversity in whole-animal models. Identifying tipping points will also help to distinguish early key events in an endocrine AOP that may be adaptive from later responses indicating loss of homeostatic function. Endocrine responses, by their nature, are variable and compensate for a variety of physical and biological stressors. Apical effects due to endocrine toxicity are likely to be general (e.g., altered development, reproduction) and difficult to attribute to a specific pathway without underlying AOP relationships anchoring the outcome to an endocrine-specific MIE. As AOP development and application continues to expand, eventually in silico and in in vitro approaches that target key events (KEs) along well defined pathways may provide sufficient information for hazard identification and risk assessments with little to no in vivo testing (e.g., MacKay et al. 2013).

The range of regulatory applications possible for a particular AOP is defined by its completeness or maturation status. While incomplete AOPs can be used in first tier screening, such as formation of chemical categories, advanced quantitative AOPs with high level of certainty can be applied in full risk assessment. Thus, AOPs provide a foundation for the design of informed approaches to testing and assessment that can strategically deploy screening level analyses to effectively focus testing resources and progressively employ more resource-intensive assays aimed at reducing the uncertainty as required by risk assessment (Tollefsen et al. 2014).

A broader use of the AOP concept in endocrine screening and testing holds the potential to improve the understanding and prediction of endocrine disruption because development of respective AOPs would help to organize and evaluate existing and new knowledge. This would allow researchers to assess confidence in the predictive relationships, as well as to identify data gaps to guide further research. Elucidation of links between bioactivity and adverse effects in individuals or populations would provide the basis for a broader and more meaningful inclusion of endocrine activity data into risk assessment frameworks. Importantly, improved mechanistic understanding would facilitate the development of alternative tests, as well as aid extrapolation across species by promoting the reciprocal use of toxicity information generated in different species (Madden et al. 2014), and focus the testing on key targets associated with a particular AOP or AOP network.

References

Allen TE, Goodman JM, Gutsell S, Russell PJ (2014) Defining molecular initiating events in the adverse outcome pathway framework for risk assessment. Chem Res Toxicol 27(12):2100–2112

Ankley GT, Bennett RS, Erickson RJ, Hoff DJ, Hornung MW, Johnson RD, Mount DR, Nichols JW, Russom CL, Schmieder PK, Serrano JA, Tietge JE, Villeneuve DL (2010) Adverse outcome pathways: a conceptual framework to support ecotoxicology research and risk assessment. Environ Toxicol Chem 29(3):730–741

Becker RA, Ankley GT, Edwards SW, Kennedy SW, Linkov I, Meek B, Sachana M, Senger H, Van Der Burg B, Villeneuve DL, Watanabe H, Barton-Maclaren TS (2015) Increasing scientific confidence in adverse outcome pathways: application of Tailored Bradford-Hill considerations for evaluating weight of evidence. Regul Toxicol Pharmacol 72(3):514–537

Browne P, Judson RS, Casey WM, Kleinstreuer NC, Thomas RS (2015) Screening chemicals for estrogen receptor bioactivity using a computational model. Environ Sci Technol 49(14):8804–8814

Burden N, Sewell F, Andersen ME, Boobis A, Chipman JK, Cronin MTD, Hutching TH (2015) Adverse outcome pathways can drive non-animal approaches for safety assessment. J Appl Tox 35:971–975

EDSTAC. Endocrine Disruptor Screening and Testing Advisory Committee Final Report (1998). http://www.epa.gov/endocrine-disruption/endocrine-disruptor-screening-and-testing-advisory-committee-edstac-final

FQPA, Food Quality Protection Act of 1996. Public Law 104-170 (1996). http://www.gpo.gov/fdsys/pkg/PLAW-104publ170/content-detail.html

Groh KJ, Carvalho RN, Chipman JK, Denslow ND, Halder M, Murphy CA, Roelofs D, Rolaki A, Schirmer K, Watanabe KH (2015) Development and application of the adverse outcome pathway framework for understanding and predicting chronic toxicity: I. Challenges and research needs in ecotoxicology. Chemosphere 120:764–777

Judson RS, Magpantay FM, Chickarmane V, Haskell C, Tania N, Taylor J, Xia M, Huang R, Rotroff DM, Filer DL, Houck KA, Martin MT, Snipes N, Richard AM, Mansouri K, Setzer RW, Knudsen TB, Crofton KM, Thomas RS (2015) Integrated model of chemical perturbations of a biological pathway using 18 in vitro high-throughput screening assays for the estrogen receptor. Tox Sci 148(1):137–154

Kleinsteuer NC, Ceger PC, Allen DG, Strickland J, Chang X, Hamm JT, Casey WM (2015) A curated database of rodent uterotrophic bioactivity. Environ Health Perspect 124(5):556–62

Knapen D, Vergauwen L, Villeneuve DL, Ankley GT (2015) The potential of AOP networks for reproductive and developmental toxicity assaydevelopment. Reprod Toxicol 56:52–55

Kramer VJ, Etterson MA, Hecker M, Murphy CA, Roesijadi G, Spade DJ, Spromberg JA, Wang M, Ankley GA (2011) Adverse outcome pathways and ecological risk assessment: bridging to population-level effects. Enviorn Tox Chem 30(1):64–76

Lacotte P, Buisson DA, Ambroise Y (2013) Synthesis, evaluation and absolute configuration assignment of novel dihydropyrimidin-2-ones as picomolar sodium iodide symporter inhibitors. Eur J Med Chem 62:722–727. doi:10.1016/j.ejmech.2013.01.043

MacKay C, Davies M, Summerfield V, Maxwell G (2013) From pathways to people: applying the adverse outcome pathway (AOP) for skin sensitization to risk assessment. ALTEX 30(4):473–486

Madden JC, Rogiers V, Vinken M (2014) Application of in silico and in vitro methods in the development of adverse outcome pathway constructs in wildlife. Philos Trans R Soc Lond Ser B Biol Sci 369(1656). pii: 20130584. doi: 10.1098/rstb.2013.0584

NRC, National Research Council (2007) Toxicity testing in the 21st century: a vision and strategy. The National Academy Press, Washington, DC

OECD (2012) OECD test no 457: BG1Luc estrogen receptor Transactivation test method for identifying estrogen receptor agonists and antagonists. http://www.oecd-ilibrary.org/environment/test-no-457-bg1luc-estrogen-receptor-transactivation-test-method-for-identifying-estrogen-receptor-agonists-and-antagonists_9789264185395-en

Patlewicz G, Simon TW, Rowlands JC, Budinsky RA, Becker RA (2015) Proposing a scientific confidence framework to help support the application of adverse outcome pathways for regulatory purposes. Regul Toxicol Pharmacol 71(3):463–477

Paul Friedman K, Watt ED, Hornung MW, Hedge JM, Judson RS, Crofton KM, Houck KA, Simmons SO (2016) Tiered high-throughput screening approach to identify thyroperoxidase inhibitors within the ToxCast phase I and II chemical libraries. Toxicol Sci 151(1):160–180

SDWA, The Safe Drinking Water Act Amendments of (1996) Public Law 104–182 (1996). http://www.gpo.gov/fdsys/pkg/PLAW-104publ182/pdf/PLAW-104publ182.pdf

Tollefsen KE, Scholz S, Cronin MT, Edwards SW, de Knecht J, Crofton K, Garcia-Reyero N, Hartung T, Worth A, Patlewicz G (2014) Applying adverse outcome pathways (AOPs) to support integrated approaches to testing and assessment (IATA). Regul Toxicol Pharmacol 70(3):629–640

US EPA, Endocrine Disruptor Screening Program Universe of Chemicals and Validation Principles (2012). http://epa.gov/endo/pubs/edsp_chemical_universe_and_general_validations_white_paper_11_12.pdf

US EPA, U. S. Environmental Protection Agency (2011) Weight-of-evidence: Evaluating results of EDSP Tier 1 screening to identify the need for Tier 2 testing. http://www.regulations.gov/#!documentDetail;D=EPA-HQ-OPPT-2010-0877-0021

US EPA. U.S. Environmental Protection Agency (2015) Use of high throughput assays and computational tools; endocrine disruptor screening program; notice of availability and opportunity for comment. Fed Regist 2015;80:35350–35355

Villeneuve DL, Crump D, Garcia-Reyero N, Hecker M, Hutchinson TH, LaLone CA, Landesmann B, Lettieri T, Munn S, Nepelska M, Ottinger MA, Vergauwen L, Whelan M (2014a) Adverse outcome pathway (AOP) development I: strategies and principles. Toxicol Sci 142(2):312–320

Villeneuve DL, Crump D, Garcia-Reyero N, Hecker M, Hutchinson TH, LaLone CA, Landesmann B, Lettieri T, Munn S, Nepelska M, Ottinger MA, Vergauwen L, Whelan M (2014b) Adverse outcome pathway development II: best practices. Toxicol Sci 142(2):321–330

Vinken M (2013) The adverse outcome pathway concept: a pragmatic tool in toxicology. Toxicology 312:158–165

Whelan M, Andersen M (2013) Toxicity pathways – from concepts to application in chemical safety assessment. Scientific and policy report by the Joint Research Centre of the European Commission. http://publications.jrc.ec.europa.eu/repository/bitstream/JRC86467/lb-na-26389-en-n%20.pdf

Chapter 3
Cell-Free Assays in Environmental Toxicology

Adeline Arini, Krittika Mittal, and Niladri Basu

Abstract Predictive toxicology requires in vitro tests that can help prioritize, screen, and evaluate a large number of chemicals (i.e., thousands) in a relatively short period of time (days to weeks). Cell-free assays represent a relatively simple in vitro tool that can characterize the interaction between test chemicals and biochemical targets, and are increasingly being used to study a range of fish and wildlife, and also screen single chemicals as well as complex mixtures of environmental samples. The purpose of this chapter is to describe cell-free assays, and propose them as a species agnostic, in vitro toxicity-testing tool of potential relevance to ecological risk assessment. In doing so, the chapter aims to show that cell-free tests are an attractive tool that can be used in predictive ecotoxicology especially considering the limited availability of test organisms (particularly species that are at-risk, difficult to maintain in captivity, etc.), lack of proven cell-based tools (e.g., cell cultures and cell lines), societal concerns over animal testing, sheer number of ecological species to study, and vast inter-species differences.

3.1 Context

Thousands of chemicals need to be evaluated for regulatory purposes. For example, large endeavours such as the European Union's Registration, Evaluation, Authorisation and Restriction of Chemicals (REACH) program, the U.S. EPA ToxCast program, and the Chemicals Management Plan (CMP) in Canada were implemented in recent years to address legislative obligations and take action on

A. Arini • K. Mittal
Faculty of Agricultural and Environmental Sciences, McGill University, Montreal H9X 3V9, QC, Canada

N. Basu (✉)
Faculty of Agricultural and Environmental Sciences, McGill University, Montreal H9X 3V9, QC, Canada

Department of Environmental Health Sciences, University of Michigan School of Public Health, Ann Arbor, MI, USA
e-mail: niladri.basu@mcgill.ca

N. Garcia-Reyero, C.A. Murphy (eds.), *A Systems Biology Approach to Advancing Adverse Outcome Pathways for Risk Assessment*,
DOI 10.1007/978-3-319-66084-4_3

chemicals believed to be harmful. However, these regulatory programs face major hurdles. Foremost, the number of chemical substances for which toxicity data are required is tremendous and backlogged (e.g., 85,000 on U.S. Toxic Substances Contract Act inventory; 23,000 under Canada's Domestic Substances List; 107,000 chemicals in EU manufactured within or imported into region in quantities exceed 1000 tons). This number continues to grow, and is substantially higher when considering the complex environmental samples (e.g., effluents) that need testing.

Historically, testing chemicals has relied on in vivo studies that use whole animals. In many respects, in vivo toxicity testing responds to the concept of "one problem, one test" (Hartung 2009), which implies that a single animal study is conducted to relate the effects of a single chemical with a single adverse outcome. A major consequence of this is that only few classes of contaminants have been subjected to intensive testing. There remains thousands of chemicals (including mixtures) for which few or no test data are available (Judson et al. 2009). In addition, these types of studies yield findings that are largely descriptive, and the work is time consuming and prohibitively costly. For example, the U.S. EPA estimates that traditional testing of a single chemical may take 4 years. and cost $1-20 M USD (Martin et al. 2012). The EU REACH program realistic case scenario calculates the need for 54 million vertebrate animals and $13.6B USD to achieve registration goals (Rovida and Hartung 2009). These realities represent major barriers to fulfilling legal obligations to manage chemicals.

The aforementioned limitations have been recognized by the U.S. National Research Council (NRC) in their document entitled "*Toxicity Testing in the 21st Century: A Vision and a Strategy*" (NRC 2007). The main outcome of this NRC document was the recommendation of a new, predictive strategy as the cornerstone of 21st century toxicity testing. This predictive strategy is based on understanding and applying in vitro toxicity assays which predict cellular level effects that can next be extrapolated to effects on individuals. It de-emphasizes the need to base assessments on animal tests, thus promoting the 3-Rs principle for humane animal research that was developed over 50 years ago (Russell et al. 1959). This new strategy harnesses recent advancements in the fields of cellular and molecular biology, toxicology, and computational biology among others. For example, advances in measurement technologies and fundamental toxicological understanding at the molecular level (i.e., transcriptomics, proteomics, metabolomics) have increased the amount and types of information available and potentially useful to risk assessors (Ankley et al. 2010). These are now contributing towards the development of New Approach Methodologies (NAMs) as discussed in a recent workshop by the European Chemicals Agency (ECHA 2016).

A major conclusion of the NRC report was the expansion and utilization of in vitro tools in chemical risk assessment. In particular, the report articulated a need to establish in vitro tests that can prioritize, screen and evaluate a large number of chemicals (i.e., thousands) in a relatively short period of time (days to weeks). Regarding in vitro tests that span a multitude of molecular, biochemical and physiological systems, the expectation is that advanced computational and bioinformatics platforms could integrate the complex data streams and predict whole organismal impacts. Such a plan lies at the heart of predictive toxicology. This is the basis of an ambitious program launched by the U.S. Environmental Protection Agency in 2007 called Toxicity Forecaster (ToxCast™) (Judson et al. 2010). As detailed elsewhere (Dix et al. 2007), ToxCast is comprised of several in vitro, automated chemical screening technologies that provide a cost-effective and rapid approach to screen for changes in biological activity in response to chemical exposure. The program has nearly 1000 high-throughput and automated assays in its repertoire that cover approximately 300 signalling pathways. The program has screened thousands of chemicals including 300 well-studied chemicals that have undergone extensive animal testing (Phase 1, Proof of Concept; (Judson et al. 2010; Martin et al. 2011; Sipes et al. 2011; Kavlock et al. 2012; Padilla et al. 2012)), >2000 chemicals from a broad range of sources including consumer products, green chemicals, and food additives (Phase 2, (Rotroff et al. 2013; Sipes et al. 2013)), and ~800 chemicals that are known or suspected endocrine disruptors (E1K library; Karmaus et al. 2016). In a recent paper, ToxCast scientists screened 10,000 chemicals (15 concentrations of each chemical in 3 independent experiments) through 30 different cell-based assays (Huang et al. 2016), and components of the testing platform are hailed to be able to screen 10,000 chemicals within a week (Attene-Ramos et al. 2013). Performing the same work in animals would have taken years and millions of dollars. Clearly the cost/performance ratio makes these attractive as tools to screen, prioritize and evaluate a large number of chemicals, and thus meet regulatory obligations as well as help satisfy societal concern.

The development of NAMS, particularly new in vitro tools for testing chemicals such as those referred to above has near-exclusively been focused on human health applications. Unfortunately they are of limited use in the ecological sciences in which many more species (and their complex interactions) are under scrutiny. Very few in vitro toxicity testing tools exist for the most standard ecotoxicological test species, and there is almost nothing for native species of ecological relevance. This is problematic since the extrapolation of results across species (i.e., from standard test species to native species of ecological relevance) introduces tremendous uncertainty, as does extrapolation from controlled laboratory tests to real-world environments (Villeneuve and Garcia-Reyero 2011). For example, native bird species can be more sensitive or respond differently to chemicals than the standard lab model (Head et al. 2008). These types of differences complicate decision-making and often necessitate additional testing.

There is a clear need to accelerate the development and application of novel in vitro toxicity testing tools for the purposes of ecological risk assessment, and this has been recognized by leading scholars in the field (Villeneuve and Garcia-Reyero 2011). As such, the purpose of this chapter is to describe cell-free assays, and propose them as a species agnostic, in vitro toxicity-testing tool of potential relevance to ecological risk assessment. The chapter describes cell-free tests and how they are conducted, and also provides examples from the literature. In doing so, the chapter aims to show that cell-free tests are an attractive tool that can be used in predictive ecotoxicology especially considering the limited availability of test organisms (particularly species that are at-risk, difficult to maintain in captivity, etc.), lack of proven cell-based tools (e.g., cell cultures and cell lines), societal concerns over animal testing, sheer number of ecological species to study, and vast inter-species differences.

3.2 Description of Cell-Free Assays

Cell-free assays are simplified in vitro platforms that can help evaluate the effects of a test chemical on a biochemical process. A number of other in vitro approaches are also employed in toxicology such as primary cell cultures and immortalized cell lines. These have the advantages of better retaining in vivo tissue-specific characteristics and cell line longevity thus in some cases facilitating the study of functional pathways (Bhogal et al. 2005) (Fig. 3.1). However, over time they tend to lose in vivo properties and cell lines are available only for a select number of species suited for laboratory studies. In comparison, while cell free platforms, typically performed in tissue homogenates, cell lysates or on purified molecules, might represent an over-simplified approach, with careful design consideration, the assays can provide complementary and useful mechanistic information on the nature of biochemical interactions (e.g., does the chemical act as an agonist or antagonist of a target receptor).

Here we briefly describe the steps involved in running a common cell-free assay, and focus on radioligand binding to a neurochemical receptor (Fig. 3.2). While assays may be permitted on other organ systems, we focus on the nervous system and draw upon examples based on previous work by our group (Basu et al. 2009; Rutkiewicz et al. 2011; Arini et al. 2016). Briefly, for receptor binding assays, cellular membranes are isolated by homogenizing cerebral tissues in a 1:10 solution of

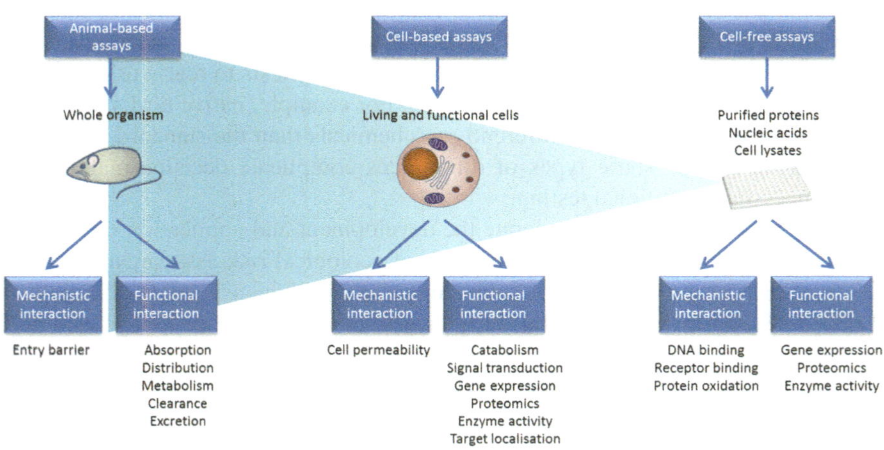

Fig. 3.1 Schematic presentation of the main differences among animal-based, cell-based and cell-free studies (Adapted from Englebienne (2005))

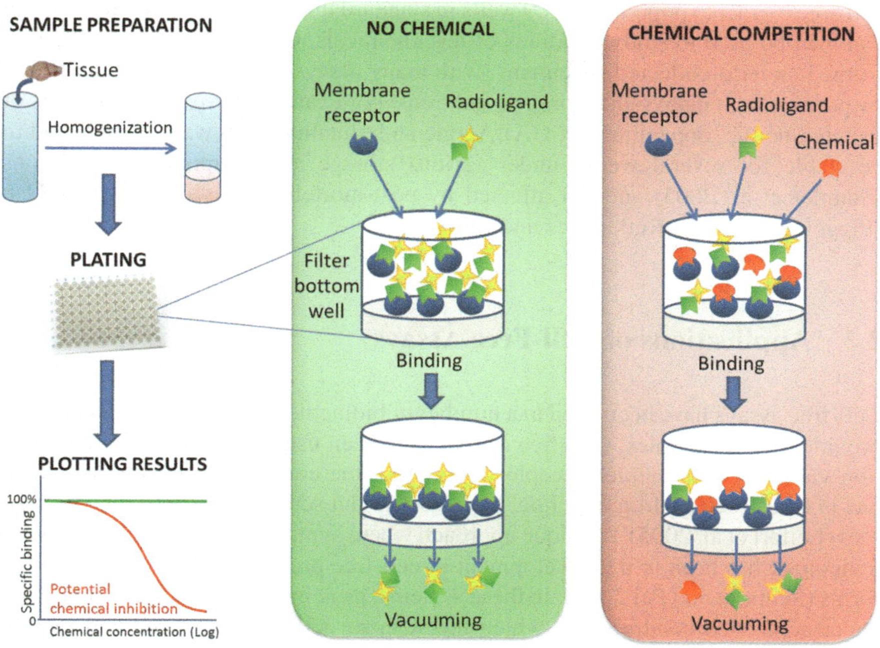

Fig. 3.2 Schematic representation of cell-free receptor binding assays, in presence or absence of a test chemical

buffer and then centrifuging the homogenate to isolate pellets, which are then washed and re-suspended frozen until use. When needed, the cellular membranes preparations are thawed and diluted to an optimal concentration, and then added to microplates that contain a glass filter bottom. The membranes are incubated with radioligands specific for the target of interest. The incubation conditions vary dependent upon the particular assay (e.g., length of incubation, temperature, buffers and assay cofactors). Following an incubation period, vacuum is applied to the well thus filtering the bound radioactive ligand (i.e., the receptor-ligand complex is trapped on the filter) from the unbound ligand that passes through the filter. The radioactivity retained by the filter provides an index of binding. Specific binding to receptors is defined as the difference in radioligand bound in the presence and absence of excess amounts of an unlabelled displacer. These assays can next be run in the presence of a test chemical to determine if that substance impairs ligand-receptor interactions. A range of biochemical parameters can be investigated, such as ligand affinity and saturation kinetics, and the inhibitory (or potentiating) effects of a test chemical on such parameters can be quantified.

A great advantage of cell-free assays is that they are amenable for use from any species from which tissue can be obtained. This is especially useful for ecological species that are difficult to maintain under laboratory conditions or for which there exists limited data. As an example, one gram of brain tissue can yield enough cell-free extract to populate ~5000 wells in standard microplates (~50 plates), which can then be used to study hundreds of test chemicals. Cell-free assays can be performed on field-collected specimens, with many assays being relatively unaffected by post-mortem delays and storage conditions. For example, several components of the cholinergic, dopaminergic, GABAergic and glutamate pathways were found to be stable for several weeks under various storage and temperature conditions (Stamler et al. 2005) and not affected by post-mortem delays of up to 36–72 h (Piggott et al. 1992; Rutkiewicz and Basu 2012).

3.3 Applications of Cell-Free Assays

Cell-free assays have been used in a number of biomedical applications and here we provide select examples. Cell-free assays have been used to study signal transduction via G-protein coupled receptors (GPCRs), the commercial interest of which lies in areas such as drug targeting, high-throughput screening systems and biosensors (Leifert et al. 2005). A unique approach where synthetic biology intersects with toxicology has been in the development of cell-free protein synthesis (CFPS) platforms (Schmidt and Pei 2011). In these systems, proteins of interest are synthesized under controlled conditions in which they can be actively monitored and rapidly sampled (Schmidt and Pei 2011). First developed with *E.coli* extracts, known as S30 extracts, a current example is Cytomim which is an *E.coli* cell-free platform can be used to produce protein therapeutics, toxins and other biochemicals that are difficult to make in vivo because of their toxicity or complexity (DeVries and Zubay 1967; Schmidt and Pei 2011). A final example are purified enzymatic systems from fungi and bacteria that have been used to determine catabolism and biodegradation of fluorinated aromatic compounds and provide information on their fate in the environment using nuclear magnetic resonance (Murphy 2007). Together, these examples showcase the breadth and versatility of cell-free platforms. Given the chapter's objective we restrict the following sections towards the application of cell-free tests towards the toxicological testing of chemicals, particularly for ecological risk assessment. For more information on synthetic biology approaches see Chap. 19.

Arguably the most concerted effort to use cell-free assays has been through by the US EPA's ToxCast program that was briefly introduced earlier. The cell-free methods in ToxCast have been performed using Novascreen from Caliper Biosciences (Judson et al. 2010; Knudsen et al. 2011). Chemicals were evaluated in approximately 300 signalling cell-free pathways: 77 G-protein coupled receptor (GPCR) binding assays; 32 CYP-450-related enzyme activity assays; enzymatic assays for 72 kinases, 22 phosphatases, 15 proteases, 6 histone deactylases (HDACs), 3 cholinesterases, and 14 other enzyme activities; 18 nuclear receptor binding assays; 20 ion channel and ligand-gated ion channel activities; and 9 transporter proteins, 2 mitochondrial pore proteins, and 2 other receptor types (Kavlock et al. 2012). First, a single concentration of test chemical was run through the assays.

Second, a concentration-response assay was conducted for all active and some selected inactive calls. Data from these assays are available online via the ToxCast Database. Toxicity signatures from ToxCast are defined and evaluated by how well these in vitro signals predict adverse outcomes in toxicity pathways relevant to human health. It is hoped that molecular initiating events, as realized via in vitro results, may be predictive of apical outcomes relevant to the whole organism. Some ToxCast studies have paid specific attention to making such in vivo and in vitro comparisons. For example, Knudsen et al. (2011) ran 292 high-throughput cell-free assays to evaluate 320 environmental chemicals. In vitro data from acetylcholinesterase assays were compared to in vivo data available in the literature for rats and humans. A qualitative association between in vitro and in vivo activity was evident for 16 of 17 (94%) chemicals studied and so the authors concluded that, to a reliable extent, in vitro generally predicted the in vivo situation. Silva et al. (2015) compared GABA(A) binding, dopamine binding and AChE activity after in vivo and in vitro exposure to two pesticides (endosulfane and methidathion). This study showed good concordance between in vitro and in vivo results for dopamine pathways with endosulfan exposure. However, in other cases in vitro results were less representative of in vivo effects. The authors showed that some in vitro assays from ToxCast resulted in false negatives in several critical endpoints. For instance, there is a strong body of evidence in the literature relating endosulfan exposure to estrogenic and anti-androgenic effects in vivo, including receptor binding, whereas endosulfan was reported as being active only in a minimal number of ToxCast assays (Silva et al. 2015). The authors suggested that the discrepancy between in vivo and in vitro responses was likely due to a lack of metabolic activation and limitations in assay design. ToxCast was designed as a collaborative effort and hence, discrepancies could also have resulted from the different analytical approaches or different assay types used by the different collaborating teams to interpret the data, and this could affect how a chemical is defined as having a positive or negative effect.

Cell-free assays have been extended to studying wild, native species not conducive to lab-based experimentation, and the outcomes of some studies are briefly reviewed here. The inhibition potential of inorganic and methyl mercury ($HgCl_2$ and MeHgCl) on muscarinic cholinergic (mACh) receptor binding from both ecological (mink, river otter) and biomedical (humans, rats, mice) tissue samples, was characterized in two brain regions (cerebral cortex and cerebellum) thus resulting in rich concentration-response data across organisms (Basu et al. 2005). The work showed that, across all species, that inorganic mercury was a more potent inhibitor of muscarinic receptor binding than organic mercury, and that the cerebellum was more sensitive than the cerebral cortex. Species-sensitivity could be determined and from most to least sensitive as: river otter > rat > mink > mouse > humans. The mean IC50 value (concentration that inhibits receptor binding by 50%) between the most and least sensitive species ranged from 5-8x. A follow-up study was performed on cortical tissues from ringed seals to show that mercurials but not several organochlorines (e.g., PCBs, toxaphene, DDT, dieldrin) inhibited muscarinic cholinergic receptor binding (Basu et al. 2006). Another follow-up study documented that the M1 muscarinic receptor subtype was more sensitive to mercury-associated inhibition that than the M2 subtype (Basu et al. 2008). Taken together, these studies demonstrate that cell-free assays are potentially useful in studying chemical-ligand interactions in native species that are otherwise difficult to study in the lab, such as marine mammals. The work demonstrates that cell-free assays may help resolve differences across species and chemicals.

Cell-free in vitro systems may also be useful in screening real-world samples, including complex mixtures. In a study concerning pulp and paper mill effluents, goldfish brains were homogenized and cell-free preparations were exposed to primary and secondary effluent extracts (Basu et al. 2009). The results showed that the extracts contained neuroactive substances that could alter the specific binding to several receptors and the activity of enzymes involved in the reproductive signalling. For instance, some extracts increased ligand-binding to Dopamine-2 (D2) and GABA(A) receptors, whereas others competed with the N-methyl-D-aspartic acid (NMDA) and muscarinic cholinergic (mACh) receptors and decreased their binding by 26–75%. Activities of the monoamine oxidase (MAO) and the acetylcholinesterase (AChE) were the most impacted with enzyme inhibition reaching 50%. The authors concluded that these cell-free assays provide a novel in vitro tool to highlight the plausible mechanism by which pulp and paper mills effluents may impair fish reproduction by interacting with neurotransmitter systems. In addition, these in vitro data were used to model potential effects at the level of the whole organism (Chap. 16). A similar approach was taken on wastewater effluents from an Area of Concern (AOC) in the Great Lakes region of North America (Arini et al. 2016). In this case two parallel approaches (in vivo and in vitro) were used to assess how the exposure to wastewater treatment plant (WWTP) effluents or to extracts targeting different classes of chemicals (steroid hormones, nonylphenols, bisphenol A) could impact neurochemistry in fathead minnow (*Pimephales promelas*). The ability of the wastewater (in vivo) or extracts (in vitro) to interact with enzymes (monoamine oxidase (MAO) and glutamine synthetase (GS)) and receptors (dopamine (D2) and N-methyl-D-aspartate receptor (NMDA)) involved in dopamine and glutamate-dependent neurotransmission were examined on brain homogenates. In vivo exposure of FHM led to significant decreases of NMDA receptor binding in females and increases of MAO activity in males (2.8–3.2-fold). In vivo and in vitro results for FHM were consistent in some cases (but not in all cases). The main correlation was found for MAO activity that increased after both in vivo and in vitro exposures to steroid hormones-targeted extracts from the WWTP.

3.4 Concluding Remarks

Cell-free assays provide a simple in vitro tool to characterize the interaction between test chemicals and biochemical targets, and ultimately these tools can be used to prioritize, screen and evaluate a large number of chemicals (i.e., thousands) in a relatively short period of time (days to weeks). Such has been shown via the U.S. EPA's ToxCast program, in which cell-free assays are an important component.

Studies more oriented towards ecological risk assessment are beginning to show that cell-free assays can be used to study a range of fish and wildlife, and also screen single chemicals and complex mixtures of environmental samples.

There are several potential advantages of cell-free assays. Cell-free assays can be developed on cell components from potentially any vertebrate, and thus are species agnostic and may be of interest for organisms that are at-risk or difficult to maintain in captivity. The data from cell-free assays can be used to inform risk assessment and to provide additional evidence for read-across to toxicologically similar chemicals. It can ultimately result in generating large databases and strengthening decision-making and environmental management.

The assays are amenable to a high degree of automation, and scalable to high-throughput screening. These types of assays can be run in a relatively rapid manner and at a fraction of the cost associated with animal bioassays. Certain cell-free assays can attain a high level of reproducibility, specificity, and sensitivity. When assays are strung together into a systems/pathway-based manner, the assay results may yield plentiful quantitative concentration-response data that may be used to develop predictive models. This information may help develop hypotheses (e.g., candidate toxicants, sensitive pathways) to be further tested via animal models and may also enable inter-species differences to be uncovered.

Cell-free assays characterize simple interactions between a molecular target and a contaminant, and such an interaction may be considered a molecular initiating event which represent the first sequence of events in an adverse outcome pathway (Landesmann et al. 2013; Ankley et al. 2010). For example, the toxic actions of domoic acid are mediated via its agonism of kainate receptors (Watanabe-Sailor et al. 2011), and so this first key molecular initiating event could be developed into a cell-free assay for the purposes of predictive ecotoxicology.

Despite the aforementioned advantages, as with any technology or method there exist limitations. Foremost among them is that the assays represent a simplistic biological system. They lack the requisite cellular machinery found in traditional in vitro methods such as cell lines and cell cultures, yet one may argue that they represent more meaningful models than can be achieved *in silico*. They lack the metabolic capacity of cells though future endeavours could aim to increase their realism via co-incubations with biological cofactors (e.g., S9 fractions). Moving forward, validation studies that enable comparisons between data from cell-free assays and physiological responses from the whole organism are required to establish these in vitro testing tools as reliable models.

References

Ankley GT, Bennett RS, Erickson RJ, Hoff DJ, Hornung MW, Johnson RD, Mount DR, Nichols JW, Russom CL, Schmieder PK, Serrano JA, Tietge JE, Villeneuve DL (2010) Adverse outcome pathways: a conceptual framework to support ecotoxicology research and risk assessment. Environ Toxicol Chem 29:730–741

Arini A, Cavallin JE, Berninger JP, Marfil-Vega R, Mills M, Villeneuve DL, Basu N (2016) In vivo and in vitro neurochemical-based assessments of wastewater effluents from the Maumee River area of concern. Environ Pollut 211:9–19

Attene-Ramos MS, Miller N, Huang R, Michael S, Itkin M, Kavlock RJ, Austin CP, Shinn P, Simeonov A, Tice RR, Xia M (2013) The Tox21 robotic platform for the assessment of environmental chemicals–from vision to reality. Drug Discov Today 18(15–16):716–723

Basu N, Stamler CJ, Loua KM, Chan HM (2005) An interspecies comparison of mercury inhibition on muscarinic acetylcholine receptor binding in the cerebral cortex and cerebellum. Toxicol Appl Pharmacol 205:71–76

Basu N, Kwan M, Chan HM (2006) Mercury but not organochlorines inhibits muscarinic cholinergic receptor binding in the cerebrum of ringed seals (Phoca hispida). J Toxicol Environ Health A 69:1133–1143

Basu N, Scheuhammer AM, Rouvinen-Watt K, Evans RD, Grochowina N, Chan LHM (2008) The effects of mercury on muscarinic cholinergic receptor subtypes (M1 and M2) in captive mink. Neurotoxicology 29:328–334

Basu N, Ta CA, Waye A, Mao J, Hewitt M, Arnason JT, Trudeau VL (2009) Pulp and paper mill effluents contain neuroactive substances that potentially disrupt neuroendocrine control of fish reproduction. Environ Sci Technol 43:1635–1641

Bhogal N, Grindon C, Combes R, Balls M (2005) Toxicity testing: creating a revolution based on new technologies. Trends Biotechnol 23:299–307

DeVries JK, Zubay G (1967) DNA-directed peptide synthesis. II. The synthesis of the alpha-fragment of the enzyme beta-galactosidase. Proc Natl Acad Sci U S A 57:1010–1012

Dix DJ, Houck KA, Martin MT, Richard AM, Setzer RW, Kavlock RJ (2007) The toxcast program for prioritizing toxicity testing of environmental chemicals. Toxicol Sci 95:5–12

ECHA (2016) New approach methodologies in regulatory science. Proceedings of a scientific workshop, Helsinki, European Chemicals Agency, April 19–20, 2016. https://echa.europa.eu/documents/10162/22816069/scientific_ws_proceedings_en.pdf/a2087434-0407-4705-9057-95d9c2c2cc57

Englebienne P (2005) High throughput screening: will the past meet the future? Front Drug Des Discovery 1:69–86

Hartung T (2009) A toxicology for the 21st century—mapping the road ahead. Toxicol Sci 109:18–23

Head JA, Hahn ME, Kennedy SW (2008) Key amino acids in the aryl hydrocarbon receptor predict dioxin sensitivity in avian species. Environ Sci Technol 42:7535–7541

Huang R, Xia M, Sakamuru S, Zhao J, Shahane SA, Attene-Ramos M, Zhao T, Austin CP, Simeonov A (2016) Modelling the Tox21 10[thinsp]K chemical profiles for in vivo toxicity prediction and mechanism characterization. Nat Commun 7:10245

Judson R, Richard A, Dix DJ, Houck K, Martin M, Kavlock R, Dellarco V, Henry T, Holderman T, Sayre P, Tan S, Carpenter T, Smith E (2009) The toxicity data landscape for environmental chemicals. Environ Health Perspect 117:685–695

Judson RS, Houck KA, Kavlock RJ, Knudsen TB, Martin MT, Mortensen HM, Reif DM, Rotroff DM, Shah I, Richard AM, Dix DJ (2010) In vitro screening of environmental chemicals for targeted testing prioritization: the toxcast project. Environ Health Perspect 118:485–492

Karmaus AL, Toole CM, Filer DL, Lewis KC, Martin MT (2016) High-throughput screening of chemical effects on steroidogenesis using H295R human adrenocortical carcinoma cells. Toxicol Sci 150(2):323–332

Kavlock R, Chandler K, Houck K, Hunter S, Judson R, Kleinstreuer N, Knudsen T, Martin M, Padilla S, Reif D, Richard A, Rotroff D, Sipes N, Dix D (2012) Update on EPA's ToxCast program: providing high throughput decision support tools for chemical risk management. Chem Res Toxicol 25:1287–1302

Knudsen TB, Houck KA, Sipes NS, Singh AV, Judson RS, Martin MT, Weissman A, Kleinstreuer NC, Mortensen HM, Reif DM, Rabinowitz JR, Setzer RW, Richard AM, Dix DJ, Kavlock RJ (2011) Activity profiles of 309 ToxCast™ chemicals evaluated across 292 biochemical targets. Toxicology 282:1–15

Landesmann B, Mennecozzi M, Berggren E, Whelan M (2013) Adverse outcome pathway-based screening strategies for an animal-free safety assessment of chemicals. ATLA Alter Lab Anim 41:461–471

Leifert WR, Aloia AL, Bucco O, Glatz RV, McMurchie EJ (2005) G-protein-coupled receptors in drug discovery: nanosizing using cell-free technologies and molecular biology approaches. J Biomol Screen 10:765–779

Martin MT, Knudsen TB, Reif DM, Houck KA, Judson RS, Kavlock RJ, Dix DJ (2011) Predictive model of rat reproductive toxicity from ToxCast high throughput screening. Biol Reprod 85:327–339

Martin MT, Knudsen TB, Judson RS, Kavlock RJ, Dix DJ (2012) Economic benefits of using adaptive predictive models of reproductive toxicity in the context of a tiered testing program. Syst Biol Reprod Med 58:3–9

Murphy CD (2007) The application of 19F nuclear magnetic resonance to investigate microbial biotransformations of organofluorine compounds. Omics: J Integr Biol 11:314–324

National Research Council (2007) Toxicity Testing in the 21st Century: A Vision and a Strategy. Washington, DC: The National Academies Press. https://doi.org/10.17226/11970.

Padilla S, Corum D, Padnos B, Hunter DL, Beam A, Houck KA, Sipes N, Kleinstreuer N, Knudsen T, Dix DJ, Reif DM (2012) Zebrafish developmental screening of the ToxCast™ phase I chemical library. Reprod Toxicol 33:174–187

Piggott M, Perry E, Sahgal A, Perry R (1992) Examination of parameters influencing [3H] MK-801 binding in postmortem human cortex. J Neurochem 58:1001–1008

Rotroff DM, Dix DJ, Houck KA, Kavlock RJ, Knudsen TB, Martin MT, Reif DM, Richard AM, Sipes NS, Abassi YA (2013) Real-time growth kinetics measuring hormone mimicry for ToxCast chemicals in T-47D human ductal carcinoma cells. Chem Res Toxicol 26:1097–1107

Rovida C, Hartung T (2009) Re-evaluation of animal numbers and costs for in vivo tests to accomplish REACH legislation requirements for chemicals—a report by the transatlantic think tank for toxicology (t (4)). ALTEX 26:187–208

Russell W. and Burch R. (1959) The Principles of Humane Experimental Technique. London UK: Methuen.

Rutkiewicz J, Nam DH, Cooley T, Neumann K, Padilla IB, Route W, Strom S, Basu N (2011) Mercury exposure and neurochemical impacts in bald eagles across several Great Lakes states. Ecotoxicology 20:1669–1676

Rutkiewicz J, Basu N (2012) Postmortem stability of brain GABAergic and glutamatergic receptors and enzymes under ecological conditions. Ecotoxicol Environ Saf 84:133–138

Schmidt M, Pei L (2011) Synthetic toxicology: where engineering meets biology and toxicology. Toxicol Sci 120:S204–S224

Silva M, Pham N, Lewis C, Iyer S, Kwok E, Solomon G, Zeise L (2015) A comparison of ToxCast test results with in vivo and other in vitro endpoints for neuro, endocrine, and developmental toxicities: a case study using Endosulfan and Methidathion. Birth Defects Res B Dev Reprod Toxicol 104:71–89

Sipes NS, Martin MT, Reif DM, Kleinstreuer NC, Judson RS, Singh AV, Chandler KJ, Dix DJ, Kavlock RJ, Knudsen TB (2011) Predictive models of prenatal developmental toxicity from ToxCast high-throughput screening data. Toxicol Sci 124:109–127

Sipes NS, Martin MT, Kothiya P, Reif DM, Judson RS, Richard AM, Houck KA, Dix DJ, Kavlock RJ, Knudsen TB (2013) Profiling 976 ToxCast chemicals across 331 enzymatic and receptor signaling assays. Chem Res Toxicol 26:878–895

Stamler CJ, Basu N, Man Chan H (2005) Biochemical markers of neurotoxicity in wildlife and human populations: considerations for method development. J Toxic Environ Health A 68:1413–1429

Villeneuve DL, Garcia-Reyero N (2011) Vision & strategy: predictive ecotoxicology in the 21st century. Environ Toxicol Chem/SETAC 30:1–8

Watanabe-Sailor K, Andersen ME, Basu N, Carvan MJ, Crofton KM, King KA, Suñol C, Tiffany-Castiglioni E, Schultz IR (2011) Defining and modeling known adverse outcome pathways. Environ Toxicol Chem 30:9–21

Chapter 4
The Fish Embryo as a Model for AOP Development

Lucia Vergauwen, Steven Van Cruchten, and Dries Knapen

Abstract Fish are routinely used for evaluating aquatic toxicity to vertebrates to set environmental quality standards. Tests using early life-stages of fish are more cost-efficient compared to tests using adult fish while maintaining the physiological relevance of a vertebrate whole-organism test system. Ethical considerations are also a driver for the use of fish embryos since they are considered alternative testing models during the early stages of development. Additionally, both in human and environmental toxicology there is a strong global interest in increasing the use of mechanistic information to support hazard assessment. The AOP (adverse outcome pathway) approach offers an interesting framework for developing mechanistically-based alternative testing methods using fish embryos. Once developed, AOPs can facilitate the identification of assays targeting key events, which have high predictive value for an adverse outcome of interest. In this chapter we first discuss what kind of information on the general biology and physiology of a fish species is important in order to use the embryonic life stage of that species as a model for AOP development, including aspects such as endocrinology, reproduction strategies, availability of genomic information, transgenic lines, and biotransformation capacity during embryonic development. Secondly, we provide an overview of strategies and examples of AOP development using fish embryos. In this context, we discuss the application of an iterative AOP development cycle, development of Fish Early Life-Stage (FELS) AOPs for developing alternative strategies for chronic toxicity testing, development of AOP networks, and development of fish AOPs for endocrine disruption.

L. Vergauwen
Zebrafishlab, Veterinary Physiology and Biochemistry, Department of Veterinary Sciences, University of Antwerp, Universiteitsplein 1, 2610 Wilrijk, Belgium

Systemic Physiological and Ecotoxicological Research (SPHERE), Department of Biology, University of Antwerp, Groenenborgerlaan 171, 2020 Antwerp, Belgium

S. Van Cruchten
Applied Veterinary Morphology, Department of Veterinary Sciences, University of Antwerp, Universiteitsplein 1, 2610 Wilrijk, Belgium

D. Knapen (✉)
Zebrafishlab, Veterinary Physiology and Biochemistry, Department of Veterinary Sciences, University of Antwerp, Universiteitsplein 1, 2610 Wilrijk, Belgium
e-mail: dries.knapen@uantwerpen.be

N. Garcia-Reyero, C.A. Murphy (eds.), *A Systems Biology Approach to Advancing Adverse Outcome Pathways for Risk Assessment*,
DOI 10.1007/978-3-319-66084-4_4

4.1 Introduction

Both vertebrate and invertebrate toxicity tests are used to provide an ecologically relevant basis for environmental quality criteria. Within the vertebrate tests, fish are valuable sentinels for evaluating aquatic toxicity since they represent a high trophic level in the aquatic food chain. However, testing for chronic fish toxicity is one of the most animal demanding areas in environmental risk assessment. Around 30 years ago, the Fish Early Life-Stage (FELS) test (OECD Testing Guideline [TG] 210, OECD 2013b; OPPTS 850.1400, USEPA 1996) was introduced as an alternative to the fish full life cycle test (McKim 1977; Woltering 1984). The latter test included all developmental life stages and assessed survival growth and reproduction. McKim et al. reviewed a set of 56 fish full life cycle tests and found that the embryo-larval and early juvenile life stages were the most, or among the most, sensitive and therefore concluded that tests with these early life stages could be useful for establishing environmental water-quality criteria. Furthermore, experiments using the early life-stages of fish are more cost-efficient compared to tests using adult fish while maintaining the physiological relevance of a vertebrate whole-organism test system. Currently, the FELS test is the primary guideline used to estimate chronic toxicity of regulated chemicals (pesticides, industrial chemicals, pharmaceuticals, food/feed additives, and cosmetics) to fish. Results obtained using these test guidelines (TG) are used to support risk assessment around the world.

Although already more cost-efficient than adult fish tests, the FELS tests are actually long-term tests. A FELS test starts with fertilized eggs and continues at least until all the control fish are swimming and free-feeding. These tests generally run for 1–2 months depending on the test species used. Therefore the FELS test is still considered as a low-throughput in vivo test method which requires high numbers of fish (Volz et al. 2011). As such, ethical considerations are driving the development of alternative test systems. For example, in Europe, toxicity testing carried out in the framework of the REACH (Registration, Evaluation, and Authorization of Chemicals) (EC 2006) legislation should be in line with the 3R principles to Replace, Reduce, and Refine the use of laboratory animals (Russel and Burch 1959). According to the EU regulation on the use of laboratory animals, independently feeding larval forms of non-human vertebrate animals, are protected (EC 2010). Fish embryos are therefore not protected until the stage of free-feeding, and are candidate models for alternative testing. For the zebrafish (*Danio rerio*), the limit was explicitly set to 5 days post fertilization (dpf) when kept at a temperature of around 28 °C (EC 2012; Straehle et al. 2012). Apart from the reproductive system, the major organ systems have developed at this age. Figure 4.1 shows different life stages of the zebrafish, including 2 and 3 day old zebrafish (eleuthero) embryos which are not protected under the European legislation on the use of laboratory animals.

On top of ethical considerations driving the development of alternative tests, both in human toxicology as well as in environmental toxicology, there is a strong global interest in increasing the use of mechanistic information to support hazard

Fig. 4.1 Photographs of zebrafish life stages. From top to bottom a zebrafish adult female, an adult male, a 2 day old embryo just before hatching, and a 3 day old eleutheroembryo (developmental phase before the larval phase, starting with hatching and ending with free-feeding) (All images copyright (The copyright holder grants permission to use these images in the original publication "Advancing Adverse Outcome Pathways for Risk Assessment", as well as for all revisions or versions, future editions, in any medium, such as in its electronic form, for all translations in any foreign language, and for distribution throughout the world.) Dries Knapen, http://zebrafish-lab.be)

assessment (Ankley et al. 2010; Krewski et al. 2010). The current FELS protocol involves only apical endpoints, including survival, hatching, overall body appearance and behavior, and final weight and length. Knowledge on the specific toxic mechanism or mode of action of chemicals is not obtained. There is an urgent need for high quality testing strategies to screen and prioritize thousands of chemicals at an acceptable cost with the maximum of relevant information.

The AOP approach offers an interesting framework for developing alternatives to the FELS test. In summary, an AOP is a detailed description of a chain of events going from a molecular initiating event (MIE, a direct interaction of a chemical with a molecular target, e.g. a hormone synthesizing enzyme) through a series of intermediate key events (KE, e.g., altered hormone levels and subsequently impaired development of a specific organ) spanning different levels of biological organization, leading to an adverse outcome (AO, e.g., reduced survival) at the individual or population level (Ankley et al. 2010; Villeneuve et al. 2014b, c). A KE is generally defined as an observable change in biological state that is necessary (but not necessarily sufficient by itself) for the progression toward a specific AO (Villeneuve et al. 2014b). Examples of KEs include changes in expression and/or function of genes, proteins, and metabolites, alterations in cellular or tissue morphology, physiological dysfunction, etc., along a causal pathway to an AO relevant to risk assessment (mainly impaired growth, survival or reproduction). Since KEs must, by definition, be measurable, there is a clear linkage between the AOP framework and assay development, particularly with respect to development of alternatives to traditional whole organism tests focused on direct observation of apical outcomes.

Once developed, AOPs can facilitate the identification of assays targeting KEs, which have high predictive value for an AO of interest. They also provide biological context for mechanistic information from existing assays, which can help increase confidence in, and utility of their results for risk assessment and regulatory decision-making. As such, AOPs could form a basis for a tiered testing strategy in which the lower Tier molecular or cellular perturbations are predictive of higher-Tier outcomes (Volz et al. 2011). AOP-specific data can be obtained at the different levels of structural and functional organization. In Tier 1 In vitro high-throughput cell-based assays can screen for molecular initiating events and the subsequent cellular responses, e.g. receptor-specific reporter assays, axonal growth assays or cell viability and functional assays using fish-specific cell lines derived from gill tissue or liver, or primary fish- cell cultures. Tier 2 could involve a short-term fish embryo test for whole-organism-based assessment of AOP-specific effects, and Tier 3 could comprise the more chronic FELS test. The low-throughput FELS (Tier 3) test would only be implemented if a chemical tested positive based on results obtained using cell-based assays (Tier 1) and alternative methods (Tier 2). By using *in silico*, in vitro and in vivo alternative tests as first medium/high-throughput systems to screen and prioritize chemicals for FELS testing, the need for long-term and costly toxicity tests requiring a large amount of animals would be reduced. The use of such a tiered testing strategy is currently considered a promising approach (Volz et al. 2011).

Due to the considerations above, fish embryos have become popular model systems in (eco)toxicology. DarT (a 48 hpf [hours post fertilization] zebrafish embryo test) was implemented to substitute fish tests in waste water evaluation in Germany (DIN 2001; ISO 2007; Nagel 2002). More recent efforts have advanced the fish embryo test as an alternative to the fish acute toxicity test (OECD TG 203; OECD 1992) for chemical registration under REACH (Embry et al. 2010; Lammer et al.

2009; Lange et al. 1995). Braunbeck and Lammer (2006) reviewed existing information to facilitate the submission of a testing guideline for the fish embryo test to the OECD. The publication of OECD TG 236, the "Fish Embryo Acute Toxicity (FET) Test" (OECD 2013a), describing a 96 h zebrafish embryo test (also called ZFET), has greatly facilitated the use of fish embryos in toxicity studies, although the FET test has not yet been officially approved as alternative to the fish acute toxicity test by regulators (Worth et al. 2014). While the Fish Acute Toxicity Test lists a set of 7 recommended species (zebrafish, fathead minnow, common carp, medaka, guppy, bluegill and rainbow trout), the FET test is currently only defined for zebrafish. There have been efforts to compare results from FET tests using zebrafish, medaka and fathead minnow. Braunbeck et al. (2005) reported that results obtained with medaka and fathead minnow embryos are generally comparable to those obtained with zebrafish embryos. They tested four compounds in the three species and found that effect concentrations differed by a maximum factor of 10 among species. This larger difference was observed for sodium dodecyl sulphate, for which the medaka was significantly less sensitive than the other two species. Except for 2,4-dinitrophenol, which was most toxic in fathead minnow, zebrafish embryos were most sensitive to all other substances. The authors also investigated some practical aspects of the standardization of fish embryo tests. Beekhuijzen et al. (2015) attempted to facilitate harmonization of the zebrafish embryo test by discussing optimal test conditions and scoring methodology. The testing guideline is currently limited to observations of lethal endpoints and hatching, while scientific research from the last decades has shown that many more sublethal toxic effects, such as molecular, biochemical and physiological responses can be effectively investigated using fish embryos. On the one hand, such more detailed measures of toxicity are necessary to describe KEs and develop AOPs. In turn, AOPs can aid in standardization of assays measuring detailed toxicity responses, and can provide the mechanistic support which can finally lead to their incorporation in new testing guidelines. For a more extensive review of the current status of **alternative** methods for regulatory toxicology, we refer to the recent Joint Research Council (JRC) report by Worth et al. (2014).

Apart from its potential for regulatory testing applications in risk assessment, the AOP framework can also aid in improving the fundamental understanding of biological processes and disruptions thereof, because it stimulates scientists to delineate a cascade of events supported by a weight of evidence approach.

4.2 Key Information on General Biology and Physiology of Fish Embryo Models

In the context of toxicity studies, the term fish embryo model is immediately linked to a small set of fish species, primarily zebrafish, fathead minnow and medaka. Together with three-spined stickleback, rainbow trout and sheepshead minnow, these are the recommended species to use in FELS and other tests (OECD 2013b; USEPA 1996). However, in these guidelines, the use of other species is not precluded. Both guidelines also provide a list of other well-documented species such as coho salmon, chinook salmon, brown trout, common carp, bluegill, channel catfish, and others. Even this extended list of well-known models is generally restricted to bony fish, while for example cartilaginous fish have unique ion-regulatory mechanisms with important implications for chemical toxicity. Also, traditionally, there has been much more focus on freshwater fish compared to saltwater fish. Teleosts form the largest group of extant vertebrates with a wide diversity and it is challenging to find adequate representation in toxicity testing.

When AOPs are developed for the purpose of risk assessment which is essentially aimed at protecting all species, it is important to consider inter-species or taxonomic applicability (Users' handbook, OECD 2015). The development of species-specific AOPs should be avoided. When considering the use of a less well-known fish model, ideally, the toxicity of a set of model compounds should be characterized, allowing comparison to established models concerning general mechanisms of toxicity. Inter-species differences are often related to the MIE. Differences in expression of molecular targets (e.g., receptors, enzymes) may lead to differences in sensitivity or even in absence of a specific toxicity mechanism in some species. Lalone et al. (2013b) showed that while reproductive capacity of fathead minnow and medaka was susceptible to an androgen receptor antagonist, *Daphnia magna* was insensitive due to the lack of a relevant homolog of the androgen receptor. Lalone et al. (2013a) described a strategy that uses molecular sequence information of molecular targets to predict which species may be more or less susceptible to a chemical with known MOA (mode of action). This approach was applied in the development of an AOP for acetylcholinesterase inhibition leading to acute mortality where sequence similarity of the enzyme acetylcholinesterase was investigated (Russom et al. 2014). Higher level organismal properties may also result in inter-species differences. Zebrafish embryos have been shown to be more sensitive than medaka embryos in some cases, and it has been suggested that this is due to the presence of a harder chorion surrounding medaka eggs (Schiller et al. 2014). Developing an AOP based on experimental evidence from a few different species thereby increases confidence and applicability. For example, exposure to a thyroid peroxidase inhibitor (thyroid peroxidase is crucial for thyroxine [T4] synthesis) was shown to result in decreased levels of T4 and impaired inflation of the anterior chamber of the swim bladder at comparable exposure concentrations in zebrafish and fathead minnow early life stages (Nelson et al. 2016; Stinckens et al. 2016). The inclusion of less straightforward fish embryo models can further improve the relevance and applicability of AOPs. Therefore, new upcoming models are welcome to contribute to AOP development and cross-species comparison. In this context, an interesting recent development is the consideration of the killifish (*Nothobranchius furzeri*) as a model for ecotoxicological testing, more specifically for rapid chronic and multigenerational toxicity testing. It has a generation time of $\leq$37 days, and produces drought resistant dormant eggs that can be stored 'on the shelf' and activated when needed (Philippe et al. 2015).

In order to interpret toxic effects and make an *a priori* estimation of taxonomic applicability of experimental findings, knowledge of the normal physiology of the species is essential. For example, chemicals may impact ion-osmoregulation including gill function, kidney function and ion pumps, the biotransformation capacity of the liver, cardiovascular performance, respiratory function, or energy metabolism among others. To interpret these effects, the availability of basic knowledge on the normal functioning of these biological processes in the species under investigation is needed. For some of these processes, important differences exist between fish species. Another important aspect is basic knowledge of the endocrinology of the species which is essential to understand the mechanisms of endocrine disrupting compounds. In the following paragraphs we will discuss what kind of information on the general biology and physiology of a fish species is important to use the embryonic life stage of that species as a model for AOP development. Because of the large variety of physiological strategies (e.g., reproduction) within the group of fish it is important to consider specific aspects of the general biology and physiology of fish species both when selecting an appropriate fish embryo model for AOP development, and when interpreting results from toxicity tests. This knowledge is important for studying the biological plausibility (a critical aspect of weight of evidence, together with empirical evidence) of a KER (key event relationship), and particularly important when the aim is to use AOPs as support to develop assays which are predictive of relevant adverse outcomes (OECD 2015).

4.2.1 Stages of Embryonic Development

Fish embryonic development is considered representative of vertebrate embryonic development in general and has therefore become an important model for developmental biology. Although fish generally develop at a faster rate compared to mammals, there are large differences between fish species in timing of important events during development, such as hatching and the onset of free-feeding. When using the embryo of a fish species to develop AOPs, basic knowledge of the embryonic development informs on which stages of development are covered in toxicity studies, and provides a basis to interpret toxicity observations. The better normal development is characterized, the more accurate AOPs can be described. For zebrafish, the stages of embryonic development were described in detail by Kimmel et al. (1995) and to date this remains the main reference in this regard. Villeneuve et al. (2014a) recently outlined a conceptual model of developmental morphological landmarks during zebrafish embryogenesis (e.g., somite formation, cardiovascular system development), that are observable during development, and therefore amenable for use in AOP development (Fig. 4.2). Moreover, these developmental landmarks were outlined with the aim of relating them to FELS AOs rather than purely describing embryonic development. For the fathead minnow, a developmental series was published, describing 32 stages during pre-hatching development (Devlin et al. 1996). The authors also provided an overview of species of which the embryonic

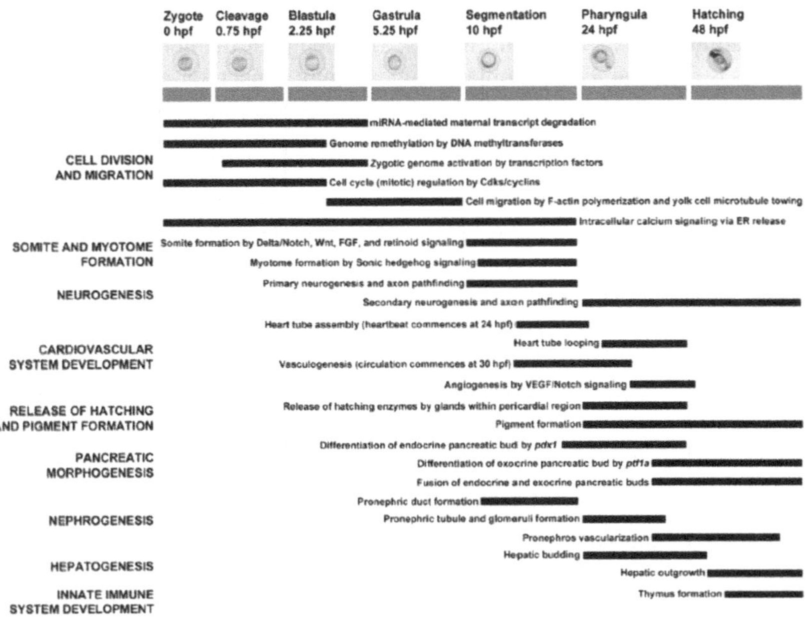

Fig. 4.2 Preliminary conceptual model of developmental landmarks during zebrafish embryogenesis. The Y-axis shows a selection of developmental landmarks, and the X-axis shows the timing during zebrafish embryonic development. *Black bars* represent the approximate duration of the events that underlie each developmental landmark (Source: Villeneuve et al. (2014a), https://creativecommons.org/licenses/by/3.0/, no changes to the original figure were made)

development had been described at the time, together with their age at hatch. For medaka, the stages of normal development were described by Iwamatsu (2004).

Fish are mostly ectothermic species and therefore the rate of development is temperature dependent. According to the Zebrafish Book (Westerfield 1995), the optimal temperature for growth and accurate developmental staging of zebrafish is 28.5 °C. The description of developmental stages published by Kimmel et al. (1995) and widely cited in zebrafish literature, including in the FET testing guideline (OECD 2013a), was recorded at 28.5 °C. The Commission Implementing Decision 2012/707/EU (EC 2012), stating that zebrafish are not protected up to 5 days post fertilization (dpf) was similarly based on a temperature of approximately 28 °C. Traditionally, OECD testing guidelines have recommended lower temperatures for zebrafish toxicity testing, e.g., 21–25 °C (OECD TG 203, OECD 1992) or 26 ± 1.5 °C (OECD TG 210, OECD 2013b). The recommended temperature for the FET test is 26 ± 1 °C (OECD 2013a). This has important implications when considering which stages of development are included in the specific time frame of a fish embryo test. A 120 hpf test at 28.5 °C, which is an alternative test according to current EU legislation, covers a substantially larger part of development than a 96 hpf test at 26 °C. The stages included in the conceptual model published by Villeneuve et al. (2014a) only comprise the first 48 h, and are therefore covered in both cases.

4.2.2 Endocrinology

Although many pathways are conserved, there are specific differences in fish compared to other vertebrates which may influence the interpretation of toxicity. For example, in mammals, the primary mineralocorticoid is aldosterone. Fish do not synthesize aldosterone, and 11-deoxycorticosterone is the suggested substitute for aldosterone in fish. The main sex hormones in zebrafish are estradiol (an estrogen), 11-ketotestosterone (an androgen) and maturation-inducing hormone (MIH, 17α,20β-dihydroxy-4-pregnen-3-one) (Tokarz et al. 2013). Comparable to mammals, estradiol regulates ovarian function and additionally, in fish, estradiol regulates vitellogenin synthesis and yolk formation. In fish, 11-ketotestosterone rather than testosterone is the primary endogenous androgen (de Waal et al. 2008). MIH which regulates maturation of oocytes, has been identified in several fish species, while it does not exist in mammals (Nagahama and Yamashita 2008).

Some hormones as well as transcripts coding for hormone receptors and steroid synthesizing enzymes are available in the early embryo through maternal transfer. This has been shown for estradiol and cortisol in zebrafish embryos, although the role of these hormones during early embryonic development is still largely unclear (Tokarz et al. 2013). Both in zebrafish and fathead minnow, maternal transfer of the thyroid hormone T4 has been shown (Chang et al. 2012; Nelson et al. 2016). This is important for AOP development using the fish embryo. We recently showed that 2-mercaptobenzothiazole (MBT), an inhibitor of thyroid peroxidase (TPO) which is essential for thyroid hormone synthesis, did not affect inflation of the posterior chamber occurring around 96 hpf in zebrafish and around 6 dpf in fathead minnow (Nelson et al. 2016; Stinckens et al. 2016), although several studies have suggested the involvement of thyroid hormones in posterior chamber inflation (Bagci et al. 2015; Heijlen et al. 2014; Jomaa et al. 2014; Liu and Chan 2002). The absence of effects on posterior chamber inflation after inhibiting thyroid hormone synthesis can possibly be explained by maternal transfer of T4 into the eggs. On the other hand, inhibition of deiodinases which are necessary for activating thyroid hormones regardless of maternal transfer, did lead to impaired posterior chamber inflation (Bagci et al. 2015; Heijlen et al. 2014).

Vitellogenin (vtg), the fish egg yolk precursor protein, is synthesized by the liver of many adult female fish and deposited in developing oocytes. Vtg synthesis is under estrogenic regulation. It is therefore a long-established biomarker for evaluating endocrine disrupting potential of chemicals in fish and vtg measurements are an important component of endocrine disruption testing (Ankley and Jensen 2014; OECD 2009a, 2011; Wheeler et al. 2005). Although the function of embryonic vtg expression is currently unknown, it has been shown that endocrine disrupting compounds already induce vtg expression in fish embryos (Schiller et al. 2014). Since different endocrine disrupting modes of action can either increase or decrease vtg levels, these AOPs can be interconnected and visualized in an AOP network (see Sect. 4.3.3).

4.2.3 Sex Determination and Differentiation

Teleosts show a high degree of diversity in sex determination and differentiation mechanisms, ranging from genetic to environmental sex determination. While some species have sex chromosomes such as mammals, many fish species have no heteromorphic chromosomes, similar to amphibian and reptile species. Additionally, many fish species are undifferentiated gonochorists, in which an indifferent gonad first develops into an ovary-like gonad which then further differentiates into either a mature ovary or a testis. In zebrafish, this differentiation process occurs between 17 and 35 dpf at a temperature of 27 °C (Sun et al. 2013). This complicates sex identification and should be taken into account when using fish embryos to study effects on sex ratio, and sex-specific traits or responses. Changes in sex ratio or intersex can be an important adverse outcome when studying endocrine disruption. If this is the intent of the study, often it will be necessary to culture the fish far beyond embryonic stages until they have developed gonads that can be histologically verified. Some species develop clearly discernible secondary sex characteristics, which may also be influenced by endocrine disruptors, namely papillary processes in male medaka and nuptial tubercles in male fathead minnow (OECD 2009b). Chemicals with endocrine modes of action may cause abnormal occurrence of secondary sex characteristics in the opposite sex. For example, androgen receptor agonists can cause the development of nuptial tubercles in female fathead minnow. In zebrafish, secondary sex characteristics are difficult to observe objectively.

For medaka, a genetic marker (DMY) is available for identifying the true genotypic sex (Urushitani et al. 2007). However, the exact mechanism of sex determination in zebrafish is not yet fully understood. Zebrafish have no heteromorphic chromosomes, and a sex-determining gene has not been identified. Recent advances suggest a polygenic sex determination mechanism, where sex is determined by the allelic combinations of several loci (Liew and Orban 2014). Additionally, environmental factors (primarily temperature) can influence sex differentiation (Uchida 2004).

4.2.4 Reproduction

Fish have a large diversity of reproductive strategies, including for example internal or external fertilization, open substrate spawners, mouth brooders, seasonal or year round reproduction. This has important implications for toxicity studies using fish embryos. Many fish species, including zebrafish, fathead minnow and medaka are oviparous, meaning that the females spawn and the eggs are externally fertilized. This offers a great advantage over viviparous animals because embryos can be collected and exposed to chemicals immediately after fertilization without further influences of the mother. Additionally, many of these species (including zebrafish, fathead minnow and medaka) produce eggs year round, while three spined sticklebacks for example are seasonal breeders. Many cartilaginous fish and also some bony fish such as the guppy (*Poecilia reticulate*) are (ovo)viviparous, meaning that the offspring develops inside the mother either with or without a placenta. This in turn offers the advantage of higher comparability to mammalian reproduction.

To actually measure reproductive failure as an adverse outcome, a mature reproductive system is needed which is not developed in the fish embryo. Efforts are underway to develop embryo tests which are predictive of reproductive effects in later life stages, such that the need for animal tests is reduced. For this purpose it is essential to document effects of endocrine disruptors in fish embryos. AOP development can subsequently aid in selecting KEs which are predictive of reproductive failure at later age.

4.2.5 Genetic Variability

For standard mammalian models (rats, mice, rabbits), strains have been developed to decrease intra- and inter-laboratory variability. For fish this is less advanced. In ecotoxicology, findings from the laboratory are extrapolated to field populations and this can be done with higher accuracy when genetic variation in the sample is high. On the other hand, reduced variation (by using clones or inbred strains) increases the precision of the results (i.e. narrow confidence limits), resulting in a trade-off between precision and accuracy (Forbes 1998). With regard to standardization of diet composition, water characteristics, lighting and temperature a similar trade-off concept is applicable. While applying standard conditions (e.g., optimal breeding temperature) provides for low variation and thus high precision, this does not take into account that natural conditions often deviate from the standards in the environment. For example, when performing toxicological experiments aimed at developing water quality criteria for chemicals in the environment, it has been proven important to investigate toxicity in conditions that deviate from standard laboratory conditions since they can influence toxicity (Vergauwen et al. 2013).

For zebrafish, a number of laboratory strains exist, which are referred to as wild-type and outbred. Tübingen is an important source of zebrafish strains, and most strains are available from the Zebrafish International Resource Center (ZIRC, https://zebrafish.org). Coe et al. (2009) assessed the genetic variation and diversity of the most commonly used wild-type strains of zebrafish (AB, TE, TL, WIK) and compared them to a sample of wild zebrafish from Bangladesh. The authors showed that genetic variation of the four laboratory strains as well as fish purchased from a commercial dealer was less than 20% of the variation found in wild fish. They also constructed a phylogram showing that wild fish form a clade separate from all laboratory strains, while they did not observe a clear grouping of any of the laboratory strains, suggesting cross-breeding between those strains. The low level of genetic variation in laboratory strains may affect behavior, fitness, susceptibility to chemicals, etc. For the fathead minnow, many laboratories originally obtained a fathead minnow stock from the Newtown EPA facility in Cincinnati (OH, USA), and routinely outbred them with wild fish according to the guidelines for the culture of fathead minnows for use in toxicity tests (USEPA 1987). Medaka has a high tolerance to inbreeding. Inbred strains have been developed which differ in behavior, body shape, brain morphology and susceptibility to mutagens (Kirchmaier et al. 2015). Both wild strains and inbred strains of medaka are available at the Japanese Medaka Stock Center (National BioResource Project Medaka, NBRP Medaka; http://www.shigen.nig.ac.jp).

4.2.6 Genomic Information

Genome projects for zebrafish (The Sanger Institute, www.sanger.ac.uk/Projects/D_rerio), medaka (National Institute of Genetics) three-spined stickleback and rainbow trout, among others, and efforts for other species such as fathead minnow (Burns et al. 2015) and carp provide genomic information. This information allows for the easy application of toxicogenomic techniques to measure changes of transcription and translation in response to a chemical insult. Such changes can be important key events, especially on the upstream part of an AOP (MIE and KEs at the cellular level). Additionally, it facilitates phylogenetic comparisons among species to assess conservation of toxicity mechanisms and therefore taxonomic applicability of AOPs (see Sect. 4.2).

In addition to DNA sequence information, it is often important to know the timing of activation of toxicity targets (e.g., enzymes, ion channels, receptors), in terms of transcription and translation, during embryonic development. This may explain susceptibility differences between life-stages which are important when the fish embryo is used as an alternative to predict toxicity at later ages.

4.2.7 Availability of Knockdowns, Knockouts and Transgenic Lines

The increasing availability of genomic information and the development of new methodologies have led to an increasing flexibility to apply knockdowns as well as genetic modifications in knockout and transgenic fish models. Lee et al. (2015) recently provided an overview of available methods in fish and their application in ecotoxicology. Zebrafish and medaka are the most popular model fish species for genomic modifications (Lee et al. 2015). The Zebrafish Mutation Project (ZMP) aims to create a knockout allele in every protein-coding gene in the zebrafish genome (https://www.sanger.ac.uk/resources/zebrafish/zmp/). Here, we provide some examples of how these techniques can benefit AOP development and application.

One of the most convincing lines of evidence for an AOP linking an MIE (for example inhibition of an enzyme) to a specific phenotype, is the use of loss of function models such as knockdowns and knockouts. Morpholino knockdowns can be designed to block RNA translation by hybridizing to the target sequence, usually resulting in incomplete loss of function (Bill et al. 2009). RNAi is another means for knockdown of specific genes (Kelly and Hurlstone 2011). Recent innovations with regard to the generation of knockouts (complete loss of function through inactivation of the gene of interest) have greatly improved their application. While earlier methodologies used random mutagenesis to create knockouts, during the last few years targeted mutagenesis methods have been developed based on engineered endonucleases such as zinc finger nucleases (ZFNs) and transcription activator-like effector nucleases (TALENs) (Bedell et al. 2012; Doyon et al. 2008). Recently, a new technique for targeted mutagenesis has emerged, the bacterial type II clustered regularly interspaced short palindromic repeats (CRISPR)/CRISPR-associated (Cas) system, which is applicable in zebrafish (Gagnon et al. 2014; Hwang et al. 2013a, b; Jao et al. 2013). The CRISPR/Cas technology can be used for the purpose of knockout as well as knock-in. Efforts are being made to apply the technology to high-throughput gene targeting and phenotyping using zebrafish (Varshney et al. 2015). Using knockdowns or knockouts, it is possible to confirm the occurrence of the KEs along multiple levels of biological organization in an AOP by specifically inducing the MIE, for example knockdown of an enzyme as a model for chemical-induced enzyme inhibition. This is much more convincing as weight of evidence for the KERs in an AOP compared to chemical exposures, since chemicals are usually suspected to target more than one MIE.

If binding to a specific receptor is thought to be the MIE of a hypothesized AOP, knockdown or knockout of the specific receptor can be used to find out whether this prohibits the progression of downstream KEs upon exposure to a chemical thought to act through this AOP. For example, polyaromatic hydrocarbons (PAHs) are often thought to cause toxicity through aryl hydrocarbon (AhR) receptor activation, similar to dioxins. Incardona et al. (2005) used AhR knockdown zebrafish to show that developmental defects induced by weathered crude oil exposure are mediated by low-molecular-weight tricyclic PAHs through an aryl hydrocarbon receptor-independent disruption of cardiovascular function and morphogenesis. Brown et al. (2015) further investigated the cardiotoxicity of weak AhR agonists using AhR and cyp1a knockdowns. Such information is highly useful for AOP development and can significantly contribute to the weight of evidence associated to the KERs in an AOP.

Many transgenic zebrafish lines have been developed in which the expression of a fluorescent protein (e.g., green fluorescent protein, GFP) is driven by the promotor of a gene of interest. As a result, localized fluorescence can be detected upon expression of the gene of interest. As a first application, this can aid in identifying and visualizing specific tissues by targeting tissue-specific genes. For example, in the vas::egfp zebrafish, expression of enhanced green fluorescent protein (eGFP) is driven by the promoter of the germ-cell-specific gene vasa (*vas*). This aids in the visualization of primordial germ cells during early embryonic development and the developing gonads later in development. Another example is the fli:egfp transgenic zebrafish line in which the fli1 promotor drives expression of eGFP in all blood vessels throughout embryogenesis (Lawson and Weinstein 2002). This model system is ideal for observing vascular defects in response to chemical exposure (Delov et al. 2014).

Transgenic fish, in addition to their role in visualization of specific cell-or tissue types, can also function as reporter lines for chemical screening purposes when the expression of fluorescent proteins is driven by the promotor of specific target genes. For example, the TSHβ:EGFP transgenic zebrafish can function as a model for screening for thyroid-disrupting chemicals (Ji et al. 2012). Gorelick and Halpern (2011) developed a zebrafish reporter line to screen for estrogen activity. This approach is especially powerful when a reporter assay can be selected based on an AOP describing the linkages between a KE that can be measured using the reporter, and the AO of interest.

4.2.8 Biotransformation Capacity During Embryonic Development

Biotransformation is the process by which an exogenous or endogenous compound is chemically modified by enzymatic activity. Biotransformation is a physiological phenomenon and is used to convert substances into those that are required for normal body function, e.g. steroidogenesis. However, biotransformation also serves as an important defence mechanism by converting toxic xenobiotics into less harmful substances that can be excreted from the body. In some cases though, biotransformation results in metabolites that are more toxic than the parent compound (Nebert and Russell 2002). This process is then called bioactivation. Formally, biotransformation reactions are classified as Phase I and Phase II reactions. Phase I reactions add a polar group to make compounds water-soluble, which is largely achieved by Cytochrome P450 (CYP) enzymes but also other enzymes can be involved such as the flavin-containing monooxygenase (FMO) system. Phase II reactions involve a covalent attachment of a small polar endogenous molecule to create a final compound of higher molecular weight (Ashauer et al. 2012). The latter is mainly achieved by glucuronidation or sulfation but other conjugation reactions can also occur. Finally, also Phase III reactions, which involve uptake and/or efflux of xenobiotics and/or their metabolites by transporters, influence the elimination of compounds. Since the toxic effects of xenobiotics depend on the achieved exposure within the organism, a thorough knowledge of Phase I, Phase II and Phase III reactions is necessary for proper risk assessment. Especially in view of AOP development, it is critical to know whether fish embryos biotransform chemicals in the same way as an adult fish.

In adult zebrafish, there has been an exhaustive characterization of CYP-mediated metabolism at the gene level, but also at the functional level. These studies suggest the presence of similar metabolic systems to those found in mammalian species (reviewed by Saad et al. 2016a). Analysis of the zebrafish genome has uncovered a total of 86 *CYP* genes that fall into seventeen categories of *CYP* gene families (Genome Reference Consortium 2015). Of these CYP families, CYP1–3 is the main xenobiotic metabolizing enzyme group. By using several well-characterized compounds in vitro and in vivo, it has been demonstrated that zebrafish clearly possess CYP1, 2 and 3 activity (Alderton et al. 2010; Chng et al. 2012; Diekmann 2013). This CYP activity often also resembles the human situation (Chng et al. 2012), although differences in metabolite profile and isoforms have been reported (Alderton et al. 2010; Diekmann 2013). Zebrafish also possess Phase II metabolic capacities as evidenced by UDP-glucuronosyltransferase (UGT) activity on testosterone (Chng et al. 2012) and activity of different sulfotransferases (SULT) (Kurogi et al. 2010; Liu et al. 2010). Also, several drug transporters have been identified in the zebrafish, e.g. 41 ABC transporters (Dean and Annilo 2005), of which some have already been characterized functionally (Fischer et al. 2013). Several xenobiotics also influence their concentrations in the organism by inhibition or induction of particular enzymes or transporters. For CYP induction, pregnane X-receptor (PXR) and aryl hydrocarbon receptor (AHR) are well known transcription factors in zebrafish and humans (Reschly and Krasowski 2006). However, constitutive androstane receptor (CAR), a third important regulatory mechanism of CYP induction in humans (Waxman 1999), is absent in zebrafish and teleost fish in general. The number of substrates that stimulate PXR also appears to be more limited in zebrafish than in humans (Ekins et al. 2008) although PXR and CYP3A are induced by a similar mechanism in both species (Bresolin et al. 2005). For AHR, only one functional AHR is detected in humans (Hahn 2002), whereas zebrafish AHR have multiple signaling members including AHR1a, AHR1b, AHR2, ARNT1, ARNT2 and two AHR repressors (Karchner et al. 2005). ARNTs are aryl hydrocarbon receptor nuclear translocators that dimerize with AHR after its translocation from the cytosol into the nucleus. As such, the high DNA binding affinity of this complex stimulates transcription of the *CYP1A1* gene and other genes (Denison and Nagy 2003). Regarding substrate affinity, AHRs bind to a broad range of aromatic and halogenated chemicals including planar halogenated aromatic hydrocarbons (pHAH) and polycyclic aromatic hydrocarbons (PAH), which are both known as environmental contaminants. Also drug transporter activity can be affected in the zebrafish by environmental chemicals, i.e. by inhibition, leading to so-called chemosensitisation (Otte et al. 2010; Scholz et al. 2008).

The available information on the biotransformation capacity of zebrafish embryos is much more limited. Most studies focus on the larval stages (96 hpf and 120 hpf), which are at the end of organogenesis and thus vital information for the early developmental stages is lacking. Knöbel et al. (2012) showed that the zebrafish embryos' toxic response, evidenced by lethality, to 38 chemicals with different physicochemical properties and mode of action was similar to the response in adult fish. However, the situation may be different for effects that are exerted by active metabolites and not by the parent compound if the required biotransformation enzymes are not present or mature yet in the embryo. The metabolic capability has been and still is a hurdle in the development of several alternative methods for animal experiments in toxicology (Spielmann et al. 2006). For CYPs, distinct spatio-temporal patterns of gene expression have been explored (Goldstone et al. 2010) and often already early peaks in expression (up to 4 hpf) are detected (Glisic et al. 2014). This probably indicates a maternal origin of the transcripts. Glisic et al. (2014) also showed that the CYP expression is inducible in 24 hpf embryos in a similar way as in adults. Indeed, atrazine exposure for up to 72 h significantly increased the CYP1A and CYP3A65 mRNA levels, albeit at a 1000 times higher concentration than present in the environment. For Phase II metabolism, it has also been shown that zebrafish embryos express all major GST isoforms (Glisic et al. 2014) and all major UGTs (Christen and Fent 2014) from very early on (4 hpf) and with clear temporal patterns. Also drug transporter transcripts have been found in very early zebrafish embryos (from 1 hpf onwards) (Fischer et al. 2013) and this was also reflected functionally as ABCB1-like efflux was inhibited in 1, 6, 12, 24 and 48 hpf embryos when using several transporter inhibitors. However, the available data are scarce on the activity of CYPs and Phase II enzymes in zebrafish embryos. So far, the only extensive study on metabolism in non-adult zebrafish was performed by Alderton et al. (2010) and they mainly focused on larvae of 168 hpf. Several compounds were tested, of which 3 were also evaluated in 72 hpf larvae. Although the 72 hpf and 168 hpf larvae were able to perform Phase I and/or Phase II metabolic reactions, only a small fraction of most of the compounds was found as metabolites in the larvae. Therefore, the authors concluded that the quantified metabolites were unlikely to contribute to observed toxicity (Alderton et al. 2010). Phase I and Phase II metabolism has also been reported by other groups in 96 hpf or older larvae (Creusot et al. 2014; Jones et al. 2010; Li et al. 2011). Otte et al. (2010) investigated earlier time points and detected CYP1A activity, assessed by an EROD assay, as early as 8 hpf. At that time point EROD activity was present in the cytoplasm of the envelope layer and in the yolk syncytial layer as well. Saad et al. (2016b) also confirmed this early EROD activity in homogenates of 5 hpf embryos and Bräunig et al. (2015) detected basal EROD activity in embryos at 24 hpf, which was clearly induced by beta-naphthoflavone after 96 h of exposure. The presence of CYP activity during early zebrafish embryonic development is not surprising as CYPs are also critical for morphogenesis, e.g. the role of CYP26 in regulating the retinoic acid concentrations in hindbrain development (Hernandez et al. 2007). However, the question remains whether the biotransformation capacity of the embryos is sufficient to bioactivate xenobiotics during the different developmental stages, in comparison to adult fish. This is still a point of controversy. Weigt et al. (2011) performed a study with several proteratogens and showed teratogenicity of the compounds. However, no analysis of biotransformation was performed and therefore this study could not answer whether these compounds were proteratogenic or teratogenic by themselves in the zebrafish. So far, allyl alcohol is the only compound for which it has been clearly demonstrated that zebrafish embryos cannot bioactivate it when exposed from 1.5 hpf until 50 hpf (Knöbel et al. 2012). Although this was due to a lack in alcohol dehydrogenase 8a activity in these embryos (Kluver et al. 2014) and not to immature Phase I or Phase II reactions, this still underlines the importance of the embryo's metabolic capacity. Therefore, co-incubation of zebrafish embryos with an exogenous metabolic activation system (MAS) has been suggested on several occasions (Busquet et al. 2008; Mattsson et al. 2012; Weigt et al. 2010). However, this complicates the assay (Pype et al. 2015) and more importantly continuous co-incubation is not possible due to embryotoxicity caused by MAS, i.e. liver microsomes and NADPH, itself (Mattsson et al. 2012). Therefore, only short and intermittent co-incubation of zebrafish embryos with MAS can be applied, which may lead to lack of exposure to metabolites during critical windows of development and consequently false negative results in the case of toxicity testing of proteratogens.

Biotransformation of xenobiotics has also been studied in the fathead minnow and medaka, but to a lesser extent. Experiments with PXR inducers clotrimazole and pregnane-16alpha-olone clearly increased the expression of PXR and CYP3A in the fathead minnow (Crago and Klaper 2011). Furthermore, CYP3A inducer rifampicin also clearly increased CYP3A activity in this species (Christen et al. 2010). CYP3A activity has also been reported for adult medakas (Kullman et al. 2004). In juvenile medakas, benzo(*a*)pyrene (BaP) induced the transcript levels of CYP1A and CYP2A together with those of glutathione-S-transferase (GST) and UGT, including their activity (Kim et al. 2014; Rhee et al. 2013). Interestingly, the pesticide aldicarb appears to be biotransformed by FMOs in adult medakas (El-Alfy and Schlenk 1998). Regarding the biotransformation capacity of medaka embryos, the earliest appearance of BaP metabolites was at 21–24 hpf in the yolk syncytial layer although diffuse metabolic activity may also have been present at this time within the yolk itself (Hornung et al. 2007). Recently, CYP1A activity could also be induced in medaka embryos by using beta-naphthoflavone (Gonzalez-Doncel et al. 2015).

4.3 AOP Development Using Fish Embryos

There can be different drivers for using fish embryos in AOP development. For example, AOP development can be driven by the need for alternative test methods to replace animal testing. Testing for chronic toxicity using FELS tests and screening for endocrine disruption using sexual development tests (OECD TG 234; OECD 2011) and fish short-term reproduction assays (OECD TG 230; OECD 2009b) require many animals and alternative methods are needed for these tests. AOPs can aid in selecting endpoints that can be measured in the fish embryo and that are predictive of AOs of ecological relevance, namely growth, survival and reproduction, which are traditionally measured in these animal tests. On the other hand, since vertebrate development, physiology and anatomy is highly conserved, data from fish (embryos) can be used as part of the weight of evidence to develop AOPs with a broader taxonomic applicability (e.g., AChE inhibition leading to acute mortality, Russom et al. 2014).

For AOP development, specific toxicity information is needed at multiple levels of biological organization, rather than only apical endpoints. Even though fish embryos are small, there is an enormous potential to measure endpoints ranging from the molecular, over the biochemical to the physiological level. New techniques are being developed or adapted from other model systems at a rapid pace.

4.3.1 Iterative AOP Development Cycle

For AOP development, it is essential to gather weight of evidence to support linkages between key events. AOP development can be visualized as an iterative process in which a hypothesized AOP is challenged experimentally and adapted accordingly until it reaches the level of detail and confidence needed for the envisaged application (left part of Fig. 4.3). Consequently, the AOP can be used to develop assays based on KEs which are predictive of the AO (right part of Fig. 4.3). Assay development can also give rise to new insights leading to updated versions of the AOP. In this way AOPs are living documents (Villeneuve et al. 2014b). This concept is facilitated by the AOPWiki (https://aopwiki.org), which allows for continuous addition and updates when new information becomes available. The fish embryo is highly amenable to such an iterative AOP development approach since quick and low-cost experiments can easily be set up to investigate specific KEs.

4.3.2 Development of FELS AOPs

AOP development can be tailored specifically to the goal of developing alternative tests for assessing chronic fish toxicity. Since the FELS test is currently the most important test to assess chronic toxicity, AOPs describing KEs measurable in the fish embryo leading up to FELS AOs would facilitate selection of predictive assays. There are two central criteria for KEs: (1) they should be measurable/observable,

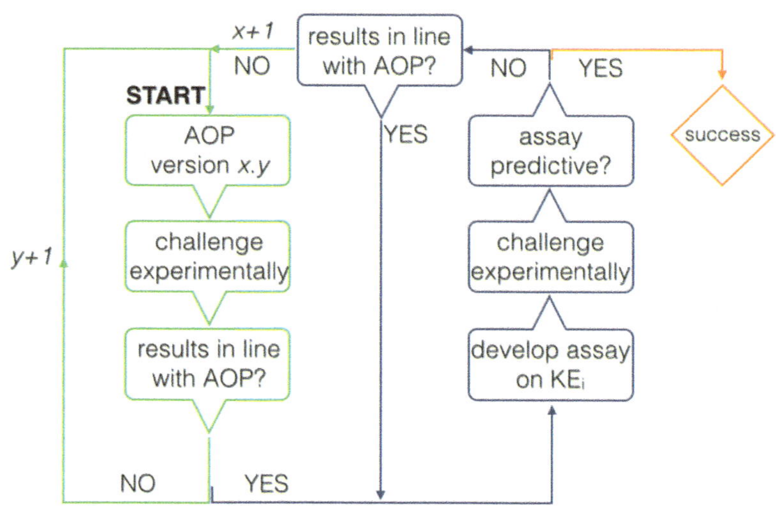

Fig. 4.3 Schematic representation of an iterative AOP development cycle. In the *green part*, experiments are aimed at developing the AOP up to the level that it can be used to develop assays based on upstream events that are predictive of downstream events (*blue part*)

and (2) they should be essential, but not necessarily sufficient for the progression from a defined biological perturbation toward a specific AO (Villeneuve et al. 2014b).

Villeneuve et al. (2014a) made some recommendations regarding the prioritization of AOP development specifically for developing alternatives to the FELS test. If the AO is directly observable in the fish embryo, there is no need to develop an AOP since a fish embryo test can directly be used to screen for this AO. If the AO is not directly observable in the fish embryo, delineating KEs leading to this AO may lead to predictive fish embryo assays. If there are no KEs that can be observed in fish embryo assays, in vitro assays may be considered. The latter two cases are considered priorities for AOP development.

Villeneuve et al. (2014b) presented an overview of common AOP development strategies. One could start by identifying ecologically relevant FELS adverse outcomes, and subsequently build AOPs delineating KEs which are measurable in fish embryos leading up to these AOs of interest, the so called top-down approach. The main AOs of regulatory relevance are growth, survival and reproduction, of which the first two are covered in the FELS test. Since these AOs are not specific and thus both are regulated by a broad array of factors, one would envision a highly complicated AOP network making it difficult to prioritize. Therefore a middle-out approach, starting from KEs that are observable in fish (not necessarily in embryos) and can be plausibly linked to relevant AOs has been applied for FELS AOP development (Groh et al. 2015; Villeneuve et al. 2014a). For this purpose, Villeneuve et al. (2014a) started by outlining a conceptual model of developmental morphological landmarks during zebrafish embryogenesis (e.g., somite formation, cardiovascular system development), to aid in identifying KEs that lead to FELS AOs. The authors used swim bladder inflation as an example KE to function as a starting point for AOP development. During early development, zebrafish undergo an embryonic-to-larval transition phase marking an important switch from yolk sac- to exogenous feeding larvae around 120 hpf. This transition includes swim bladder inflation (posterior chamber, around 96 hpf), structural and functional maturation of the mouth and gastrointestinal tract, and resorption of the yolk sac (Liu and Chan 2002). Impaired posterior chamber inflation is not directly lethal, but it impacts growth and survival especially in natural habitats where swimming capacity is essential for foraging and predator avoidance (Czesny et al. 2005; Villeneuve et al. 2014a). Later during development (around 21 dpf for zebrafish) the anterior swim bladder chamber inflates, which has an additional role in hearing (Lechner and Ladich 2008; Popper 1974). Since inflation of the posterior chamber is observed at the border of legal limitations with regard to alternative testing, and inflation of the anterior chamber can only be observed long after the time frame of the FET test, they are a priority for AOP development. By selecting swim bladder inflation as a KE, the biological pathways of interest have been narrowed down. Subsequently, a more thorough study of the biological pathways leading to the normal formation and function of the swim bladder allows for hypothesizing how chemicals can disrupt these processes. In zebrafish, these formation processes have been studied in detail (Robertson et al. 2007; Teoh et al. 2010; Winata et al. 2009; Yin et al. 2011). Several studies have suggested the involvement of thyroid hormones in posterior chamber inflation (Jomaa et al. 2014; Liu and Chan 2002). Therefore, we hypothesized that thyroid disruption can impair swim bladder inflation. Based on a literature search, a putative AOP leading from thyroid peroxidase (TPO) inhibition to impaired swim bladder inflation was constructed.

During the AOP development phase, knowledge gaps are usually identified. The logical next steps for AOP development can include targeted experimental studies that are set up to address these gaps and increase confidence in the hypothesized AOP. This process can be iterated until the AOP is sufficiently developed to serve its purpose (Fig. 4.3). In the thyroid example, assays were developed to measure the KEs along this AOP and these assays were applied to test the hypothesized AOP in two fish species, the zebrafish and the fathead minnow (Nelson et al. 2016; Stinckens et al. 2016). The two species were exposed to 2-mercaptobenzothiazole (MBT), an environmentally relevant TPO inhibitor. Whole-body T4 decreased upon MBT exposure. Anterior chamber inflation was impaired, and there was a clear relationship between T4 levels and the anterior chamber surface in zebrafish. The absence of effects on posterior chamber inflation was not expected, but can possibly be explained by maternal transfer of T4 into the eggs (see Sect. 4.2.2). Deiodinase (ID) type 1 (ID1) and type 2 (ID2) are essential to activate T4 (including maternally derived T4) into its biologically active form, T3. If the inflation process of the posterior swim bladder chamber is indeed mediated by thyroid hormones, but maternal T4 transfer is sufficient to compensate for TPO inhibition, we can assume that TPO inhibitors do not impair posterior chamber inflation, while ID inhibitors do (Stinckens et al. 2016). This has led to a new version of the hypothesized AOP. In zebrafish deiodinase knockdown studies we indeed showed impaired posterior chamber inflation (Bagci et al. 2015; Heijlen et al. 2014). When sufficiently developed, this AOP can be used for the selection of assays predictive for thyroid mediated effects on swim bladder inflation, leading to reductions in growth and survival.

Another example of applying a middle-out approach for FELS AOP development was given by Groh et al. (2015). The authors selected growth as the FELS AO of interest, and subsequently selected the KE 'reduction in food intake' as starting point for middle-out AOP development. To ensure relevance of the AOP under development, the authors applied four criteria to the selection of this KE: (1) the process underlying the KE should be important for growth regulation, (2) the biological pathways underlying the KE should be highly conserved among species, (3) the KE should be susceptible to many chemicals, (4) the KE should be induced at environmentally relevant chemical concentrations. Subsequently, the authors identified impaired locomotion as a KE that is already observable in fish embryos and plausibly linked to reduced food intake. From then on the authors developed a selection of case studies in which AOPs are delineated.

Volz et al. (2011) also used three case studies to delineate FELS AOPs as a basis for a tiered testing approach to reduce the need for FELS tests. The AOPs were based on three reference chemicals with known MOA: 2,3,7,8-tetrachlorodibenzo-p-dioxin (TCDD)–induced cardiotoxicity, chlorpyrifos (CPF)–mediated inhibition of neurite outgrowth, and linear alkylbenzene sulfonate (LAS)-induced gill toxicity and narcosis. Although such case studies are valuable starting points, it is important to note that the goal of AOP development is not to develop AOPs that are chemical specific, but to develop AOPs that reflect mechanisms through which several (classes of) chemicals elicit toxicity (Villeneuve et al. 2014b). In a next step, evidence can be assembled to support generalization over larger groups of chemicals.

4.3.3 Development of AOP Networks

A single AOP is considered as a pragmatic unit for AOP development and not as a complete biological representation of toxicological processes encompassing all possible molecular, biochemical and physiological components involved. Consequently, individual AOPs are generally conceptualized as a "linear" construct, without converging or diverging pathways connected to it. However, it is recognized that a single AOP may not capture all events that contribute to any relevant toxic effect. In the example of thyroid disruption (see Sect. 4.3.2), it became clear that thyroid disruptors impact swim bladder inflation, with an important distinction among specific subtypes of TH disrupting compounds (e.g., TPO inhibitors vs ID inhibitors). In such cases, several MIEs can converge in the same downstream KEs. AOP networks are defined as sets of AOPs sharing at least one common element, and are capable of more realistically representing potential chemical effects. They provide information on interactions between AOPs and have the potential to reveal previously unknown links between biological pathways. Analysis of these AOP networks can aid the prioritization of assay development, whether the goal is to develop a single assay with predictive utility of multiple outcomes, or development of assays that are highly specific for a particular mode of action (Knapen et al. 2015). In Knapen et al. (2015) we provided an example of an AOP network for reproductive and developmental toxicity in fish that was built based on the five relevant AOPs that were available for fish in the AOP Wiki (AOP Nos. 21, 23, 25, 29 and 30). This way, we illustrated how AOP networks can be used for assay development and refinement (Fig. 4.4). In this example, reduced estradiol synthesis in granulosa cells is linked to two different MIEs, while reduced vitellogenin synthesis in hepatocytes is linked to three different MIEs, meanwhile reduced testosterone concentration in theca cells is uniquely linked to androgen receptor agonism. While all three KEs lead to, and can potentially be used to predict, the same AO of decreased female fecundity in terms either of egg production or embryonic survival, they have varying specificity with respect to the MIE triggering the chain of events. In general, AOP networks therefore offer the potential to guide the development of assays with different degrees of specificity for toxicological mode(s) of action, being indicative of either a very specific MIE or, alternatively, of clusters of mechanistically related MIEs. This type of assay development logic may be particularly useful for differential screening of compounds with unknown molecular targets, e.g. in the context of Integrated Approaches to Testing and Assessment (IATA, Tollefsen et al. 2014), in which sequential elimination of possible mechanisms may be quickly achieved

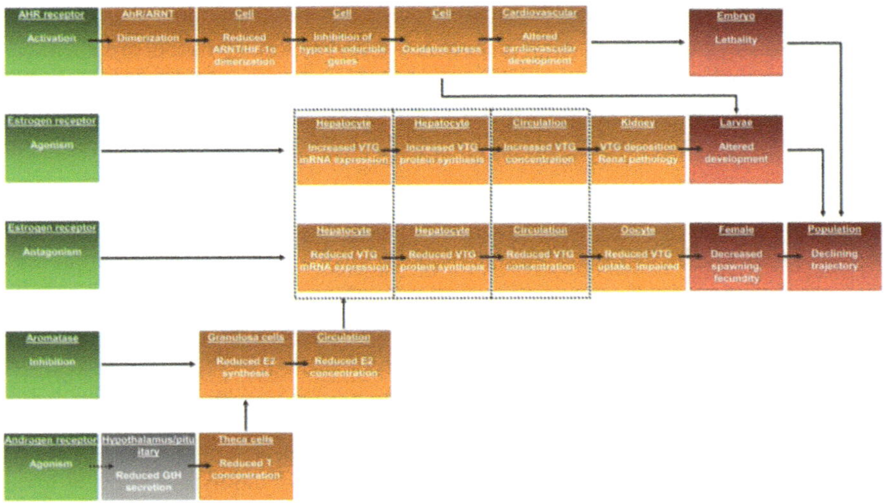

Fig. 4.4 Example of an AOP network based on the five reproductive and developmental toxicity-related AOPs that were available for fish in the AOP Wiki (Jan. 2015). MIEs are indicated in *green*, KEs in *orange*, and AOs in *red*, as per the AOP Wiki template. The *dotted squares* indicate KEs that are defined as changes in opposite direction (increase versus decrease) of the same biological component. *AHR* aryl hydrocarbon receptor, *GtH* gonadotrope hormone, *T* testosterone, *VTG* vitellogenin, *E2* estradiol. KE descriptions have been directly derived from the AOP Wiki whenever possible. In some cases, slight modifications of descriptions were necessary to generate a reusable KE in this specific network. This figure illustrates the AOP network approach but does not make any assumptions about the scientific validity of the underlying AOPs. AOPs, and hence the depicted AOP network, may be subject to change before they are formally finalized (Source: Reprinted from Knapen et al. (2015) with permission from Elsevier)

using assays probing strategically chosen KEs in an AOP network. Such an approach has, for example, already been implicitly implemented to some extent in the Fish Sexual Developmental Test (OECD TG 234, OECD 2011).

4.3.4 Development of Fish AOPs for Endocrine Disruption

When specifically developing AOPs for endocrine disruption, such as those in Fig. 4.4, a number of aspects should be considered. There have been many different definitions of endocrine disrupting chemicals (EDCs), some very strict and others very broad. For regulatory applications, EDCs are currently widely defined as agents that cause alterations in reproduction or development through direct effects on the vertebrate hypothalamic–pituitary–thyroidal or hypothalamic–pituitary–gonadal (HPG) axes (USEPA 1998). Both USEPA and OECD have developed tiered testing frameworks to screen for endocrine disrupting potential at low levels of biological organization using non-animal tests before proceeding to long-term tests to observe the AO. The selection of appropriate exposure concentrations is essential to avoid confounding effects of systemic toxicity on endocrine endpoints, and thereby avoid false positives. For this reason, the concept of maximum tolerated concentration (MTC) has been adopted (Hutchinson et al. 2009; Wheeler et al. 2013). The MTC is the highest concentration at which no mortality or signs of morbidity (e.g., feeding inhibition, abnormal behavior, morphology or color) are observed. Beyond this MTC a specific toxicity observation cannot be attributed to a test chemical since the general health of the organism has been compromised. Therefore, the test concentrations for tests assessing potential endocrine activity in fish should be below this MTC.

Ankley et al. (2009) provided a conceptual model of how the hypothalamic–pituitary–gonadal axis regulates fish reproduction and where/how chemicals with different MOAs can disrupt these pathways. This model has been important for the development of AOPs leading from aromatase inhibition, estrogen receptor (ant) agonism and androgen receptor (ant)agonism to reproductive impairment. These AOPs for endocrine disruption in fish are among the most advanced AOPs developed thus far (current knowledge is integrated in the AOP network in Fig. 4.4).

These AOPs have been developed based on experimental data from adult fish tests. In this respect, many of the aspects discussed in Sect. 4.2 are of particular importance. Since the reproductive system is not yet developed in fish embryos, reproductive dysfunction cannot be directly measured in embryos. Attempts are being made to identify KEs that are already measurable in fish embryos and that eventually lead to and thus are predictive of endocrine relevant AOs. A project currently on the OECD workplan in relation to endocrine disruptor testing and assessment called 'zebrafish embryo assay for the detection of endocrine active substances acting through the estrogen receptor' (EASZY), aims to detect endocrine active substances acting through human ER, using transgenic cyp19a1b-GFP zebrafish embryos (Carvalho et al. 2014). Schiller et al. (2014) showed that transcription of common endocrine disruption markers such as aromatase and vtg responded to exposure to endocrine disrupting chemicals in zebrafish and medaka embryos, and that the responses were generally comparable to those in later life stages.

Chemicals causing developmental outcomes are sometimes included in the group of EDCs (see definition above). While fish embryos are used to investigate mechanisms of toxicity which also occur in later life stages, they are obviously especially useful to investigate disruptions of development. An example is the AOP for AhR receptor activation leading to altered cardiovascular development and embryo toxicity (Fig. 4.4).

4.4 Conclusion

Both vertebrate and invertebrate toxicity tests are used to provide an ecologically relevant basis for environmental quality criteria. Within the vertebrate tests, fish are valuable sentinels for evaluating aquatic toxicity since they represent a high trophic level in the aquatic food chain. Experiments using the early life-stages of fish are more cost-efficient compared to tests using adult fish while maintaining the physiological relevance of a vertebrate whole-organism test system. Ethical considerations are also a driver of the use of fish embryos since they are considered alternative testing models during early development.

The AOP approach offers an interesting framework for developing alternative testing methods using fish embryos. For AOP development, specific toxicity information is needed at multiple levels of biological organization, rather than only apical endpoints. Even though fish embryos are small, there is an enormous potential to measure endpoints ranging from the molecular, over the biochemical to the physiological level. As a better fundamental understanding of fish biology under both normal and chemical exposure conditions becomes available, the fish embryo is becoming increasingly useful for AOP development. Once developed, AOPs can facilitate the identification of assays targeting key events, which have high predictive value for an adverse outcome of interest.

References

Alderton W, Berghmans S, Butler P, Chassaing H, Fleming A, Golder Z, Richards F, Gardner I (2010) Accumulation and metabolism of drugs and CYP probe substrates in zebrafish larvae. Xenobiotica 40:547–557

Ankley GT, Bencic DC, Breen MS, Collette TW, Conolly RB, Denslow ND, Edwards SW, Ekman DR, Garcia-Reyero N, Jensen KM, Lazorchak JM, Martinovic D, Miller DH, Perkins EJ, Orlando EF, Villeneuve DL, Wang R-L, Watanabe KH (2009) Endocrine disrupting chemicals in fish: developing exposure indicators and predictive models of effects based on mechanism of action. Aquat Toxicol 92:168–178

Ankley GT, Bennett RS, Erickson RJ, Hoff DJ, Hornung MW, Johnson RD, Mount DR, Nichols JW, Russom CL, Schmieder PK, Serrano JA, Tietge JE, Villeneuve DL (2010) Adverse outcome pathways: a conceptual framework to support ecotoxicology research and risk assessment. Environ Toxicol Chem 29:730–741

Ankley GT, Jensen KM (2014) A novel framework for interpretation of data from the fish short-term reproduction assay (FSTRA) for the detection of endocrine-disrupting chemicals. Environ Toxicol Chem 33:2529–2540

Ashauer R, Hintermeister A, O'Connor I, Elumelu M, Hollender J, Escher BI (2012) Significance of xenobiotic metabolism for bioaccumulation kinetics of organic chemicals in Gammarus pulex. Environ Sci Technol 46:3498–3508

Bagci E, Heijlen M, Vergauwen L, Hagenaars A, Houbrechts AM, Esguerra CV, Blust R, Darras VM, Knapen D (2015) Deiodinase knockdown during early zebrafish development affects growth, development, energy metabolism, motility and phototransduction. PLoS One 10:e0123285

Bedell VM, Wang Y, Campbell JM, Poshusta TL, Starker CG, Krug RG II, Tan W, Penheiter SG, Ma AC, Leung AYH, Fahrenkrug SC, Carlson DF, Voytas DF, Clark KJ, Essner JJ, Ekker SC (2012) In vivo genome editing using a high-efficiency TALEN system. Nature 491:114–U133

Beekhuijzen M, de Koning C, Flores-Guillen M-E, de Vries-Buitenweg S, Tobor-Kaplon M, de Waart B, Emmen H (2015) From cutting edge to guideline: a first step in harmonization of the zebrafish embryotoxicity test (ZET) by describing the most optimal test conditions and morphology scoring system. Reprod Toxicol 56:64–76

Bill BR, Petzold AM, Clark KJ, Schimmenti LA, Ekker SC (2009) A primer for morpholino use in zebrafish. Zebrafish 6:69–77

Braunbeck T, Bottcher M, Hollert H, Kosmehl T, Lammer E, Leist E, Rudolf M, Seitz N (2005) Towards an alternative for the acute fish LC50 test in chemical assessment: the fish embryo toxicity test goes multi-species – an update. Altex-Alternativen Zu Tierexperimenten 22:87–102

Braunbeck T, Lammer E (2006) Draft detailed review paper on fish embryo toxicity assays. UBA contract number 203 85 422. Umweltbundesamt–German Federal Environment Agency, Dessau. 136 pp

Bresolin T, de Freitas RM, Celso Dias BA (2005) Expression of PXR, CYP3A and MDR1 genes in liver of zebrafish. Comp Biochem Physiol C Toxicol Pharmacol 140:403–407

Brown DR, Clark BW, Garner LVT, Di Giulio RT (2015) Zebrafish cardiotoxicity: the effects of CYP1A inhibition and AHR2 knockdown following exposure to weak aryl hydrocarbon receptor agonists. Environ Sci Pollut Res 22:8329–8338

Bräunig J, Schiwy S, Broedel O, Muller Y, Frohme M, Hollert H, Keiter SH (2015) Time-dependent expression and activity of cytochrome P450 1s in early life-stages of the zebrafish (Danio rerio). Environ Sci Pollut Res Int. doi:10.1007/s11356-11015-14673-11356

Burns FR, Cogburn AL, Ankley GT, Villeneuve DL, Waits E, Chang Y-J, Llaca V, Deschamps SD, Jackson RE, Hoke RA (2015) Sequencing and *de novo* draft assemblies of a fathead minnow (Pimephales promelas) reference genome. Environ Toxicol Chem 9999:1–6

Busquet F, Nagel R, von Landenberg F, Mueller SO, Huebler N, Broschard TH (2008) Development of a new screening assay to identify proteratogenic substances using zebrafish danio rerio embryo combined with an exogenous mammalian metabolic activation system (mDarT). Toxicol Sci 104:177–188

Carvalho RN, Arukwe A, Ait-Aissa S, Bado-Nilles A, Balzamo S, Baun A, Belkin S, Blaha L, Brion F, Conti D, Creusot N, Essig Y, Ferrero VEV, Flander-Putrle V, Furhacker M, Grillari-Voglauer R, Hogstrand C, Jonas A, Kharlyngdoh JB, Loos R, Lundebye A-K, Modig C, Olsson P-E, Pillai S, Polak N, Potalivo M, Sanchez W, Schifferli A, Schirmer K, Sforzini S, Sturzenbaum SR, Softeland L, Turk V, Viarengo A, Werner I, Yagur-Kroll S, Zounkova R, Lettieri T (2014) Mixtures of chemical pollutants at European legislation safety concentrations: how safe are they? Toxicol Sci 141:218–233

Chang J, Wang M, Gui W, Zhao Y, Yu L, Zhu G (2012) Changes in thyroid hormone levels during zebrafish development. Zool Sci 29:181–184

Chng HT, Ho HK, Yap CW, Lam SH, Chan ECY (2012) An investigation of the bioactivation potential and metabolism profile of zebrafish versus human. J Biomol Screen 17:974–986

Christen V, Caminada D, Arand M, Fent K (2010) Identification of a CYP3A form (CYP3A126) in fathead minnow (Pimephales promelas) and characterisation of putative CYP3A enzyme activity. Anal Bioanal Chem 396:585–595

Christen V, Fent K (2014) Tissue-, sex- and development-specific transcription profiles of eight UDP-glucuronosyltransferase genes in zebrafish (Danio rerio) and their regulation by activator of aryl hydrocarbon receptor. Aquat Toxicol 150:93–102

Coe TS, Hamilton PB, Griffiths AM, Hodgson DJ, Wahab MA, Tyler CR (2009) Genetic variation in strains of zebrafish (Danio rerio) and the implications for ecotoxicology studies. Ecotoxicology 18:144–150

Crago J, Klaper RD (2011) Influence of gender, feeding regimen, and exposure duration on gene expression associated with xenobiotic metabolism in fathead minnows (Pimephales promelas). Comp Biochem Physiol C Toxicol Pharmacol 154:208–212

Creusot N, Brion F, Piccini B, Budzinski H, Porcher JM, Ait-Aissa S (2014) BFCOD activity in fish cell lines and zebrafish embryos and its modulation by chemical ligands of human aryl hydrocarbon and nuclear receptors. Environ Sci Pollut Res Int. doi:10.1007/s11356-014-3882-8

Czesny SJ, Graeb BDS, Dettmers JM (2005) Ecological consequences of swim bladder noninflation for larval yellow perch. Trans Am Fish Soc 134:1011–1020

de Waal PP, Wang DS, Nijenhuis WA, Schulz RW, Bogerd J (2008) Functional characterization and expression analysis of the androgen receptor in zebrafish (Danio rerio) testis. Reproduction 136:225–234

Dean M, Annilo T (2005) Evolution of the ATP-binding cassette (ABC) transporter superfamily in vertebrates. Annu Rev Genomics Hum Genet 6:123–142

Delov V, Muth-Koehne E, Schaefers C, Fenske M (2014) Transgenic fluorescent zebrafish Tg(fli1:EGFP)(y1) for the identification of vasotoxicity within the zFET. Aquat Toxicol 150:189–200

Denison MS, Nagy SR (2003) Activation of the aryl hydrocarbon receptor by structurally diverse exogenous and endogenous chemicals. Annu Rev Pharmacol Toxicol 43:309–334

Devlin E, Brammer J, Puyear R, McKim J (1996) Prehatching development of the fathead minnow Pimephales promelas rafinesque. In: EPA/600/R-96/079. USEPA

Diekmann HH, A. (2013) ADMETox in zebrafish. Drug Discov Today Dis Model 10:e31–e35

DIN (2001) DIN 38415-6. Suborganismische Testverfahren (Gruppe T) Teil 6: Giftigkeit gegenüber Fischen Bestimmung der nicht akut giftigen Wirkung von Abwasser auf die Entwicklung von Fischeiern über Verdünnungsstufen (T 6). [German standard methods for the examination of water, waste water and sludge–Subanimal testing (group T)–Part 6: Toxicity to fish. Determination of the non-acute-poisonous effect of waste water to fish eggs by dilution limits (T6)]. Beuth Verlag, Berlin

Doyon Y, McCammon JM, Miller JC, Faraji F, Ngo C, Katibah GE, Amora R, Hocking TD, Zhang L, Rebar EJ, Gregory PD, Urnov FD, Amacher SL (2008) Heritable targeted gene disruption in zebrafish using designed zinc-finger nucleases. Nat Biotechnol 26:702–708

EC (2006) Regulation No 1907/2006 of the European Parliament and of the Council of 18 December 2006 concerning the Registration, Evaluation, Authorisation and Restriction of Chemicals (REACH), establishing a European Chemicals Agency

EC (2010) EU Parliament and Council Directive 2010/63/EU of 22 September 2010 on the protection of animals used for scientific purposes

EC (2012) EU Commission Implementing Decision 2012/707/EU of 14 November 2012 establishing a common format for the submission of the information pursuant to Directive 2010/63/EU of the European Parliament and of the Council on the protection of animals used for scientific purposes

Ekins S, Reschly EJ, Hagey LR, Krasowski MD (2008) Evolution of pharmacologic specificity in the pregnane X receptor. BMC Evol Biol 8:103

El-Alfy A, Schlenk D (1998) Potential mechanisms of the enhancement of aldicarb toxicity to Japanese medaka, Oryzias latipes, at high salinity. Toxicol Appl Pharmacol 152:175–183

Embry MR, Belanger SE, Braunbeck TA, Galay-Burgos M, Halder M, Hinton DE, Leonard MA, Lillicrap A, Norberg-King T, Whale G (2010) The fish embryo toxicity test as an animal alternative method in hazard and risk assessment and scientific research. Aquat Toxicol 97:79–87

Fischer S, Kluver N, Burkhardt-Medicke K, Pietsch M, Schmidt AM, Wellner P, Schirmer K, Luckenbach T (2013) Abcb4 acts as multixenobiotic transporter and active barrier against chemical uptake in zebrafish (Danio rerio) embryos. BMC Biol 11:69

Forbes V (1998) Sources and implications of variability in sensitivity to chemicals for ecotoxicological risk assessment. Arch Toxicol Suppl 20:407–418

Gagnon JA, Valen E, Thyme SB, Huang P, Ahkmetova L, Pauli A, Montague TG, Zimmerman S, Richter C, Schier AF (2014) Efficient mutagenesis by Cas9 protein-mediated oligonucleotide insertion and large-scale assessment of single-guide RNAs. PLoS One 9:e98186

Genome Reference Consortium (2015) http://www.ensembl.org/Danio_rerio/Info/Index?db=core

Glisic B, Hrubik J, Fa S, Dopudj N, Kovacevic R, Andric N (2014) Transcriptional profiles of glutathione-S-Transferase isoforms, Cyp, and AOE genes in atrazine-exposed zebrafish embryos. Environ Toxicol. doi:10.1002/tox.22038

Goldstone JV, McArthur AG, Kubota A, Zanette J, Parente T, Jonsson ME, Nelson DR, Stegeman JJ (2010) Identification and developmental expression of the full complement of cytochrome P450 genes in zebrafish. BMC Genomics 11:643

Gonzalez-Doncel M, Carbonell G, San Segundo L, Sastre S, Beltran EM, Fernandez-Torija C (2015) Stage-dependent ethoxyresorufin-O-deethylase (EROD) in vivo activity in medaka (Oryzias latipes) embryos. Chemosphere 135:108–115

Gorelick DA, Halpern ME (2011) Visualization of estrogen receptor transcriptional activation in zebrafish. Endocrinology 152:2690–2703
Groh KJ, Carvalho RN, Chipman JK, Denslow ND, Halder M, Murphy CA, Roelofs D, Rolaki A, Schirmer K, Watanabe KH (2015) Development and application of the adverse outcome pathway framework for understanding and predicting chronic toxicity: II. A focus on growth impairment in fish. Chemosphere 120:778–792
Hahn ME (2002) Aryl hydrocarbon receptors: diversity and evolution. Chem Biol Interact 141:131–160
Heijlen M, Houbrechts A, Bagci E, Van Herck S, Kersseboom S, Esguerra C, Blust R, Visser T, Knapen D, Darras V (2014) Knockdown of type 3 iodothyronine deiodinase severely perturbs both embryonic and early larval development in zebrafish. Endocrinology 155:1547–1559
Hernandez RE, Putzke AP, Myers JP, Margaretha L, Moens CB (2007) Cyp26 enzymes generate the retinoic acid response pattern necessary for hindbrain development. Development 134:177–187
Hornung MW, Cook PM, Fitzsimmons PN, Kuehl DW, Nichols JW (2007) Tissue distribution and metabolism of benzo[a]pyrene in embryonic and larval medaka (Oryzias latipes). Toxicol Sci 100:393–405
Hutchinson TH, Boegi C, Winter MJ, Owens JW (2009) Benefits of the maximum tolerated dose (MTD) and maximum tolerated concentration (MTC) concept in aquatic toxicology. Aquat Toxicol 91:197–202
Hwang WY, Fu Y, Reyon D, Maeder ML, Kaini P, Sander JD, Joung JK, Peterson RT, Yeh J-RJ (2013a) Heritable and precise zebrafish genome editing using a CRISPR-Cas system. PLoS One 8:e68708
Hwang WY, Fu Y, Reyon D, Maeder ML, Tsai SQ, Sander JD, Peterson RT, Yeh JRJ, Joung JK (2013b) Efficient genome editing in zebrafish using a CRISPR-Cas system. Nat Biotechnol 31:227–229
Incardona JP, Carls MG, Teraoka H, Sloan CA, Collier TK, Scholz NL (2005) Aryl hydrocarbon receptor-independent toxicity of weathered crude oil during fish development. Environ Health Perspect 113:1755–1762
ISO (2007) Water quality – determination of the acute toxicity of waste water to zebrafish eggs (Danio rerio). ISO 15088
Iwamatsu T (2004) Stages of normal development in the medaka Oryzias latipes. Mech Dev 121:605–618
Jao L-E, Wente SR, Chen W (2013) Efficient multiplex biallelic zebrafish genome editing using a CRISPR nuclease system. Proc Natl Acad Sci U S A 110:13904–13909
Ji C, Jin X, He J, Yin Z (2012) Use of TSH beta:EGFP transgenic zebrafish as a rapid in vivo model for assessing thyroid-disrupting chemicals. Toxicol Appl Pharmacol 262:149–155
Jomaa B, Hermsen SAB, Kessels MY, van den Berg JHJ, Peijnenburg AACM, Aarts JMMJG, Piersma AH, Rietjens IMCM (2014) Developmental toxicity of thyroid-active compounds in a zebrafish embryotoxicity test. Altex-Altern Anim Exp 31:303–317
Jones HS, Panter GH, Hutchinson TH, Chipman JK (2010) Oxidative and conjugative xenobiotic metabolism in zebrafish larvae in vivo. Zebrafish 7:23–30
Karchner SI, Franks DG, Hahn ME (2005) AHR1B, a new functional aryl hydrocarbon receptor in zebrafish: tandem arrangement of ahr1b and ahr2 genes. Biochem J 392:153–161
Kelly A, Hurlstone AF (2011) The use of RNAi technologies for gene knockdown in zebrafish. Brief Funct Genomics 10:189–196
Kim BM, Rhee JS, Jeong CB, Lee SJ, Lee YS, Choi IY, Lee JS (2014) Effects of benzo[a]pyrene on whole cytochrome P450-involved molecular responses in the marine medaka Oryzias melastigma. Aquat Toxicol 152:232–243
Kimmel CB, Ballard WW, Kimmel SR, Ullmann B, Schilling TF (1995) Stages of embryonic development of the zebrafish. Dev Dyn 203:253–310
Kirchmaier S, Naruse K, Wittbrodt J, Loosli F (2015) The genomic and genetic toolbox of the teleost medaka (Oryzias latipes). Genetics 199:905–918

Kluver N, Ortmann J, Paschke H, Renner P, Ritter AP, Scholz S (2014) Transient overexpression of adh8a increases allyl alcohol toxicity in zebrafish embryos. PLoS One 9:e90619

Knapen D, Vergauwen L, Villeneuve DL, Ankley GT (2015) The potential of AOP networks for reproductive and developmental toxicity assay development. Reprod Toxicol 56:52–55

Knöbel M, Busser FJM, Rico-Rico Á, Kramer NI, Hermens JLM, Hafner C, Tanneberger K, Schirmer K, Scholz S (2012) Predicting adult fish acute lethality with the zebrafish embryo: relevance of test duration, endpoints, compound properties, and exposure concentration analysis. Environ Sci Technol. doi:10.1021/es301729q

Krewski D, Acosta D Jr, Andersen M, Anderson H, Bailar JC III, Boekelheide K, Brent R, Charnley G, Cheung VG, Green S Jr, Kelsey KT, Kerkvliet NI, Li AA, McCray L, Meyer O, Patterson RD, Pennie W, Scala RA, Solomon GM, Stephens M, Yager J, Zeise L, Staff Comm Toxicity Testing A (2010) Toxicity testing in the 21st century: a vision and a strategy. J Toxicol Environ Health-Part B-Crit Rev 13:51–138

Kullman SW, Kashiwada S, Hinton DE (2004) Analysis of medaka cytochrome P450 3A homotropic and heterotropic cooperativity. Mar Environ Res 58:469–473

Kurogi K, Dillon J, Nasser A, Liu M-Y, Williams FE, Sakakibara Y, Suiko M, Liu M-C (2010) Sulfation of drug compounds by the zebrafish cytosolic sulfotransferases (SULTs). Drug Metab Lett 4:62–68

LaLone CA, Villeneuve DL, Burgoon LD, Russom CL, Helgen HW, Berninger JP, Tietge JE, Severson MN, Cavallin JE, Ankley GT (2013a) Molecular target sequence similarity as a basis for species extrapolation to assess the ecological risk of chemicals with known modes of action. Aquat Toxicol 144:141–154

LaLone CA, Villeneuve DL, Cavallin JE, Kahl MD, Durhan EJ, Makynen EA, Jensen KM, Stevens KE, Severson MN, Blanksma CA, Flynn KM, Hartig PC, Woodard JS, Berninger JP, Norberg-King TJ, Johnson RD, Ankley GT (2013b) Cross-species sensitivity to a novel androgen receptor agonist of potential environmental concern, spironolactone. Environ Toxicol Chem 32:2528–2541

Lammer E, Carr GJ, Wendler K, Rawlings JM, Belanger SE, Braunbeck T (2009) Is the fish embryo toxicity test (FET) with the zebrafish (Danio rerio) a potential alternative for the fish acute toxicity test? Comp Biochem Physiol C-Toxicol Pharmacol 149:196–209

Lange M, Gebauer W, Markl J, Nagel R (1995) Comparison of testing acute toxicity on embryo of zebrafish, brachydanio rerio and RTG-2 cytotoxicity as possible alternatives to the acute fish test. Chemosphere 30:2087–2102

Lawson ND, Weinstein BM (2002) In vivo imaging of embryonic vascular development using transgenic zebrafish. Dev Biol 248:307–318

Lechner W, Ladich F (2008) Size matters: diversity in swimbladders and Weberian ossicles affects hearing in catfishes. J Exp Biol 211:1681–1689

Lee O, Green JM, Tyler CR (2015) Transgenic fish systems and their application in ecotoxicology. Crit Rev Toxicol 45:124–141

Li ZH, Alex D, Siu SO, Chu IK, Renn J, Winkler C, Lou S, Tsui SK, Zhao HY, Yan WR, Mahady GB, Li GH, Kwan YW, Wang YT, Lee SM (2011) Combined in vivo imaging and omics approaches reveal metabolism of icaritin and its glycosides in zebrafish larvae. Mol BioSyst 7:2128–2138

Liew WC, Orban L (2014) Zebrafish sex: a complicated affair. Brief Funct Genomics 13:172–187

Liu YW, Chan WK (2002) Thyroid hormones are important for embryonic to larval transitory phase in zebrafish. Differentiation 70:36–45

Liu T-A, Bhuiyan S, Liu M-Y, Sugahara T, Sakakibara Y, Suiko M, Yasuda S, Kakuta Y, Kimura M, Williams FE, Liu M-C (2010) Zebrafish as a model for the study of the phase II cytosolic sulfotransferases. Curr Drug Metab 11:538–546

Mattsson A, Ulleras E, Patring J, Oskarsson A (2012) Albendazole causes stage-dependent developmental toxicity and is deactivated by a mammalian metabolization system in a modified zebrafish embryotoxicity test. Reprod Toxicol 34:31–42

McKim JM (1977) Evaluation of tests with early life stages of fish for predicting long-term toxicity. J Fish Res Board Can 34:1148–1154

Nagahama Y, Yamashita M (2008) Regulation of oocyte maturation in fish. Develop Growth Differ 50:S195–S219
Nagel R (2002) DarT: the embryo test with the zebrafish Danio rerio – a general model in ecotoxicology and toxicology. Altex-Alternativen Zu Tierexperimenten 19:38–48
Nebert DW, Russell DW (2002) Clinical importance of the cytochromes P450. Lancet 360:1155–1162
Nelson K, Schroeder A, Ankley G, Blackwell B, Blanksma C, Degitz S, Flynn K, Jensen K, Johnson R, Kahl M, Knapen D, Kosian P, Milsk R, Randolph E, Saari T, Stinckens E, Vergauwen L, Villeneuve D (2016) Impaired anterior swim bladder inflation following exposure to the thyroid peroxidase inhibitor 2-mercaptobenzothiazole part I: fathead minnow. Aquat Toxicol 173:192–203
OECD (1992) OECD guidelines for the testing of chemicals. No.203: Fish, Acute Toxicity Test
OECD (2009a) OECD guidelines for the testing of chemicals. No.229: Fish Short Term Reproduction Assay
OECD (2009b) OECD guidelines for the testing of chemicals. No.230: 21-day Fish Assay: A Short-Term Screening for Oestrogenic and and Androgenic Activity, and Aromatase Inhibition
OECD (2011) OECD guidelines for the testing of chemicals. No.234: Fish Sexual Development Test
OECD (2013a) OECD guidelines for the testing of chemicals. No.236: Fish Embryo Acute Toxicity (FET) Test
OECD (2013b) OECD guidelines for the testing of chemicals. No. 210: Fish, Early-life Stage Toxicity Test
OECD (2015) Users' handbook supplement to the guidance document for developing and assessing AOPs. p https://aopkb.org/common/AOP_Handbook.pdf
Otte JC, Schmidt AD, Hollert H, Braunbeck T (2010) Spatio-temporal development of CYP1 activity in early life-stages of zebrafish (Danio rerio). Aquat Toxicol 100:38–50
Philippe C, Grégoir A, De Boeck G, Van Breendonck L (2015) Are short-term tests telling the whole truth? Nothobranchius furzeri as a new model in ecotoxicology. In: SETAC Europe: Barcelona
Popper AN (1974) Response of swim bladder of goldfish (Carassius auratus) to acoustic stimuli. J Exp Biol 60:295–304
Pype C, Verbueken E, Saad MA, Casteleyn CR, Van Ginneken CJ, Knapen D, Van Cruchten SJ (2015) Incubation at 32.5 degrees C and above causes malformations in the zebrafish embryo. Reprod Toxicol 56:56–63
Reschly EJ, Krasowski MD (2006) Evolution and function of the NR1I nuclear hormone receptor subfamily (VDR, PXR, and CAR) with respect to metabolism of xenobiotics and endogenous compounds. Curr Drug Metab 7:349–365
Rhee JS, Kim BM, Choi BS, Choi IY, Wu RS, Nelson DR, Lee JS (2013) Whole spectrum of cytochrome P450 genes and molecular responses to water-accommodated fractions exposure in the marine medaka. Environ Sci Technol 47:4804–4812
Robertson GN, McGee CAS, Dumbarton TC, Croll RP, Smith FM (2007) Development of the swimbladder and its innervation in the zebrafish, Danio rerio. J Morphol 268:967–985
Russel W, Burch R (1959) The principles of humane experimental technique. Methuen, London
Russom CL, LaLone CA, Villeneuve DL, Ankley GT (2014) Development of an adverse outcome pathway for acetylcholinesterase inhibition leading to acute mortality. Environ Toxicol Chem 33:2157–2169
Saad M, Cavanaugh K, Verbueken E, Pype C, Casteleyn C, Van Ginneken C, Van Cruchten S (2016a) Xenobiotic metabolism in the zebrafish: a review of the spatiotemporal distribution, modulation and activity of cytochrome P450 families 1 to 3. J Toxicol Sci 41:1–11
Saad M, Verbueken E, Pype C, Casteleyn C, Van Ginneken C, Maes L, Cos P, Van Cruchten S (2016b) In vitro CYP1A activity in the zebrafish: temporal but very low metabolite levels during organogenesis and lack of gender differences in the adult stage. Reproductive Toxicology accepted

Schiller V, Zhang X, Hecker M, Schaefers C, Fischer R, Fenske M (2014) Species-specific considerations in using the fish embryo test as an alternative to identify endocrine disruption. Aquat Toxicol 155:62–72

Scholz S, Fischer S, Gundel U, Kuster E, Luckenbach T, Voelker D (2008) The zebrafish embryo model in environmental risk assessment – applications beyond acute toxicity testing. Environ Sci Pollut Res Int 15:394–404

Spielmann H, Seiler A, Bremer S, Hareng L, Hartung T, Ahr H, Faustman E, Haas U, Moffat GJ, Nau H, Vanparys P, Piersma A, Sintes JR, Stuart J (2006) The practical application of three validated in vitro embryotoxicity tests. The report and recommendations of an ECVAM/ZEBET workshop (ECVAM workshop 57). Altern Lab Anim 34:527–538

Stinckens E, Vergauwen L, Schroeder A, Maho W, Blackwell B, Witters H, Blust R, Ankley G, Covaci A, Villeneuve D, Knapen D (2016) Impaired anterior swim bladder inflation following exposure to the thyroid peroxidase inhibitor 2-mercaptobenzothiazole part II: Zebrafish. Aquat Toxicol 173:204–217

Straehle U, Scholz S, Geisler R, Greiner P, Hollert H, Rastegar S, Schumacher A, Selderslaghs I, Weiss C, Witters H, Braunbeck T (2012) Zebrafish embryos as an alternative to animal experiments-a commentary on the definition of the onset of protected life stages in animal welfare regulations. Reprod Toxicol 33:128–132

Sun D, Zhang Y, Wang C, Hua X, Zhang XA, Yan J (2013) Sox9-related signaling controls zebrafish juvenile ovary-testis transformation. Cell Death Dis 4:e930

Teoh P-H, Shu-Chien AC, Chan W-K (2010) Pbx1 is essential for growth of zebrafish swim bladder. Dev Dyn 239:865–874

Tokarz J, Moeller G, de Angelis MH, Adamski J (2013) Zebrafish and steroids: what do we know and what do we need to know? J Steroid Biochem Mol Biol 137:165–173

Tollefsen KE, Scholz S, Cronin MT, Edwards SW, de Knecht J, Crofton K, Garcia-Reyero N, Hartung T, Worth A, Patlewicz G (2014) Applying adverse outcome pathways (AOPs) to support integrated approaches to testing and assessment (IATA). Regul Toxicol Pharmacol 70:629–640

Uchida D (2004) An aromatase inhibitor or high water temperature induce oocyte apoptosis and depletion of P450 aromatase activity in the gonads of genetic female zebrafish during sex-reversal. Comp Biochem Physiol A Mol Integr Physiol 137:11–20

Urushitani H, Katsu Y, Kato Y, Tooi O, Santo N, Kawashima Y, Ohta Y, Kisaka Y, Lange A, Tyler C, Johnson R, Iguchi T (2007) Medaka (Oryzias latipes) for use in evaluating developmental effects of endocrine active chemicals with special reference to gonadal intersex (testis-ova). Environ Sci 14:211–233

USEPA (1987) Guidelines for the culture of fathead minnows Pimephales promelas for use in toxicity tests. EPA/600/3–87/001

USEPA (1996) Ecological effects test guidelines. OPPTS 850.1400: Fish Early-Life Stage Toxicity Test

USEPA (1998) Endocrine disruptor screening and testing advisory committee (EDSTAC) final report. Office of Prevention, Pesticides and Toxic Substances, US Environmental Protection Agency, Washington, DC. http://www.epa.gov/scipoly/oscpendo/pubs/edspoverview/finalrpt.htm

Varshney GK, Pei W, LaFave MC, Idol J, Xu L, Gallardo V, Carrington B, Bishop K, Jones M, Li M, Harper U, Huang SC, Prakash A, Chen W, Sood R, Ledin J, Burgess SM (2015) High-throughput gene targeting and phenotyping in zebrafish using CRISPR/Cas9. Genome Res 25:1030–1042

Vergauwen L, Hagenaars A, Blust R, Knapen D (2013) Temperature dependence of long-term cadmium toxicity in the zebrafish is not explained by liver oxidative stress: evidence from transcript expression to physiology. Aquat Toxicol 126:52–62

Villeneuve D, Volz DC, Embry MR, Ankley GT, Belanger SE, Leonard M, Schirmer K, Tanguay R, Truong L, Wehmas L (2014a) Investigating alternatives to the fish early-life stage test: a strategy for discovering and annotating adverse outcome pathways for early fish development. Environ Toxicol Chem 33:158–169

Villeneuve DL, Crump D, Garcia-Reyero N, Hecker M, Hutchinson TH, LaLone CA, Landesmann B, Lettieri T, Munn S, Nepelska M, Ottinger MA, Vergauwen L, Whelan M (2014b) Adverse outcome pathway (AOP) development I: strategies and principles. Toxicol Sci 142:312–320

Villeneuve DL, Crump D, Garcia-Reyero N, Hecker M, Hutchinson TH, LaLone CA, Landesmann B, Lettieri T, Munn S, Nepelska M, Ottinger MA, Vergauwen L, Whelan M (2014c) Adverse outcome pathway development II: best practices. Toxicol Sci 142:321–330

Volz DC, Belanger S, Embry M, Padilla S, Sanderson H, Schirmer K, Scholz S, Villeneuve D (2011) Adverse outcome pathways during early fish development: a conceptual framework for identification of chemical screening and prioritization strategies. Toxicol Sci 123:349–358

Waxman DJ (1999) P450 gene induction by structurally diverse xenochemicals: central role of nuclear receptors CAR, PXR, and PPAR. Arch Biochem Biophys 369:11–23

Weigt S, Huebler N, Braunbeck T, von Landenberg F, Broschard TH (2010) Zebrafish teratogenicity test with metabolic activation (mDarT): effects of phase I activation of acetaminophen on zebrafish Danio rerio embryos. Toxicology 275:36–49

Weigt S, Huebler N, Strecker R, Braunbeck T, Broschard TH (2011) Zebrafish (Danio rerio) embryos as a model for testing proteratogens. Toxicology 281:25–36

Westerfield M (1995) The zebrafish book: a guide for the laboratory use of zebrafish (Brachydanio rerio), 3rd edn. University of Oregon Press, Eugene

Wheeler JR, Gimeno S, Crane M, Lopez-Juez E, Morritt D (2005) Vitellogenin: a review of analytical methods to detect (anti) estrogenic activity in fish. Toxicol Mech Methods 15:293–306

Wheeler JR, Panter GH, Weltje L, Thorpe KL (2013) Test concentration setting for fish in vivo endocrine screening assays. Chemosphere 92:1067–1076

Winata CL, Korzh S, Kondrychyn I, Zheng W, Korzh V, Gong Z (2009) Development of zebrafish swimbladder: the requirement of hedgehog signaling in specification and organization of the three tissue layers. Dev Biol 331:222–236

Woltering DM (1984) The growth-response in fish chronic and early life stage toxicity tests – a critical-review. Aquat Toxicol 5:1–21

Worth A, Barosso J, Bremer S, Burton J, Casati S, Coecke S, Corvi R, Desprez B, Dumont C, Gouliarmou V, Goumenou M, Gräpel R, Griesinger C, Halder M, Janusch Roi A, Kienzler A, Madia F, Munn S, Nepelska M, Paini A, Price A, Prieto P, Rolaki A, Schäffer M, Triebe J, Whelan M, Wittwehr C, Zuang V (2014) Alternative methods for regulatory toxicology – a state-of-the-art review. In: JRC science and policy reports

Yin A, Korzh S, Winata CL, Korzh V, Gong Z (2011) Wnt signaling is required for early development of zebrafish swimbladder. PLoS One 6:e18431

Chapter 5
Invertebrate Model Species in AOP Development

Geoff Hodges, Steve Gutsell, Nadine Taylor, Erica Brockmeier, Emma Butler, Cecilie Rendal, and John Colbourne

Abstract In this chapter, we present the use of invertebrate model species in the development of adverse outcome pathways (AOPs), its challenges, and the current state of invertebrate toxicity studies. Invertebrates can contribute significantly towards the development of robust AOPs, providing many advantages over the use of vertebrate species. This includes a generally shorter life cycle allowing for chronic and full life cycle toxicity tests, and a wide array of powerful molecular genetic tools such as genome sequences, genomic engineering including gene knock-outs, and comprehensive bioinformatics databases. Currently, the most robustly developed invertebrate model species for toxicity testing include *Daphnia*, *Caenorhabditis elegans*, plus members of the *Drosophila* genus. The potential use of these and other invertebrate organisms for assessing chemical risk for most animals (including vertebrate species) is evaluated via a comparative phylogenetic approach to ecotoxicological testing, seeking to discover the evolutionary origins and distribution of toxicity pathways across the internal branches of the animal phylogeny. Comparative –omics data from cellular and developmental studies suggest a high degree of conservation in regulatory pathways in fly, worm and human. By comparing –omics studies between vertebrates and invertebrate species in toxicology, we begin to also discover coherence in pathway level responses, indicating potentially numerous overlapping responses to specific stressors, even across species that have different physiologies and ecological niches. At present, only a small number of invertebrate AOPs are informed by evidence. Perhaps the most robust of these is the Acetylcholinesterase inhibition (AChE) AOP for pesticides. We present

G. Hodges (✉) • S. Gutsell • E. Butler • C. Rendal
Safety & Environmental Assurance Centre, Unilever, Colworth Science Park, Sharnbrook, Bedfordshire MK44 1LQ, UK
e-mail: geoff.hodges@unilever.com

N. Taylor • J. Colbourne
School of Biosciences, University of Birmingham, Birmingham B15 2TT, UK

E. Brockmeier
University of Liverpool, Institute of Integrative Biology, Crown Street, Liverpool L69 7ZB, UK

N. Garcia-Reyero, C.A. Murphy (eds.), *A Systems Biology Approach to Advancing Adverse Outcome Pathways for Risk Assessment*,
DOI 10.1007/978-3-319-66084-4_5

a case study of using the AOP framework for risk assessment and discuss how the use of models, such as those using Dynamic Energy Budget theory linked to populations, can enhance the use of AOPs for understanding and predicting chemical risk.

5.1 Introduction (Key Challenges)

Conventional methods of Environmental Risk Assessment (ERA) largely rely on short-term acute toxicity tests carried out in the laboratory on various model species that occupy different trophic levels; these studies are supported occasionally by long-term chronic toxicity tests. Assessment factors, to account for the uncertainties in extrapolating from laboratory data to the natural environment, are then used to derive toxicity thresholds. However, since these factors lack a mechanistic/causal basis and do not quantify variation both within and among species, they have limited potential for quantitatively estimating cross-species toxicity thresholds and are, therefore, set to be protective rather than predictive. Economic as well as ethical factors also influence the type of testing typically conducted for ERA; new regulations are phasing out the use of vertebrate animal testing (Council Directive 2010/63/EU), while the costs of conducting chronic toxicity studies on whole vertebrate organisms are also prohibitive (REACH 2006).

The Adverse Outcome Pathway (AOP) approach looks at the effect of chemical perturbation from the Molecular Initiating Event (MIE) through several Key Events (KEs) which exhibit responses at various levels of biological organisation (e.g., cellular, tissue, organ, etc.), which will lead to an Adverse Outcome (AO) that is relevant to risk assessment (Ankley et al. 2010). The individual organism level AOs commonly used in ecotoxicology describe impacts on survival, growth and reproduction. As more AOPs and KEs are documented, it may become possible to predict and assess potential AOs using information from lower levels of biological organisation as part of a new approach to ERA based on mechanistic understanding. As a minimum the AOP framework provides a way to collect, organise and integrate information from multiple sources to enable a safety decision to be made based on risk assessment and for communication of that information in a biologically plausible manner (Burden et al. 2015; Perkins et al. 2015).

Given that invertebrates account for at least 95% of all known animal species and are critical to ecosystem structure and function (Verslycke et al. 2007), investigations using these key species are vital when obtaining data from which to develop AOPs. Invertebrates offer many advantages over vertebrates, including their generally short life cycle, large brood sizes and the ease with which large numbers of individuals may be studied. The use of invertebrates in toxicity testing is often cost-effective, as thousands of organisms can be housed in single testing facilities. Many invertebrates are longstanding model species for biomedical and basic research, and their genome biology is very well understood for associating genetic diversity with their phenotypic effects. For example, the genome size of *Caenorhabditis elegans* is only 100 megabases, with predicted protein products exhibiting 40% homology between *C. elegans* and humans; many *C. elegans* genes having similar functions with human proteins (The *C. elegans* Sequencing Consortium 1998). In fact, functional annotations and associations among human genes are most often first discovered in these systems (Williams and Auwerx 2015). Their discovery and predictive power have yet to be fully aimed at supporting AOP development. Moreover, a systems toxicology approach using a carefully chosen panel of such model species, which includes a large swath of animal diversity, will reveal commonalities by virtue of our common ancestry, thereby enabling cross-species predictions based on phylogenetic principles.

This chapter will explore the benefits and limitations of using invertebrates as a model species for AOP development from all perspectives, including:

- Appropriate invertebrate species selection
- Existing AOPs and KEs based on invertebrate data
- The importance of test design
- How invertebrate AOPs or KEs could be incorporated into risk assessment in the future

5.2 Species Selection

5.2.1 Model Species

Model organisms can be defined in multiple ways. In general, model organisms have particular experimental advantages: they can be easy to maintain in a laboratory setting, are amenable to genetic analysis and manipulation, are used to understand the genetic basis of disease, or are keystone species in ecology or representatives of biodiversity (http://genome.wellcome.ac.uk/doc_WTD020803.html). Model organisms can therefore, serve to understand the human condition (biomedical models) and to understand how ecosystems function (ecological models). Model organisms with significant research communities and characterized genomes have particular value in research. Over the past decade, there has been a growing appreciation of the democratization of genomics resulting in a new class of "emerging" model species that are beginning to reveal genome diversity, the context dependency of gene functions and their products, and the level of genetic variation for environmentally relevant traits (Feder and Mitchell-Olds 2003; Tagu et al. 2014). From a comparative biology perspective approach, the origins and conservation of genes, and their functional associations, can be mapped onto the animal phylogeny to infer homology, thereby increasing confidence in cross species extrapolation. By focussing attention on research dedicated to discovering the molecular underpinnings for fitness related responses to environmental conditions bridges an artificial divide between human- and eco-toxicology. Given these developments, invertebrates are certainly important for the discovering AOPs in risk assessment.

The National Institute of Health maintains a list of approved model organisms for use in biomedical research (http://www.nih.gov/science/models/). Of these, currently there are three invertebrate animal genera: *Daphnia, Drosophila* and *Caenorhabditis*. These three species are commonly used in comparative genomics, toxicology and/or ecotoxicology studies. These taxa offer a wealth of historical data and biological knowledge from over a century of genetic investigations.

Daphnia species are small freshwater crustaceans, ubiquitous worldwide and integral to the pelagic ecosystem (Lampert 2011). Fundamentally, *Daphnia* are a long-established model species with an important role in determining chemical safety criteria around the world, and is the most commonly used system for ecotoxicological testing worldwide (Shaw et al. 2008). Highlights of the *Daphnia* system are: (1) reproduction by cyclical parthenogenesis; genotypes can be maintained indefinitely in a clonal fashion (Lampert 2011). It is possible to self or outcross lineages, thereby experimentally dissecting the relative contributions of genes and environment for toxicity by partitioning the variance between and among clones, with biological replication involving fixed genetic backgrounds. (2) A draft sequence assembly and annotation of the *Daphnia* genome (Colbourne et al. 2011). (3) A transparent carapace that allows for imaging gene expression by fluorescence-based assays in whole animals; (Gorokhova and Kyle 2002; Paul et al. 1998). (4) Methods for reverse-genetic testing, including RNAi-based gene knockdown, CRISPR/CAS9-based gene knockout, and transformation system (Kato et al. 2011, 2012; Nakanishi et al. 2014). (5) Multiple mutation accumulation lines for obtaining direct estimates of the mutational rates and spectra (Seyfert et al. 2008). (6) A bioinformatics database (wFleaBase.org) modelled after Drosophila's FlyBase.org. (7) A large number of overlapping genes with human, more than any other sequenced invertebrate (Colbourne et al. 2011). As it stands, few species can rival *Daphnia* for possessing key biological attributes, research tools and infrastructure, as well as the support of a global research community; the Daphnia Genomics Consortium. The resources of this consortium can be used to discover how genomes and environments, including chemical stressors, interact. This is a key invertebrate species in all current environmental risk assessment paradigms and as such is perfect for merging known chemical effects with the testing paradigms being suggested.

Drosophila melanogaster (fruit fly, a dipteran insect) is a model organism that has been utilized for over a century in the field of genetics and has numerous biological research tools available (FlyBase.org). The striking conservation of >60% of human disease genes makes it an important model for neurological diseases, cancers, heart disease, metabolic diseases and diabetes, and responses to infection by pathogens (www.flydiseasemodels.blogspot.com). The physiological attributes of *Drosophila* including a brain, a beating heart, a tubular network analogous to lungs, an osmoregulatory/excretory system analogous to kidneys and many other aspects of physiology and homeostasis make it an excellent model species for the development of AOPs that are highly applicable across species.

Caenorhabditis elegans (a nematode worm) is a much-used and long-established model system for obtaining integrated information on the cellular, developmental, and molecular aspects of the effects of toxicants on growth and development, as well as gene expression. There is a wealth of knowledge available on *C. elegans* biology; an exceptionally detailed database on the cell and developmental biology as well as gene and protein expression patterns and regulation (WormBase). *C. elegans* is a practical and powerful species for toxicological testing having the added value of being able to observe all of the somatic cells in the living organism (http://www.wormatlas.org). There is a conservation of neurophysiological components; shared genetic networks and developmental programs between nematodes and vertebrates make it an excellent model for these systems in particular. Several studies document that responses in *C. elegans* following chemical exposure appear to be predictive of developmental shifts or neurological damage in vertebrates (Leung et al. 2008). As a result of the evolutionarily conserved nature of signal transduction and stress-response pathways, it is likely that responses elicited in *C. elegans* will be applicable to understanding similar processes in higher organisms, including humans.

Utilizing invertebrate models experimentally will help to reduce our reliance on animal-based methods, positively impacting animal welfare whilst elucidating mechanistic information that can aid in cross-species extrapolation (Burden et al. 2015). The short-generation time and ease of handling of invertebrates in the laboratory makes them amenable to high-throughput screening approaches for assessing the cellular and molecular responses to chemicals that can take advantage of the myriad of technical advances that have occurred over recent years. Although risk assessment approaches have traditionally been based on the use of in vivo data generation supported by in silico methods, there has been a recent shift, in the US in particular, to incorporate these new in vitro (cell-based) alternative approaches (Ankley et al. 2008). AOPs represent one mechanism for helping to collate and interpret these data in combination with more traditional apical endpoints (e.g., growth, reproduction, mortality) for assessments.

5.2.2 Using Phylogenetic Approaches to Maximize the Use of Invertebrate Models and Existing Data in Risk Assessment

The AOP is a conceptual framework for obtaining data on early mechanistic events, leading to toxicity by a chemical, that can be linked to eventual adverse outcomes at many levels of biological organization. By virtue of the rapidity and cost effectiveness of in vitro and computational approaches, there is a growing database of the toxicological potential of chemicals based on their disruption of pathways that are integral to cellular functions (USEPA 2015). Yet despite the richness of this database for risk assessment, these results do not reflect the complexity of whole organisms including metabolic capacity, complex interactions among cells within tissues, tissues within organs, organs within individuals, and individuals within populations and under varying ecological settings. Cell-based assays also suffer from genetic homogeneity, frequent aneuploidy or loss of specific functions and adaptation through homeostatic mechanisms. In effect, all biological models suffer from inherent variation in their responses to environmental conditions because of natural genetic variation caused by mutation, genetic drift and natural selection, even among populations of the same species.

Variability in the response to chemicals is well known and derives from differences in genetics, epigenetics, life histories (including development, sex), ecological setting and lifestyles (Barata et al. 2002; De Coninck et al. 2014). Accounting for innate variation among biological systems requires diversity in the assays used to discover toxicity pathways – preferably discovering mechanisms that are shared by evolutionary descent among many biological systems that altogether predict the chemical effects on the vast majority of untested organisms, including humans.

A program of comparative, multiomics chemical screening research programme using both invertebrate and vertebrate test species for discovering AOPs that are built from evolutionarily conserved molecular mechanisms of toxic responses to compounds is being pursued by a grassroots Consortium for Environmental Omics and Toxicology (CEOT), which both widens the set of processes investigated for potentially new mechanistic insights, and draws knowledge from genetic variation as part of the AOP discovery process. By utilizing a suite of research-intensive experimental organisms that are recognized biomedical model species, and by including data from cell lines, researchers are enhancing their studies to include a much broader range of potential adverse outcomes that complement higher throughput in vitro experimental data. The molecular responses to hundreds, then thousands, of chemically induced perturbations, measured by genome-wide RNA profiling and non-targeted metabolomics, are extracted and combined into co-responsive networks of genes and metabolites that show reproducible correlative structure across many samples and test conditions. Machine learning approaches are then used to relate the different omics data types, including forms of sparse regression and feature selection that put forward candidate pathways of toxicological relevance. The co-expression networks identified in this way are predicted to participate in the same metabolic reactions in different species. This comparative approach is a far more powerful notion than merely relying on shared sequence similarity to infer functional gene-homology. The fact that genes share a common evolutionary ancestor is important, but does not guarantee they retain similar biological roles. Yet by identifying clusters of genes that influence the same metabolic processes, this research generalizes the notion of gene homology to homology at the level of networks that function in the same way. These shared co-functioning networks need not be composed entirely of evolutionarily conserved genes, which is important given the very large evolutionary distance among the test species. Several studies have already suggested that pathways or biological processes (when discovered) are more likely than genes to be functionally conserved among most animals. Examples include DNA repair (Taylor and Lehmann 1998) and the decoding chromatin state and epigenetic information (Gerstein et al. 2014)

Using a phylogenetic approach that includes invertebrates is, therefore, necessary to identify the evolutionary origins and preservation of toxicological pathways starting from the base of the animal phylogeny, which can be useful to predict the susceptibility of a large swath of animal diversity to chemicals (Burgess-Herbert and Euling 2013). By studying the toxicological responses of organisms that vary in their phylogenetic relatedness, adverse outcomes are discovered from chemical exposure that are potentially shared due to the inheritance of toxicological pathways (KEs) from a common ancestor, thereby being predictive of the outcome in untested species along the same evolutionary branch. Connecting (eco)toxicity and sensitivity data to the phylogeny of the tested species can potentially provide an *a priori* prediction of a species' sensitivity to contaminants (Larras et al. 2014) and a comparative approach which overlays a species' trait values onto phylogenetic trees. This can then be used to determine whether species possessing similar traits attributable to a shared history, or convergence, can help in extrapolating these findings across species (Hammond et al. 2012).

5.3 AOP-Relevant Test Design (Benefits of Invertebrates Over Other Sentinel Species)

To experimentally discover the molecular mechanisms involved in a toxicological response, including MIEs, KEs and key event relationships (KER), test species such as those listed above requires well-developed molecular platforms, powerful biological research tools and support by large model species research communities. –Omics can be used as one tool in an integrated approach (weight of evidence; WoE) in combination with the readily available in silico or in vitro data to support the categorisation of chemicals by their KEs. There is still immense work to be done in identifying and/or further elucidating the molecular mechanisms or relevant KEs. This is particularly the case when trying to elucidate the molecular-level responses of model species in order to predict survival, growth and reproduction following chronic exposure to a chemical. The desired output is a new decision-making tool that includes a suite of newly identified KEs that are predictive of chronic phenotypic responses to chemical exposure, and which are indicative of a specific Mode of Action (MoA)/AOP.

There are several practical benefits that are associated with the choice of invertebrate systems in the context of –omics experiments. The availability of genome sequence and gene annotation information enables more comprehensive analyses of processes informed by –omics experiments which are often derived by gene set enrichment analysis (Subramanian et al. 2005) and determination of co-regulated pathways and biological processes (http://geneontology.org). When comparing genome sequences available for invertebrates and vertebrates, there is currently more invertebrate sequence information (378 entries versus 311, as of June 2016) in the National Center for Biotechnology Information (NCBI) genome database (www.ncbi.nlm.nih.gov/genome). Due to the size of many experimental invertebrates, molecular samples are often obtained from whole organisms, although tissue dissections can also be conducted. Commercially-available microarrays currently include 13 vertebrate and 3 invertebrate species. However, platforms are now available which allow custom microarrays from sequence information to be made, opening up additional opportunities for developing and using –omics tools for invertebrate species. In addition, the advent of RNASeq technologies have enabled other approaches for measuring differential gene expression in species where no prior genome sequence is available (Nookaew et al. 2012).

As there is a crucial need in AOP development to address both the acute and chronic effects of toxicant exposure (Patwardhan and Ghaskadbi 2013), invertebrates provide a solution to addressing long-term effects as well as impacts within multiple life-cycles in a shorter experimental time frame. Such chronic exposure is particularly relevant for ERA in which chemical exposure is often at low concentrations over extended duration. Due to the shorter life cycles of invertebrate test species (Buikema and Cairns 1980), complete toxicity assessments on multiple life stages and windows of sensitivity are possible. In addition, laboratory and testing space requirements are greatly reduced due to the size and life cycle length of invertebrates, with a reduction in over 50% of the space and facilities required for toxicity testing as compared to toxicity testing in fish (Buikema and Cairns 1980).

Considerations of sample quality control are required for the appropriate use of –omics tools. For microarray experiments, RNA quality must be accurately assessed in order to ensure samples are not degraded. While RNA extraction procedures are easily available and provide reliable methods for invertebrate tissues (Stevanik et al. 2013; Santiago-Vazquez et al. 2006; Spade et al. 2010), several species exhibit a 'hidden break' in the 28S subunit of their ribosomal RNA (Ishikawa 1977). This makes assessing RNA quality using methods such as RNA Integrity Number (RIN) evaluation using tools such as the Agilent BioAnalyzer more difficult. Researchers must keep this in mind when assessing RNA quality and have an SOP available for both RNA extraction and appropriate quality control assessment for their invertebrate organism of interest.

In terms of biomass required, genome-wide studies of differential gene expression currently utilize low amounts of RNA, so a small number of these organisms are sufficient for RNA-based studies. For example, four *Daphnia magna* at 5 days of age were sufficient as a pooled number of samples for RNA extraction (Taylor et al. 2010). In metabolomics analyses, individual *D. magna* samples were collected for both whole-body homogenates as well as hemolymph samples (approximately 1 μl per organism), with extracts able to provide metabolite profiles using Fourier transform ion cyclotron resonance mass spectrometry (FT-ICR MS) (Taylor et al. 2010).

For homogenization methods, difficulties may also arise with sequence-based –omics studies, for example RNA-seq (Wang et al. 2009) if whole-body preps are utilized. Carry-over of gut and intestine contents during whole-body homogenizations may skew these results to sequences found in internal bacterial contents. If whole body preparations are to be utilized, appropriate depuration methods should be used to avoid additional biases in the sequence analysis.

5.3.1 Pathways Inferred from –Omics Studies: A Comparison Between Invertebrate and Vertebrate Test Systems

When thinking about the use of invertebrate test systems and identification of MIEs and KEs for AOPs, an important question is if results from invertebrate experiments are biologically relevant or analogous to the responses of vertebrates.

Here we present a comparison of –omics analyses comparing vertebrate and invertebrate test species and how biological results obtained relate between these two systems.

5.3.1.1 Case Study 1: Nickel's Impact on Global Gene Expression Via Microarray Analysis

Nickel is a ubiquitous earth metal but is found in concentrated and potentially toxic concentrations in some locations due to industrial activities such as mining and smelting. Toxic responses in humans range from dermatitis to cancer (Vandenbrouck et al. 2009). Several groups have used –omics tools to evaluate the toxic mechanism of action of this chemical, as knowledge of how nickel toxicity is elicited in non-mammalian species is currently not well-understood.

Vandenbrouck et al. (2009) exposed a single clone of *Daphnia magna* to four waterborne concentrations of $NiCl_2$ (0.125, 0.5, 1, and 2 mg/L), which were below the EC50 of immobility at 48 h. Exposures for RNA extraction and microarray analysis were conducted for 96 h and additional growth measurements on *D. magna* were also assessed. RNA from a pool of 45 *D. magna* was extracted using TRIzol and a custom *D. magna* cDNA with 2445 genes related to life stage, moulting processes, and metabolism was used. After background correction and normalisation, significance analysis of microarrays (SAM) (Tushers et al. 2001) was used with a 5% false discovery rate (FDR) cut-off to determine differential gene expression. Blast2GO was used to determine differential regulation at the biological process level (www.blast2go.com).

Results demonstrated a dose-dependent increase in genes related to the *D. magna* cuticle, as well as several genes related to chitin-steroid metabolism and protein metabolism. Several ribosomal protein genes were downregulated in a dose-dependent manner, as well as genes related to lipid, oxygen, and ATP transport. Physiological data collected from this experiment showed a significant decrease of cellular energy allocation and consumed energy after 96 h at the three highest doses of Ni2+ (Vandenbrouck et al. 2009). This physiological response related to the dose-dependent changes in several metabolic processes, including decreases in oxygen and ATP transport genes.

In a separate study, Mohamed et al. (2014) exposed Mediterranean mussel (*Mytilus galloprovinciali*) to Ni. Mussels of 5–6 cm in length from aquaculture were acclimatized and exposed to 135 µg/L Ni via semi-static renewal for 4 days at both a preferred and high water temperature (18 °C and 26 °C). RNA was extracted from digestive glands of female mussels, and lysosomal membrane stability (LMS) was also assessed as a physiological measure of bivalve stress. A custom cDNA array of 1748 mussel sequences (Mytarray V1.1) was used with a two-colour labelling. Linear model for microarray analysis (Limma) (Smyth 2004) was used to determine differential gene expression and Blast2GO for gene ontology annotation.

In the single-dose exposure, there was a significant decrease in lysosomal membrane stability from the control, which is a characteristic response of bivalves to several toxic stressors (both biotic and abiotic) (Svendsen et al. 2004). Focusing on results from the 18 °C exposure, as other studies were also conducted under normal conditions for the test organism, pathways that were over-represented in the differential gene sets include response to chemical stimulus, ribosome biogenesis, cell development, cellular catabolic process, and chitin metabolic process (Table 2, Mohamed et al. 2014). While the primary focus of this paper was to determine differential responses between Ni-exposure and temperature stress, the data revealed overlapping gene expression responses with *D. magna*, specifically for genes and pathways related to ribosome synthesis and chitin. Chitin forms part of both the exoskeleton of arthropods and the shells of molluscs. Its differential regulation may be a response to the chemical stressor and as an attempt to maintain homeostasis during xenobiotic stress (Rodriguez-Serrano et al. 2009), and the presence in this over-represented pathways in both *D. magna* and *M. galloprovinciali* exposures demonstrated the usability of these types of data to infer responses in other species of invertebrates.

When considering how these pathways relate to exposures conducted in vertebrate organisms, one example from Bougas et al. (2013) exposed juvenile yellow perch (*Perca flavescens*) to two concentrations of nickel. The low concentration used in this study (68.5 μg/L) was an observed concentration in lakes from the Sudbury region in Canada whilst the high concentration (542 μg/L) was five times the low concentration and was known to result in significant metal accumulation for the exposure duration. Juveniles were exposed for 45 days with weekly monitoring of dissolved metal concentrations. Kidneys were collected upon completion of the exposure to determine internal metal concentrations. Livers were used for RNA extraction and microarray analysis using a 1000 probe custom chip with probe selection conducted to represent genes related to metabolism and known metal exposure responses, as well as a set of genes found differentially expressed between two lakes in the region but not associated with metal exposure. A mixed ANOVA was used with a multiple testing correction to determine differential gene expression, and Blast2GO was used for biological process analysis.

There were significantly increased concentrations of Ni in the kidneys over the controls in both the low and high treatment groups. However, only the high dose of Ni resulted in significant differential gene expression over the controls. Focusing again on experiments conducted at a normal physiological temperature only, high dose nickel exposure resulted in the following enriched processes and functional categories: translation, ribosome biogenesis, iron binding, structural constituent of ribosomes, and cellular homeostasis (Table 1, Bougas et al. 2013). Genes related to ribosome biogenesis appear once again, and can also be found enriched in mouse cells treated with Ni (Lu et al. 2010). It was proposed by the authors that this decrease may be an adaptive response to overall decreased protein-level metabolism during Ni exposures.

While these studies do not demonstrate convincing evidence for overlapping biological pathways between vertebrate and invertebrate species, they do give some insight into the potential for such studies to provide information on the concordance of processes exhibited between these species after nickel exposure. This could be true even in organisms with vastly different physiologies and using different exposure scenarios (acute versus chronic). Although individual gene responses will differ, focusing on similarities at the pathway level can provide a more holistic insight into how chemicals are interfering with normal processes during toxicant exposure, demonstrating that results are relevant and coherent between invertebrate and vertebrate systems. Indeed, early investigations from the modENCODE consortium, (providing an encyclopedia of genomic functional elements in the model organisms *C. elegans* and *D. melanogaster*) (www.modencode.org), at comparing the genome-wide expression of human, worm and fly, discovered co-expression modules that are shared across these animals, often functionally associated with cellular and developmental processes (Gerstein et al. 2014). As expected, regulatory modules that are shared among species were enriched by orthologous genes, yet those modules that are most conserved also contained the greatest number of interacting genes (Gerstein et al. 2014). This key finding is reinvigorating the use of experimental model species for understanding animal biology and the human condition. It also provides a platform for the next big and transformative set of experiments that combine genomics with toxicology using model invertebrate species to discover overlapping pathway-level biological responses shared among invertebrates to vertebrates that can aid at identifying Mode of Action (MoA), which is useful in the context of ERA.

5.4 Current State of AOPs

Invertebrates play an important role in the functioning of most ecosystems and represent about 95% of the metazoan diversity (GIGA 2014). Yet despite this fact, and the many discussed benefits of using invertebrate species as model organisms (including the ability to investigate multigenerational and sublethal endpoints with relative ease, plus the comparative wealth of existing in vivo data), there is a surprising dearth of activity in using invertebrates in the development of AOPs. This and the overall pace of discovery is reflected in the low numbers of full or partial AOPs that are currently available or being developed as part of activities driven by such organisations as the OECD, EPA etc. Here we review the current state of several AOPs available through the AOP Wiki (as a joint initiative between the European Commission – DG Joint Research Centre (JRC) and U.S. Environmental Protection Agency (EPA)) (https://aopkb.org/aop-wiki/index.php/Main_Page).

Of the AOPs in the wiki at this time, only a small number currently contain evidence from and, therefore, are applicable to invertebrates. This perhaps reflects the relatively early development of the majority of AOPs and also the limited knowledge in being able to apply them across species. Since the value of any AOP is enhanced when it is applicable for multiple species, this value is further increased if it is applicable across multiple taxa. Therefore, it is perhaps not surprising that few of the AOPs that currently are being developed are for invertebrate species exclusively.

5.4.1 Case Study 2: Summary of AOP Linking Acetylcholinesterase Inhibition (AChE) to Acute Mortality Based on Multi Data Approaches

Perhaps the most robust of the developing pathways currently is the AOP linking Acetylcholinesterase (AChE) inhibition to acute mortality (Russom et al. 2014) which is described under AOP16 in the AOP wiki. AChE is found in many types of conducting tissue including nerves and muscle, central and peripheral tissues, and motor and sensory fibres, but it is primarily found in the blood, brain and muscles. Its primary function is to hydrolyze the neurotransmitter Acetylcholine (ACh). AChE contains both an anionic and an esteratic site (Quinn 1987). During neurotransmission ACh is released from the nerve into the synaptic cleft and binds to ACh receptors, relaying the signal from the nerve. The signal is stopped when AChE hydrolyzes ACh. There are extensive datasets describing the impact of AChE inhibiting chemicals on the mortality in multiple species as exemplified by the prevalence of data which continues to become available linking organophosphates such as Chlorpyrifos to adverse effects in organisms. A simple search of the USEPA ECOTOX database (www.epa.gov/ecotox) for example, reveals over 3700 recorded values for this compound over the last 10 years. However, even for such a well-studied toxicity pathway(s) where there is significant evidence to support the link between acetylcholinesterase inhibition and acute toxicity, the lack of quantification to allow prediction of apical endpoints from in vitro or in vivo measurements highlights the scale of the challenge in developing and using AOPs in risk assessment.

Much of the existing literature considering AChE inhibition has limited taxonomic coverage. However, the review of the literature and AOP development by Russom et al. (2014) considers the biological conservation of the MIE and evidence supporting linkage of the MIE to AOs across a wide range of ecologically relevant taxa at different life stages. In addition, the authors present a chemical category approach to AOP development, considering toxicity data from a diversity of organophosphate and carbamate insecticides that act via inhibition of AChE.

To help define the taxonomic domain of applicability, and to predict relative intrinsic susceptibility to organophosphate and carbamate chemicals, Russom et al. (2014) utilized a comparative method developed by LaLone et al. (2013) to identify AChE as a potential toxicological target across a greater swath of animal diversity. This method identifies homologs, or genes that are shared across genomes by evolutionary descent. The applied logic is that the greater the level of protein sequence conservation, the greater the probability that they also retain their functions, which are preserved by natural selection (the orthology–function conjecture; Gabaldón and Koonin 2013). Although this premise is often true for identified orthologs (genes that are shared because of speciation), those that are identified as paralogs (genes that are shared because of duplication) often diverge in their functions. Therefore, the reliability of methods at predicting toxicity by comparative genomics, especially comparing both vertebrates and invertebrates, continues to improve as more genome sequences populate pre-computed databases that reconcile a gene tree with the corresponding species tree, for example OrthoDB (www.orthodb.org). Orthology analysis has understandably become an important sub-discipline of bioinformatics (Kriventseva et al. 2015; Dessimoz et al. 2012) in which attempts are made to identify orthologous genes which have descended from a single gene from the last common ancestor (Fang et al. 2010; Koonin 2005). Our search for AChE homologs among the vertebrates and invertebrates within the OrthoDB V8 database of orthologous groups for major clades (Kriventseva et al. 2015) uncovers 395 genes in 169 animal species (out of 173); the gene is found as single copy in only 35 species (gene ortholog group EOG8H1C6C at the Metazoa level). Phylogenetic reconstruction reveals an ancient gene duplication event of the AChE gene prior to the origins of Mammalia, resulting in most species having at least two copies in their genomes (tree not shown). Not surprising, AChE among the invertebrates have a more complex history of gene duplication and deletion along evolutionary lineages that include the arthropods and echinoderms through to nematodes and trematodes. However, despite such complexity, the tools and databases now available to researchers provide them with improved opportunities to interrogate and demonstrate the true breadth of the applicable domain of species for a given AOP; a trend which will increase as genome sequences become available.

The MIE for AChE inhibition is triggered by the interaction of the chemical with the anionic site of the enzyme, blocking the site for acetylcholine (ACh) and resulting in a build-up of ACh at synapses (KE2) and unregulated excitation (KE3) at neuromuscular junctions, preganglionic neurotransmitters and postganglionic nerve endings in the autonomous nervous system and neurotransmitters in the brain and CNS (Fig. 5.1). Significantly, protein sequence alignments of the AChE enzyme related to the MIE are relatively well-conserved across vertebrates and invertebrates, suggesting that the manner of the chemical interaction may be similar across a wide range of species and taxa (Russom et al. 2014). Hence, the likely domain of applicability ranges from invertebrate classes branchiopoda, insecta, arachnida, cephalopoda, ascidiacea, trematoda, gastrapoda as well as amphibia, mammalia and avia, thereby reflecting the ubiquitous nature of ACHE in all life stages in vertebrates and invertebrates.

5.4.1.1 Evidence from Invertebrates Supporting the AChE KEs and KERs

Supporting evidence for developing the AChE AOP was found in existing empirical data from both the ECOTOX database and the scientific literature (Russom et al. 2014). Studies preferentially selected were those which reported endpoints indicative of neurotoxicity such as seizure activity, muscle responses, heart, respiration rates, etc. The ECOTOX database yielded significant numbers of studies reporting physiological responses in terrestrial and aquatic organisms, only 39 of which reported results that were associated with at least 3 KEs, and only 2 studies which provided results supporting all 4 KEs (Fig. 5.1). The open literature also provided a substantial number of studies linking AChE inhibition with downstream KEs.

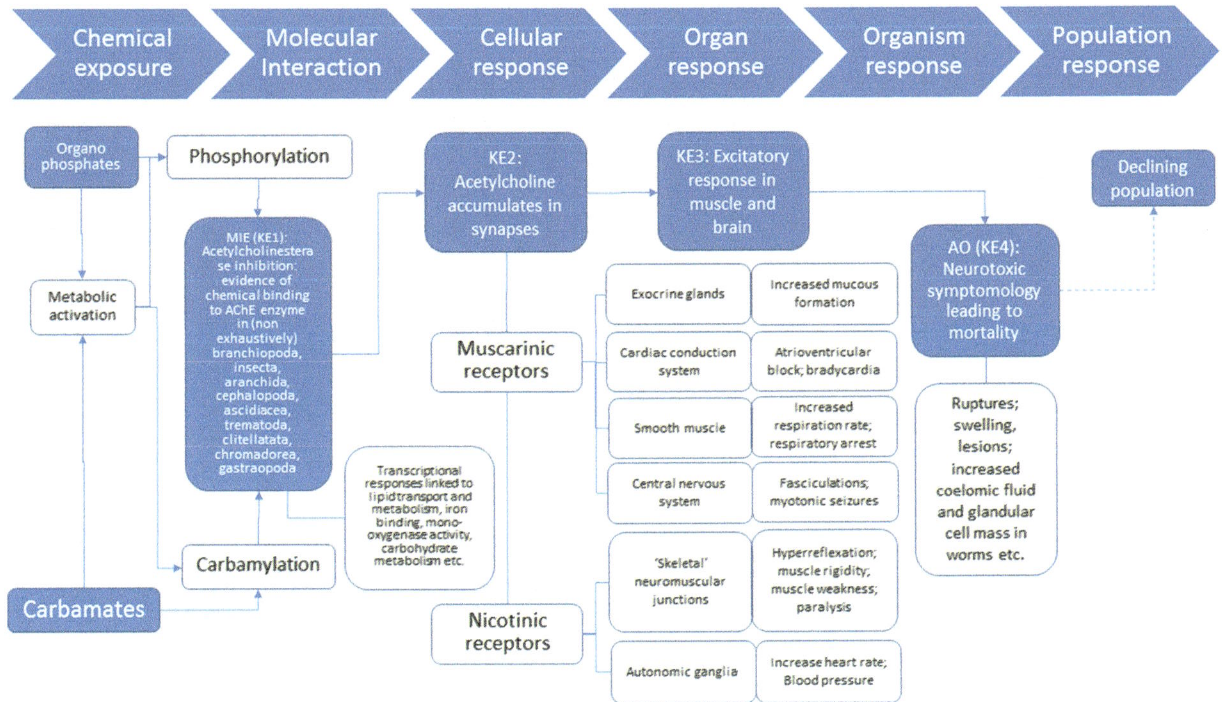

Fig. 5.1 Proposed adverse outcome pathway (AOP) for acetylcholinesterase (AChE) inhibition leading to acute lethality highlighting invertebrate specific attributes. Key events (KEs) are in bold letters, starting with the inhibition of AChE, the molecular initiating event (MIE), and proceeding to the accumulation of acetylcholine in cholinergic synapses (KE2),excitatory response in tissues (KE3), and lethality, the adverse outcome (AO) of regulatory significance. Dashed arrow indicates a plausible connection for which sufficient direct empirical evidence was not available (Copyright and copying: Copyright©2014 SETAC. All rights reserved. No part of the publication may be reproduced, stored or transmitted in any form or by any means without the prior permission in writing from the copyright holder. Authorization to copy items for internal and personal use is granted by the copyright holder for libraries and other users registered with their local Reproduction Rights Organization (RRO), e.g., Copyright Clearance Center (CCC), 222 Rosewood Drive, Danvers, MA 01923 (www.copyright.com), provided the appropriate fee is paid directly to the RRO. This consent does not extend to other kinds of copying such as copying for general distribution, for advertising or promotional purposes, for creating new collective works, or for resale. Special requests should be addressed to permissionsuk@wiley.com) (Modified from Russom et al. (2014))

Lu et al. (2012) were able to provide evidence linking the MIE to KE4 (Fig. 5.1) use gene silencing techniques. They were able to determine that AChE produced by the TcAce 1 gene was responsible for cholinergic neurotransmissions, while AChE produced by the TcAce2 gene is involved in noncholinergic activities including growth and reproduction in the red flour beetle *Tribolium castaneum*.

Empirical evidence linking KE2 to KE4 (Fig. 5.1) has been found in a single study using the earthworm *Eisenia foetida* (Reddy and Rao 2008). The study using profenofos (PFF) demonstrated a link between increased AChE levels (measured using the method described by Ellman et al. 1961) with body ruptures, lesions, excessive formation of glandular cell mass and disintegration of muscles causing internal coelomic pressure leading to mortality.

Some studies have attempted to quantify links between KEs for the ACHe AOP. Barata et al. (2004) exposed *Daphnia magna* to organophosphorus and carbamate pesticides in order to assess the inhibition and subsequent recovery patterns of both AChE and carboxylesterase (CbE), and related these patterns to individual observed effects. Time course experiments were conducted using two concentrations (the 24 h LC50 and 50% of the 24 h EC50) over a 48 h exposure period followed by a 72 h recovery period to determine the concentration of each tested compound (Chlorpyrifos, Malathion and Carbofuran) which caused a 50% inhibition of AChE and ChE (IC_{50}), as well as the kinetics of inhibition and recovery. Results indicated that AChE inhibition levels were greater than 50%. More specifically, 56% and 80% AChE inhibition was needed to impair survival to 10% and 50% respectively.

Similar attempts to quantify links between AChE inhibition levels as an indicator of KE2 linked to mortality (KE4) have been made using *Caenorhabditis elegans* as the test species (Rajini et al. 2008). Specifically, the nematodes were exposed to a number of organophosphorus (OP) insecticides for 4 h exposures over a range of concentrations. AChE levels were determined using the Ellman et al. (1961) method and linked to both acute lethal and sublethal behavioural effects. All OPs studied produced significant toxicity at greater than 50% AChE inhibition.

Other studies using the freshwater shrimp (*Paratya australiensis*) and the common shrimp (*Palaemon serratus*) have also established a link between AChE inhibition following exposure to lethal concentrations OPs, demonstrating between 70% and 100% inhibition of AChE at lethal doses (Abdullah et al. 1994; Bocquene and Galgani 1991). Abdullah et al. also reported >40% reduction in AChE levels resulting from sublethal concentrations of 0.1–10 μg/L of profenofos test chemical. Similarly studies with other invertebrate species investigating AChE inhibition in midge larvae (*Chironomus riparius*) (Detra and Collins 1991) and the freshwater gastropod Chilina gibbosa (Bianco et al. 2013) established strong links between KE2 and KE4.

A number of other studies have shown links between the MIE and KE2, including a study using the speckled shrimp *Metapendeus monoceros* (Reddy et al. 1990). The authors reported a significant reduction in AChE activity concurrent with an increase in ACh levels in nervous tissues.

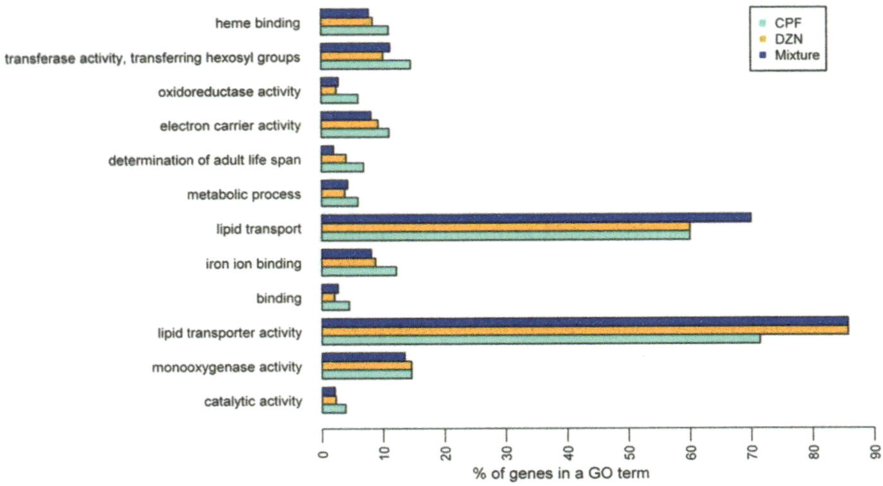

Fig. 5.2 Differential gene expression in C. elegans after chlorpyrifos (CPF), diazinon (DZN) exposures, and mixture OP exposures (Adapted from Vinuela et al. (2010) (Figure 2 in original manuscript))

5.4.1.2 –Omics Data Supporting the AChE Pathway

In the nematode *C. elegans*, Viñuela et al. (2010) evaluated differential gene expression after chlorpyrifos exposure. *C. elegans* were exposed from hatching to the L3 stage (72 h) to 0.5 mg/L of chlorpyrifos, a dose below the EC50 for reproduction (3.5 mg/L) as well as growth (14 mg/L). Whole body RNA extracts were hybridised to a whole-genome *C. elegans* array developed by the Genome Sequencing Center at Washington University. Differential gene expression was determined using rank product analysis with control of the false positive rate at 5%. Gene ontology information was obtained from Wormbase (http://www.wormbase.org) and over-representation of Gene Ontology (GO) terms was determined using a hypergeometric test (p-value cut-off <0.01).

Chlorpyrifos exposure resulted in the differential regulation of 551 genes, with key enriched GO terms in the categories of lipid transporter activity, lipid transport, mono-oxygenase activity, immunity, transferase activity, iron binding, and electron carrier activity (Fig. 5.2, Viñuela et al. 2010). There was also a strong expression of genes within the daf-16 pathway, such as glutathione S-transferase P 10 (gst-10) which are involved in phase II detoxification. Vitellogenins were increased by chlorpyrifos exposure, as well as genes involved in the Insulin Growth Factor (IGF) pathway.

In the marine bivalve *Mytilus galloprovincialis*, Dondero et al. (2011) conducted exposures to chlorpyrifos for 4 days to 0.77 mg/L, equivalent to the EC50 for lysosomal membrane stability. Pathways with significant enrichment following chlorpyrifos exposure include ion binding, transmembrane receptor activity, catabolic processes, carbohydrate metabolism, iron oxidase, and oxidoredictase

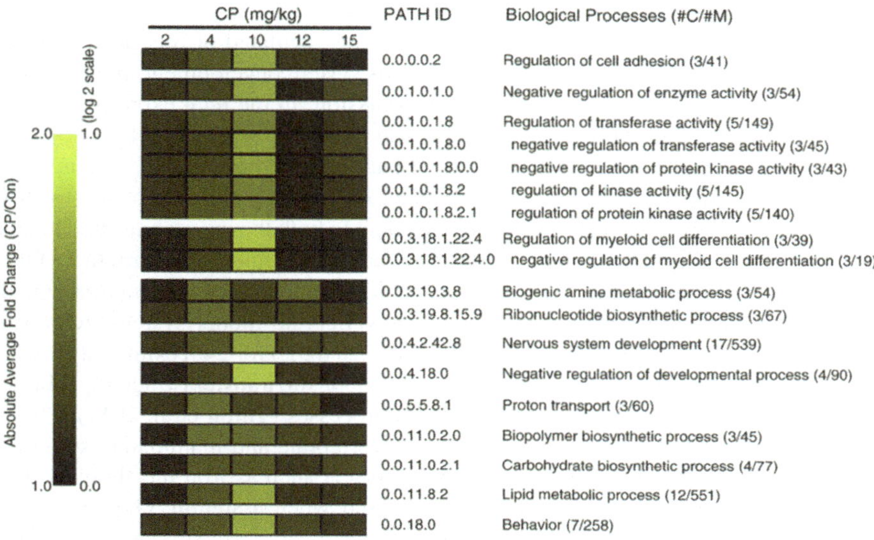

Fig. 5.3 Significant gene ontology terms associated with chlorpyrifos treatment in the maternal rat brain (Adapted from Moreira et al. (2010))

(Dondero 2011). Chitinase activity was also found in catabolic process enriched genes, which is consistent with results found in *C. elegans* following chlorpyrifos exposures (Vinuela et al. 2010). Impacts on iron binding and iron-related processes are also found in both of these systems, which are supported by findings that iron levels have an impact on regulating acetylcholine receptor expression in rats (Han and Kim 2015). Over-representation of genes within the iron binding and iron oxidase pathways may be a concordant toxic response between nematodes and mussels in response to OP stressors.

In a study by Moreira et al. (2010) looking at the maternal and fetal effects of chlorpyrifos exposure, C57BL/6 mice (*Mus musculus*) dams were treated with up to 15 mg/kg/day of chlorpyrifos via subcutaneous injection, with the highest doses (10, 12, and 15 mg/kg/day) resulting in decreased AChR activity in maternal brains. RNA was extracted from fetal and maternal brains and microarray analysis conducted using the affymetrix mouse whole genome 430 2.0 platform, which has over 39,000 transcripts. Differential gene expression was determined using the Limma analysis and MAPPFinder was used to determine enriched biological processes.

Focusing on the maternal brain data, enriched GO processes include regulation of transferase activity, lipid metabolism, carbohydrate biosynthesis, proton transport, and myeloid cell differentiation (Fig. 5.3, Moreira 2010). Additional specific biological information can be inferred from the Kyoto Encyclopedia of Genes and Genomes (KEGG) pathway analysis, with the following pathways significantly enriched by chlorpyrifos exposure in maternal brains: adherens junction, axon guidance, ErbB signaling, GnRH signaling, and Jak-STAT signaling. Focusing on genes that had a peak upregulation at 10 mg/kg, genes related to lipid and phosphorus metabolism are prevalent, and genes that are downregulated at this dose include developmental processes, cell adhesion, nervous system development, lipid metabolism, secretory pathways, and synaptic transmission. In the fetal exposures, 5 out of 43 genes related to oxidative stress were also present.

Whilst there was not complete overlap of pathways described in *C. elegans* and *M. musculus* (Vinuela et al. 2010; Moreira et al. 2010) overlapping pathways related to lipid and carbohydrate metabolism are prevalent in both the invertebrate and mammalian systems. The lack of other overlapping pathways could also an artefact of the experimental design such as the nature of the dosing methods used (environmental exposures versus direct injection). In another vertebrate model system, zebrafish (*Danio rerio*) exposed to chlorpyrifos (35, 88 or 220 μg/L) for 24 h exhibited enrichments in pathways related to morphogenesis, metabolism, transferase activity, kinase activity, cell growth/replication, and catabolic processes (Tilton et al. 2011). These pathways related to what was found in both invertebrate and mammalian systems, and while some specific pathways may not intersect such as iron metabolism, this could be due to varying genes and methods used when assigning GO categories between model and non-model organisms. These pathway-level similarities provide support to broader comparisons in the context of risk assessment as they provide support for similarities in key events and toxic modes of action.

One of the strengths of the AOP approach in ERA is in the value of combined evidence. Information provided by invertebrates combined with data obtained from mammalian, fish, and avian species using all-available evidence from in vitro, in silico and in vivo as well as gene expression information to support the KEs and KERs can be integrated to consider the adverse impacts of (in particular) chemical stressors on organisms. This is particularly the case when considering chemicals where there is a paucity of such data. However, even for chemicals for which target effects are known and for which significant evidence exists to support the pathway identification, the value of an AOP approach has the potential to significantly enhance current approaches in identifying pathways across species for ERA purposes. This is exemplified by the example of acetylcholinesterase inhibition, in which the network of AOPs as a whole, including the indirect KERs, supports the potential utility of in vitro or short-term in vivo measures of acetylcholinesterase inhibition for identifying chemicals with potential to cause systemic neurotoxicity at sub-narcotic concentrations. For example, Gong et al. (2010) demonstrated that hexanitrohexaazaisowurtzitane exhibited strong neurotoxic behaviour in the earthworm *Eisenia fetida* using a range of information including behavioural observations and neurobiochemical and eletrophysiological measurements. However, at present, while these approaches can in some cases provide strong evidence for the activation of a particular pathway, quantitative understanding is not sufficiently complete to accurately predict apical outcomes or potency from in vitro measurements alone. In addition, well-known chemical initiators of these AOPs are known to require metabolic activation, suggesting that chemical-specific ADME and toxicokinetic considerations will be strong determinant of quantitative outcomes along these AOPs (Groh et al. 2015).

5.4.2 Other AOPs Using Invertebrates

The case study of AChE exemplifies the need for multiple data sources to identify, support and quantify the pathway. Given the complexity of responses to a chemical assault within an organism or the complexity of responses given the multiple stressors to which organisms are exposed in the environment, single data types alone are unlikely to be able to fully characterize subsequent impacts on the individual following exposure to a chemical toxin. Such a problem was considered in a recent study by De Coninck et al. (2014), where the authors limited the stressors to two. The authors investigated the possibilities and limitations of using a genome-wide transcription based approach to consider the impact of two stressors, cadmium and microcystis (producing mycrocystin neurotoxin), on two genotypes of *Daphnia pulex* isolated from two populations; one population was exposed to high levels of cadmium (tolerant) for over a century and the other was exposed to naturally occurring low a levels (sensitive). The authors were able to interrogate the effects of mixture components and genotypes, both independently and in combination, to identify interaction responses which contributed to tolerance in individuals. They identified oxidative stress and polyunsaturated fatty acid metabolism-related pathways, as well as trypsin and neurexin IV gene-families as candidates for the underlying causes of genotypic differences in tolerance to microcystis. However, the approaches were less successful in linking gene expression results from single chemical exposure to organismal responses. The study thus demonstrated the potential value of the technique for better understanding and extricating pathways, but also highlighted that additional techniques and information would be needed to understand key events quantitatively and links to phenotypic and population relevant endpoints for application in risk assessment, particularly after exposure to multiple stressors.

The study of neurotoxicity is of course not limited to AChE inhibitors alone. Thousands of chemicals are known or thought to have neurotoxic properties and have been studied in an environmental toxicology context. Other key neurotransmitter pathways, in addition to the cholinergic pathway, which can be impacted and have been studied include the Dopaminergic (DA), Serotonergic, GABAergic and Glutamatergic acid pathways (Basu 2015). Invertebrates lend themselves well to investigating some of these pathways more than others. The inhibition of gamma aminobutyric acid (GABA) receptor is well-studied in vertebrates but perhaps less so in invertebrates. However, ionotropic GABA receptors (iGABARs) have also been described in many different phyla of invertebrates such as social amoeba (*Dictyostelium discodeum*), cnidarians, mollusks, annelids, arthropods, nematodes, and chordates (as described in the AOP under development in the AOP wiki). As such, the described AOP has potential broad relevance across invertebrate as well as vertebrate taxa.

Perhaps more poignant is the current limited representation of invertebrate data supporting other developing AOPs in the wiki. For example, there are currently two AOPs under development related to N-methyl-D-aspartate receptors (NMDAR). These are focussed primarily on mammalian brain development. The NMDA receptor is a glutamate receptor and ion channel protein found in nerve cells and there is significant evidence that such receptors are present in invertebrate species as well as mammalian species. The genes involved in ionotropic glutamate receptors have been found to be significantly activated in daphnids (Toyota et al. 2015). Genes are predicted to encode the subunits of an NMDA-type (NMDAR) iGluR necessary for memory retention in *C. elegans* (Kano et al. 2008) and a partial cDNA encoding the leech NR1 subunit of the NMDA receptor (HirNR1) has also been identified (Grey et al. 2009). The inclusion of such information in developing AOPs will significantly broaden their applicability across taxa.

The new developing AOP describing the alkylation of DNA in male pre-meiotic germ cells leading to heritable mutations provides empirical evidence across the KEs and KERs when it includes invertebrate data. Of particular relevance for the broad applicability of the AOP is that data are reported for multiple species to support this indirect KER showing that a variety of O-alkylating agents cause male germ cell mutations in many species including invertebrates such as *Drosophila* (Stilwell et al. 2006; Raymond-Delpech et al. 2005).

5.5 Application to Environmental Risk Assessment

This section aims to illustrate how the AOP framework can be used in ERA with the methods and tools that are currently available. Full AOPs are not currently necessary to complete a robust risk assessment because the framework can still provide valuable insights, even when only individual segments of the AOP are addressed. The following is a conceptual approach to show the value of the AOP framework in risk assessment. The approach here is limited to consider invertebrates. Figure 5.4 provides an overview of the AOP framework and the tools that are currently available to support the use of AOPs in environmental risk assessment.

5.5.1 QSARs

There is a long history of use of Quantitative Structure-Activity Relationships (QSAR) in ERA for the prediction of hazard data. Many hundreds of models have been developed to predict the aquatic toxicity of chemicals. The most widely accepted QSAR models for prediction of aquatic toxicity are based on MoA. Therefore, to apply QSARs to a novel chemical, it is first necessary to assign that chemical to a MoA. This is often the most difficult step in applying QSARs and is,therefore, the largest potential source of error. An inaccurate assignment of MoA can lead to a QSAR prediction several orders of magnitude different to the true value.

Although there are numerous QSARs developed that are based on invertebrate test data, the current methods for predicting MoA from chemical structure are based predominantly on insights from fish toxicity tests (Verhaar et al. 1992; Russom et al. 1997). It is unclear to what extent these MoA classifications are applicable across other species, including invertebrates. Despite this, once a MoA is assigned, it is

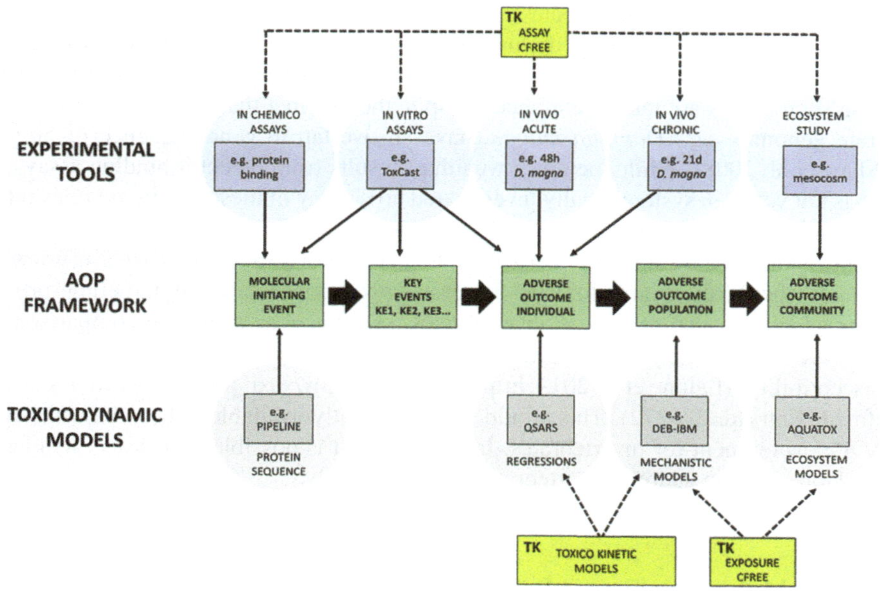

Fig. 5.4 AOP toolbox. The AOP framework is given in the *green* boxes. Lab based tools are given as *blue boxes*. Toxicodynamic tools are given in *grey*. Toxicokinetic tools are given in *yellow*

generally applied across all species being assessed. The information captured in AOPs and KEs related to aquatic toxicity, combined with application of phylogenetic insights, will allow more informed predictions to be made across species by highlighting commonalities and, more importantly, differences between AOPs in different species. It is anticipated that the grouping of chemicals for read across/QSAR approaches will need to be updated in light of such developments as the existing MoA classifications may not be sufficient to accommodate this level of detail.

The development of AOPs (and QSARs) for sub-lethal AOs is a major need in the area of ecotoxicology. As discussed above, invertebrate species are ideal candidates for such work. As the network of pathways underlying sub-lethal effects is unveiled, the use of QSARs to predict KEs at lower levels of biological organisation may become necessary.

5.5.2 *In Vitro Assays*

While the majority of receptor binding assays are currently designed to inform issues related to human health and pharmacology, these assays also have the potential to provide valuable insights for the identification of MoA and molecular initiating events in other species in cases where targets are conserved. Endocrine disruption is a good example of this; despite the fact that the majority of research in this area has focused on vertebrate species, some of the most robust studies showing the adverse effects of endocrine disruption have been carried out on invertebrates (Oetken et al. 2004). For invertebrate risk assessment and AOP development, results from such assays are rarely referenced despite the fact that that human and invertebrate genomes e.g., *Daphnia magna*, have an overlap in gene sequence of 56% (Shaw et al. 2008). While there is a wealth of results from receptor binding assays, it has not yet been systematically investigated how many of these assays are relevant to a broader range of species, including invertebrates. What is required is a review of how many of the commonly performed human health assays e.g., ToxCast (www.epa.gov/chemical-research/toxicity-forecaster-toxcasttm-data), are targeting proteins which are also present in invertebrates. For this gene sequence comparison, tools are available, e.g., SeqAPASS: Sequence alignment to predict across-species susceptibility) (Lalone et al. 2013; https://cfpub.epa.gov/si/si_public_record_report.cfm?dirEntryId=276372). This would allow currently available data to be used in AOP development for invertebrates. In the future, it is possible that assays will be available that can address invertebrate proteins directly.

5.5.3 Molecular Target Sequence Analysis

For all new human pharmaceuticals coming on to market, it became mandatory in the EU in 2006 to conduct chronic endpoint studies on *Daphnia*, algae and early life stage fish (Gunnarsson et al. 2008). However, this strategy alone is not predictive for all wildlife species. The challenge of addressing the difficulties of assessing the impact of chemical exposure on multiple species using data from a few is prominent in the environmental risk assessment of all chemicals. It is an impossible task in environmental risk assessment to conduct toxicity tests on every wildlife species to ensure complete safety, let alone to address the complexity of potential ecological interactions influencing sensitivity to toxicants. For this reason, tools for species extrapolation are essential. One such method is to make use of results obtained from the comparative genomics research community at detecting homologous elements across all sequenced genomes and using molecular phylogenies and functional genomics data to understand the functional preservation of these elements that include genes. Annotated draft genome and transcriptome sequences are known for many species and their numbers are increasing exponentially (Reddy et al. 2014). For example, Gunnarsson et al. (2008) observed conservation of 1318 human drug targets across 16 species. Remarkably this included 61% target conservation in *Daphnia magna*, highlighting the significant conservation of 'human' targets across lower trophic level invertebrates. Searching for target conservation can lead to an increased efficiency in toxicity testing by identifying appropriate test organisms (Ankley et al. 2007).

There are two main types of methods used to identify gene orthologs among species: *(1)* orthology by sequence similarity searches that use local alignment algorithms such as BLAST or Smith-Waterman (Altschul et al. 1990; Smith and Waterman 1981) for all pairwise sequence comparisons; *(2)* phylogeny-based searches that aim to delineate speciation from gene duplication by comparing the gene tree with the corresponding species tree (Goodman et al. 1979). Although the methods for predicting the functional conservation of genes are regularly evaluated against benchmarks that minimize known biases and pitfalls (Gabaldón et al. 2009; Dessimoz et al. 2012), the toxicology community has developed a custom tool (SeqAPASS) (Lalone et al. 2013), which take the peptide sequence for a known target protein and aligns it with protein sequences from a publicly available sequence databank. Orthologs are presumed using Blast and the output of this process is a qualitative prediction of susceptible taxonomic groups. Despite the early stage of the development of SeqAPASS and a need to evaluate its performance alongside the >30 established comparative genomic databases (http://questfororthologs.org/), several publications underline the potential benefits of such approaches (Lalone et al. 2013; Schreiber et al. 2011; Russom et al. 2014).

There are limitations of using such tools in isolation. Only a limited number of species across relevant taxonomic groups have genome maps which can limit the applicability. Additionally, these tools tend to use a two dimensional protein structures and very little is known about the three dimensional structures and how the docking site might vary on a similar protein across different species. To consider three dimensional binding interactions is very time consuming, and the current tools are not yet powerful enough to identify sensitive species with high levels of confidence. Therefore, the molecular target sequence analysis tools can currently only be used to identify vulnerable species groups and guide risk assessors in the selection of test species.

Molecular –omics technologies now exist that can revolutionise testing and environmental monitoring. These –omics technologies offer insight into the mechanisms of toxicity by measuring the expression of 10,000s of genes and the levels of 1000s of metabolites in an organism. These molecular data can then be used in AOPs to link from the MIE or key event to the phenotypic responses (e.g., growth and reproduction) of organisms to pollutants. By understanding the gene expression or changes in the metabolome in response to a given group of chemicals with a known MoA, the resulting molecular signature can be used to identify similar responses in other species. One of the main challenges which currently is preventing the full scale use of such technologies in elucidating pathways is the current lack of a cohesive approach to interpreting and differentiating gene expression data which connect to KEs (leading to an AO by definition) from those connected to more incidental events in a pathway.

5.5.4 *In Vivo*

5.5.4.1 Acute

There is a wealth of acute toxicity data for invertebrates and these data are usually fast, easy, and relatively cheap to generate without the ethical dilemmas associated with vertebrate species. In addition, the use of invertebrates also avoids the legal and regulatory constraints imposed on vertebrate species. However, the data mainly come from a very narrow range of universally-recognized model species (i.e. *D. magna, C. elegans* etc.) and provide limited information on toxicity mechanisms due to the limited range of endpoints which are traditionally recorded such as immobility and survival.

While acute data are often criticised as too crude to provide population relevant, mechanistic insights, it must be acknowledged that when generated carefully, these values can provide valuable supporting evidence for AOPs through a better understanding of MoA. This evidence is enhanced when combined with other data such as gene expression. Toxicokinetic concepts such as chemical activity and critical body burden can help to define the MIE and related AOP. For example, Thomas et al. (2015) illustrated how the use of the chemical activity framework can distinguish between baseline toxicants and specifically acting chemicals. The approach utilizes an activity threshold for acute toxicity, where chemicals that exhibit activities in the range of 0.01–0.1 can be classed as acute baseline toxicants. Lower activities imply a more specific mode of action. This approach is similar to the critical body burden approach where chemicals that have an internal concentration of 2–8 mM at acute EC50 are generally classed as baseline toxicants (van Wezel et al. 1996). Careful consideration of exposure concentrations is critical for this type of analysis as nominal concentrations can misrepresent the actual free concentration in the medium. Processes such as partitioning into organic compartments, evaporation, precipitation, degradation and metabolism may significantly reduce the concentration of chemical which is available for uptake by the organism. Such processes, if present, typically lead to an underestimation of the toxicity of the chemical. Modelling approaches have been developed to help identify potential exposure-related issues in advance of experimental work (Armitage et al. 2014) and advanced passive exposure techniques have been developed to help overcome some of these issues (Kramer et al. 2010).

5.5.4.2 Chronic

Despite some of the potentially useful aspects associated with acute toxicity data, there is currently a call for more chronic data for invertebrate species. The use of chronic data over acute data increases the confidence in risk assessments and allows the use of reduced assessment factors. From an AOP perspective, however, there is valuable information which is not captured when experiments are limited to standard test protocols. For instance, physiological and behavioural effects may provide insights into key events or adverse outcomes. Furthermore, information on MIE can be captured by taking samples for transcriptomics analyses. Such additional information can support modelling and extrapolation both upwards and downwards in the AOP. In short, we can obtain a stronger mechanistic understanding from the results of these assays. A standard chronic toxicity assay may completely miss the adverse event (i.e. reduction of offspring survival) or fail to identify important key events (i.e. inhibition of growth).

5.5.5 Use of Mechanistic Effect Models in AOPs

When assessing the potential for adverse effects in the environment following exposure to a chemical, there is a need in ERA firstly to be able to determine the nature of the adverse effect in relevant species, secondly at what concentration or dose that impact occurs (the tipping point) and lastly how relevant that impact is, given the actual exposure in the environment under consideration. AOPs provide a valuable approach to be able to understand relevant mechanisms across relevant species at an individual level. Critically, however, for use in ERA there is a need to be able to extrapolate these findings to understanding impacts on populations. Predictive systems models are under development which can begin to account for some of the complexity of chemical impacts on populations, communities and ecosystems. Models such as Dynamic Energy Budget linked with Individual based models (DEB-IBM) attempt to extrapolate from individual level effects to population effects (Martin et al. 2013). When combined with the AOP framework, mechanistic effect models (at the sub individual level) even have the potential to be used to link chemical effects at different levels of biological organisation based on an understanding of the chemical MoA (Groh et al. 2015); qualitative links have been established, for example, between DEB models and transcriptomics data (Wren et al. 2011; Swain et al. 2010).

DEB modelling is a toxicodynamic modelling approach that relies on DEB theory to link either steady-state or time-varying concentrations of a chemical to the effects on the life history of an organism (Kooijman 2009). In simple terms, DEB considers the total energy intake of an organism, and maps out how the organism will allocate the available energy to the main processes in the life history: growth, maintenance, reproduction, and maturation. Chemicals can either affect the direction of the energy flow, or can affect the energy cost of one or more of the life history processes. DEB uses measured time point data from toxicity experiments (growth, reproduction, survival) to determine the *physiological mode of action* (pMoA) – for example costs for growth, costs for reproduction, maintenance costs, and decrease in feeding ability. The pMoA differs from a traditional MoA in that it only considers the effect of a chemical on the life history traits of an organism – or in other words – pMoA provides the population relevant effects of a chemical on an organism (Martin et al. 2013). Therefore, although DEB modelling occurs at the individual level, predicted DEB parameters are suitable for extrapolating to the population level.

An example of this type of extrapolation for potential use in ERA is given in Martin et al. (2013) where a comprehensive dataset for *Daphnia magna* was used to identify the physiological mode of action (pMoA) and the subsequent dynamic population level effects of 3,4-dichloroaniline. The data consisted of a 42 day chronic exposure study to five continuous concentrations and a control with monitored growth, reproduction and survival. The DEB analysis identified the most likely pMoA to be either embryonic hazard or reproduction costs. It was not possible to definitively distinguish between these two pMoAs – but as the dose-response curve fit with embryonic hazard as a pMoA resulted in the highest likelihood this was used in the further analysis. Linking the DEB parameters, chemical stress level and pMoA to an Individual Based Model (IBM) resulted in accurate predictions of dynamic population level effects of the stressor.

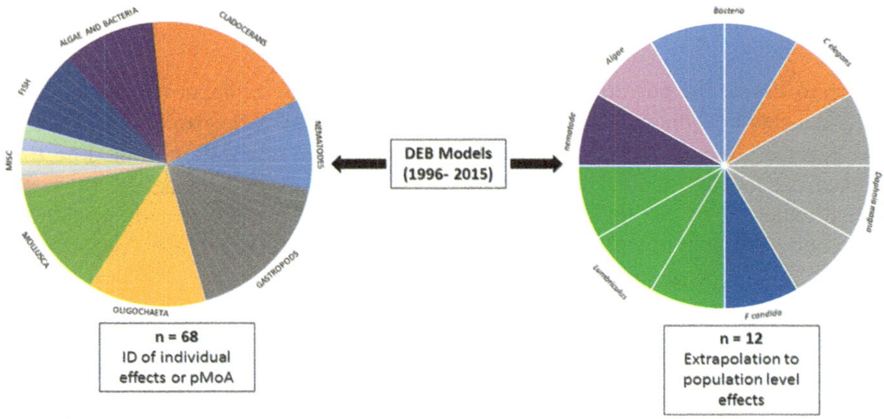

Fig. 5.5 An overview of the species distribution in peer reviewed publications relating to Dynamic Energy Budget theory

DEB models can be parameterized for any species, although initial parameterization of a new species usually requires extensive measurements. However, of the 68 papers identified through the website of the department of Theoretical Biology of the VU University of Amsterdam (http://www.bio.vu.nl/thb/) and the debtox information page of Tjalling Jager (http://www.debtox.info/index.html), which deal with the impact of chemical stressors on the individual level, 55 were based on data from invertebrate studies. The most frequently used species were *Daphnia magna* (n = 12), *Lymnaea stagnalis* (n = 11), *Mytilus edulis* (n = 8), *Caenorhabditis elegans* (n = 6) and *Eisenia andrei* (n = 6) (Fig. 5.5). Invertebrates lend themselves to DEB modelling because of their relatively short life-cycle, ease of culturing, and feasibility of measuring the necessary DEB parameters. Whilst the examples of the extension of DEB theory to link AOPs to population level effects is not yet prevalent, for invertebrates at least, DEB provides a tangible opportunity to enable AOPs potentially to be integrated into an environmental risk assessment approach considering population and potentially community level impacts.

5.6 Conclusion

Traditional ERA approaches used to understand the impacts of chemical exposure rely heavily on short-term acute and/or chronic in-vivo toxicity tests using various model species combined with a variety of assessment factors to derive toxicity thresholds. However, since these factors lack a mechanistic basis, they have limited potential for quantitatively estimating cross-species toxicity thresholds. AOPs provide a real opportunity to create a future framework for ERA based on a mechanistic, exposure driven understanding at its core. They also enable the exploitation of the wealth of knowledge and data of traditionally discrete disciplines such as Toxicology and (Eco)toxicology, Systems Toxicology and Environmental Genomics in helping to define such mechanistic challenges in ERA as cross-species KE homology.

In such a framework, invertebrates offer some unique advantages over vertebrates in the development of AOPs. Their short life cycle and relative ease in which large numbers of organisms can be studied provides valuable opportunities to study the impact of chemical exposure at environmentally relevant concentrations over chronic time spans. Relatedly, economic as well as ethical factors also influence the type of testing typically conducted for ERA; for example, the time and labour required to perform chronic toxicity studies on whole vertebrate organisms as well as the call for a reduction of animals used for testing limits their use (REACH 2006). Invertebrates also lend themselves to genomic and phylogenetic investigation and allow the study of sublethal key events. This is facilitated by the fact that the basic biology of many invertebrates is well-understood, with a number of model species with fully mapped genomes. Perhaps one of the primary potential limitations of using invertebrates is that they are not biologically representative of vertebrates due to differences in their physiology. It is certainly true that differences will be observed ranging from individual gene responses to depuration processes and for certain pathways the presence or absence of receptors will drive differential responses across species. Yet, such differences are also observed even among more closely related vertebrate species. However, despite this, focussing on similarities rather than the differences still allows significant conclusions to be drawn between species. Importantly, there is strong conservation in drug targets between humans and invertebrates, thereby demonstrating the applicability of using invertebrate species as a model for potential effects on vertebrates. Comparative genomic methods and databases including SeqAPASS may help to increase the ability to read across the results of pathways in one species to other species for such cross species extrapolation. Finally, and perhaps most importantly, the use of invertebrate models begins to allow us to consider realistic options for extrapolating from individual to population effects. Through the development and use of such models as those using Dynamic Energy Budgets linked with population models, the potential for extrapolating sub-individual events described by an AOP to population level effects relevant for ERA decision making becomes a genuine possibility.

References

Abdullah AR, Kumar A, Chapman JC (1994) Inhibition of acetylcholinesterase in the Australian freshwater shrimp (*Paratya australiensis*) by profenofos. Environ Toxicol Chem 13(11):1861–1866

Altschul SF, Gish W, Miller W, Myers EW, Lipman DJ (1990) Basic local alignment search tool. J Mol Biol 215(3):403–410

Ankley GT, Bennett RS, Erickson RJ, Hoff DJ, Hornung MW, Johnson RD, Mount DR, Nichols JW, Russom CL, Schmieder PK, Serrano JA, Tietge JE, Villeneuve DL (2010) Adverse outcome pathways: a conceptual framework to support ecotoxicology research and risk assessment. Environ Toxicol Chem 29(3):730–741

Ankley GT, Brooks BW, Huggett DB, Sumpter JP (2007) Repeating history: pharmaceuticals in the environment. Environ Sci Technol 41(24):8211–8217

Ankley GT, Miracle A, Perkins EJ, Daston GP (eds) (2008) Genomics in regulatory ecotoxicology: applications and challenges. CRC Press, Boca Raton

Armitage JM, Wania F, Arnot JA (2014) Application of mass balance models and the chemical activity concept to facilitate the use of in vitro toxicity data for risk assessment. Environ Sci Technol 48:9770–9779

Barata C, Markich SJ, Baird DJ, Taylor G, Soares AM (2002) Genetic variability in sublethal tolerance to mixtures of cadmium and zinc in clones of Daphnia magna Straus. Aquat Toxicol 60(1–2):85–99

Barata C, Solayan A, Porte C (2004) Role of B-esterases in assessing toxicity of organophosphorus (chlorpyrifos, malathion) and carbamate (carbofuran) pesticides to *Daphnia magna*. Aquat Toxicol 66(2):125–139

Basu N (2015) Applications and implications of neurochemical biomarkers in environmental toxicology. Environ Toxicol Chem 34(1):22–29

Bianco K, Yusseppone MS, Otero S, Luquet C, Rios de Molina MC, Kristoff G (2013) Cholinesterases and neurotoxicity as highly sensitive biomarkers for an organophosphate insecticide in freshwater gastropod (*Chilina gibbosa*) with low sensitivity carboxylesterases. Aquat Toxicol 144–145:26–35

Bocquené G, Galgani F (1991) Acetylcholinesterase activity in the common prawn (*Palaemon serratus*) contaminated by carbaryl and phosalone: choice of a method for detection of effects. Ecotoxicol Environ Saf 22(3):337–344

Bougas B, Normandeau E, Pierron F, Campbell PGC, Bernatchez L, Couture P (2013) How does exposure to nickel and cadmium affect the transcriptome of yellow perch (*Perca flavescens*) – results from a 1000 candidate-gene microarray. Aquat Toxicol 142–143:355–364

Buikema AL, Cairns J (1980) Aquatic invertebrate bioassays. ASTM International, West Conshohocken

Burden N, Sewell F, Andersen ME, Boobis A, Chipman JK, Cronin MTD, Hutchinson TH, Kimber I, Whelan M (2015) Adverse outcome pathways can drive non-animal approaches for safety assessment. J Appl Toxicol 35(9):971–975

Burgess-Herbert SL, Euling SY (2013) Use of comparative genomics approaches to characterize interspecies differences in response to environmental chemicals: challenges, opportunities, and research needs. Toxicol Appl Pharmacol 271(3):372–385

Colbourne JK, Pfrender ME, Gilbert D, Thomas WK, Tucker A, Oakley TH, Tokishita S, Aerts A, Arnold GJ, Basu MK (2011) The ecoresponsive genome of *Daphnia pulex*. Science 331(6017):555–561

Council Directive 2010/63/EU of 22 September 2010 on the protection of animals used for scientific purposes [2010] OJ L276/33

De Coninck DIM, Asselman J, Glaholt S, Janssen CR, Colbourne JK, Shaw JR, De Schamphelaere KAC (2014) Genome-wide transcription profiles reveal genotype-dependent responses of biological pathways and gene-families in daphnia exposed to single and mixed stressors. Environ Sci Technol 48(6):3513–3522

Dessimoz C, Gabaldón T, Roos DS, Sonnhammer ELL, Herrero J, Altenhoff A, Apweiler R, Ashburner M, Blake J, Boeckmann B, Bridge A, Bruford E, Cherry M, Conte M, Dannie D, Datta R, Domelevo Entfellner J, Ebersberger I, Galperin M, Joseph J, Koestler T, Kriventseva E, Lecompte O, Leunissen J, Lewis S, Linard B, Livstone MS, Lu H-C, Martin M, Mazumder R, Messina D, Miele V, Muffato M, Perrière G, Punta M, Rouard M, Schmitt T, Schreiber F, Silva A, Sjölander K, Škunca N, Stanley E, Szklarczyk R, Thomas P, Uchiyama I, Van Bel M, Vandepoele K, Vilella AJ, Yates A, Zdobnov E (2012) Toward community standards in the quest for orthologs. Bioinformatics 28(6):900–904

Detra RL, Collins WJ (1991) The relationship of parathion concentration, exposure time, cholinesterase inhibition and symptoms of toxicity in midge larvae (*Chironomidae: Diptera*). Environ Toxicol Chem 10(8):1089–1095

Dondero F, Banni M, Negri A, Boatti L, Dagnino A, Viarengo A (2011) Interactions of a pesticide/heavy metal mixture in marine bivalves: a transcriptomic assessment. BMC Genomics 12:195

Ellman GL, Courtney KD, Andres V Jr, Featherstone RM (1961) A new and rapid colorimetric determination of acetylcholinesterase activity. Biochem Pharmacol 7(2):88–90

Fang G, Bhardwaj N, Robilotto R, Gerstein MB (2010) Getting started in gene orthology and functional analysis. PLoS Comput Biol 6(3):1–8

Feder ME, Mitchell-Olds T (2003) Evolutionary and ecological functional genomics. Nat Rev Genet 4:649–655

Gabaldón T, Koonin EV (2013) Functional and evolutionary implications of gene orthology. Nat Rev Genet 14:360–366

Gabaldón T, Dessimoz C, Huxley-Jones J, Vilella AJ, Sonnhammer ELL, Lewis S (2009) Joining forces in the quest for orthologs. Genome Biol 10(9):403. doi:10.1186/gb-2009-10-9-403

Gerstein MB, Rozowsky J, Yan KK, Wang D, Cheng C, Brown JB, Davis CA, Hillier L, Sisu C, Li JJ, Pei B, Harmanci AO, Duff MO, Djebali S, Alexander RP, Alver BH, Auerbach R, Bell K, Bickel PJ, Boeck ME, Boley NP, Booth BW, Cherbas L, Cherbas P, Di C, Dobin A, Drenkow J, Ewing B, Fang G, Fastuca M, Feingold EA, Frankish A, Gao G, Good PJ, Guigó R, Hammonds A, Harrow J, Hoskins RA, Howald C, Hu L, Huang H, Hubbard TJ, Huynh C, Jha S, Kasper D, Kato M, Kaufman TC, Kitchen RR, Ladewig E, Lagarde J, Lai E, Leng J, Lu Z, MacCoss M, May G, McWhirter R, Merrihew G, Miller DM, Mortazavi A, Murad R, Oliver B, Olson S, Park PJ, Pazin MJ, Perrimon N, Pervouchine D, Reinke V, Reymond A, Robinson G, Samsonova A, Saunders GI, Schlesinger F, Sethi A, Slack FJ, Spencer WC, Stoiber MH, Strasbourger P, Tanzer A, Thompson OA, Wan KH, Wang G, Wang H, Watkins KL, Wen J, Wen K, Xue C, Yang L, Yip K, Zaleski C, Zhang Y, Zheng H, Brenner SE, Graveley BR, Celniker SE, Gingeras TR, Waterston R (2014) Comparative analysis of the transcriptome across distant species. Nature 512:445–448

Giga Community of Scientists (2014) The global invertebrate genomics alliance (GIGA): developing community resources to study diverse invertebrate genomes. J Hered 105(1):1–18

Gong P, Basu N, Scheuhammer AM, Perkins EJ (2010) Neurochemical and electrophysiological diagnosis of reversible neurotoxicity in earthworms exposed to sublethal concentrations of CL-20. Environ Sci Pollut Res 17(1):181–186

Goodman M, Czelusniak J, Moore GW, Romero-Herrera AE, Matsuda G (1979) Fitting the gene lineage into its species lineage, a parsimony strategy illustrated by cladograms constructed from globin sequences. Syst Biol 28(2):132–163

Gorokhova E, Kyle M (2002) Analysis of nucleic acids in *Daphnia*: development of methods and ontogenetic variations in RNA-DNA content. J Plankton Res 24(5):511–522

Grey KB, Moss BL, Burrell BD (2009) Molecular identification and expression of the NMDA receptor NR1 subunit in the leech. Invertebr Neurosci 9(1):11–20

Groh KJ, Carvalho RN, Chipman JK, Denslow ND, Halder M, Murphy CA, Roelofs D, Rolaki A, Schirmer K, Watanabe KH (2015) Development and application of the adverse outcome pathway framework for understanding and predicting chronic toxicity: I. Challenges and research needs in ecotoxicology. Chemosphere 120:764–777

Gunnarsson L, Jauhiainen A, Kristiansson E, Nerman O, Larsson DGJ (2008) Evolutionary conservation of human drug targets in organisms used for environmental risk assessments. Environ Sci Technol 42(15):5807–5813

Hammond JI, Jones DK, Stephens PR, Relyea RA (2012) Phylogeny meets ecotoxicology: evolutionary patterns of sensitivity to a common insecticide. Evol Appl 5(6):593–606

Han M, Kim J (2015) Effect of dietary iron loading on recognition memory in growing rats. PLoS One 10(3):e0120609

Ishikawa H (1977) Evolution of ribosomal RNA. Comp Biochem Physiol B 58(1):1–7

Kano T, Brockie PJ, Sassa T, Fujimoto H, Kawahara Y, Iino Y, Mellem JE, Madsen DM, Hosono R, Maricq AV (2008) Memory in *Caenorhabditis elegans* is mediated by NMDA-type ionotropic glutamate receptors. Curr Biol 18(13):1010–1015

Kato Y, Shiga Y, Kobayashi K, Tokishita S, Yamagata H, Iguchi T, Watanabe H (2011) Development of an RNA interference method in the cladoceran crustacean *Daphnia magna*. Dev Genes Evol 220(11–12):337–345

Kato Y, Matsuura T, Watanabe H (2012) Genomic integration and germline transmission of plasmid injected into crustacean *Daphnia magna* eggs. PLoS One 7(9):e45318

Kooijman B (2009) Dynamic energy budget theory for metabolic organisation, 3rd edn. Cambridge University Press, Cambridge, UK

Koonin EV (2005) Orthologs, paralogs and evolutionary genomics. Annu Rev Genet 39:309–338

Kramer NI, Busser FJM, Oosterwijk MTT, Schirmer K, Escher BI, Hermens JLM (2010) Development of a partition-controlled dosing system for cell assays. Chem Res Toxicol 23:1806–1814

Kriventseva EV, Tegenfeldt F, Petty TJ, Waterhouse RM, Simão FA, Pozdnyakov IA, Ioannidis P, Zdobnov EM (2015) OrthoDB v8: update of the hierarchical catalog of orthologs and the underlying free software. Nucleic Acids Res 43(Database issue):D250–D256. doi:10.1093/nar/gku1220

LaLone CA, Villeneuve CL, Burgoon LD, Russom CL, Helgen HW, Berninger JP, Tietge JE, Severson MN, Cavallin JE, Ankley GT (2013) Molecular target sequence similarity as a basis for species extrapolation to assess the ecological risk of chemicals with known modes of action. Aquat Toxicol 144–145:141–154

Lampert W (2011) In: Kinne O (ed) Daphnia: development of a model organism in ecology and evolution, Excellence in ecology series. International Ecology Institute, Luhe

Larras F, Keck F, Montuelle B, Rimet F, Bouchez A (2014) Linking diatom sensitivity to herbicides to phylogeny: a step forward for biomonitoring? Environ Sci Technol 48(3):1921–1930

Leung MC, Williams PL, Benedetto A, Au C, Helmcke KJ, Aschner M, Meyer JN (2008) *Caenorhabditis elegans*: an emerging model in biomedical and environmental toxicology. Toxicol Sci 106(1):5–28

Lu X, Lu H, Zhao L, Yang Y, Lu Z (2010) Genome-wide pathways analysis of nickel ion-induced differential genes expression in fibroblasts. Biomaterials 31(8):1965–1973

Lu Y, Park Y, Gao X, Zhang X, Yoo J, Pang Y, Jiang H, Zhu KY (2012) Cholinergic and non-cholinergic functions of two acetylcholinesterase genes revealed by gene-silencing in Tribolium castaneum. Sci Rep 2(288):1–7

Martin BT, Jager T, Nisbit RM, Preuss TG, Hammers-Wirtz M, Grimm V (2013) Extrapolating ecotoxicological effects from individuals to populations: a generic approach based on dynamic energy budget theory and individual-based modeling. Ecotoxicology 22(3):574–583

Mohamed B, Hajer A, Susanna S, Caterina O, Flavio M, Hamadi B, Aldo V (2014) Transcriptomic responses to heat stress and nickel in the mussel *Mytilus galloprovincialis*. Aquat Toxicol 148:104–112

Moreira EG, Yu X, Robinson JF, Griffith W, Hong SW, Beyer RP, Bammler TK, Faustman EM (2010) Toxicogenomic profiling in maternal and fetal rodent brains following gestational exposure to chlorpyrifos. Toxicol Appl Pharmacol 245(3):310–325

Nakanishi T, Kato Y, Matsuura T, Watanabe H (2014) CRISPR/Cas-mediated targeted mutagenesis in *Daphnia magna*. PLoS One 9(5):e98363

Nookaew I, Papini M, Pornputtpong N, Scalcinati G, Fagerberg L, Uhlen M, Nielsen J (2012) A comprehensive comparison of RNA-Seq-based transcriptome analysis from reads to differential gene expression and cross-comparison with microarrays: a case study in Saccharomyces cerevisiae. Nucleic Acids Res 2012:1–14

Oetken M, Bachmann J, Schulte-Oehlmann U, Oehlmann J (2004) Evidence for endocrine disruption in invertebrates. Int Rev Cytol 236:1–44

Patwardhan V, Ghaskadbi S (2013) Invertebrate alternatives for toxicity testing: hydra stakes its claim. Proceedings of animal alternatives in teaching, toxicity testing, and medicine, Bhubaneswar, 2012: 69–76

Paul RJ, Colmorgen M, Pirow R, Chen YH, Tsai MC (1998) Systemic and metabolic responses in *Daphnia magna* to anoxia. Comp Biochem Physiol Part A 120(3):519–530

Perkins EJ, Antczak P, Burgoon L, Falciani F, Garcia-Reyero N, Gutsell S, Hodges G, Kienzler A, Knapen D, McBride M, Willett C (2015) Adverse outcome pathways for regulatory applications: examination of four case studies with different degrees of completeness and scientific confidence. Toxicol Sci 148(1):14–25

Quinn DM (1987) Acetylcholinesterase: enzyme structure, reaction dynamics, and virtual transition states. Chem Rev 87(5):955–979

Rajini PS, Melstrom P, Williams PL (2008) A comparative study on the relationship between various toxicological endpoints in *Caenorhabditis elegans* exposed to organophsophorus insecticides. J Toxicol Environ Health A 71(15):1043–1050

Raymond-Delpech V, Matsuda K, Sattelle BM, Rauh JJ, Sattelle DB (2005) Ion channels: molecular targets of neuroactive insecticides. Invertebr Neurosci 5(3–4):119–133
REACH (2006). Regulation (EC) No 1907/2006 of The European Parliament and of The Council In Registration, Evaluation, Authorisation and Restriction of Chemicals (REACH), establishing a European Chemicals Agency, amending Directive 1999/45/EC and repealing Council Regulation (EEC) No 793/93 and Commission Regulation (EC) No 1488/94 as well as Council Directive 76/769/EEC and Commission Directives 91/155/EEC, 93/67/EEC, 93/105/EC and 2000/21/EC. The European Parliament and The Council of The European Union
Reddy MS, Jayaprada P, Rao KVR (1990) Impact of methyl parathion and malathion on cholinergic and non-cholinergic enzyme systems of penaeid prawn, Metpenaeus monoceros. Biochem Int 22:769–780
Reddy NC, Rao JV (2008) Biological response of earthworm, *Eisenia foetida* (Savigny) to an organophosphorus pesticide, profenofos. Ecotoxicol Environ Saf 71(2):574–582
Reddy TBK, Thomas A, Stamatis D, Bertsch J, Isbandi M, Jansson J, Mallajosyula J, Pagani I, Lobos E, Kyrpides N (2014) The genomes OnLine database (GOLD) v.5: a metadata management system based on a four level (meta)genome project classification. Nucl Acids Res. doi:10.1093/nar/gku950
Rodriguez-Serrano M, Romero-Puertas MC, Pazmino DM, Testillano PS, Risueno MC, del Rio LA, Sandalio LM (2009) Cellular response of pea plants to cadmium toxicity: cross talk between reactive oxygen species, nitric oxide, and calcium. Plant Physiol 150(1):229–243
Russom CL, Bradbury SP, Broderius SJ, Hammermeister DE, Drummond RA (1997) Predicting modes of toxic action from chemical structure: acute toxicity in the fathead minnow (*Pimephales promelas*). Environ Toxicol Chem 16(5):948–967
Russom CL, Lalone CA, Villeneuve DL, Ankley GT (2014) Development of an adverse outcome pathway for acetylcholinesterase inhibition leading to acute mortality. Environ Toxicol Chem 33(10):2157–2169
Santiago-Vázquez LZ, Ranzer LK, Kerr RG (2006) Comparison of two total RNA extraction protocols using the marine gorgonian coral *Pseudopterogorgia elisabethae* and its symbiont *Symbiodinium* sp. Electron J Biotechnol 9(5):598–603
Schreiber R, Gündel U, Franz S, Küster A, Rechenberg B, Altenburger R (2011) Using the fish plasma model for comparative hazard identification for pharmaceuticals in the environment by extrapolation from human therapeutic data. Regul Toxicol Pharmacol 61(3):261–275
Seyfert AL, Cristescu ME, Frisse L, Schaack S, Thomas WK, Lynch M (2008) The rate and spectrum of microsatellite mutation in *Caenorhabditis elegans* and *Daphnia pulex*. Genetics 178(4):2113–2121
Shaw JR, Pfrender ME, Eads BD, Klaper R, Callaghan A, Sibly RM, Colson I, Jansen B, Gilbert D, Colbourne JK (2008) Daphnia as an emerging model for toxicological genomics. In: Hogstrand C, Kille P (eds) Comparative Toxicogenomics, vol 2. Elsevier Science, Amsterdam, pp 165–219
Smith TF, Waterman MS (1981) Identification of common molecular subsequences. J Mol Biol 147:195–197
Smyth GK (2004) Linear models and empirical bayes methods for assessing differential expression in microarray experiments. Stat Appl Genet Mol Biol 3(1):3
Spade DJ, Griffitt RJ, Liu L, Brown-Peterson NJ, Kroll KJ, Feswick A, Glazer RA, Barber DS, Denslow ND (2010) Queen conch (*Strombus gigas*) testis regresses during the reproductive season at nearshore sites in the Florida keys. PLoS One 5(9):e12737
Stevanik DJ, Wolenski FS, Friedman LE, Gilmore TD, Finnerty JR (2013) Isolation of DNA, RNA and protein from the starlet sea anemone *Nematostella vectensis*. Nat Protoc 8(5):892–899
Stilwell GE, Saraswati S, Littleton JT, Chouinard SW (2006) Development of a *Drosophila* seizure model for in vivo high-throughput drug screening. Eur J Neurosci 24(8):2211–2222
Subramanian A, Tamayo P, Mootha VK, Mukherjee EBL, Gillette MA, Paulovic A, Pomeroy SL, Golub TR, Lander ES, Mesirov JP (2005) Gene set enrichment analysis: a knowledge-based approach for interpreting genome-wide expression profiles. Proc Natl Acad Sci 102(43):15545–15550
Svendsen C, Spurgeon DJ, Hankard PK, Weeks JM (2004) A review of lysosomal membrane stability measured by neutral red retention: is it a workable earthworm biomarker? Ecotoxicol Environ Saf 57(1):20–29

Swain S, Wren JF, Stürzenbaum SR, Kille P, Morgan AJ, Jager T, Jonker MJ, Hankard PK, Svendsen C, Owen J, Hedley BA, Blaxter M, Spurgeon DJ (2010) Linking toxicant physiological mode of action with induced gene expression changes in *Caenorhabditis elegans*. BMC Syst Biol 4:32

Tagu D, Colbourne JK, Nègre N (2014) Genomic data integration for ecological and evolutionary traits in non-model organisms. BMC Genomics 15:490

Taylor EM, Lehmann AR (1998) Conservation of eukaryotic DNA repair mechanisms. Int J Radiat Biol 74(3):277–286

Taylor NS, Weber RJM, White TA, Viant MR (2010) Discriminating between different acute chemical toxicities via changes in the daphnid metabolome. Toxicol Sci 118(1):307–317

The *C. elegans* Sequencing Consortium (1998) Genome sequence of the Nematode *C. elegans*: a platform for investigating biology. Science 282(5396): 2012–2018

Thomas P, Dawick J, Lampi M, Lemaire P, Presow S, van Egmond R, Arnot JA, Mackay D, Mayer P, Burgos MG (2015) Application of the activity framework for assessing aquatic ecotoxicology data for organic chemicals. Environ Sci Technol 49(20):12289–12296

Tilton FA, Tilton SC, Bammler TK, Beyer RP, Stapleton PL, Scholz NL, Gallagher EP (2011) Transcriptional impact of organophosphate and metal mixtures on olfaction: copper dominates the chlorpyrifos-induced response in adult zebrafish. Aquat Toxicol 102(3–4):205–215

Toyota K, Miyakawa H, Yamaguchi K, Shigenobu S, Ogino Y, Tatarazako N, Miyagawa S, Iguchi T (2015) NMDA receptor activation upstream of methyl farnesoate signaling for short day-induced male offspring production in the water flea, *Daphnia pulex*. BMC Genomics 16(1):186

Tushers VG, Tibshirani R, Chu G (2001) Significance analysis of microarrays applied to the ionizing radiation response. Proc Natl Acad Sci 98(9):5116–5121

USEPA (2015) ToxCast & Tox21 summary files from invitrodb_v2. Retrieved from http://www2.Epa.Gov/chemical-research/toxicity-forecaster-toxcasttm-data on October 28, 2015. Data released. October 2015

Van Wezel AP, de Vries DA, Sijm D, Opperhuizen A (1996) Use of the lethal body burden in the evaluation of mixture toxicity. Ecotoxicol Environ Saf 35(3):236–241

Vanderbrouck T, Soetaert A, van der Ven K, Blust R, de Coen W (2009) Nickel and binary metal mixture responses in *Daphnia magna*: molecular fingerprints and (sub)organismal effects. Aquat Toxicol 92(1):18–29

Verhaar HJM, Van Leeuwen CJ, Hermens JLM (1992) Classifying environmental pollutants 1: structure-activity relationships for prediction of aquatic toxicity. Chemosphere 25(4):471–491

Verslycke T, Ghekiere A, Raimondo S, Janssen C (2007) Mysid crustaceans as standard models for the screening and testing of endocrine-disrupting chemicals. Ecotoxicology 16(1):205–219

Viñuela A, Snoek LB, Riksen JAG, Kammenga JE (2010) Genome-wide gene expression analysis in response to organophosphorus pesticide chlorpyrifos and diazinon in *C. elegans*. PLoS One 5(8):e12145

Wang Z, Gerstein M, Snyder M (2009) RNA-Seq: a revolutionary tool for transcriptomics. Nat Rev 10(1):57–63

Williams EG, Auwerx J (2015) The convergence of systems and reductionist approaches in complex trait analysis. Cell 162(1):23–32

Wren JF, Kille P, Spurgeon DJ, Swain S, Sturzenbaum SR, Jager T (2011) Application of physiologically based modelling and trascriptomics to probe the systems toxicology of aldicarb for *Caenorhabditis elegans* (Maupas 1900). Ecotoxicology 20(2):397–408

Chapter 6
Non-model Species in Ecological Risk Assessment

Markus Hecker

Abstract Ecological risk assessors are increasingly recognizing the need for objectively characterizing the sensitivity of specific ecological receptors of interest to environmental contaminants. Current testing strategies in support of ecological risk assessments primarily rely on extensive animal testing, and on extrapolation from standard laboratory model species to native species of relevance in local ecosystems. In addition to the huge costs and large numbers of animals needed, it has been shown that these approaches are often not adequately predictive, and thus, protective of organisms of interest. This chapter reviews the current challenges and developments in ecological and chemical risk assessment of non-model ecological species with specific reference to the current paradigm shift in toxicity testing from classic empirical live animal testing approaches to alternative concepts. The status and applicability of (high-throughput) in vitro systems, predictive toxicity-pathway models such as adverse outcome pathways (AOPs), quantitative structure-activity relationship (QSARs), and computational approaches are discussed in context with their potential to address current uncertainties in cross-species extrapolation of chemical hazards and associated regulatory needs. Specifically, comparative 'omics and systems biology approaches are increasingly seen as powerful tools for cross-species extrapolation based on the assumption that structural and functional similarities or differences of specific molecular targets or pathways are likely to be one of the main drivers of the intrinsic sensitivity of organisms to contaminants. However, there are a number of uncertainties that remain to be addressed before these approaches and associated computational tools such as USEPA's SeqAPASS tool become a viable option in non-model species risk assessment. Main concerns include the limited number of mature toxicity pathways currently available, their limited taxonomic application and their mostly qualitative nature. Furthermore, large data gaps exist with regard to toxicodynamic and toxicokinetic properties of chemicals in ecological species that determine target site concentrations, and that are critical factors influencing intrinsic sensitivity. The chapter concludes by

M. Hecker (✉)
School of the Environment & Sustainability and Toxicology Centre,
University of Saskatchewan, 44 Campus Drive, Saskatoon, SK S7N 5B3, Canada
e-mail: markus.hecker@usask.ca

N. Garcia-Reyero, C.A. Murphy (eds.), *A Systems Biology Approach to Advancing Adverse Outcome Pathways for Risk Assessment*,
DOI 10.1007/978-3-319-66084-4_6

providing a theoretical road map for future research building on the current promising advances in the field of ecotoxicogenomics and computational biology combined with alternative testing approaches using in vitro systems and early life stage animal tests to anchor pathways to species-specific biological outcomes.

6.1 Introduction

Legislation in North America, Europe and other parts of the world, such as the Canadian Environmental Protection Act (CEPA), the European Union's Registration, Evaluation and Authorization of Chemicals (REACH), the U.S. Safe Drinking Water Act (SDWA), and the U.S. Toxic Substances Control Act (TSCA), mandates the assessment of risks of chemicals to wildlife and human health. Under these programs, regulators and industry are faced with the challenge to assess the toxicological risks associated with an ever-increasing number of chemicals used by society, and that are ultimately released, either intentionally or unintentionally, into the environment. Assessment of the ecological risks of exposure to a certain pollutant or a complex mixture of contaminants follows a standardized approach (e.g. Suter et al. 2000) that aims to assess the probability of an adverse ecological outcome such as a fish kill or the impact on the fitness of a population or community of interest.

One of the key steps in ecological risk assessments of chemicals is the characterization of the hazard of the chemical or mixtures of concern to biological receptors of interest. This is based on establishing so called toxicity reference values (TRVs) that are derived by identifying threshold or effective concentrations such as No Observable Adverse Effect Levels (NOAELs)/Lowest Observable Adverse Effect Levels (LOAECs) or concentrations at which a certain effect level occurs (e.g. LC_x or EC_x), respectively. To be of regulatory relevance, these TRVs are based on measurements of apical outcomes including survival, reproduction, growth and development (Calow et al. 1997; Suter 2004). Current testing strategies for the determination of TRVs rely on extensive animal testing using selected model species that are easy to culture in the laboratory and for which standardized testing protocols exist. Data derived from these standard laboratory model species is then extrapolated to predict the potential risks to native receptors of interest in local ecosystems or to humans. These strategies pose a daunting challenge to regulators and industry, as the number of chemical substances for which toxicity data are required under the above legislations is tremendous (e.g., 23,000 under Canada's Chemical Management Plan (CMP), 84,000 under the US Toxic Substances Control Act (TSCA), and 107,000 under the European Union's Registration, Evaluation, Authorization of Chemicals (REACH)). In addition to the ethical concerns surrounding the use of large numbers of live animals needed to fulfill current testing requirements, the associated time and monetary costs are prohibitive. For example, the U.S. EPA estimates that it would take 3–4 years and one to 20 million US dollars to test a single chemical under their current testing mandates, and it was estimated that testing costs resulting from the European Union REACH program would amount to approximately 9.5 billion Euro and require 54 million animals (Rovida and Hartung 2009).

In addition to the above-mentioned economic and ethical concerns with respect to current hazard assessment approaches, there is a great amount of uncertainty associated with their application in ecological risk assessment. The assessment of the hazard of a contaminant to an ecological receptor of interest predominantly relies on data generated with acute (short-term) in vivo toxicity studies with a few selected laboratory model species and, to a much lesser extent, on sub-chronic or chronic experiments trying to characterize the sublethal effects or the mode of action of a chemical. These data are then extrapolated to native species of relevance or interest in local ecosystems (further termed as "non-model species"). There is great uncertainty associated with extrapolation from acute studies that mostly deal with effects on survival to field scenarios representing long-term exposures at low concentrations. Uncertainty also exists with regard to the extrapolation from model laboratory species to non-model species and requires the application of safety factors, which sometimes can be as great as several orders of magnitude, to ensure risk assessments are sufficiently protective (CEC 1996; Forbes and Calow 2002). While such safety factors typically tend to provide a relatively "safe" answer, in many situations they are likely to significantly overestimate the true risk that a chemical poses to a receptor species. This can result in unrealistic hazard assessments (Chapman et al. 1998) that could trigger unnecessary remediation measures, which themselves can have significant impacts on the environment and are very costly. In other cases, extrapolation from acute to chronic data or among species or taxonomic classes is not acceptable (Allard et al. 2010). Given that data regarding acute toxicity is based solely on mortality, it ignores the intricacies of toxic responses associated with equally important apical responses such as reproduction, development and growth, and which may be much more sensitive endpoints. Similarly, it was shown that with increasing taxonomic distance the application of species extrapolation tools such as interspecies correlation models become increasingly uncertain, resulting in unacceptable error rates (Rainmondo et al. 2008).

In summary, knowledge about the hazards that environmental chemicals pose to the diversity of non-model species is critical to enable objective ecological risk assessments. However, traditional species extrapolation approaches that primarily rely on live animal testing and extrapolation from laboratory model species are insufficient for reliably predicting the sensitivity of the vast diversity of ecological species to the large number of chemicals that need to be tested under current legislations. This book chapter reviews the current challenges associated with ecological risk assessment of non-model species and summarizes recent developments and approaches to address these challenges. Particularly, this paper examines recent strategies such as functional omics, the adverse outcome pathway (AOP) concept, computational methods, and in vitro testing that are proposed as the path forward to address current limitations of both chemical and ecological risk assessment practices, and specifically, how they may help overcome the barriers associated with species extrapolation.

6.2 Factors Contributing to Uncertainty in Non-model Species Risk Assessment

6.2.1 Extrapolation from Standard Laboratory Models to Native Species

Considering the multiplicity of organisms inhabiting earth's ecosystems, the wide range of their susceptibilities to environmental pollutants, and the limited amount of species-specific toxicological information available, accurate prediction of risks of chemical contaminants to ecosystems represents a huge challenge to risk assessors. For example, assessment of contaminant risks to cold freshwater systems is routinely based on selected biotests with one or two species of algae, a few species of crustaceans, midge larvae and certain model species of fishes, assuming that they are reasonable representatives of aquatic communities. However, they often ignore other key elements of these ecosystems such as microbial communities, other species of insect larvae, worms, snails, amphibians, native fishes, etc. Similar limitations are associated with terrestrial risk assessment strategies/guidelines (Fernández et al. 2006).

Traditionally, it has been assumed that organisms from the same class or family would have comparative sensitivity to environmental pollutants, and thus, the use of standard laboratory organisms would be sufficiently protective of wildlife species (e.g. a rainbow trout would allow to predict sensitivity of a sturgeon or lake trout, or a chicken would be predictive of an eagle). However, within the past decade it has become apparent that these assumptions often are not true, and that such extrapolations are either not sufficiently protective or vastly overestimate the sensitivity of phylogenetically related organisms to chemical toxicants (Allard et al. 2010; Sanderson and Solomon 2009; Vardy et al. 2013). A second uncertainty stems from the large diversity of species across ecosystems. Site-specific risk assessments are based on the selection of receptors of concern from a list of species that are likely to be exposed at the site of concern. The receptor(s) of concern should represent one (or a few) species that are considered sensitive to the stressor of interest, potentially threatened or endangered, or ecologically significant, etc. However, selection of the appropriate receptor is often hampered by a lack of information that is available for the majority of species in an ecosystem of interest, including their specific physiological traits that would either render them vulnerable or tolerant to certain contaminants (see Sect. 6.3).

To fill some of the existing data gaps for non-standard test species (i.e. native species of concern) the risk assessment community often relies on short-term (mostly acute) toxicity tests that typically assess the effects of high concentrations of a chemical on survival of the test species. These tests are quickly completed (typically within 48 to 96 h), and generate comparable data based on standardized test protocols. The most common use of these data is for the construction of species sensitivity distributions (SSDs) that aim to derive hazard concentrations protective of the majority species in an ecosystem of concern (Rainmondo et al. 2008). However, there are multiple drawbacks with this approach that render it of limited usefulness for risk assessments of non-model species (Allard et al. 2010; Baird and Van den Brink 2007):

1. Acute studies typically use mortality as the sole endpoint, which, depending on the contaminant tested, may result in an unacceptable underestimation of risk (even after application of large uncertainty factors) for certain chemical groups with very specific modes of action (e.g. some endocrine disruptors, immunotoxicants, etc.);
2. They mostly ignore toxicokinetic and toxicodynamic (accumulation, distribution, metabolization and elimination [ADME]) properties of a contaminant in a given test species, and which are critical especially for bioaccumulative or biologically (metabolically) active compounds;
3. They mostly focus on one life-stage (typically early life stages) that is considered most sensitive to the exposure with contaminants. However, data obtained with an early life-stage test does not allow for conclusions to be made with respect to critical biological functions unique to adulthood such as reproduction;
4. They do not consider life-traits/-history of the target species, and which determine when, for how long and through what route organisms may be exposed; and
5. They are limited to species that can be maintained for certain periods or cultured in the laboratory. As a consequence, special interest species such as some endangered species that are difficult to obtain or to maintain under laboratory conditions can often not be assessed.

Increasingly, attempts have been made to develop and use specialized assays for some non-common test organisms (e.g. Doering et al. 2014; Dwyer et al. 2005; Fairchild et al. 2005; Vardy et al. 2011, 2013; Wang et al. 2007), as well as to investigate the sub-chronic or chronic effects of contaminants on organisms of interest with the aim to generate information on more subtle effects (e.g. reproduction, development, behavior). In doing so, it has become apparent that complete representation of the diversity of organisms in hazard assessments using standard ecotoxicological approaches will not be feasible using current standard testing approaches. Even if increased resources were to be made available for the testing of additional target receptors of interest, there are a number of ethical and logistical hurdles that will be difficult to overcome. This is particularly true for species of special interest such as endangered species and long-lived species.

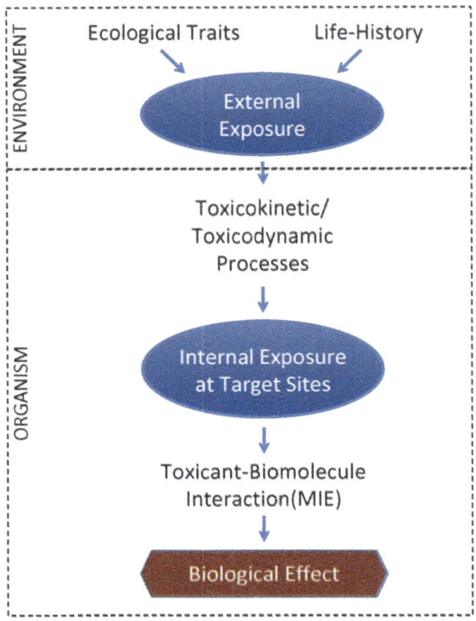

Fig. 6.1 Factors affecting the susceptibility of a species to environmental contaminants

6.2.2 Causes for Differences in Species Sensitivity to Contaminants

As discussed in the previous sections, there are a number of uncertainties with strategies used to extrapolate from standard laboratory model species to native species of relevance to ecological risk assessments. These uncertainties are a function of the great differences in the inherent sensitivity of each species to contaminants, which is rooted in their unique physiology, life-history, ecological niche, evolutionary traits, adaptation and differences in ADME of contaminants (Fig. 6.1). The relevance of biological or life history traits and their role in predicting species sensitivity in context with ecological risk assessments have already been thoroughly reviewed elsewhere (e.g. Baird and Van den Brink 2007; Calow et al. 1997), and will not be discussed in great detail here. It is acknowledged that the understanding of the life history and ecological traits of a species such as generation time, time to maturity, ecological niche (e.g. benthic vs pelagic), etc., are important when conducting ecological risk assessments as it will determine the likelihood of external exposure during critical life stages or periods, as well as the route and duration of exposure. However, for any exposure to result in an adverse effect, the chemical or mixture of chemicals of concern has to interfere with the function or structure of certain bio-molecules (also termed the molecular initiating event [MIE]) that lead to the alteration of normal physiological functioning of an organism, and which ultimately manifests as reduced fitness. Therefore, the key to understanding the sensitivity of a species to a contaminant or a chemical mixture is knowledge about its specific mechanism of action. A second critical factor that determines the sensitivity of a species is the ADME properties of a compound that determine internal exposure at the molecular target sites of a toxicant (Jager et al. 2011; Nichols et al. 1990).

6.2.3 Chemical Modes of Action and Their Role in Determining Species-Specific Sensitivity

It has been recognized that the sensitivity of an organism to a toxicant is often driven by the MIE that then initiates a cascade of downstream events ultimately causing an adverse biological effect. Inter-species differences in sensitivity to chemical exposure may then arise because of differences in molecular targets, which have undergone changes throughout evolution (ECETOC 2007; Gunnarsson et al. 2008; Celander et al. 2010). In cases where biological properties such as basic cellular structures and functions of cells are highly conserved throughout evolution, chemicals that interact with such processes are assumed to be comparable in their toxicity among diverse species (Ashauer and Escher 2010; Vaal et al. 1997). Examples include persistent hydrophobic chemicals that have the tendency to accumulate in cell membranes, leading to alterations of the structure and functioning of these membranes (i.e. non-polar narcosis; Rand 1995). In contrast, a large number of chemicals elicit toxic mechanisms that are highly specific to certain physiological traits, and thus, to certain groups of organisms to which these traits are unique or where they play important biological roles. Examples include disruption of functions such as photosynthesis in plants and algae or the production of egg yolk proteins in oviparous animals. Some xenobiotics only interact with very specific molecules such as receptors or enzymes while others are rather unspecific and affect multiple processes simultaneously or are changed in their toxicological properties through metabolic processes. While chemicals that interact only with specific biomolecules such as receptor proteins allow for the relatively simple categorization of chemicals through approaches like quantitative structure-activity relationships (QSAR) as well as the development of risk assessment tools such as relative potency factors (RePs) as in the case of dioxin-like chemicals (DLCs; e.g. Van den Berg et al. 2006), reliable effect assessment of chemicals that act via non-specific or multiple mechanisms is a greater challenge (Escher and Hermens 2002).

The best-characterized examples of the specific interactions of a substance with a certain target molecule is demonstrated by chemicals that are ligands of nuclear receptors. For example, studies have shown that the MIE that is likely to determine in vivo sensitivity of certain vertebrate groups to DLCs is sensitivity of activation of the arylhydrocarbon receptor (AhR). It was demonstrated that key amino acids in the ligand binding domain of the AhR determine affinity of binding and are the molecular basis for differences in sensitivity to DLCs among strains of mice (*Mus musculus*) (Pandini et al. 2007) and among species of birds (Farmahin et al. 2013; Head et al. 2008; Karchner et al. 2006). Similarly, key amino acids in the ligand-binding domain of the AhR drive differences in sensitivity to activation between

AhR1α and AhR1β of the African clawed frog (*Xenopus laevis*) (Odio et al. 2013) and between AhR1a and AhR2 of zebrafish (*Danio rerio*) (Fraccalvieri et al. 2013). In addition to studies of the AhR, some studies have attempted to link differences in primary structure of the estrogen receptor (ER) with binding affinity of xenoestrogens among selected mammals, birds and fish (e.g. Matthews et al. 2000; Toyahama et al. 2015). Interestingly, a thorough review of the pertinent ecotoxicological literature revealed that with a few exceptions, including those discussed above, there is almost a complete lack of published research that investigated the role of the structure of molecular targets and how it may drive species-specific sensitivity to xenobiotics. This is surprising given the importance of the MIE and associated molecule-toxicant interactions, especially when considering that identification of specific molecular targets is common practice in the fields of drug discovery and development of pesticides. One of the main reasons for this may be rooted in our lack of understanding of the physiology of many of the species native to ecosystems of concern. While there is a thorough coverage of the physiology of standard laboratory vertebrate species such as mice, chicken, quail, zebrafish, rainbow trout, and many insect species because of their role in pharmacological and agricultural research, very little is known about the myriad of species inhabiting the diverse ecosystems of our planet (e.g. LaFont 2000). However, the current advances in "big data" 'omics and systems biology are likely to greatly stimulate research into the basic biology of non-model species in ecotoxicology.

6.2.4 Toxicokinetic and Toxicodynamic Considerations in Predicting Species Sensitivity

One of the main limitations of the mechanism-based characterization of species sensitivities is the lack of the ability to integrate reliable metrics of the internal exposure concentrations of the chemical of concern. Toxicologists have long recognized that the ability of an organism to cope with a toxic insult is a function of its physiological "machinery" that deals with the ADME of a chemical. For example, in a review by Wang and Rainbow (2008) it was shown that metal bioaccumulation was a function of different ion regulation (e.g. ion-channels or – pumps) and detoxification (e.g. induction of metallothionein and heat-shock proteins) strategies among different fresh- and saltwater species of fishes. Another example is the difference in inducibility of P450 phase I metabolic enzymes among different species of mammals, and which significantly affects the genotoxic/mutagenic potency as well as elimination of certain drugs (e.g. tamoxifen) and PAHs (e.g. benzo(a)pyrene) (Lewis et al. 1998). For example, organisms with a slower metabolism may be at lesser risk from exposure to compounds that need to be metabolically activated to elicit toxicity (e.g. certain PAHs require metabolic activation to become genotoxic). However, these organisms may be at greater risk from exposure to chemicals that are metabolically detoxified because of their slower metabolic activity. Other examples of factors that significantly influence the internal exposure of organisms are surface to volume ratio, lipid content, and life span, with the latter two being of great relevance with regard to bioaccumulative contaminants (reviewed in Escher and Hermens 2002).

Increasingly sophisticated models are being developed to predict the exposure concentration at the biological target site of an organism. The simplest practitioner models include toxicokinetic (TK) and toxicodynamic (TD) models. TK models can predict the time course of a toxicant's concentration at the site of toxic action, including ADME processes (i.e. what happens with the chemical in an organism). TD models can predict the dynamics of a toxicant's interactions with a biological target site and the resulting effects. TD models can be simple mathematical descriptions of the kinetics of induction of toxicological effects, e.g. the study of carcinogenesis, but ideally are based on a quantitative description of the underlying mechanisms of toxicity (Jager et al. 2011). Traditional TK modeling approaches are more pragmatic and assume organisms to consist of a single homogeneously mixed compartment that accumulates and eliminates chemicals at specific rates (Barron 1990). By far, the most comprehensive and sophisticated state-of-the-art models in this field are physiologically based TK (PBTK) models. In this type of model, organs and tissue groups (e.g. liver and kidney) are explicitly represented by their weight, their lipid and water content and the rate at which they are perfused by the circulating blood (Nichols et al. 1990). Thus, PBTK models are capable of more precisely predicting the internal concentration of chemicals in an animal's body and in specific organs at any time post exposure.

As discussed above, significant progress has been made in context with developing models that can predict the ADME properties of chemicals in organisms. One of the main limitations of the routine application of TK, TD or PBTK models in ecological risk assessment is the lack of information on the physiological parameters for species other than the current standard laboratory models. To be useful, these models will have to be calibrated for each species of interest based on its specific physiological properties such as lipid and water content of target tissues, metabolic activity, cardiac output and distribution of blood flow among tissues (Ashauer et al. 2011). Interestingly, a number of recent studies found that ADME properties developed for humans were reasonably predictive of pharmacological responses in fish (Perkins et al. 2013). For example, the ratio of the acutely toxic and therapeutic drug doses derived from pharmacological studies with mammals was shown to be predictive of chronic toxicity of the same pharmaceuticals in fish (Berninger and Brooks 2010). Also, the use of species-specific in vitro (liver) models showed good promise with regard to deriving parameters such as hepatic clearance rates that can be used to parameterize species-specific PBTK models (Brinkmann et al. 2014; Han et al. 2007). In this context, recent developments in 3D tissues cultures and "mini-organs" are anticipated to advance the way by which these models can be parametrized and calibrated and will likely increase their accuracy.

6.3 Novel Developments and Tools

The limitations of traditional ecological risk assessment approaches to address the hazard assessment needs for the tens of thousands of chemicals mandated by current legislations in Europe, North America, and some Asian countries have led to a paradigm shift in toxicity testing (Villeneuve et al. 2014a, b). As a consequence, toxicity testing in the twenty-first century is moving from classic approaches that use empirical live animal testing to alternative concepts including (high-throughput) in vitro systems, predictive toxicity-pathway models such as adverse outcome pathways (AOPs), quantitative structure-activity relationship (QSARs), and computational approaches (ECHA 2015; Kavlock et al. 2012; NCR 2007; Villeneuve and Garcia-Reyero 2011). This section explores the utility of these novel concepts and approaches to advance our ability to reliably predict sensitivity of non-model species.

6.3.1 Toxicity Pathway-Based Approaches in Cross-Species Extrapolation

Considering differences in the conservation and evolution of molecular targets of chemicals, their biological context/applicability, and the potential that inter-species differences in susceptibility to environmental pollutants may arise at multiple levels of organization, it becomes apparent that there is unlikely to be one approach that can be used to characterize and predict species sensitivity across all phyla and taxa. However, as recently reviewed by Perkins et al. (2013), there is evidence that understanding the specific molecular perturbations caused by a chemical that can be linked to an adverse outcome, and the conservation of these perturbations across certain species or organism groups, represents a promising starting point for cross-species extrapolation. Within animal groups, most fundamental pathways such as development, reproduction, stress response, etc. tend to be highly conserved (Adamska et al. 2007; Ankley and Johnson 2004; Rand-Weaver et al. 2013; Simmons et al. 2009; Vallee et al. 2008). This is also likely to be true for the majority of the macromolecules that regulate these pathways, such as receptors, enzymes and other functional proteins that share common ancestral genes. In fact, decades of pharmacological research demonstrated that non-mammalian model species such as the zebrafish and even invertebrates such as *Drosophila melanogaster* express highly conserved molecular pathways directly applicable to humans and other mammals (reviewed in Perkins et al. 2013), clearly showing the applicability of highly conserved pathways to extrapolate among even very distantly related species (Garcia-Reyero et al. 2011). It needs to be noted, however, that gene function and conservation is less and less conserved with increasing evolutionary distance, especially for processes that do not have a shared evolutionary history. Examples include pathways involved with development of skeletal structures or reproductive processes in vertebrates vs invertebrates, or the restriction of certain metabolic pathways, such as the glycan pathway, to metazoans due to its specific role in processes associated with multicellularity (Peregrin-Alvarez et al. 2009).

The recent advent of advanced, high-content and –throughput 'omics and bioinformatics has made it possible to efficiently and reliably probe entire organismal systems and to describe the complete molecular machinery driving these systems within relatively short timeframes. In context with ecological risk assessment of chemicals, 'omics are increasingly seen as a powerful tool to characterize toxicity pathways among species, and to identify conserved pathways or molecular targets affected by contaminants (Perkins et al. 2013). One framework that has gained significant attention with regard to characterizing toxicity pathways of chemicals is that of the adverse outcome pathway (AOP). AOPs organize and evaluate biologically plausible and empirically-supported links among different levels of biological organization (Ankley et al. 2010; Villeneuve et al. 2014a). They systematically link molecular-level perturbations (e.g. the MIE) to an adverse outcome of regulatory relevance through the characterisation of a series of Key Events (KEs). One of the most common starting points within an AOP to investigate the intrinsic sensitivity of an organism to a chemical is to compare the molecular structures (e.g. amino-acid sequences) at the MIE (see Sect. 6.2.1; LaLone et al. 2013a). For example, it has been demonstrated within vertebrate classes such as birds, mammals and fish that the large differences in the sensitivity to DLCs are likely to be driven by small differences in critical amino acid residues in the ligand-binding pocket of the AhR (Karchner et al. 2006; Doering et al. 2014). Comparable relationships are hypothesized to also occur for other molecular targets, particularly receptors such as the ER (Matthews et al. 2000; Toyahama et al. 2015). However, although current computational cross-species extrapolation approaches such as the SeqAPASS model (Sect. 6.2.1) assume a direct link between structure of target molecules and susceptibility to certain contaminants, there is little direct evidence for similar relationships for molecules other than the above discussed receptors, highlighting the need for additional research in this field.

While information on the sequence and functional homology of a molecular target provides valuable insights for explaining the differential susceptibility of organisms to contaminants, differences in the role of the MIE and/or KEs in downstream biological functions can provide additional information. For example, while the MIE of binding of agonists to the estrogen receptor may be highly conserved between oviparous and viviparous animals, the role of KEs such as production of vitellogenin represents a critical outcome in one class of organisms (e.g. fishes) but not in another (e.g. mammals) resulting in very different biological outcomes (Fig. 6.2). In other cases, interspecies differences in susceptibility to certain contaminants may arise through changes in less defined structural features such as cellular membranes or even physiological processes that are only indirectly related to the interaction with the target site, and for which structural sequence information may not provide plausible explanations to the differences observed. This applies in particular to species differences in catabolism or metabolism of endogenous and exogenous ligands, where differences in ADME properties influence the

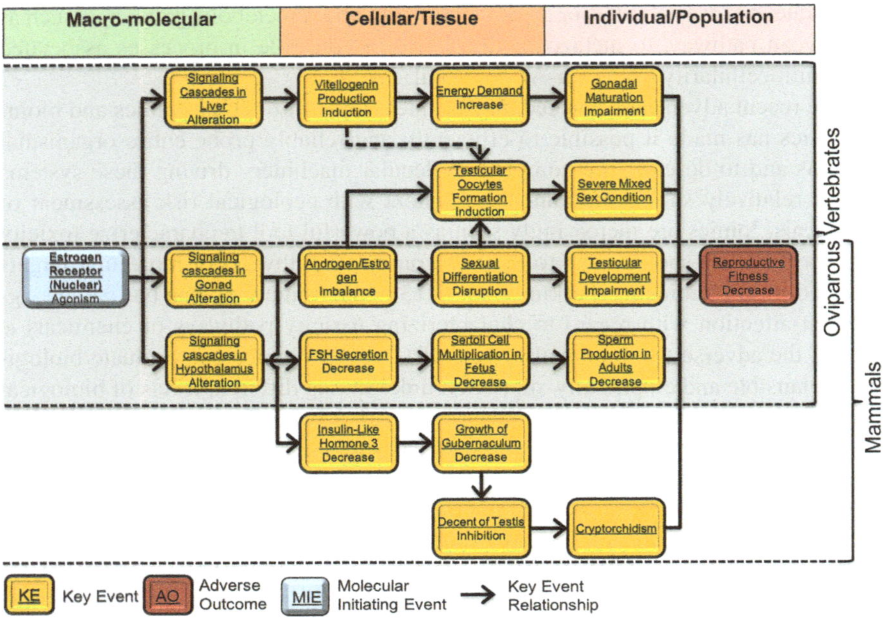

Fig. 6.2 Examples of hypothetical adverse outcome pathways of nuclear estrogen receptor mediated estrogenic responses in oviparous vertebrates and mammals leading to reproductive impairment in adult organisms

concentrations at the biological targets (e.g. critical body residue), and thus, become a driver in susceptibility considerations (ECETOC 2007; Escher et al. 2011). Therefore, to be realistic and useful to regulators, cross-species extrapolation models that use conserved molecular targets or toxicity pathways to predict the sensitivity of an organism to contaminants need to integrate the TD and TK properties of these chemicals in a given species. This is supported by a recent review of data generated by USEPA's Endocrine Disruptor Screening Program (EDSP) that compared the responses to 12 model compounds between the 21 day fathead minnow reproductive assay and a selection of rat assays (uterotrophic, Hershberger, and male and female pubertal assays) (Ankley and Gray 2013). The authors confirmed that the effects of potent (xeno)estrogens on ER-mediated pathways that are highly conserved among vertebrates were in general predictable among species, even considering the different physiological manifestations in oviparous fish versus viviparous mammals. However, when comparing findings obtained for a weak ER agonist, bisphenol A, two of the three rat assays produced negative results when compared to the fish assay that tested positive. The likely cause for these differences was that due to the oral dosing route used for rats compared to waterborne exposure of fish, in rats most of the bisphenol A was likely to have been cleared from the system by first-pass hepatic metabolism, further highlighting the role of ADME as well as types of exposures in cross-species extrapolation. Interestingly, effects of chemicals mediated through the AR or disruption of sex hormone steroidogenesis were less variable when comparing outcomes of the fathead minnow with the rat assay. Overall, however, these data suggest that highly conserved biological pathways such as hormone receptor-mediated or steroid synthesis-mediated reproductive functions can be used as a starting point in AOP-based cross species hazard assessments.

Despite evidence that identification of conserved toxicity pathways might be useful for cross species extrapolations of toxicity, there are a number of remaining uncertainties that need to be addressed before approaches such as conserved toxicity pathways will be truly useful for ecological risk assessors. The major shortcoming with regard to using AOPs in this context is the lack of maturity – or the existence – of AOPs for most chemicals of regulatory concern and a strong bias in availability of AOPs toward vertebrates. This represents a particular concern when considering bacteria, plants and invertebrates. It is, however, acknowledged that clearer definition and differentiation of AOP applicability for a given regulatory need may assist in using data even if the AOP has not been fully developed (Tollefsen et al. 2014). Another main concern includes the mostly qualitative nature of approaches such as the AOP framework, which limits the prediction of the sensitivity across species as this is inadvertently linked to exposure concentrations (internal and external). As reviewed by Perkins et al. (2013), factors such as exposure routes for e.g. terrestrial (typically non-continues oral exposures) and aquatic (typically continuous immersion) organisms render the comparison among these species difficult given the differences in ADME properties solely associated with exposure route and frequency. Similarly, approaches are needed to integrate toxicokinetic and –dynamic properties into current qualitative toxicity pathway models in order to successfully use this approach for cross-species sensitivity assessments. Also, it needs to be considered that each of the examples provided above are based on standard model species such as zebrafish, fathead minnow, rainbow trout, rat, mouse, etc., for which extensive knowledge of their basic physiology, basal metabolic activity, etc. as well as large toxicological datasets for many chemicals are available. However, no such information is available for the vast majority of the non-model species of interest. This is particularly true for invertebrates, but even for large groups of vertebrates, such as the fishes for which there are greater than 30,000 species and very little information is available. For example, while being considered of great priority in context with ecological risk assessments in North America and Asia due to their endangered status, we know very little about the vulnerability to environmental contaminants of ancient fishes such as sturgeons that are evolutionary far removed from most of the modern teleosts. This is concerning as recent studies have shown that some sturgeon species tend to be unique in their responses to certain contaminants. For example, white sturgeon (*Acipenser transmontanus*) were shown to be among the most sensitive species of fishes to the exposure with selected heavy metals (e.g. copper) during certain early life stages, and which renders the protectiveness of current water quality standards for these organisms questionable (Little et al. 2012; Vardy et al. 2011, 2013). One hypothesis for this high sensitivity is a blunted ability to mount a cellular stress response against metal ions as demonstrated by a very low inducibility of functions involved with amelioration of toxicity including expression of metallothioneins that sequester metals as well as genes that mediate anti-oxidant responses (Doering et al. 2015; Tang et al. 2016). This lack of compensatory response is likely to have a significant impact on the TK/TD properties of metals in sturgeon, affecting internal exposure to metals at target tissues.

6.3.2 Species Read Across Approaches Using Conserved Molecular Targets: SeqAPASS

As discussed in Sect. 6.2, conservation of molecular targets of chemicals with specific modes of action is likely to provide some information regarding the intrinsic susceptibility of a species. This assumption has been extensively used in the development of pharmaceuticals for human and veterinary use as well as in the development of pesticides to target specific pest species. This section provides a review of a recent initiative by the USEPA that explores the concept of molecular target conservation as a tool to predict the susceptibility to chemicals with specific modes of action across phylogenetic taxa (LaLone et al. 2013b). Specifically, this initiative investigates the utility of protein sequences/structure similarity in predicting the likelihood of susceptibility of any species for which sequence information is available, and which is termed Sequence Alignment to Predict Across Species Susceptibility (SeqAPASS). SeqAPASS is a computational tool that aligns the sequence of the functional molecule representing a MIE, such as a receptor or enzyme, which has been shown to trigger an adverse effect. It relies on existing information on the amino acid sequence of proteins and their structure. Initial case studies have shown the promise of this tool to predict species susceptibility across taxa for selected pharmaceuticals and pesticides (LaLone et al. 2013a). For example, a comparison of the susceptibility to estrogenic chemicals based on ER sequence similarities among animal taxa using protein sequence similarity analysis showed a high degree of conservation among vertebrate species when using the human receptor as query (LaLone et al. 2013b; Fig. 6.3). Furthermore, invertebrates were predicted to be generally less sensitive, which is in accordance with the absence of functional ERs in many invertebrates. Furthermore, when comparing the intrinsic susceptibility predictions derived from the SeqAPASS analysis for aquatic species with empirical toxicity data these showed a good correlation.

The advantage of the SeqAPASS tool is that it utilizes relatively underutilized and continuously expanding resources of data that aim to predict chemical susceptibility across a broad range of taxa ranging from humans to viruses. However, while existing databases within the National Centre for Biotechnology Information (NCBI) are rapidly expanding and include more-and-more information on non-model species, they are still far from providing a representative overview of the animal and plant kingdoms. The SeqAPASS approach also allows the characterization of chemicals that interact with multiple molecular targets by combining queries from different data sets. As discussed in Sect. 6.2, current cross-species extrapolation

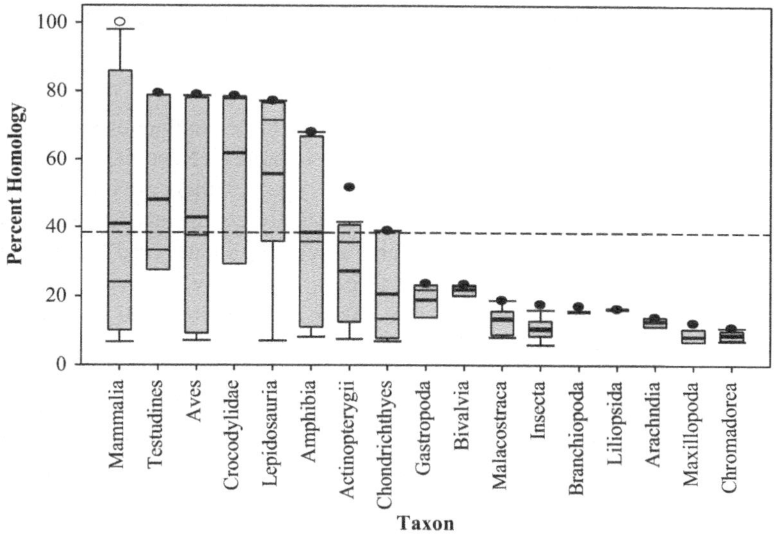

Fig. 6.3 SeqAPASS analysis output of the distribution of homolog candidate proteins across species within select taxa based on percent similarity to the human estrogen receptor (ER). *The open circle* represents the human (Homo sapiens) ER, and *solid circles* represent the species with the highest percent similarity within the specified taxonomic group. Box plots: the *top* and *bottom* of the *box* represent the 75th and 25th percentiles, respectively, and the *top* and *bottom* whiskers on plot represent the 90th and 10th percentiles, respectively. *Small black dots* indicate outliers representing the 95th and 5th percentiles. The mean and median values for each taxon are represented by horizontal thick and thin *black lines* on the *box*, respectively. In some cases, *lines* representing mean and median overlap and are displayed as a single horizontal thick *black line*. If <3 species represent a taxon, only maximum and mean values are shown. - - - indicates the cut-off for intrinsic susceptibility predictions (based on ortholog analysis), with those above the line predicted to be susceptible (Figure courtesy of Carlie LaLone and modified from LaLone et al. (2013b))

approaches often lack the inclusion of TK/TD properties. First attempts to combine available mammalian ADME data such as drug clearance rates for approximately 1200 different human and veterinary drugs from 100 drug classes using a probabilistic distribution approach with the above-discussed SeqAPASS model showed promise in prioritizing certain drug classes with regard to their ecotoxicological risks (LaLone et al. 2013b). This demonstrates the potential of this model to be integrated with other approaches important for a more realistic assessment of cross-species sensitivities to contaminants. However, given the significant differences in ADME properties between mammals and other vertebrates and invertebrates, much additional work is needed to better characterize the ADME properties of chemicals across species from all taxa that would be required to expand the above approach for use with species other than current mammalian model organisms. Also, factors such as life stage, life history, ability to mount a compensatory response, sex and other critical factors driving an organism's response to contaminant exposure are not considered by SeqAPASS.

6.3.3 Alternatives to Live Animal Testing to Inform Cross-Species Extrapolation

As discussed in the previous sections, there are multiple challenges with assessing the effects of contaminants to native species of interest. Beyond any molecular, biochemical and physiological differences across species, there are challenges associated with maintaining wild species under laboratory conditions, including ethical concerns when working with live animals, especially endangered species (which can be of particular interest with regard to their sensitivity to environmental contaminants), and high investments in time, labour and cost involved with traditional in vivo assays. Therefore, alternatives to animal testing are increasingly used as tools to investigate the toxicity of chemicals. These include (1) computational *in silico* models such as Quantitative Structure-Activity Relationships (QSARs), (2) in vitro tests either using sub-cellular components, immortalized cell lines or primary cell and tissue cultures, as well as (3) toxicity testing with early life stages of oviparous organism that are not considered live animals until they have depleted their yolk-sacs, including *in ovo* assays with bird embryos or fish embryos prior to swim up (Hartung and Hoffmann 2009; Knight 2008).

6.3.4 In Silico *Approaches*

Per definition, in toxicology *in silico* refers to any methodology that involves computer-based planning, analysis, evaluation or prediction of toxicological information (Hartung and Hoffmann 2009). For the purpose of this chapter, the focus is those current computational approaches that are directly applicable to cross-species sensitivity extrapolation. Toxicological *in silico* methodologies use experimental data derived from in vivo or in vitro tests to identify commonalities and patterns among chemicals or biological targets (molecular through populations/communities) that can be used to develop computational models to predict their hazard potential or susceptibility, respectively. The most common *in silico* approaches utilized in chemical risk assessment are QSARs, which are used to predict the potential toxicological hazard of a chemical based on its structure and physicochemical properties (Bradbury 1994). To be useful in cross-species extrapolations, however, knowledge about the actual target sites of the toxicants of interest is required. The integration of QSARs with other predictive tools such as the above-discussed SeqAPASS has the promise to identify certain taxa that may be particularly vulnerable to certain contaminants based on predicted chemical-target site interactions. Other important computational approaches in species sensitivity extrapolation are species sensitivity distribution (SSD) tools such as interspecies correlation models (Barron et al. 2012). These models primarily rely on acute toxicity data to develop computational models based e.g. on QSAR properties. However, Barron et al. (2012) found that such models were associated with significant variability, which limited their predictive power. Finally, *in silico* approaches include a number of other current concepts discussed elsewhere in this book chapter, including the SeqAPASSS model and USEPA's ToxCast program.

6.3.5 In Vitro *Approaches*

In vitro assays are increasingly used to inform the potential toxicological risks of chemicals because they often have great sensitivity to low concentrations, great specificity of response, high throughput, and have a lesser cost than in vivo assays (Gray et al. 1997). Additionally, in vitro assays require much fewer numbers of animals compared to in vivo assays, which, for reasons outlined above, is becoming increasingly important in toxicity testing. The relevance and utility of stable cell line-based in vitro approaches in support of chemical hazard assessment and prioritization including the use of high-throughput screening in vitro batteries to identify specific MIEs have been extensively reviewed in the past (see e.g. Kavlock et al. 2012; Perkins et al. 2013). Because they do not directly apply to cross-species extrapolation, with the few exceptions discussed below, they will not be discussed in detail here. Instead, this section focuses on the use of in vitro tests as an alternative tool to generate and test hypothesis pertaining to the sensitivity of non-model species to environmental contaminants.

One method for in vitro testing that has shown potential with regard to identifying species-specific sensitivities to environmental contaminants are tissue explants or primary cell cultures (Beitel et al. 2014, 2015; Eisner et al. 2015). Such approaches are beneficial as tissues maintain some of their natural functions (e.g. paracrine interactions) outside their natural environment, as most of the necessary machinery required for the cell- or tissue-specific function is present (Powlin et al. 1998; Gray et al. 1997). Therefore, test systems using tissue explants could be used to identify species-specific responses to the exposure with contaminants. Recent studies confirmed the potential of species-specific tissue explant assays to predict relative sensitivity of selected key events (KEs) in native fish species to certain environmental contaminants including endocrine disruptors and DLCs (Beitel et al. 2014, 2015; Eisner et al. 2015). While absolute sensitivities (threshold or effective concentrations) were not directly comparable, Eisner et al. (2015) demonstrated that the relative potencies of six DLCs to four evolutionary distinct fish species determined using in vitro liver explants were directly correlated with embryo toxicity data. Similar correlations in relative responses occurred when three different fish species were exposed to estrogenic compounds (reviewed in Beitel et al. 2015). Furthermore, Beitel et al. (2014) showed that tissues in primary culture were representative of seasonal fluctuations in reproductive endocrine functioning (i.e. steroid synthesis) in vivo, and allowed for identification of the most sensitive stage of the reproductive cycle of three fish species native to North America. Therefore, it was hypothesized that primary tissue cultures have the potential to identify most sensitive phases or life stages of an organism. These initial data are promising with regard to the future potential of tissue culture assays to help in identifying species that may be particularly vulnerable to the exposure with chemicals of concern. However, significant work is still required to confirm and validate the predictive power of species-specific tissue explant assays for in vivo effects for different classes of contaminants and among greater numbers of diverse ecological species.

Stable cell lines have also been used as tools to characterize and compare MIEs among different species, and to establish baseline information on the responsiveness of certain target molecules such as receptors to the exposure with contaminants. One of the best-described examples in the ecotoxicological literature is the use of green monkey (COS-7) cells to characterize the role of the AhR in mediating sensitivity of birds and fishes to DLCs (Doering et al. 2014; Farmahin et al. 2014; Karchner et al. 2006). Transfection of COS-7 cells with AhRs from different species of birds, and more recently fishes, was used successfully to categorize the in vivo potency of variety of DLCs to activate AhR signalling (Farmahin et al. 2014; Doering et al. 2014). In birds, COS-7 cells were used in conjunction with site-directed mutagenesis studies replacing key amino acid in the ligand-binding domain of the receptor to characterize the specific MIE driving species sensitivity, and based on which the sensitivity of any bird species of interest can now be predicted. Similar research is currently ongoing to characterise the role of amino acids in the ligand binding domain of the AhR in determining sensitivity of fishes to the exposure with DLCs. Successful completion of this work would represent a critical milestone in advancing risk assessment of these priority pollutants across the greater than 30,000 species of fishes inhabiting our planet. Initial efforts are also currently underway to characterize the molecular basis for the differences in potency of environmental estrogens among fishes (Toyahama et al. 2015). Considering the potential of these in vitro based tools in combination with recent advances in 'omic technologies that enable quick and inexpensive identification of the specific molecular composition (e.g. gene or protein sequence information) of molecular targets of interest, it is anticipated that similar approaches will become routine practice in the future to elucidate specific MIEs that inform the sensitivity of different ecological species.

It should be acknowledged, however, that although numerous advantages exist with regard to the potential of in vitro assays to predict sensitivity to contaminants across species, there are remaining uncertainties regarding their use as surrogates for in vivo assays. For example, ADME properties are often not, or only partially, accounted for by in vitro assays. This can lead to false positive or false negative results in cases of chemicals that are rapidly metabolized or that require metabolic activation, respectively (Gray et al. 1997). Also, in vitro systems do not represent organismal feedback systems and interactions among organs and tissues. Therefore, it is unlikely that in vitro approaches will completely replace live animal testing in the near future. However, it is anticipated that alternative tests will increasingly be used in chemical prioritization and the identification of MIEs and associated molecular toxicity pathways, and in combination with computational modeling such as currently applied in USEPA's ToxCast Program, will provide powerful tools to advance our understanding of species-specific modes or chemical action while significantly reducing the need for in vivo testing.

6.3.6 Early Life Stage Testing

Another alternative testing approach that is increasingly applied in ecotoxicological testing is the use of embryonic life-stages of oviparous animals, including fish and birds. These embryos are not considered "live animals" under many legislations (Knight 2008), although the exact life-stages at which these organisms are considered sentient animals differ among countries (Strähle et al. 2012). Embryos of egg-laying organisms are a powerful alternative to in vitro assays as they are representative of the intricacies and complexity of whole organisms. While some embryos may not have completely developed organ systems during very early stages, it could be shown that they seem to present most of the molecular regulatory networks driving adult physiological functions. For example, transcriptional analyses of zebrafish embryos exposed to estrogenic chemicals revealed effects on the expression of genes and pathways indicative of disruption of potential downstream events associated with estrogen signalling, steroid hormone production, and neurodevelopment, regardless of the fact that some of these processes are not expressed in larval fish (Schiller et al. 2013; Vosges et al. 2010). Zebrafish embryos have also been shown to express complete pathways for other key physiological functions including thyroid signalling and cardiovascular system development (Hill 2005; Thienpont et al. 2011). Similarly, studies with early tadpole stages of the African clawed frog (*Xenopus laevis*) around the time of sexual differentiation exposed to EE2 showed that these early stages expressed molecular pathways whose disruption was indicative of later effects on biological functions including metamorphosis, gonadal development and growth (Tompsett et al. 2013). However, the vast majority of research with early life stages has been conducted with very few model species such as the zebrafish and the chicken, and it remains to be demonstrated whether effects on embryonic stages of other non-model ecological receptors are similarly predictive of biological effects that manifest in adult organisms.

Regardless of whether assays with early life stages is useful to predict effects at later life stages, with some exceptions, embryos are thought to be among the most sensitive life-stages to the exposure with contaminants (Mohammed 2013). This is because most organ systems are developing during this time, a process that may be vulnerable to toxic insults, the low volume to surface ratio allows fast uptake and distribution of contaminants, the lack of efficient metabolism and clearance mechanisms, and low fat reserves that may sequester lipophilic contaminants. Therefore, toxic effects in embryos are considered to be a conservative proxy of the sensitivity of a species to the exposure with pollutants with the exception of adult-specific functions such as reproduction. Early life stage tests with most oviparous animals are usually completed within a few days to weeks depending on the species of interest, and they represent a great opportunity to assess species that typically cannot be tested under controlled conditions, such as endangered or long-lived organisms. This is particularly true for fishes and amphibians that produce hundreds to thousands of eggs, and for which standard culture methods can be easily adopted. For example, Vardy et al. (2013) demonstrated that early life stage studies with embryos of rainbow trout (*Oncorhynchus mykiss*), fathead minnow (*Pimephales promelas*) and white sturgeon represented a reliable tool to predict the sensitivity to metals among fish species. Also, two comparative studies conducted with shortnose (*A. brevirostum*) and Atlantic (*A. oxyrichnus*) sturgeon (Chambers et al. 2012) and with shovelnose (*Saphirhynchus platorynchus*) and pallid (*S. albus*) sturgeon (Buckler et al. 2015) showed the utility of early life stages of endangered species to predict their sensitivity to DLCs. Given the relative ease of obtaining and culturing of fish, amphibian and bird embryos, and the relatively low cost and time investment needs to conduct early life stage studies, it is surprising that there only have been a few efforts to use them for cross-species extrapolation in ecotoxicology. As early development across fishes, amphibians or birds is highly comparable within each of these taxonomic groups, and considering the availability of standardized guidelines for assessing endpoints in embryos (e.g. OECD 2013) they represent a highly promising tool in comparative ecotoxicology.

6.4 Conclusions

The ability to assess the hazards chemicals may pose to the vast diversity of ecological species is increasingly becoming a necessity in ecological risk assessment. It also is apparent that traditional testing approaches using live animals will not be able to address these needs given the economic and ethical restrictions associated with them. Acknowledging these limitations, and motivated by the current paradigm shift from empirical testing to systems- and pathway-based approaches in the field of human health and ecological risk assessment (NRC 2007; Villeneuve et al. 2014a), regulators, scientists and industry are currently exploring novel concepts and methodologies to enable the prediction of the toxicological risks across species and taxonomic groups. Particularly, the recognition that specific molecular targets representing MIEs or KEs, or partial or entire toxicity pathways can be conserved among or within taxonomic groups has resulted in a focus on comparative 'omics as predictive tools for cross-species extrapolation (Brockmeier et al. 2017). One platform that is increasingly used in this context for identifying key processes that can drive sensitivity of a species are AOPs because they cover all levels of biological organization, provide biological context and inform the chemical and taxonomic applicability of the toxicity pathway. However, there are a number of uncertainties that remain to be addressed before AOPs and associated tools such as the above-discussed SeqAPASS tool and other *in silico* methodologies become a viable option in non-model species risk assessment, including the limited number of mature AOPs currently available, their limited taxonomic application (virtually no AOPs exist for microorganisms, invertebrates and plants) and their mostly qualitative nature. Furthermore, large data gaps exist with regard to ADME properties of chemicals in ecological species that determine target site concentrations, and are a critical factor influencing intrinsic sensitivity.

Therefore, there is the need to expand our current knowledge of the MIEs and KEs for the large number of chemicals of environmental concern in addition to their ADME properties across diverse species. A tiered approach exploiting (1) existing knowledge about AOPs along with available sequence and functional homology information when the MIE and KEs are well-characterized, and (2) generation of custom *de novo* gene or protein sequence information to expand the knowledge base may provide a feasible solution to rapidly characterize the taxonomic applicability domain for certain pathways. In parallel, development and validation of models that allow predicting the ADME properties of environmental contaminants across diverse taxonomic groups are needed to enable prioritization of organism groups with regard to their risk of internal exposure to contaminant groups of concern (e.g. organisms with low metabolic activity are less likely to be at risk to the exposure with chemicals that require metabolic activation). Based on this information, expert-curated species similarity maps can be constructed with the goal of identifying so called "forecaster species", which allow extrapolation of sensitivity to similar species that cannot be investigated due to feasibility or ethical reasons (e.g. endangered species). The proposed approach is similar to the current practice in risk assessment of dioxin-like contaminants for birds where species are categorized as chicken-like (highly sensitive), pheasant-like (moderately sensitive), and quail-like (not sensitive) (Karchner et al. 2006). Subsequent confirmation of these predictions to identify the main drivers of species-specific differences in susceptibility is proposed by a combination of (1) advanced *in silico* modeling of MIEs or KEs across species using tools such as SeqAPASS (LaLone et al. 2013b), (2) use of targeted high-throughput in vitro assays following principles currently used in drug discovery (see e.g. Doering et al. 2014; Farmahin et al. 2013), (3) conduct species-specific tissue explant assays to generate hypotheses for mechanism-specific sensitivity (Beitel et al. 2015; Eisner et al. 2015), and (4) develop of *embryo toxicity* tests for oviparous vertebrates that allow anchoring to an AO using relatively economical higher-throughput systems while addressing animal welfare concerns. However, in vivo whole organism tests will still be required to validate the predictions in cases where the approaches proposed above are not adequate.

Acknowledgements I would like to thank Steve Wiseman for his critical review and helpful comments. Also, I greatly appreciate Carlie LaLone introducing me to the SeqAPASS tool, and providing the examples included in this book chapter.

References

Adamska M, Degnan SM, Green KM, Adamski M, Craigie A, Larroux C, Degnan BM (2007) Wnt and TGF-β expression in the sponge *Amphimedon queenslandica* and the origin of metazoan embryonic patterning. PLoS One:e1031. doi:10.1371/journal.pone.0001031

Allard P, Fairbrother A, Hope BK, Hull RN, Johnson MS, Kapustka L, Mann G, McDonald B, Sample BE (2010) Recommendations for the development and application of wildlife toxicity reference values. Integr Environ Assess Manag 6(1):28–37. doi:10.1897/IEAM_2009-010.1

Ankley GT, Johnson RD (2004) Small fish models for identifying and assessing the effects of endocrine-disrupting chemicals. ILAR J 45:469–483

Ankley GT, Bennett RS, Erickson RJ, Hoff DJ, Hornung MW, Johnson RD, Mount DR, Nichols JW, Russom CL, Schmieder PK, Serrano JA, Tietge JE, Villeneuve DL (2010) Adverse outcome pathways: a conceptual framework to support ecotoxicology research and risk assessment. Environ Toxicol Chem 29(3):730–741

Ankley GT, Gray LE (2013) Cross-species conservation of endocrine pathways: a critical analysis of tier 1 fish and rat screening assays with 12 model chemicals. Environ Toxicol Chem 32:1084–1087.

Ashauer R, Escher BI (2010) Advantages of toxicokinetic and toxicodynamic modelling in aquatic ecotoxicology and risk assessment. J Environ Monit 12(11):2056–2061. doi:10.1039/c0em00234h. Epub 2010 Sep 23

Ashauer R, Agatz A, Albert C, Ducrot V, Galic N, Hendriks J, Jager T, Kretschmann A, O'Connor I, Runach MN, Nyman A-M, Schmitt W, Stadnicka J, Van den Brink PJ, Preuss TG (2011) Toxicokinetic-toxicodynamic modeling of quantal and graded sublethal endpoints: a brief discussion of concepts. Environ Toxicol Chem 30:2519–2524

Baird DJ, Van den Brink PJ (2007) Using biological traits to predict species sensitivity to toxic substances. Ecotoxicol Environ Saf 67(2):296–301

Barron MG (1990) Bioconcentration. Will water-borne organic chemicals accumulate in aquatic animals? Environ Sci Technol 24:1612–1618

Barron MG, Jackson CR, Awkerman JA (2012) Evaluation of in silico development of aquatic toxicity species sensitivity distributions. Aquat Toxicol 116-117:1–7

Beitel SC, Doering JA, Patterson SE, Hecker M (2014) Assessment of the sensitivity of three north American fish species to disruptors of steroidogenesis using in vitro tissue explants. Aquat Toxicol 152:273–283

Beitel SC, Doering JA, Eisner BK, Hecker M (2015) Comparison of the sensitivity of four native Canadian fish species to 17-α Ethinylestradiol, using an in vitro liver explant assay. Environ Sci Pollut Res. doi:10.1007/s11356-015-5101-7

Berninger JP, Brooks BW (2010) Leveraging mammalian pharmaceutical toxicology and pharmacology data to predict chronic fish responses to pharmaceuticals. Toxicol Lett 193:69–78. doi:10.1016/j.toxlet.2009.12.006

Bradbury S (1994) Predicting modes of toxic action from chemical structure: an overview. SAR QSAR Environ Res 2(1–2):89–104

Brinkmann M, Eichbaum K, Buchinger S, Reifferscheid G, Bui T, Schäffer A, Hollert H, Preuss TG (2014) Understanding receptor-mediated effects in rainbow trout: in vitro–in vivo extrapolation using physiologically based toxicokinetic models. Environ Sci Technol 48:3303–3309

Buckler J, Candrl JS, McKee MJ, Papoulias DM, Tillitt DE, Galat DL (2015) Sensitivity of shovelnose sturgeon (*Scaphirynchus platorynchus*) and pallid sturgeon (*S. albus*) early life stages to 3,3′,4,4′5-pentachlorobiphenyl and 2,3,7,8-tetrachlorodibenzy-p-dioxin exposure. Environ Toxicol Chem 34(6):1417–1424

Brockmeier EK, Hodges G, Hutchinson TH, Butler E, Hecker M, Tollefsen KE, Garcia-Reyero N, Kille P, Becker D, Chipman K, Colbourne J, Collette TW, Cossins A, Cronin M, Graystock P, Gutsell S, Knapen S, Katsiadaki I, Lange A, Marshall S, Owen SF, Perkins EJ, Plaistow S, Schroeder A, Taylor D, Viant M, Ankley G, Falciani F (2017) The role of omics in the application of adverse outcome pathways for chemical risk assessment. Toxicol Sci 158:252–262

Calow P, Sibly RM, Forbes V (1997) Risk assessment on the basis of simplified life-history scenarios. Environ Toxicol Chem 16(9):1983–1989

CEC (Commission of the European Communities) (1996) Technical guidance documents of in support of the Commission Directive 93/67 EEC on risk assessment for new substances and the Commission Regulation (EC) no. 1488/94 on risk assessment of existing substances. Parts I through IV, Brussels, Commission of the European Communities

Celander MC, Goldstone JV, Denslow ND, Iguchi T, Kille P, Meyerhoff RD, Smith BA, Hutchinson TH, Wheeler JR (2010) Species extrapolation for the 21st century. Environ Toxicol Chem 30(1):52–63

Chambers RC, Davis DD, Habeck EA, Roy NK, Wirgin I (2012) Toxic effects of PCB126 and TCDD on shortnose sturgeon and Atlantic sturgeon. Environ Toxicol Chem 31(10):2324–2337

Chapman PM, Fairbrother A, Brown D (1998) A critical evaluation of safety (uncertainty) factors for ecological risk assessment. Environ Toxicol Chem 17:99–108

Doering JA, Wiseman S, Beitel SC, Giesy JP, Hecker M (2014) Identification and expression of aryl hydrocarbon receptors (AhR1and AhR2) provide insight in an evolutionary context regarding sensitivity of white sturgeon (*Acipenser transmontanus*) to dioxin-likecompounds. Aquat Toxicol 150:27–35

Doering J, Beitel S, Eisner B, Heide T, Hollert H, Giesy J, Hecker M, Wiseman S (2015) Identification and response to metals of metallothionein in two ancient fishes: white sturgeon (*Acipenser transmontanus*) and lake sturgeon (*Acipenser fulvescens*). Comp Biochem Physiol 171:41–48

Dwyer FJ, Mayer FL, Sappington LC, Buckler DR, Bridges CM, Greer IE, Hardesty DK, Henke CE, Ingersoll CG, Kunz JL, Whites DW, Augspurger T, Mount DR, Hattala K, Neuderfer GN (2005) Assessing contaminant sensitivity of endangered and threatened aquatic species: I. Acute toxicity of five chemicals. Arch Environ Contam Toxicol 48:143–154

ECETOC (European Centre for Ecotoxicology and Toxicology of Chemicals) (2007) Intelligent testing strategies in ecotoxicology: mode of action approach for specifically acting chemicals. Technical report 102, Brussels

ECHA (2015) Read-across assessment framework (RAAF). Eur Chem Agency (ECHA). May 2015, Helsinki, Finland: 38. doi:10.2823/69353

Eisner BK, Doering JA, Beitel SC, Wiseman S, Raine JC, Hecker M (2015) Cross-species comparison of relative potencies and relative sensitivities of fishes to dibenzo-*p*-dioxins, dibenzofurans, and polychlorinated biphenyls in vitro. Environ Toxicol Chem. doi:10.1002/etc.3173

Escher BI, Hermens JLM (2002) Modes of action in ecotoxicology: their role in body burdons, species sensitivity, QSARs, and mixture effects. Environ Sci Technol 36(20):4201–4217

Escher B, Ashauer R, Dyer S, Hermens JL, Lee JH, Leslie HA, Mayer P, Meador JP, Warne MS (2011) Crucial role of mechanisms and modes of toxic action for understanding tissue residue toxicity and internal effect concentrations of organic chemicals. Integr Environ Assess Manag 7:28–49

Fairchild JF, Allert AL, Sappington LC, Waddell B (2005) Chronic toxicity of un-ionized ammonia to early life stages of endangered Colorado pikeminnow (*Ptychocheilus Lucius*) and razorback sucker (*Xyrauchen texanus*) compared to the surrogate fathead minnow (*Pimephales promelas*). Arch Environ Contam Toxicol 49:378–384

Farmahin R, Manning GE, Crump D, Wu D, Mundy LJ, Jones SP, Hahn ME, Karchner SI, Giesy JP, Bursian SJ, Zwiernik MJ, Fredricks TB, Kennedy SW (2013) Amino acid sequence of the ligand-binding domain of the aryl hydrocarbon receptor 1 predicts sensitivity of wild birds to effects of dioxin-like compounds. Toxicol Sci 131:139–152

Farmahin R, Jones SP, Crump D, Hahn ME, Giesy JP, Zwiernik MJ, Bursian SJ, Kennedy SW (2014) Species-specific relative AHR1 binding affinities of 2,3,4,7,8-pentachlorodibenzofuran explain avian species differences in its relative potency. Comp Biochem Physiol C 161:21–25

Fernández MD, Vega MM, Tarazona JV (2006) Risk-based ecological soil quality criteria for the characterization of contaminated soils. Combination of chemical and biological tools. Sci Total Environ 366:466–484

Forbes VE, Calow P (2002) Extrapolation in ecological risk assessment: balancing pragmatism and precaustion in chemical controls legislation. Bioscience 52:249–257

Fraccalvieri D, Soshilov AA, Karchner SI, Franks DG, Pandini A, Bonati L, Hahn ME, Denison MS (2013) Comparative analysis of homology models of the ah receptor ligand binding domain: verification of structure-function predictions by site-directed mutagenesis of a nonfunctional receptor. Biochemistry 52:714–725

Garcia-Reyero N, Habib T, Pirooznia M, Gust KA, Gong P, Warner C, Wilbanks M, Perkins E (2011) Conserved toxic responses across divergent phylogenetic lineages: a meta-analysis of the neurotoxic effects of RDX among multiple species using toxicogenomics. Ecotoxicology 20(3):580–594

Gray EL, Kelce WR, Wiese T, Lyl R, Caido K, Cook J, Klinefelter G, Desaulniers D, Wilson E, Zacharewski T, Waller C, Foster P, Laskey J, Reel J, Laws S, McLachlan J, Breslin W, Cooper R, Di Giulio R, Johnson R, Purdy R, Mihaich E, Safe S, Sonnenschein C, Welshons W, Miller R, McMaster S, Colborn T (1997) Endocrine screening methods workshop report: detection of estrogenic and androgenic hormonal and Antihormonal activity for chemicals that act via receptor or steroidogenic enzyme mechanisms. Reprod Toxicol 11(5):719–750

Gunnarsson L, Jauhiainen A, Kristiansson E, Nerman O, Larsson DGJ (2008) Evolutionary conservation of human drug targets in organisms used for environmental risk assessments. Environ Sci Technol 42:5807–5813

Han X, Nabb DL, Mingoia RT, Yang CH (2007) Determination of xenobiotic intrinsic clearance in freshly isolated hepatocytes from rainbow trout (*Oncorhynchus mykiss*) and rat and its application in bioaccumulation assessment. Environ Sci Technol 41:3269–3276. doi:10.1021/es0626279

Hartung T, Hoffmann S (2009) Food for thought … on in silico methods in toxicology. ALTEX 26:155–166

Head JA, Hahn ME, Kennedy SW (2008) Key amino acids in the aryl hydrocarbon receptor predict dioxin sensitivity in avian species. Environ Sci Technol 42:7535–7541

Hill AJ (2005) Zebrafish as a model vertebrate for investigating chemical toxicity. Toxicol Sci 86:6–19. doi:10.1093/toxsci/kfi110

Jager T, Albert C, Preuss TG, Ashauer R (2011) General unified threshold model for survival – a toxicokinetic-toxicodynamic framework for ecotoxicology. Environ Sci Technol 45:2529–2540

Karchner SI, Franks DG, Kennedy SW, Hahn ME (2006) The molecular basis for differential dioxin sensitivity in birds: role of the aryl hydrocarbon receptor. Proc Nat Acad Sci U S A 103(16):6252–6257

Kavlock R, Chandler K, Houck K, Hunter K, Judson R, Kleinstreuer N, Knudsen T, Martin M, Padilla S, Reif D, Richard A, Rotroff D, Sipes N, Dix D (2012) Update on EPA's ToxCast program: providing high throughput decision support tools for chemical risk management. Chem Res Toxicol 25:1287–1302

Knight A (2008) Non-animal methodologies within biomedical research and toxicity testing. ALTEX 25:213–231

LaFont R (2000) The endocrinology of invertebrates. Ecotoxicology 9:41–57

LaLone CA, Berninger JP, Villeneuve DL, Ankley GT (2013a) Leveraging existing data for prioritization of the ecological risks of human and veterinary pharmaceuticals to aquatic organisms. Philos Trans R Soc B 369:2014.0022

Lalone CA, Villeneuve DL, Burgoon LD, Russom CL, Helgen HW, Berninger JP, Tietge JE, Severson MN, Cavallin JE, Ankley GT (2013b) Molecular target sequence similarity as a basis for species extrapolation to assess the ecological risk of chemicals with known modes of action. Aquat Toxicol 144–145:141–154

Lewis DFV, Ioannides C, Parke DV (1998) Cytochromes P450 and species differences in xenobiotic metabolism and activation of carcinogens. Environ Health Perspect 106:633–641

Little EE, Calfee RD, Linder G (2012) Toxicity of copper to early-life stage Kootenai River white sturgeon, Columbia River white sturgeon, and rainbow trout. Arch Environ Contam Toxicol 63:400–408

Matthews J, Celius T, Halgren R, Zahcharewski T (2000) Differential estrogen receptor binding of estrogenic substances: a species comparison. J Steroid Biochem Mol Biol 74:223–234

Mohammed A (2013) Why are early life stages of aquatic organisms more sensitive to toxicants than adults? In: Gowder S (ed) New insights into toxicity and drug testing. InTech. doi:10.5772/55187

Nichols JW, McKim JM, Andersen ME et al (1990) A physiologically based toxicokinetic model for the uptake and disposition of waterborne organic chemicals in fish. Toxicol Appl Pharmacol 106:433–447

NRC (National Research Council) (2007) Toxicity testing in the 21st century: a vision and a strategy. National Academies Press, Washington, DC. http://www.nap.edu/openbook.php?record_id=11970

Odio C, Holzman SA, Denison MS, Fraccalvieri D (2013) Specific ligand binding domain residues confer low dioxin responsiveness to AhR1β of *Xenopus laevis*. Biochemistry 52(10):1746–1754
OECD (2013) Test no. 210: fish, early-life stage toxicity test. OECD guidelines for the testing of chemicals, Section 2. OECD Publishing, Paris. doi:10.1787/9789264203785-en
Pandini A, Denison MS, Song Y, Soshilov AA, Bonati L (2007) Structural and functional characterization of the aryl hydrocarbon receptor ligand binding domain by homology modeling and mutational analysis. Biochemistry 47:696–708
Peregrin-Alvarez J, Sanford C, Parkinson J (2009) The conservation and evolutionary modularity of metabolism. Genome Biol 10:R63
Perkins EJ, Ankley GT, Crofton KM, Garcia-Reyero N, LaLone CA, Johnson MS, Tietge JE, Villeneuve DL (2013) Current perspectives on the use of alternative species in human health and ecological hazard assessments. Environ Health Perspect 121:1002–1010. doi:10.1289/ehp.1306638
Powlin SS, Cook JC, Novac S, O'Connor JC (1998) Ex vivo and in vitro testes and ovary explants: utility for identifying steroid biosynthesis inhibitors and comparison to a tier I screening battery. Toxicol Sci 46:61–74
Rainmondo S, Vivian DN, Delos C, Barron M (2008) Protectiveness of species sensitivity distribution hazard concentrations for acute toxicity used in endangered species risk assessment. Environ Toxicol Chem 27(12):2599–2607
Rand GM (1995) Fundamentals of aquatic toxicology: effects, environmental fate, and risk assessment, 2nd edn. CRC Press, Boca Raton, pp 50–53. isbn:1-56032-091-5
Rand-Weaver M, Margiotta-Casaluci L, Patel A, Panter GH, Owen SF, Sumpter J (2013) The read-across hypothesis and environmental risk assessment of pharmaceuticals. Environ Sci Technol 47:11384–11395
Rovida C, Hartung T (2009) Re-evaluation of animal numbers and costs for in vivo tests to accomplish REACH legislation requirements for chemicals – a report by the transatlantic think tank for toxicology. ALTEX 26:187–208
Sanderson H, Solomon K (2009) Contaminants of emerging concern challenge ecotoxicology. Environ Toxicol Chem 28:1359–1360
Schiller V, Wichmann A, Kriehuber R, Muth-Koehne E, Giesy JP, Hecker M, Fenske M (2013) Studying the effects of genistein on gene expression of fish embryos as an alternative testing approach for endocrine disruption. Comp Biochem Physiol C 157:41–53
Simmons SO, Fan CY, Ramabhadran R (2009) Cellular stress response pathway system as a sentinel ensemble in toxicological screening. Toxicol Sci 111:202–225. doi:10.1093/toxsci/kfp140
Strähle U, Scholz S, Geisler R, Greiner P, Hollert H, Rastegar S, Schumacher A, Selderslaghs I, Weiss C, Witters H, Braunbeck T (2012) Zebrafish embryos as an alternative to animal experiments—a commentary on the definition of the onset of protected life stages in animal welfare regulations. Reprod Toxicol 33:128–132
Suter GW (2004) Bottom-up and top-down integration of human and ecological risk assessment. J Toxicol Environ Health A 67(8–10):779–790
Suter GW, Efroymson RA, Sample BE, Jones DS (2000) Ecological risk assessment for contaminated sites. Lewis, Boca Raton
Tang S, Doering JA, Sun J, Beitel SC, Patterson S, Crawford S, Giesy JP, Wiseman SB, Hecker M (2016) Mechanistic investigations into the life-stage specific differences in sensitivity of white sturgeon (*Acipenser transmontanus*) to copper or cadmium. Environ Sci Technol 50:9717–9726
Thienpont B, Tingaud-Sequeira A, Prats E, Barata C, Babin PJ, Raldúa D (2011) Zebrafish eleutheroembryos provide a suitable vertebrate model for screening chemicals that impair thyroid hormone synthesis. Environ Sci Technol 45:7525–7532. doi:10.1021/es202248h
Tollefsen KE, Scholz S, Cronin MTC, Edwards SC, de Knecht J, Crofton K, Garcia-Reyero N, Hartung T, Worth A, Patlewiczi G (2014) Applying adverse outcome pathways (AOPs) to support integrated approaches to testing and assessment (IATA). Regul Toxicol Pharmacol 70:629–640
Tompsett AR, Wiseman S, Higley E, Giesy JP, Hecker M (2013) Effects of exposure to 17α-ethynylestradiol during sexual differentiation on the transcriptome of the African clawed frog (*Xenopus laevis*). Environ Sci Technol 47:4822–4828

Toyahama S, Miyagawa S, Lange A, Ogino Y, Mizutani T, Tatarazako N, Katsu Y, Ihara M, Tanaka H, Ishibashi H, Kobayashi T, Tyler CR, Iguchi T (2015) Understanding the molecular basis for differences in responses of fish estrogen receptor subtypes to environmental estrogens. Environ Sci Technol 49(12):7439–7447

Vaal M, van der Wal JT, Hoekstra J, Hermens J (1997) Variation in the sensitivity of aquatic species in relation to the classification of environmental pollutants. Chemosphere 35:1311–1327

Vallee M, Aiba K, Piao Y, Palin MF, Ko MSH, Sirard MA (2008) Comparative analysis of oocyte transcript profiles reveals a high degree of conservation among species. Reproduction 135:439–448. doi:10.1530/REP-07-0342

Van den Berg M et al (2006) The 2005 World Health Organization re-evaluation of human and mammalian toxic equivalency factors for dioxins and dioxin-like compounds. Toxicol Sci 93(2):223–241

Vardy DW, Tompsett AR, Sigurdson JL, Doering JA, Zhang XW, Giesy JP, Hecker M (2011) Effects of subchronic exposure of early life stages of white sturgeon (*Acipenser transmontanus*) to copper, cadmium and zinc. Environ Toxicol Chem 30:2497–2505

Vardy DW, Oellers J, Doering JA, Hollert H, Giesy JP, Hecker M (2013) Sensitivity of early life stages of white sturgeon, rainbow trout, and fathead minnow to copper. Environ Toxicol Chem 22:139–147

Villeneuve DL, Garcia-Reyero N (2011) Vision & strategy: predictive ecotoxicology in the 21st century. Environ Toxicol Chem 30(1):1–8. doi:10.1002/etc.396

Villeneuve DL, Crump D, Garcia-Reyero N, Hecker M, Hutchinson T, LaLone CA, Landesmann B, Lettieri T, Munn S, Nepelska M, Ottinger MA, Vergauwen L, Whelan M (2014a) Adverse outcome pathway (AOP) development I: strategies and principles. Toxicol Sci 142:312–320

Villeneuve DL, Crump D, Garcia-Reyero N, Hecker M, Hutchinson T, LaLone CA, Landesmann B, Lettieri T, Munn S, Nepelska M, Ottinger MA, Vergauwen L, Whelan M (2014b) Adverse outcome pathway (AOP) development II: best practices. Toxicol Sci 142:321–330

Vosges M, Le Page Y, Chung B-C, Combarnous Y, Porcher JM, Kah O, Brion F (2010) 17α-ethinylestradiol disrupts the ontogeny of the forebrain GnRH system and the expression of brain aromatase during early development of zebrafish. Aquat Toxicol 99:479–491. doi:10.1016/j.aquatox.2010.06.009

Wang W-X, Rainbow PS (2008) Comparative approaches to understand metal bioaccumulation in aquatic animals. Comp Biochem Physiol C 148:315–323

Wang N, Ingersoll CG, Hardesty DK, Ivey CD, Kunz JL, May TW, Dwyer FJ, Roberts AD, Augspurger T, Kane CM, Neves RJ, Barnhart MC (2007) Acute toxicity of copper, ammonia, and chlorine to glochidia and juveniles of freshwater mussels (Unionidae). Environ Toxicol Chem 26:2036–2047

Chapter 7
Green Algae and Networks for Adverse Outcome Pathways

Anze Zupanic, Smitha Pillai, Diana Coman Schmid, and Kristin Schirmer

Abstract If adverse outcome pathways (AOPs) are to become the new standard predictive tool for chemical risk assessment in ecotoxicology, substantial effort will be required to construct AOPs for exposures to different chemical groups making sure that we have enough representation of different test species to adequately cover the tree of life. This should include plants, which have not yet received sufficient attention from the AOP community. In this chapter, we present *Chlamydomonas reinhardtii*, a unicellular green microalga that serves as a model organism for, among others, photosynthesis and the circadian rhythm. We review *C. reinhardtii* as a model organism for ecotoxicology and summarize different publicly available genomic and OMICS resources for the species. We also present a new putative AOP for *C. reinhardtii* exposed to silver, constructed based on integration of transcriptomic and proteomic datasets. Finally, we present the current state-of-the-art bioinformatics procedures that can be used for constructing AOPs from OMICS type of datasets and evaluate whether the approaches are suitable for *C. reinhardtii*.

A. Zupanic (✉) • S. Pillai • D. Coman Schmid
Department of Environmental Toxicology, Eawag, Swiss Federal Institute for Aquatic Science and Technology, 8600 Dübendorf, Switzerland
e-mail: anze.zupanic@eawag.ch

K. Schirmer
Department of Environmental Toxicology, Eawag, Swiss Federal Institute for Aquatic Science and Technology, 8600 Dübendorf, Switzerland

EPF Lausanne, School of Architecture, Civil and Environmental Engineering, 1015 Lausanne, Switzerland

Department of Environmental Systems Science, ETH Zürich, 8092 Zürich, Switzerland

N. Garcia-Reyero, C.A. Murphy (eds.), *A Systems Biology Approach to Advancing Adverse Outcome Pathways for Risk Assessment*,
DOI 10.1007/978-3-319-66084-4_7

7.1 Introduction

The Adverse Outcome Pathway (AOP) is a knowledge-based toxicological framework that covers the adverse effects of chemicals across multiple levels of biological organization, from the molecular initiating event (MIE) to key events (KE) at the level a of cell or tissue and to the final adverse outcome (AO) of ecological relevance on the individual organism or population level (Ankley et al. 2010). The goal of AOP developers is that AOPs become a predictive tool for chemical risk assessment and thus become an important part in environmental regulatory decision-making (Groh et al. 2015; Villeneuve et al. 2014). The extent to which this can be achieved depends, among other aspects, on the validity of the assumptions behind the AOP concept. For one, AOPs are not chemical specific, but rather general enough to account for toxic actions of whole classes of chemicals. For example, while around 60% of all organic industrial chemicals are thought to be characterized as narcotics, their action in fish seem to fall under two different AOPs, one based on changes in cellular metabolism and one on more specific damage to the gill epithelium (Ankley et al. 2010; Perkins et al. 2015; Volz et al. 2011). Another assumption behind AOPs is that toxic responses are conserved among similar species and therefore these species share AOPs for at least some MIEs. In this way it should be possible to use AOPs for across species extrapolation, as was already shown for the neurotoxic effects of cyclotrimethylenetrinitramine (Garcia-Reyero et al. 2011). If these assumptions are valid, then a finite number of AOPs would be able to describe a vast majority of the toxic effects of environmental chemicals and the only limitation for the utility of the AOP concepts would be the amount of work researchers could dedicate to AOP construction and the diversity of the constructed AOPs.

Until the beginning of 2016, approximately 50 different AOP related projects have been registered with the OECD (Jan 19, 2016[1]). Most of the projects involve developing AOPs for humans, rodents and fish, while not a single one deals with adverse outcomes of chemicals on plants. For the AOP framework to become successful, it is vital that the AOP assortment of species includes a much wider diversity, ideally covering every branch of the tree of life. Among plant species, it would make the most sense to start with model plant species with the most available biological knowledge, such as the annual *Arabidopsis thaliana* or the unicellular green alga *Chlamydomonas reinhardtii*. Unicellular organisms, such as *C. reinhardtii*, could serve as especially attractive targets for AOP development because (1) they are easier to work with and more amenable to high throughput data collection, and (2) the lack of the higher multicellular organization would make the developed AOPs shorter and more tractable. It is not yet clear how useful AOPs of unicellular plants would be for understanding chemical toxicity to multicellular ones, but since many biological processes are shared in the plant kingdom it is reasonable to assume some level of across species extrapolation to be possible.

---

[1] http://www.oecd.org/env/ehs/testing/listsofprojectsontheaopdevelopmentprogrammeworkplan.htm.

In this chapter, we introduce *C. reinhardtii* as a model organism for ecotoxicology and describe available resources for toxicological and OMICS types of data. We present a hypothetical AOP for *C. reinhardtii* exposed to non-essential metals, which we developed based on a time course exposure of the alga to silver and the resulting combined transcriptomics/proteomics dataset. We conclude with a vision for future development of AOPs for unicellular plant organisms based on high-throughput OMICS.

7.2 The Alga *C. reinhardtii* as a Model Species

C. reinhardtii is a biflagellate unicellular green alga which is commonly found in soil and freshwater. The alga is a heterothallic species that reproduces sexually or asexually and can grow robustly under photoautotrophic, mixotrophic and heterotropic conditions. It is about 10 μm in diameter with a glycoprotein rich cell wall, a large single chloroplast, a nucleus, an eyespot which senses light, a pyrenoid which stores starch and two anterior flagella for motility (Fig. 7.1). *C. reinhardtii* can be cultured easily in the lab with a short generation time of 8–12 h and is amenable to genetic manipulation, with a vast array of functional mutants available. It has three genomes: the nuclear, the mitochondrial and the chloroplastic, which have been sequenced (Merchant et al. 2007). Many strains, both lab generated and isolated

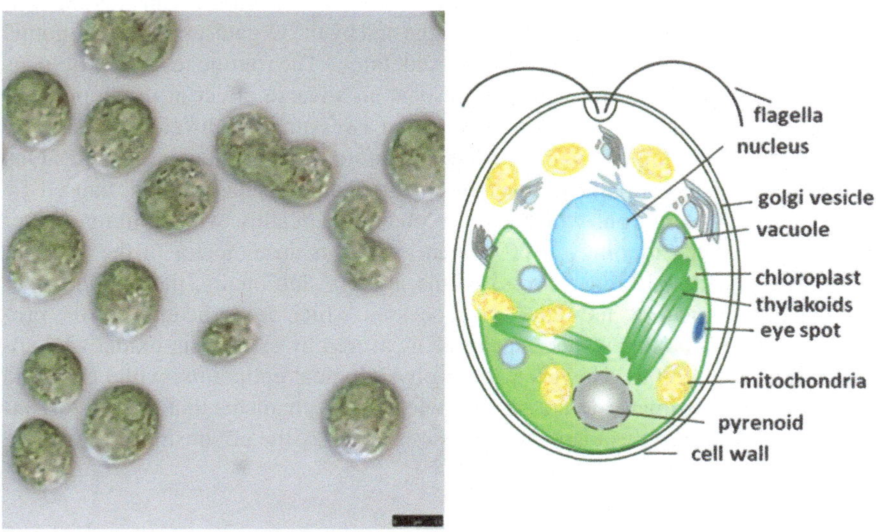

Fig. 7.1 *Chlamydomonas reinhardtii* and a schematic representation of a cell. Scale bar = 7.5 μm (Source: Department of Environmental Toxicology, Eawag)

from the environment, are available at the *Chlamydomonas* Resource Center.[2] Extensive work in the last decade has produced a fairly well annotated genome, with the most recent version at JGI v5.5,[3] including nearly 17,800 protein-coding loci (Blaby et al. 2014). *C. reinhardtii* is a representative of the lineage which evolved to early plants but it also shares features of animal cells, retained from the last eukaryotic common ancestor (Cross and Umen 2015), making it a greatly amenable model system. In the recent past *C. reinhardtii* has been used to study many fundamental cellular processes including chloroplast biogenesis, photosynthesis (Houille-Vernes et al. 2011), circadian rhythm (Matsuo and Ishiura 2008), flagellar assembly and motility (Pazour and Witman 2000), DNA methylation, metabolism, and sex determination (Harris 2001). Advances in genetic manipulation has further allowed *C. reinhardtii* to be explored for the development of sustainable algal biofuels and bioproducts (Scranton et al. 2015).

7.2.1 The Alga C. reinhardtii as a Model Species for Ecotoxicology

Aquatic systems are sinks for accumulating toxicants and primary producers, such as microalgae, comprise the base of the food chain from which effects can be propagated to higher trophic levels. Therefore, the estimation of bioaccumulation and toxicity to primary producers are important for accurate risk assessment and *C. reinhardtii* is an excellent model in this regard. Indeed, *C. reinhardtii* is routinely used in ecotoxicological risk assessment as one of the standard organisms for testing effects of toxicants in fresh water, as suggested by the Organisation for Economic Co-operation and Development (OECD) guidelines.[4] The routine tests mostly focus on the inhibition of growth, which represents an adverse effect at the population level and requires at least 24 h. A more detailed assessment of different physiological endpoints is used to study the mechanisms of toxicity (Nestler et al. 2012). One such physiological endpoint is the inhibition of photosynthetic yield, caused by herbicides such as diuron, which can be quantified readily within minutes after exposure. Another endpoint is ATP content, which is an indicator of the viability and physiological state (e.g., stress) of the algae. Additionally, the regulation or disturbance of oxidative and reductive processes, which are indicators for the production of reactive oxygen species, can be measured by estimating oxidative damage (Sarkar et al. 2005). However, these physiological endpoints, while easy and relatively fast to estimate, do not reveal molecular mechanisms that precede organism level changes nor the adaptive responses that allow the organism to recover.

---

[2] http://www.chlamycollection.org/.

[3] https://phytozome.jgi.doe.gov/pz/portal.html.

[4] http://www.oecd.org/chemicalsafety/testing/oecdguidelinesforthetestingofchemicals.htm.

The advancement of quantitative OMICS technologies, such as transcriptomics for abundance of RNA molecules, proteomics for abundance of proteins, and metabolomics for abundance of internal metabolites, allows for measurement of adaptive and toxic effects in a time and concentration dependent fashion at different molecular levels. These different levels are likely to cover a larger proportion of chemical-related KEs; therefore their integration with the traditionally measured physiological responses could, in principle, be used to propose relatively complete AOPs. To our knowledge, OMICS based AOPs have thus far not been proposed for *C. reinhardtii* nor for any other ecotoxicologically relevant plant species. In the next section, we will review the available OMICS datasets for *C. reinhardtii* and discuss different approaches that could lead to AOP development from OMICS data. Although the focus in the chapter is on *C. reinhardtii*, most of the proposed approaches are general enough to be used for other species.

7.2.2 OMICS Resources for *C. reinhardtii*

The most mature and broadly used OMICS is transcriptomics, which is based on either microarray or sequencing technology. In the last decade, the number of eco-toxicological studies utilizing transcriptomics has consistently risen (Schirmer et al. 2010) as has the number of publicly available transcriptomic datasets in public repositories. However, despite its amenability to transcriptomics, publicly available studies in *C. reinhardtii* are still relatively rare, with only 28 stress related transcriptomic datasets available for environmental and chemical related stress (a list can be found in Table 7.1). Among the available studies, those looking at the effects of insufficiency or excess of metabolic resources, such as nitrogen or sulphur, or essential metals, dominate, while those looking at chemical stress have only started appearing in the last couple of years. We expect that the number of transcriptomic studies will substantially increase in the future, enabling the use of integrative, network-based methods for the discovery of the mechanisms of toxicity and the definition of AOPs (Perkins et al. 2011) (see also below).

Other OMICS technologies have not yet achieved a similar technological maturity, and consequently there have been fewer studies in general. For *C. reinhardtii*, only mass spectroscopy based proteomics and metabolomics studies can be found in double figures, if all publications, not only toxicological ones, are taken into consideration (e.g., (Kleessen et al. 2015; Schmollinger et al. 2014)). Unfortunately, as public repositories for these technologies have only recently started appearing (e.g., the PRIDE repository for proteomics has only come online this year (Vizcaino et al. 2016)), normally the only way to obtain the respective datasets is by contacting the authors of the respective publications where they appeared. Hopefully, in the near future the number of all OMICS studies and their availability will increase for all species.

Table 7.1 Transcriptomic, stress-related datasets for *C. reinhardtii* published by March 1, 2016 in GEO (http://www.ncbi.nlm.nih.gov/geo/) and ArrayExpress (http://www.ebi.ac.uk/arrayexpress/)

Stressor	Repository ID	Platform	References
Cerium dioxide nanoparticles	E-MTAB-2454	Microarray	Taylor et al. (2016)
Heat shock	E-GEOD-20859	Microarray	Voss et al. (2011)
Heat shock, hemin and mg-protoporphyrin	E-GEOD-20861	Microarray	Voss et al. (2011)
Oxidative stress	E-GEOD-30646	Microarray	Fischer et al. (2012)
Electrophilic stress	E-GEOD-30646	Microarray	Fischer et al. (2012)
Silver, silver nanoparticles	E-GEOD-48677	Microarray	Pillai et al. (2014)
Light irradiation	E-GEOD-56800	Microarray	Mettler et al. (2014)
Rotifer predation	E-MEXP-3562	Microarray	Becks et al. (2012)
Sulphur stress	E-GEOD-33039	Microarray	Toepel et al. (2011)
Sulphur starvation	E-GEOD-33040	Microarray	Toepel et al. (2011)
Nitrogen starvation	E-GEOD-33041	Microarray	Toepel et al. (2011)
Sulphur starvation	E-SMDB-2992	Microarray	Zhang et al. (2004)
Sulphur starvation	E-GEOD-17970	RNA-seq	Gonzalez-Ballester et al. (2010)
Nitrogen starvation	E-GEOD-24365	RNA-seq	Miller et al. (2010)
Copper	E-GEOD-25124	RNA-seq	Castruita et al. (2011)
Oxidative stress	E-GEOD-33548	RNA-seq	Fischer et al. (2012)
CO_2	E-GEOD-33927	RNA-seq	Fang et al. (2012)
Nitrogen starvation	E-GEOD-34585	RNA-seq	Boyle et al. (2012)
Oxidative stress	E-GEOD-34826	RNA-seq	Urzica et al. (2012)
Iron starvation	E-GEOD-35305	RNA-seq	Boyle et al. (2012)
Zinc starvation	E-GEOD-41096	RNA-seq	Malasarn et al. (2013)
Iron starvation	E-GEOD-44611	RNA-seq	Urzica et al. (2013)
Nitrogen starvation	E-GEOD-51602	RNA-seq	Blaby et al. (2014)
Phosphate starvation	E-GEOD-56505	RNA-seq	No publication available
Sulfur starvation	E-MTAB-1329	RNA-seq	Toepel et al. (2011)
Phosphate starvation	E-MTAB-2556	RNA-seq	No publication available
UV-B stress	E-GEOD-68739	RNA-seq	No publication available
Zinc starvation	E-GEOD-58786	RNA-seq	No publication available

Lists with differentially expressed transcripts (based on the limma algorithm, (Ritchie et al. 2015)) for each individual dataset can be retrieved from http://www.eawag.ch/en/department/utox/projekte/integrative-network-toxicogenomics/

7.3 How to Use OMICS for AOPs?

7.3.1 From Gene Expression Through Pathways of Toxicity to AOPs

While there are many potential methods for including OMICS datasets into the construction of AOPs, a concrete path from OMICS to AOPs has not yet been presented for any species. Most often in ecotoxicological studies, individual

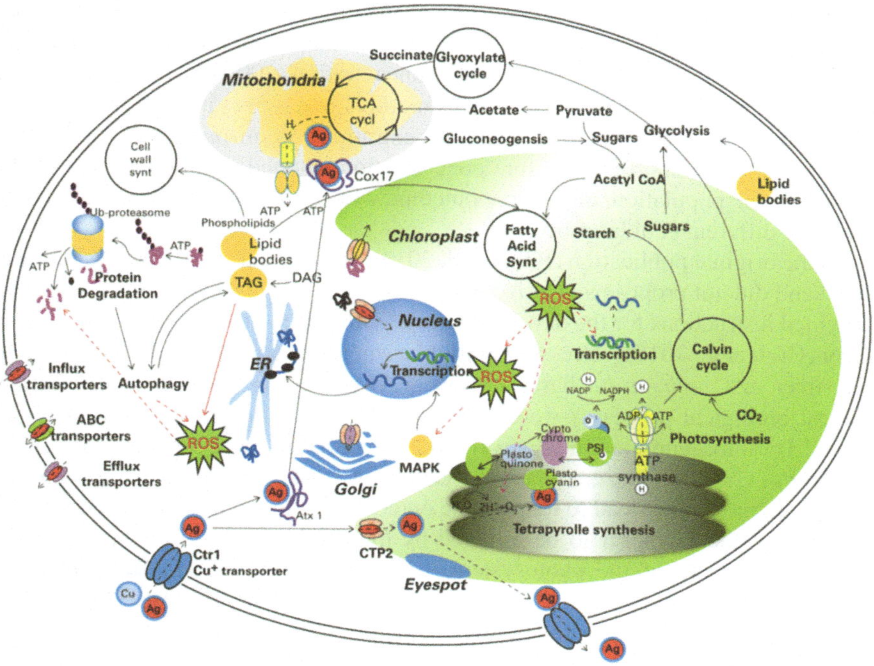

Fig. 7.2 Schematic representation of the effects of silver in *C. reinhardtii* (Pillai et al. 2014)

transcriptomics datasets are analyzed to identify differentially expressed transcripts in response to a stressor. This gene-centric approach is leveraged by pathway-centric analyses, in which a priori known molecular pathways are tested for overrepresentation of differentially expressed transcripts. The identified pathways of toxicity are then combined with the available knowledge of adverse outcomes for that particular stressor, and finally used to inform AOP development.

For example, in one of our studies, the responses of *C. reinhardtii* at the transcript, protein and physiological levels on exposure to silver was estimated in a concentration and time dependent manner (Pillai et al. 2014). We identified pathways of toxicity based on the perturbed expression of the transcripts and the proteins and linked them to the observable physiological outcome. This enabled us to put forward a conceptual mechanistic hypothesis for toxicity of silver on *C. reinhardtii* (Fig. 7.2). Silver is suggested to be transported into the cells by copper transporters. In the cells, silver is distributed via the copper chaperones and this concomitantly elicits several effects. Silver binds to thiol groups of proteins causing mis-folding and damage, it replaces copper in key proteins of the electron transport chain and photosystem and regulates the expression of proteins leading to inhibition of ATP and photo-synthesis. The disturbance of the electron transport chain in the photosystem leads to increased reactive oxygen species (ROS) production which causes peroxidation of lipids and membrane damage. As a defense mechanism against the oxidative stress, the algae mount an antioxidant response. At lower concentrations of silver (10 and 100 nM), the antioxidant response is seemingly adequate for the recovery of algae. At higher concentration (200 nM), in addition to the antioxidant response, the efflux mechanism probably removes intracellular silver as observed by decreasing intracellular concentrations, confirming a detoxification process. Nonetheless, this exposure concentration resulted in the inhibition of the growth of the population, an adverse outcome which integrates the many effects of silver on different cellular processes.

In the original publication (Pillai et al. 2014), we stopped at the toxicity pathway stage and did not propose a putative AOP. Since then, several excellent papers have provided instructions for the development of AOPs for ecotoxicological risk assessment (Groh et al. 2015). Here, we exploit the suggested strategy to develop an AOP for silver in *C. reinhardtii* based on our previous results. On exposure to silver, several biological processes were affected within durations spanning minutes to hours. The bioaccumulation of silver over the duration of two generations of *C. reinhardtii* (24 h) and the cumulative effect on key processes such as photosynthesis and energy utilisation lead to the adverse outcome of growth inhibition. The effects on photosynthesis appear to be initiated by the displacement of copper by silver from plastocyanin of the photosystem, which we define as an MIE (Fig. 7.3). Consequently, plastocyanin is misfolded (KE) and its function lost. This impairs photosystem II and the process of photosynthesis (KE) and leads to production of ROS (KE), which is followed by oxidative damage (KE). All of these KEs were observed not only at the physiological level, but also at the molecular level, in the form of dysregulation of transcripts and proteins. More importantly, the effects of silver on individual cells had an adverse outcome on the growth of the population. This preliminary AOP is mapped for defined acute exposure concentrations. It would be valuable to incorporate modelling approaches which would allow prediction of adverse effects for a range of environmentally relevant silver concentrations.

7.3.2 From Reverse Engineering Gene Co-expression Networks (GCNs) to AOPs

Another approach is to build AOPs not from a single OMICS dataset, but to leverage in house and publicly available OMICS datasets by using data-driven algorithms. A popular method for integration of OMICS is gene co-expression networks (GCNs), which was successfully used on single studies interrogating transcriptomes under different environmentally relevant conditions (Perkins et al. 2011; Williams et al. 2011). Construction of GCNs relies on the guilt by association assumption: genes with a correlated (linearly or non-linearly) expression patterns across several different conditions have a higher probability of participating in the same biological processes and of being under common transcriptional regulatory programs. Consistently, genes participating in the same pathways are usually found in densely correlated

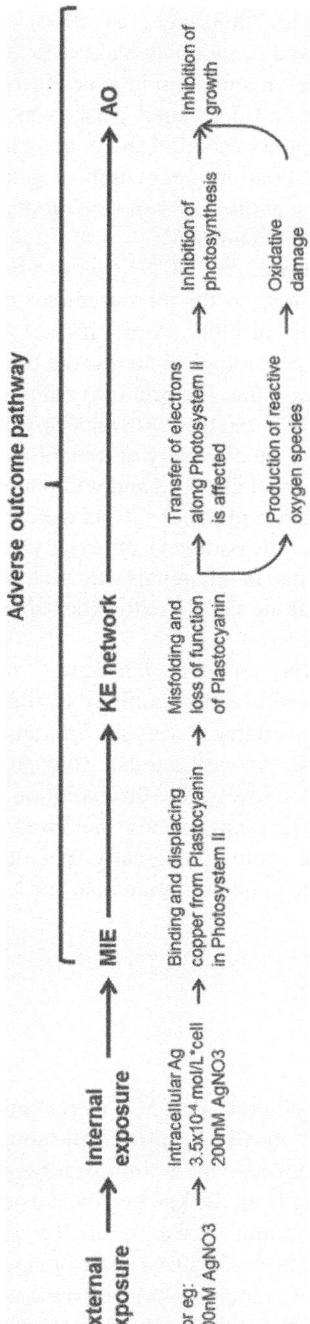

Fig. 7.3 A simple schematic representation of the AOP for silver in *C. reinhardtii* based on the comprehensive work presented in Pillai et al. (2014). *MIE* molecular initiating event, *KE* key event

regions of the GNC, i.e. in co-expressed modules (Hansen et al. 2014; Heyndrickx and Vandepoele 2012; Stuart et al. 2003; Wei et al. 2006).

GCNs can be used in a focused fashion when a specific toxicological pathway is of interest and prior knowledge on some guide genes involved in the pathway are available. In this approach, to find correlated genes which might have a similar function one can query the neighbourhood of the guide genes in the co-expression network. After experimental validation, annotation of genes with unknown functions is enabled and the subsequent discovery of new biomarkers that can be used in future assays covering key events in an AOP.

The algorithms used for constructing GCNs can also be applied for integration with adverse outcomes, just as long as the measurements of these are available for all the included studies/datasets. In other words, if there are several studies with measured transcriptomics and physiological states, the transcripts and the physiological states both feature in the same (correlation) network (Garcia-Reyero et al. 2014). The use of multiple OMICS together with data-driven GCN inference has the distinct advantage that it enables the discovery of toxicologically relevant genes and mechanisms, making use of all available data and without requiring comprehensive gene functional annotation. In a recent paper, GCNs were used in combination with text and database mining to infer the pathways of toxicity of MPTP (a chemical that elicits Alzheimer like symptoms in mammals) in mice (Maertens et al. 2015). Therefore it is possible to combine GCNs with other approaches as well in AOP development.

To date, the GCN approach has not yet been used in *C. reinhardtii* for ecotoxicological purposes. However, there are two publicly available databases that have compiled some of the publicly available *C. reinhardtii* transcriptome datasets (Aoki et al. 2016; Zheng et al. 2014). As well another study reports the use of GCNs inferred from multiple datasets to investigate the evolution of light-dependent gene regulatory modules across several plant species, including *C. reinhardtii* (Romero-Campero et al. 2013). As more ecotoxicologically specific datasets become available, the GCN approach will become an option also for *C. reinhardtii* based AOP construction.

7.4 Causal Networks

Another network approach that could lead to AOP development are causal networks. Although these have until now mostly been used in human toxicological studies where more data is available, they also hold promise for ecotoxicology. Causal networks are normally built starting from the known adverse outcome, such as a known human disease caused by an environmental exposure, for example lung cancer after exposure to tobacco smoke (Titz et al. 2016). After manually selecting scientific articles covering a known adverse outcome and with a molecular mechanistic focus, text mining is used to find sentences that feature causal connections between genes, proteins, metabolites, etc. (Fig. 7.4). The recognized connections are collected into

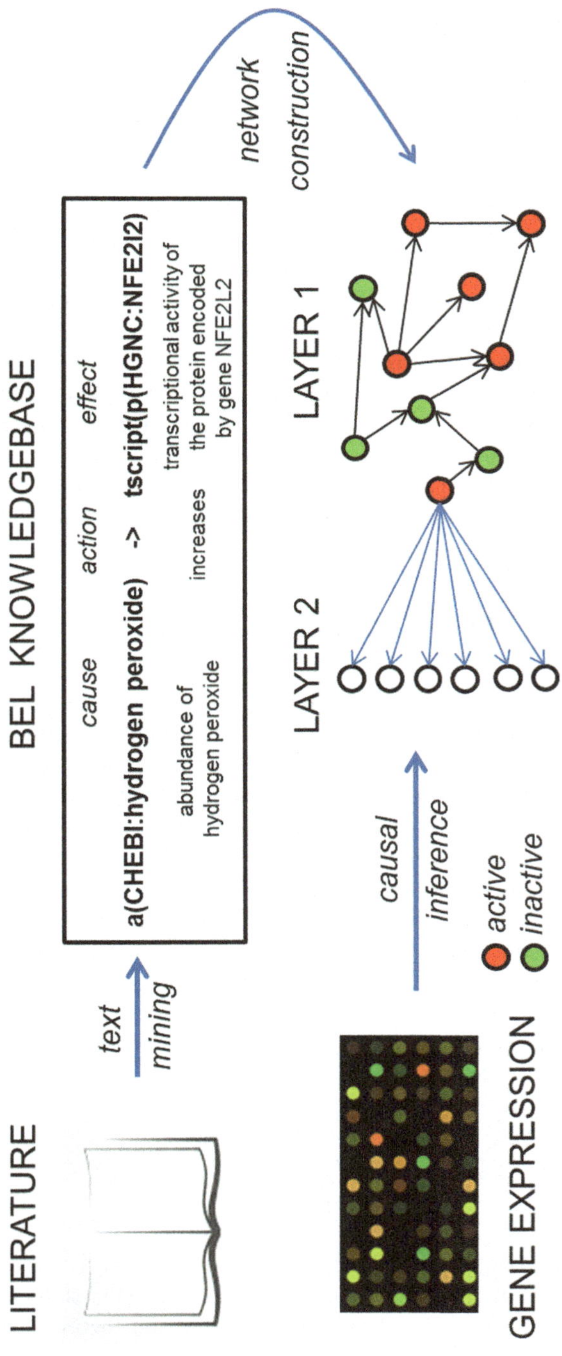

Fig. 7.4 Finding candidate MIEs and KEs by using causal modeling (Figure provided by Dr. Julia Hoeng and adapted by the authors)

a knowledge database written in BEL (biological expression language) and then joined together into a causal network that describes what is currently known about the molecular processes behind the adverse outcome. Every node in the causal network is then connected to several nodes of a second layer network that comprises all known downstream targets of that node. For example, if a causal node is a transcription factor, all its known targets would be in the second layer network. Finally, transcript abundance measurements of the second layer are used to infer the activity of the causal node, based on the transcript abundance of their downstream targets. Those causal nodes that are active under different exposures would therefore be prime targets for KE and KE assay development.

The quality of constructed causal networks depends greatly on the amount and quality of information and data available for a specific adverse outcome and species (Boue et al. 2015). Therefore it is probably not yet possible to construct them for ecotoxicological purposes, except for the most studied adverse outcomes in the most well studied species. For *C. reinhartdii*, the approach might work for inhibition of growth after chemical exposure. Other data mining based methods, such as frequent itemset mining (Oki and Edwards 2016) require even more information to be available and will be difficult to use for plant AOPs in the near future.

7.5 Towards Mechanistic Computable AOPs

Finally, although not yet attempted for AOP development, it would be possible to integrate OMICS datasets with mechanistic mathematical modelling. A potential approach of this is using genome-scale metabolic reconstructions and gene expression (either transcriptomic or proteomic) data to predict adverse metabolic phenotypes and metabolic toxicity biomarkers. Two successful examples are prediction of drug effects on growth of cancer cells (Folger et al. 2011) and prediction of growth of *C. reinhardtii* exposed to different nutrient conditions (Imam et al. 2015). An advantage of the metabolic modelling approach is the possibility of discovering KEs at the level of endogenous metabolism, which has so far been missing in the proposed AOPs. When the reconstruction of signalling or gene regulatory pathways on the genome scale also becomes possible we will be one step closer to whole cell models, which could be directly used as quantitative AOPs (Hyduke and Palsson 2010).

7.6 Conclusion

For AOPs to become one of the main tools for environmental risk assessment, their development will have to move from the current mammal and fish focus to the other branches of the tree of life. The AOP community should therefore try to involve researchers that are working on non-vertebrate ecological and ecotoxicological model organisms and beyond. OMICS technologies and meta-OMICS integrative analyses will undoubtedly play an important role in the effort to enrich the information about the chemical sensitivity of the model and non-model species and should be considered as the backbone of future AOP construction.

References

Ankley GT, Bennett RS, Erickson RJ, Hoff DJ, Hornung MW, Johnson RD, Mount DR, Nichols JW, Russom CL, Schmieder PK et al (2010) Adverse outcome pathways: a conceptual framework to support ecotoxicology research and risk assessment. Environ Toxicol Chem 29(3):730–741

Aoki Y, Okamura Y, Ohta H, Kinoshita K, Obayashi T (2016) ALCOdb: gene coexpression database for microalgae. Plant Cell Physiol 57(1):e3

Becks L, Ellner SP, Jones LE, Hairston NG Jr (2012) The functional genomics of an eco-evolutionary feedback loop: linking gene expression, trait evolution, and community dynamics. Ecol Lett 15(5):492–501

Blaby IK, Blaby-Haas CE, Tourasse N, Hom EF, Lopez D, Aksoy M, Grossman A, Umen J, Dutcher S, Porter M et al (2014) The Chlamydomonas genome project: a decade on. Trends Plant Sci 19(10):672–680

Boue S, Talikka M, Westra JW, Hayes W, Di Fabio A, Park J, Schlage WK, Sewer A, Fields B, Ansari S et al (2015) Causal biological network database: a comprehensive platform of causal biological network models focused on the pulmonary and vascular systems. Database (Oxford) 2015:bav030

Boyle NR, Page MD, Liu B, Blaby IK, Casero D, Kropat J, Cokus SJ, Hong-Hermesdorf A, Shaw J, Karpowicz SJ et al (2012) Three acyltransferases and nitrogen-responsive regulator are implicated in nitrogen starvation-induced triacylglycerol accumulation in Chlamydomonas. J Biol Chem 287(19):15811–15825

Castruita M, Casero D, Karpowicz SJ, Kropat J, Vieler A, Hsieh SI, Yan W, Cokus S, Loo JA, Benning C et al (2011) Systems biology approach in Chlamydomonas reveals connections between copper nutrition and multiple metabolic steps. Plant Cell 23(4):1273–1292

Cross FR, Umen JG (2015) The Chlamydomonas cell cycle. Plant J 82(3):370–392

Fang W, Si Y, Douglass S, Casero D, Merchant SS, Pellegrini M, Ladunga I, Liu P, Spalding MH (2012) Transcriptome-wide changes in Chlamydomonas Reinhardtii gene expression regulated by carbon dioxide and the CO2-concentrating mechanism regulator CIA5/CCM1. Plant Cell 24(5):1876–1893

Fischer BB, Ledford HK, Wakao S, Huang SG, Casero D, Pellegrini M, Merchant SS, Koller A, Eggen RI, Niyogi KK (2012) SINGLET OXYGEN RESISTANT 1 links reactive electrophile signaling to singlet oxygen acclimation in Chlamydomonas reinhardtii. Proc Natl Acad Sci U S A 109(20):E1302–E1311

Folger O, Jerby L, Frezza C, Gottlieb E, Ruppin E, Shlomi T (2011) Predicting selective drug targets in cancer through metabolic networks. Mol Syst Biol 7:501

Garcia-Reyero N, Habib T, Pirooznia M, Gust KA, Gong P, Warner C, Wilbanks M, Perkins E (2011) Conserved toxic responses across divergent phylogenetic lineages: a meta-analysis of the neurotoxic effects of RDX among multiple species using toxicogenomics. Ecotoxicology 20(3):580–594

Garcia-Reyero N, Ekman DR, Habib BT, Villeneuve DL, Collette TW, Bencic DC, Ankley GT, Perkins EJ (2014) Integrated approach to explore the mechanisms of aromatase inhibition and recovery in fathead minnows (Pimephales promelas). Gen Comp Endocrinol 203:193–202

Gonzalez-Ballester D, Casero D, Cokus S, Pellegrini M, Merchant SS, Grossman AR (2010) RNA-seq analysis of sulfur-deprived Chlamydomonas cells reveals aspects of acclimation critical for cell survival. Plant Cell 22(6):2058–2084

Groh KJ, Carvalho RN, Chipman JK, Denslow ND, Halder M, Murphy CA, Roelofs D, Rolaki A, Schirmer K, Watanabe KH (2015) Development and application of the adverse outcome pathway framework for understanding and predicting chronic toxicity: I challenges and research needs in ecotoxicology. Chemosphere 120:764–777

Hansen BO, Vaid N, Musialak-Lange M, Janowski M, Mutwil M (2014) Elucidating gene function and function evolution through comparison of co-expression networks of plants. Front Plant Sci 5:394

Harris EH (2001) Chlamydomonas as a model organism. Annu Rev Plant Physiol Plant Mol Biol 52:363–406

Heyndrickx KS, Vandepoele K (2012) Systematic identification of functional plant modules through the integration of complementary data sources. Plant Physiol 159(3):884–901

Houille-Vernes L, Rappaport F, Wollman FA, Alric J, Johnson X (2011) Plastid terminal oxidase 2 (PTOX2) is the major oxidase involved in chlororespiration in Chlamydomonas. Proc Natl Acad Sci U S A 108(51):20820–20825

Hyduke DR, Palsson BO (2010) Towards genome-scale signalling network reconstructions. Nat Rev Genet 11(4):297–307

Imam S, Schauble S, Valenzuela J, Lopez Garcia de Lomana A, Carter W, Price ND, Baliga NS (2015) A refined genome-scale reconstruction of Chlamydomonas metabolism provides a platform for systems-level analyses. Plant J 84(6):1239–1256

Kleessen S, Irgang S, Klie S, Giavalisco P, Nikoloski Z (2015) Integration of transcriptomics and metabolomics data specifies the metabolic response of Chlamydomonas to rapamycin treatment. Plant J 81(5):822–835

Maertens A, Luechtefeld T, Kleensang A, Hartung T (2015) MPTP's pathway of toxicity indicates central role of transcription factor SP1. Arch Toxicol 89(5):743–755

Malasarn D, Kropat J, Hsieh SI, Finazzi G, Casero D, Loo JA, Pellegrini M, Wollman FA, Merchant SS (2013) Zinc deficiency impacts CO2 assimilation and disrupts copper homeostasis in Chlamydomonas reinhardtii. J Biol Chem 288(15):10672–10683

Matsuo T, Ishiura M (2008) Chlamydomonas reinhardtii, a new model system for studying the circadian clock. Tanpakushitsu Kakusan Koso 53(14):1873–1880

Merchant SS, Prochnik SE, Vallon O, Harris EH, Karpowicz SJ, Witman GB, Terry A, Salamov A, Fritz-Laylin LK, Marechal-Drouard L et al (2007) The Chlamydomonas genome reveals the evolution of key animal and plant functions. Science 318(5848):245–250

Mettler T, Muhlhaus T, Hemme D, Schottler MA, Rupprecht J, Idoine A, Veyel D, Pal SK, Yaneva-Roder L, Winck FV et al (2014) Systems analysis of the response of photosynthesis, metabolism, and growth to an increase in irradiance in the photosynthetic model organism Chlamydomonas reinhardtii. Plant Cell 26(6):2310–2350

Miller R, Wu G, Deshpande RR, Vieler A, Gartner K, Li X, Moellering ER, Zauner S, Cornish AJ, Liu B et al (2010) Changes in transcript abundance in Chlamydomonas reinhardtii following nitrogen deprivation predict diversion of metabolism. Plant Physiol 154(4):1737–1752

Nestler H, Groh KJ, Schonenberger R, Behra R, Schirmer K, Eggen RI, Suter MJ (2012) Multiple-endpoint assay provides a detailed mechanistic view of responses to herbicide exposure in Chlamydomonas Reinhardtii. Aquat Toxicol 110-111:214–224

Oki NO, Edwards SW (2016) An integrative data mining approach to identifying adverse outcome pathway signatures. Toxicology 350:49–61

Pazour GJ, Witman GB (2000) Forward and reverse genetic analysis of microtubule motors in Chlamydomonas. Methods 22(4):285–298

Perkins EJ, Chipman JK, Edwards S, Habib T, Falciani F, Taylor R, Van Aggelen G, Vulpe C, Antczak P, Loguinov A (2011) Reverse engineering adverse outcome pathways. Environ Toxicol Chem 30(1):22–38

Perkins EJ, Antczak P, Burgoon L, Falciani F, Garcia-Reyero N, Gutsell S, Hodges G, Kienzler A, Knapen D, McBride M et al (2015) Adverse outcome pathways for regulatory applications: examination of four case studies with different degrees of completeness and scientific confidence. Toxicol Sci 148(1):14–25

Pillai S, Behra R, Nestler H, Suter MJ, Sigg L, Schirmer K (2014) Linking toxicity and adaptive responses across the transcriptome, proteome, and phenotype of Chlamydomonas reinhardtii exposed to silver. Proc Natl Acad Sci U S A 111(9):3490–3495

Ritchie ME, Phipson B, Wu D, Hu Y, Law CW, Shi W, Smyth GK (2015) Limma powers differential expression analyses for RNA-sequencing and microarray studies. Nucleic Acids Res 43(7):e47

Romero-Campero FJ, Lucas-Reina E, Said FE, Romero JM, Valverde F (2013) A contribution to the study of plant development evolution based on gene co-expression networks. Front Plant Sci 4:291

Sarkar N, Lemaire S, Wu-Scharf D, Issakidis-Bourguet E, Cerutti H (2005) Functional specialization of Chlamydomonas reinhardtii cytosolic thioredoxin h1 in the response to alkylation-induced DNA damage. Eukaryot Cell 4(2):262–273

Schirmer K, Fischer BB, Madureira DJ, Pillai S (2010) Transcriptomics in ecotoxicology. Anal Bioanal Chem 397(3):917–923

Schmollinger S, Muhlhaus T, Boyle NR, Blaby IK, Casero D, Mettler T, Moseley JL, Kropat J, Sommer F, Strenkert D et al (2014) Nitrogen-sparing mechanisms in Chlamydomonas affect the transcriptome, the proteome, and photosynthetic metabolism. Plant Cell 26(4):1410–1435

Scranton MA, Ostrand JT, Fields FJ, Mayfield SP (2015) Chlamydomonas as a model for biofuels and bio-products production. Plant J 82(3):523–531

Stuart JM, Segal E, Koller D, Kim SK (2003) A gene-coexpression network for global discovery of conserved genetic modules. Science 302(5643):249–255

Taylor NS, Merrifield R, Williams TD, Chipman JK, Lead JR, Viant MR (2016) Molecular toxicity of cerium oxide nanoparticles to the freshwater alga Chlamydomonas reinhardtii is associated with supra-environmental exposure concentrations. Nanotoxicology 10(1):32–41

Titz B, Boue S, Phillips B, Talikka M, Vihervaara T, Schneider T, Nury C, Elamin A, Guedj E, Peck MJ et al (2016) Effects of cigarette smoke, cessation, and switching to two heat-not-burn tobacco products on lung lipid metabolism in C57BL/6 and Apoe−/− mice-an integrative systems toxicology analysis. Toxicol Sci 149(2):441–457

Toepel J, Albaum SP, Arvidsson S, Goesmann A, la Russa M, Rogge K, Kruse O (2011) Construction and evaluation of a whole genome microarray of Chlamydomonas reinhardtii. BMC Genomics 12:579

Urzica EI, Casero D, Yamasaki H, Hsieh SI, Adler LN, Karpowicz SJ, Blaby-Haas CE, Clarke SG, Loo JA, Pellegrini M et al (2012) Systems and trans-system level analysis identifies conserved iron deficiency responses in the plant lineage. Plant Cell 24(10):3921–3948

Urzica EI, Vieler A, Hong-Hermesdorf A, Page MD, Casero D, Gallaher SD, Kropat J, Pellegrini M, Benning C, Merchant SS (2013) Remodeling of membrane lipids in iron-starved Chlamydomonas. J Biol Chem 288(42):30246–30258

Villeneuve DL, Crump D, Garcia-Reyero N, Hecker M, Hutchinson TH, LaLone CA, Landesmann B, Lettieri T, Munn S, Nepelska M et al (2014) Adverse outcome pathway (AOP) development I: strategies and principles. Toxicol Sci 142(2):312–320

Vizcaino JA, Csordas A, Del-Toro N, Dianes JA, Griss J, Lavidas I, Mayer G, Perez-Riverol Y, Reisinger F, Ternent T et al (2016) 2016 update of the PRIDE database and its related tools. Nucleic Acids Res 44(D1):D447–D456

Volz DC, Belanger S, Embry M, Padilla S, Sanderson H, Schirmer K, Scholz S, Villeneuve D (2011) Adverse outcome pathways during early fish development: a conceptual framework for identification of chemical screening and prioritization strategies. Toxicol Sci 123(2):349–358

Voss B, Meinecke L, Kurz T, Al-Babili S, Beck CF, Hess WR (2011) Hemin and magnesium-protoporphyrin IX induce global changes in gene expression in Chlamydomonas reinhardtii. Plant Physiol 155(2):892–905

Wei H, Persson S, Mehta T, Srinivasasainagendra V, Chen L, Page GP, Somerville C, Loraine A (2006) Transcriptional coordination of the metabolic network in Arabidopsis. Plant Physiol 142(2):762–774

Williams TD, Turan N, Diab AM, Wu H, Mackenzie C, Bartie KL, Hrydziuszko O, Lyons BP, Stentiford GD, Herbert JM et al (2011) Towards a system level understanding of non-model organisms sampled from the environment: a network biology approach. PLoS Comput Biol 7(8):e1002126

Zhang Z, Shrager J, Jain M, Chang CW, Vallon O, Grossman AR (2004) Insights into the survival of Chlamydomonas reinhardtii during sulfur starvation based on microarray analysis of gene expression. Eukaryot Cell 3(5):1331–1348

Zheng HQ, Chiang-Hsieh YF, Chien CH, Hsu BK, Liu TL, Chen CN, Chang WC (2014) AlgaePath: comprehensive analysis of metabolic pathways using transcript abundance data from next-generation sequencing in green algae. BMC Genomics 15:196

Chapter 8
Neurobehavioral Analysis Methods for Adverse Outcome Pathway (AOP) Models and Risk Assessment

Francisco X. Mora-Zamorano, Jeremy K. Larson, and Michael J. Carvan III

Abstract The emerging use of neurobehavioral analysis techniques in toxicology promotes the implementation of neurobehavior, a powerful integrator of molecular, physiological, and environmental stimuli, in the development of Adverse Outcome Pathway (AOP) models. In recent years, zebrafish have been extensively investigated for their potential as a model organism in behavioral toxicology due to their low maintenance cost and similarities with rodent behavior and physiology. This chapter will review: (1) the beneficial role of neurobehavioral assays in the development of AOPs; (2) the diverse neurobehavioral endpoints to be considered in the evaluation of neurotoxicity and; (3) the challenges of integrating neurobehavioral outcomes into AOP development. Discussion of the many neurobehavioral screening assays that have been adapted from rodents to zebrafish is included. Furthermore, this chapter will review studies in which behavioral phenotypes and neurophysiological outcomes have been anchored to specific molecular initiating events induced by a chemical exposure. Although the study of the genetic and physiological basis of behavior is still nascent, there are many noteworthy studies that have enabled the creation of AOP models for the prediction of how chemical exposure affects the behavior of individuals in a population and, in turn, how these alterations can affect population dynamics.

8.1 Introduction

Neurobehavior is the study of an organism's behavior and how it relates to the function of its nervous system. This powerful experimental endpoint serves as an integrator of the diverse and complex internal (e.g., chemical, molecular, cellular, physiological) and external (e.g., environmental) stimuli encountered by an

F.X. Mora-Zamorano • J.K. Larson • M.J. Carvan III (✉)
School of Freshwater Sciences, University of Wisconsin - Milwaukee, Milwaukee, WI, USA
e-mail: carvanmj@uwm.edu

N. Garcia-Reyero, C.A. Murphy (eds.), *A Systems Biology Approach to Advancing Adverse Outcome Pathways for Risk Assessment*,
DOI 10.1007/978-3-319-66084-4_8

organism. Therefore, the knowledge gained from neurobehavioral studies contribute significantly to the understanding of gene-environment interactions, physiology, and ecology. Behavior is the output of molecular and cellular level events (i.e. gene expression, neurotransmitter signaling, and neurodevelopment) that enable an organism to respond to its surroundings (e.g., locomotion, foraging, escape predators, seek shelter, grow and reproduce). Importantly, such complex responses can be compromised via environmental insult, such as exposure to a neurotoxicant. Neurotoxicity is defined by the United States Environmental Protection Agency (US EPA) as "an adverse change in the structure and/or function of the central and/or peripheral nervous system measured at the neurochemical, behavioral, neurophysiological or anatomical levels" (Tilson et al. 1995). It is estimated that nearly 30% of all commercially used chemicals (currently >80,000 chemicals) possess neurotoxic properties (Basu 2015). However, over the past four decades, few of these chemicals have been extensively studied and characterized as neurotoxicants (Bal-Price et al. 2015).

Neurobehavioral screening assays (NBSAs) provide an excellent platform to identify the effects of a large number of potentially neurotoxic compounds. Some NBSAs are particularly useful in toxicology and pharmacology due to their robustness and the possibility of being automated and implemented in a high throughput manner (Reif et al. 2015). NBSAs can also be a quite sensitive approach to identify neurotoxic effects that would be otherwise too subtle to be elucidated by anatomical or histological screens (Detrich III et al. 2009). However, the organism-level data obtained from NBSAs is much more powerful for risk assessment purposes when coupled with the knowledge of the mechanisms that mediate neurobehavior, as well as the possible broader implications of behavior alterations. In such a scenario, the data gathered by NBSAs allows the possibility of making meaningful predictions of how chemical-induced behavioral alterations observed in individual organisms can affect higher levels of biological organization (i.e., communities and populations). Hence, the continued development and adoptions of a systems biological approach (i.e., Adverse Outcome Pathway [AOP]) will significantly improve our ability to make such predictions from neurobehavioral-derived data obtained in the laboratory. The AOP is a methodological framework that utilizes the knowledge of the biological effects of molecular, cellular and organism-level events to predict the potential adverse outcomes at higher levels of biological organization (Fig. 8.1) (Landesmann et al. 2013; Ankley et al. 2010).This approach de-emphasizes the apical adverse outcomes at the organismal level and higher, and focuses more on the effects on initiating and intermediate measurable key events of biological organization, which can be mechanistically linked to apical adverse effects of broad ecological impact. This chapter will review: (1) the beneficial role of neurobehavioral assays in the development of AOPs; (2) the diverse neurobehavioral endpoints to be considered in the evaluation of neurotoxicity and; (3) the challenges of integrating neurobehavioral outcomes into AOP development.

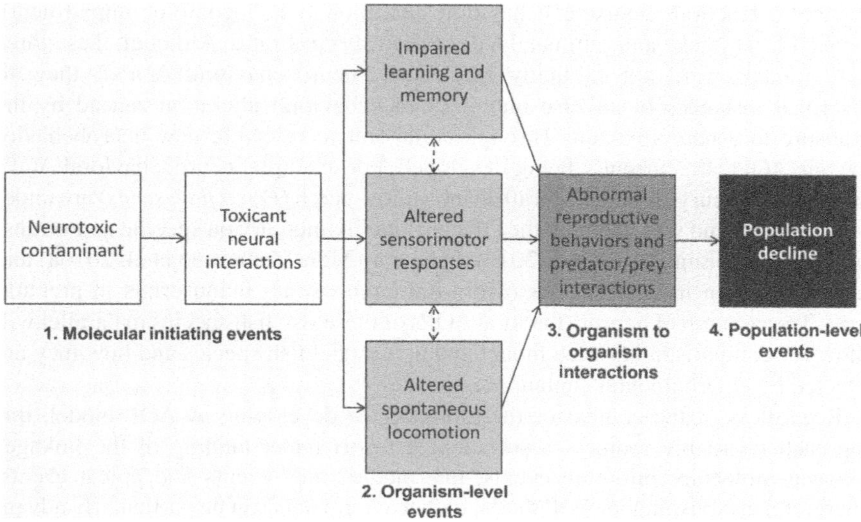

Fig. 8.1 A conceptual neurobehavioral AOP diagram. Exposure to a neurotoxicant can initiate molecular and/or cellular consequences that may impact different levels of behavioral complexity and lead to an altered interaction between organisms (i.e., interactions required for reproduction and survival), which could ultimately result in population decline due to a failure to reproduce, vulnerability to predators and/or starvation

8.2 Neurobehavioral AOPs

For the purpose of clarity, this chapter will define neurobehavioral AOPs as any AOP model that makes use of NBSA-derived data to make predictions of adverse outcomes that may branch from behavioral alteration. As discussed extensively by Bal-Price and collaborators (Bal-Price et al. 2015), the development of AOP models specific to neurotoxicity outcomes is nascent. To illustrate this point, individual-based models (also referred to as "agent-based" models) have existed for roughly four decades and the development of these models specifically for the study of population dynamics has been in progress for two decades (DeAngelis and Grimm 2014). However, neurobehavior has only recently been considered and implemented in such models to predict population-wide outcomes of neurotoxic exposure. One example of such implementation is the use of behavioral data from methylmercury exposed Atlantic croaker to predict impaired survival skills and, consequently, population decline (Alvarez Mdel et al. 2006).

A recent review (Groh et al. 2015) discussed three case studies of AOP models of growth impairment. The molecular initiating events in these three models were associated with exposure to pyrethroids, cadmium and selective serotonin reuptake inhibitors, all of which have known neurological effects (Eriksson and Fredriksson 1991; Isbister et al. 2004; Wang and Du 2013). In all cases, the models indicated that the effects of chemical exposure would initiate a cascade of physiological and metabolic reactions, causing behavioral alteration (i.e., locomotor impairment), reduced food intake and, ultimately, decreased survival rates. Although these three AOP models were not explicitly labeled as "neurobehavioral AOPs", they all described instances of adverse outcomes of behavioral alteration caused by the exposure to a neurotoxicant. Through collaborative research, new neurobehavior specific AOPs are currently being developed. For example, neurobehavioral AOPs for methylmercury-exposed zebrafish and yellow perch (*Perca flavescens*) are under development, and will integrate the effects of methylmercury on specific neurotransmitters in the brain (Arini et al. 2016), behavior (Mora-Zamorano et al. 2016a) and gene expression in both species (Mora-Zamorano et al.; manuscripts in preparation). The creation of neurobehavioral AOPs of these two fish species in parallel will allow for comparison between model and non-model fish species and how they are affected by environmental contaminants.

Regardless of the context or the rationale for developing an AOP model, this approach invariably requires a sufficient *a priori* understanding of the linkages between molecular initiating events, intermediate key events and apical events. Therefore, establishing an AOP model on behavioral data can be particularly advantageous, given that behavior inherently encompasses the molecular, physiological and anatomical implications of neurotoxicant exposure. A neurobehavioral AOP model consists of three essential levels of biological organization: (1) molecular-level (i.e., toxicant-induced alteration in cell physiology/biochemistry), (2) organism-level (i.e., behavioral alteration), and (3) population-level (i.e., apical outcomes). It has been previously discussed that behavioral responses can be classified into three hierarchical tiers: (1) basic motor responses (2) sensorimotor responses, and (3) learning and memory (Tierney 2011). In this chapter, a "fourth tier" will be added to this proposed hierarchy: (4) organism-organism (i.e., predator-prey) interactions. The advantage of organizing behaviors by their complexity is that it facilitates the implementation of behavioral data into AOP models. For example, spontaneous locomotor activity is a very fundamental behavior, while courtship, prey capture and predator avoidance are much more complex (Scott and Sloman 2004). Ideally, a neurobehavioral AOP would link different levels of behavioral complexity in a manner whereby fundamental behaviors predict the outcome of more complex ones. Although neurobehavior is a practical and integrative endpoint of neurotoxicity, it is also important to acknowledge its potential pitfalls. The most obvious weakness is the fact that neurobehavior alone does not provide insight into the vast and complex mechanisms that modulate it. More often than not, molecular mechanisms that are very distinct from each other will render behavioral outcomes that are practically indistinguishable (Groh et al. 2015).

8.3 Model Organisms in Neurobehavioral Studies

Model organisms belonging to a wide variety of taxa have been employed for neurobehavioral analysis for decades. Invertebrates such as the nematode *C. elegans* and the fruit fly (*D. Melanogaster*) are two examples of model organisms that have been extensively utilized for such purposes (Benzer 1967; Cronin et al. 2006). Among vertebrates, rodents have been historically the traditional animal models for neurotoxicity screening (Eddins et al. 2010). As mammalian models, it is generally accepted that neurotoxic effects observed in rodents are frequently predictive of similar effects in humans (Bal-Price et al. 2015). Nonetheless, exclusively utilizing rodents to assess the neurotoxicity of the thousands of commercially available chemicals is proving to be prohibitively costly and time consuming (Perkins et al. 2013). Another widely utilized vertebrate model is the zebrafish (*Danio rerio*), which has recently emerged as an increasingly popular model in toxicology (Miklósi and Andrew 2006). Despite being a non-mammalian model, orthologs for 70% of all human genes have been identified in this model organism (Howe et al. 2013), thus making this species well-suited to assess potential human and environmental hazards. This chapter will focus primarily on zebrafish neurobehavior assays, and assays developed originally developed for rodents that have now been adapted for zebrafish research.

8.4 Neurobehavioral Endpoints and General Considerations

Neurotoxicants are capable of altering behavior via diverse mechanisms (e.g., affect perception by altering the senses, alter gait and locomotion, modulate emotional states and impair cognition; Fig. 8.2). To date, several detailed reviews have been published summarizing behavioral assays that were designed to target a specific mode of alteration (e.g., visual deficit, locomotor activity, startle responses, anxiety and learning). The examples described herein are not all inclusive, but rather a consolidation of some of the most widely utilized assays in neurobehavioral toxicology studies. Literature featuring the diverse types of neurobehavioral assays and their documented application to ecotoxicology research are summarized in Table 8.1.

Prior to performing behavioral assays there are a number of important facets that must be considered. Firstly, although developmental toxicity and teratogenesis are not neurobehavioral endpoints, it is critical to recognize that overt morphological abnormalities caused by chemical exposure can drastically affect behavioral output. Therefore, it is highly advisable to perform a developmental toxicity screen prior to any neurobehavioral assessment. This action will ensure that the origin of any observed behavioral abnormalities stems from altered nervous system function rather than independent morphological defects. The most acknowledged approach to carry out a toxicity screen consists of observing a cohort of organisms and identifying the proportion of individuals in the cohort that present one or more morphological abnormalities. For the sake of consistency, a scoring rubric can be implemented to assign a value to the severity of morphological abnormalities, such as the early life stage toxicity (ELS-tox) score (Heiden et al. 2005). In zebrafish, scoring rubrics for developmental toxicity include observations of embryo mortality, blood circulation, somite formation, pigmentation, body morphology, swim

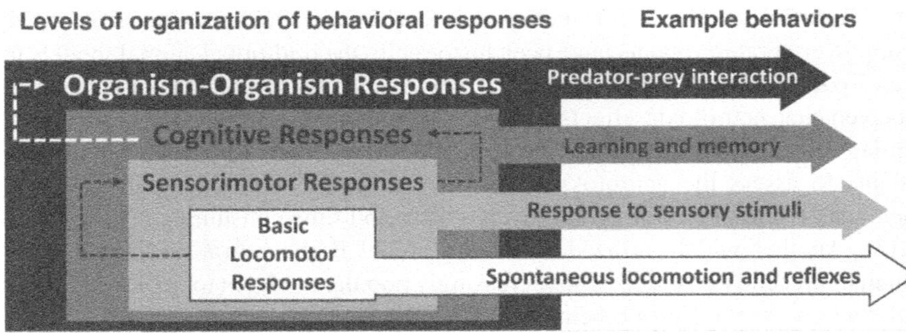

Fig. 8.2 Behavioral outputs can originate from different levels of behavioral complexity, and the interactions between these levels. Spontaneous locomotor output and reflexes are fundamental behavioral outputs that require relatively "primitive" anatomical structures to be evoked (e.g., muscles, spinal cord and hindbrain). Sensorimotor responses and the processes of learning and memory are more complex phenomena. However, all the aforementioned processes are required for an organism to interact with other organisms and with the environment

bladder inflation and the presence of yolk sac edemas (Truong et al. 2011). It is also important to consider that circadian rhythms can affect the way organisms behave depending on the time of day (MacPhail et al. 2009), thus the duration of light/dark cycles and/or disruption of normal illumination may impact experimental outcomes. The developmental stage at which organisms are exposed to a neurotoxicant can also alter the outcome (Weis and Weis 1995b). It is also not advisable to feed organisms immediately prior to behavioral analysis, as this can affect the behavior of zebrafish (Clift et al. 2014; Hurd et al. 1998).

8.4.1 Neuromotor Development

Locomotor activity is by far the most prominently documented and well understood behavioral endpoint. It is also the most fundamental behavioral output that can be plausibly linked to more complex behaviors such as foraging (Groh et al. 2015). Neuromotor and reflex development are critical milestones during the early ontogeny of an organism. Likewise, these first manifestations of locomotor output are arguably the earliest behavioral endpoints that can be observed and measured in a developing organism.

8.4.1.1 Spontaneous Activity

Spontaneous tail flicks or contractions can be quantified in zebrafish embryos as early as 30-hours post-fertilization (hpf) (Saint-Amant and Drapeau 1998). This endpoint can be analyzed by empirical observation or video recording the embryos

Table 8.1 Selected references describing the methodologies for different behavioral assays, as well as the applications of these approaches in ecotoxicology

Behavioral endpoint	Selected references	Examples of neurotoxic compounds assessed
Spontaneous activity	Budick and O'Malley (2000)	Chlorpyrifos; (Selderslaghs et al. 2010)
Photomotor response (PMR)	Kokel et al. (2011)	Trimethyltin chloride; (Chen et al. 2011)
Visual motor response (VMR)	Emran et al. (2008)	Perfluorooctane sulfonate; (Spulber et al. 2014)
Visual startle, avoidance and escape response	Orger et al. (2009), Neuhauss (2003)	Methylmercury; (Weber et al. 2008)
Optomotor response (OMR)	Orger et al. (2009), Neuhauss (2003)	PCB 1254; (Zhang et al. 2015)
Optokinetic response (OKR)	Orger et al. (2009), Neuhauss (2003)	[a]N/A
Touch response	Budick and O'Malley (2000)	Domoic acid; (Tiedeken et al. 2005)
Olfactory-evoked locomotion	Lindsay and Vogt (2004)	[a]N/A
Olfactory conditioning	Braubach et al. (2009)	[a]N/A
Open field, diving test and scototaxis	Maximino et al. (2011)	[b]Benzodiazepines; (Maximino et al. 2011)
Thigmotaxis	Schnörr et al. (2012)	PCBs; (Gonzalez et al. 2016)
Habituation	Best et al. (2008)	Chlorpyrifos; (Eddins et al. 2010)
Plus maze	Sison and Gerlai (2010)	[a]N/A
Spatial alternation test	Williams et al. (2002)	Methylmercury; (Smith et al. 2010)
Predator avoidance	Luca and Gerlai (2012)	Methylmercury; (Alvarez Mdel et al. 2006)
Prey capture	Budick and O'Malley (2000)	Methylmercury; (Mora-Zamorano et al. 2016a)

[a]N/A indicates there are no known applications of the methodology in the field of ecotoxicology thus far

[b]Although benzodiazepines are more amenable to be considered a chemical of pharmacological relevance, the study cited represents a proof-of-concept with the potential of being utilized with environmental contaminants

and manually counting the number of tail flicks thereafter. However, automated analysis of tail flicks can also be achieved with a commercial system such as the DanioScope suite, offered by Noldus Information Technology (Wageningen, Netherlands). Alternatively, activity can be analyzed with a custom ImageJ (Schneider et al. 2012a) macro designed for such purpose, as suggested by Kokel and collaborators (2010). Tail flicks in early embryonic stages of the zebrafish can be induced via exposure to a brief and intense flash of light, this phenomenon has been referred to as the photomotor response (PMR) (Kokel et al. 2010). The PMR has been characterized and documented in zebrafish embryos between 30 and 42 hpf (Kokel et al. 2010). Upon exposure to a sudden flash of light, the embryos exhibit a robust sequence of tail bends that lasts 5–7 s, after which the embryos enter a refractory period (>15 s) where they fail to respond to another flash of light. Although the neurological basis of this response is seldom understood, the PMR offers many advantages as a behavioral screening paradigm. The fish can be screened as early as 1.5 days post-fertilization when they are still inside of their chorions, and the small size of the embryos facilitates the use of 96-well microtiter plates to image as many as 8–10 embryos per well, allowing for high throughput data acquisition. This approach has been successfully utilized for the high-throughput screening of the effects of many psychotropic compounds (Kokel and Peterson 2011).

Spontaneous activity can also be induced in zebrafish embryos as early as 36 hpf via an acute exposure to an aqueous solution of nicotine (30–240 μM) (Mora-Zamorano et al. 2016b). Nicotine induces a characteristic burst of activity by acting as an agonist of nicotinic acetylcholine receptors in the spinal cord (Thomas et al. 2009). This behavioral paradigm has been previously utilized for genetic (Petzold et al. 2009) and drug (Schneider et al. 2012b) screening. More recently, however, this nicotine-evoked locomotor response (NLR) has been utilized to assess the effects of methylmercury on the locomotor activity of 48 hpf zebrafish embryos (Mora-Zamorano et al. 2016b). One advantage of this assay is that the NLR can be directly anchored to the neurological basis of this behavior; the locomotor output induced by nicotine is known to be mediated by the spinal cord (Thomas et al. 2009).

8.4.1.2 Development of Neuromotor Control

Fine motor control and coordination need to be developed in order for an organism to interact with its environment. In zebrafish, the locomotor pattern of embryos becomes mature once they reach 5 days of age (Lambert et al. 2012). In contrast, rat pups acquire an adult-like locomotor pattern roughly between 13 and 16 days post-birth (Geisler et al. 1993). Both zebrafish and rodents can be observed throughout their development to identify possible delays in neuromotor milestones. To analyze neuromotor development in zebrafish, the embryos are often videotaped and their locomotor activity is tracked either manually by the observer or using machine vision algorithms that allow the automation of the analysis process. Commercially available systems to perform locomotion analysis in zebrafish include the Noldus DanioVision and the Viewpoint ZebraBox (Ahmad et al. 2012). However, it is also possible to analyze locomotor activity with freely available software such as ImageJ or Ctrax (Branson et al. 2009). The use of ImageJ for activity analysis in zebrafish larvae has been extensively described in previous studies (Colwill and Creton 2011a; Creton 2009; Richendrfer and Creton 2015; Richendrfer and Créton 2013). The Ctrax motion tracking algorithm has been successful at quantifying the activity of developing zebrafish embryos from 3 to 5 days of age (Lambert et al. 2012), and it has also been utilized to assess the effects of methylmercury on the swimming behavior of zebrafish larvae (Mora-Zamorano et al. 2016a). In rodents, the analysis of neuromotor development often involves observing the first manifestations of activity in the pups, such as crawling, rearing and grooming (Geisler et al. 1993). Later in development, rodents can be assessed for their performance in crossing rods of different widths, pivoting and forelimb grip strength (Dubovický et al. 2008).

8.4.2 Vision

Towards the end of the 1990s a number of research groups were interested in analyzing the genetic basis of vision. As a result, several behavioral assays have been developed in both rodents and zebrafish to screen for visual acuity and impairment (Fig. 8.3) (Neuhauss 2003; Prusky et al. 2000). Recent studies support that environmental toxicants can yield vision abnormalities and subsequent behavioral abnormalities in zebrafish via dysregulated expression of photoreceptor cell-related genes (Zhang et al. 2015), delayed retinal neurodifferentiation (Sun et al. 2016b), as well as altered retinal morphology and electrophysiology (Weber et al. 2008). Examples of the vision assays used in such studies are described below.

8.4.2.1 Visual Motor Response

The visual motor response (VMR) assay was originally developed to quantify the locomotor output of multiple zebrafish larvae in response to sudden changes in light intensity (Emran et al. 2008). This assay consists of recording the larvae in a multiple-well plate that is placed on top of a light source. The recording apparatus contains an infrared (IR) sensitive camera and a source of IR light to allow for video recording in the dark. Additionally, the experimental setting is often surrounded by a chamber that impedes the entrance of extraneous light. The experimenter can then program the light source within the chamber to turn on or off; the duration of the light and dark periods can be varied, as well as the number of times that these light-dark cycles are repeated. Deeti et al. have recently described use of the VMR assay to evaluate the safety of human oculotoxic drugs, thus highlighting its potential inclusion in future high-throughput approaches (Deeti et al. 2014). The VMR has only been extensively documented in the zebrafish, however, there have been recent efforts to evaluate this behavioral paradigm in a non-model fish species (yellow perch; *Perca flavescens*). Interestingly, the behavior exhibited by yellow perch was the opposite of that observed in zebrafish, in other words, yellow perch exhibit increased swimming activity during light periods, while zebrafish do so in dark periods (Mora-Zamorano et al.; manuscript in preparation). This illustrates the

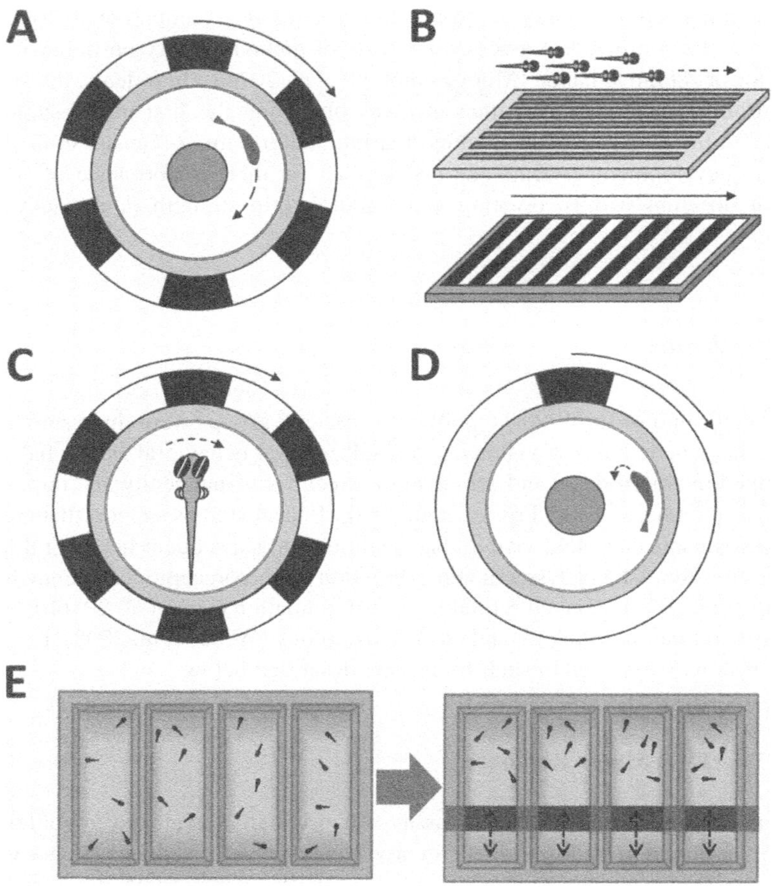

Fig. 8.3 Among the approaches to assess sensorimotor function in zebrafish, assays for vision are some of the most varied and documented. For instance, the optomotor response has been characterized in both (**a**) adult and (**b**) larval zebrafish, in both cases, the organisms swim in the perceived direction of moving parallel lines. (**c**) The optokinetic response has also been observed in adult and larval zebrafish (only larval zebrafish depicted in image), in this assay the experimenter records the saccadic movements of the eyes in response to a moving grating of bars. (**d**) Visual avoidance assays are another approach to assess vision, one variation of these experiments in adult zebrafish are performed by placing the fish in a vessel surrounded by a rotating drum with a single *black bar*, which the fish avoids on each rotation. (**e**) A similar avoidance assay can be performed in larval zebrafish placed on top of a computer monitor displaying a single moving bar on one side of the recording arena, the larvae swim away from the visual stimulus and aggregate to the side of the vessel where there is no stimulus

strong opinion of the authors that baseline behavioral data need to be thoroughly evaluated prior to the assessment of toxicant effects, with the understanding that activation of the same molecular pathway in two species may result in diametric behaviors that only make sense after exploring life history differences.

8.4.2.2 Visual Startle, Avoidance and Escape Response

A variety of startle and avoidance responses can be evoked with different kinds of visual stimuli in both adult and larval zebrafish. For instance, a robust escape response can be evoked in adult zebrafish by a black stripe inscribed on the inner surface of a white rotating drum (Li and Dowling 1997). Similar escape maneuvers have been reported in adult zebrafish presented with an expanding dot on an LCD monitor, which presumably mimics a fast approaching predator (Ahmed et al. 2012; Luca and Gerlai 2012). In zebrafish larvae, the visual startle response can be triggered by a sudden change in light intensity (Colwill and Creton 2011a, b), however, in contrast with vibrational or touch-evoked startle responses, the visual startle response in zebrafish larvae has been shown to not involve the Mauthner cell (Easter Jr and Nicola 1996; Portugues and Engert 2009). The visual startle response has been extensively studied by Burgess and Granato; this group has developed a software tool (FLOTE software) to assess visual startle as well as other kinds of startle responses in multiple zebrafish larvae simultaneously (Burgess et al. 2009). Larval zebrafish are also known to actively avoid animations of moving dots and bars (Colwill and Creton 2011b), which has led to the development of high-throughput methodologies to assess avoidance behavior (Richendrfer and Créton 2013).

8.4.2.3 Optomotor Response

The optomotor response (OMR) is an innate behavior that has been observed primarily in fish and insects. The OMR can be triggered by presenting a pattern of moving vertical stripes, which is commonly achieved by means of a rotating drum, but can also be presented on a computer monitor or with an LCD projector. Fish and insects react to this stimulus by moving in the direction of the perceived movement of the stripes, presumably in an attempt to adjust their trajectory in accordance to an environment that appears to be moving. The OMR assay has been widely employed to perform behavioral screens in adult and larval zebrafish (Neuhauss 2003; Li and Dowling 1997; Orger et al. 2009), and there are reports that this assay successfully elicits head movements in rodents (Abdeljalil et al. 2005).

8.4.2.4 Optokinetic Response

The optokinetic response (OKR) is a characterized by a series of saccadic eye movements that occur in response to moving objects while the head remains stationary. This response has been observed in primates (Miles 1993) including humans (Howard and Simpson 1989), as well as rodents (Scudder 2009), and in zebrafish (Neuhauss 2003; Orger et al. 2009). Although this response has been well described, it has not been extensively exploited in toxicology studies.

8.4.3 Acoustic and Vibrational Startle

Both rodents and zebrafish exhibit a robust response to acoustic and vibrational stimuli, both of which can be triggered by a speaker or a mechanical hammer. These stimuli are often used to evaluate non-associative learning (habituation) in both rodents and zebrafish (Wolman et al. 2011; Best et al. 2008; Pilz and Schnitzler 1996), as will be discussed below in more detail. However, both rodents and zebrafish are well-established models for studying hearing loss and the screening of ototoxic compounds (Buck et al. 2012; Bang et al. 2002; Goldey et al. 1995). In zebrafish, these assays consist of housing individual fish in tanks to which a sound or vibration is delivered with a loud speaker. In order to increase the throughput, multiple fish are often videotaped simultaneously by housing them in several contiguous cubicles, after which the startle responses of the fish are analyzed (Bang et al. 2002; Goldey et al. 1995; Bailey et al. 2013). In rodents, the procedure is very similar to the aforementioned; organisms are housed in an acoustic startle chamber and the startle stimulus is delivered by a loud speaker. Afterwards, the startle response of the organism is videotaped for later analysis (Goldey et al. 1995). Studies have focused on elucidating the molecular and neurophysiological mechanisms (e.g., the potential role of vesicular glutamate transporter 3 (Obholzer et al. 2008) and glycinergic signaling (Hirata et al. 2011) of the acoustic startle response (Bhandiwad et al. 2013; Lin et al. 2015; Burgess and Granato 2007; Tanimoto et al. 2009) and the optimization of experimental techniques (Bhandiwad and Sisneros 2016; Zeddies and Fay 2005). Coffin and Ramcharitar have extensively reviewed chemically-induced ototoxicity in fish and its impact of neurobehavior (Coffin and Ramcharitar 2016).

8.4.4 Touch Response

The touch response has been extensively documented in zebrafish embryos and adults, and it represents one of the most fundamental behavioral assays available to induce a robust locomotor response. A touch response test is a very simple procedure that involves empirically observing or recording the organism to be tested with a high speed camera (500–1000 frames per second) while touching the tail or the head of the fish with a dissecting needle; a water jet to the tail or head may also be used to evoke the touch response (Budick and O'Malley 2000). This straightforward assay is suitable for quick preliminary screens to identify locomotor abnormalities linked to neurodevelopmental defects (Patton and Zon 2001).

8.4.5 Olfaction and Taste

Perhaps the least studied of the senses in both zebrafish and rodents are olfaction and taste. In zebrafish, the chemosensory system is developed within the first 7 days post-fertilization (dpf), presumably to support feeding upon yolk depletion (Lindsay and Vogt 2004). The olfactory bulb in zebrafish is sensitive to five classes of odorants: amino acids, bile salts, steroids, prostaglandins, and nucleotides (Bhinder and Tierney 2012). Very few assays have been developed to assess olfactory-evoked behavior in zebrafish; however, olfactory cues have been utilized in classical conditioning experiments (Lindsay and Vogt 2004), both in rodents and in zebrafish (Braubach et al. 2009). Other assays have focused in characterizing the locomotor activity of zebrafish embryos after the addition of a variety of amino acids of which L-alanine was reported to evoke a subtle yet consistent increase in the locomotor output of larvae (Bhinder and Tierney 2012). Furthermore, brief exposure (from 80 to 83 hpf) of transgenic hsp70/eGFP zebrafish to waterborne cadmium (125 μM) induced gene expression of heat-shock protein 70 (i.e., a biomarker of cellular stress) in the olfactory neurons with concomitant observation of significant cell death in this neuron type within cadmium-exposed wild-type zebrafish larvae compared to control (Blechinger et al. 2007).Using an identical exposure paradigm, juvenile (50 dpf) zebrafish exposed to 125 μM cadmium showed a significant decrease in dashing activity (i.e., rapid bursts of apparently disoriented swimming) and a significant increase in the time required to initiate a response to an alarm substance stimulus compared to control fish. Therefore, cadmium-induced toxicity of the olfactory system can alter predator avoidance behaviors in teleosts. Nathan et al. (2015) have suggested that the neuropeptide kisspeptin1 may regulate the odorant (alarm substance) evoked fear response in zebrafish via 5-HT1A and 5-HT2 serotonin receptors.

In regard to taste, the molecular mechanisms of gustation in fish are relatively unknown (Okada 2015). Similarly, research focused on investigating the effects of environmental contaminants on neurobehavior via modulation of the gustatory pathway in teleosts is scant. Vendrell-Llopis and Yaksi have revealed that taste stimuli of different categories evoked different neural activity in the brainstem of semi-restrained juvenile *Tg[elval3:GCaMP5]* zebrafish as analyzed using a two-photon microscope (Vendrell-Llopis and Yaksi 2015). Results also showed that the zebrafish yielded weak behavioral responses (as assessed via angular tail speed and tail-beat frequency) upon ingestion of amino acids relative to the heightened locomotor output observed in the zebrafish that ingested sour and bitter taste stimuli. Thus, the modulation of neurobehavior in zebrafish is dependent on the category of the taste stimuli.

8.4.6 Activity and Emotional Reactivity

The activity of an organism can serve as a proxy to identify emotional states such as fear, anxiety, aggression or lethargy. One the simplest assays to perform for stress and anxiety analysis is the open field test, which has been well-documented in both rodents and zebrafish (Stewart et al. 2012; Cryan and Holmes 2005). During the open field test, stressed organisms tend to spend more time around the edges of the arena (thigmotaxis) and will avoid entering the center region of a brightly lit open field (Champagne et al. 2010). In larval zebrafish, thigmotaxis has also been linked to anxiety. Larvae that tend to spend more time close to the edges of an arena are considered to be exhibiting an anxiety-like behavior (Schnörr et al. 2012).

A substantial number of assays have been developed to analyze stress and anxiety in rodents. In addition to the open field test, there is also the light-dark preference paradigm, in which the organism is placed in an open field and provided with a dark box to hide. Other assays include the elevated plus maze and the elevated zero maze. As their names imply, the elevated plus maze is a maze shaped like a plus symbol of which two arms are enclosed and the other two are open. The maze itself is elevated well above the ground level by means of a base. Similarly, the zero maze is shaped like the number zero and is divided into four quadrants, two of which are enclosed and the others are open, this maze is also elevated from the ground level. In such assays, rodents will exhibit stress and anxiety by seeking shelter in a dark and enclosed area, such as the dark box in the light-dark preference test or by avoiding the risk of falling over the open sections of the plus or the zero mazes (Cryan and Holmes 2005).

In zebrafish, one variation of the open field test is the diving test (Egan et al. 2009). In this assay, adult fish are placed in a deep and narrow tank, which enhances the ability of the experimenter to observe spatial preference in the vertical axis. Fish that spend more time in the bottom of the tank are considered to be exhibiting more stress and anxiety than fish that spend more time in the top portion of the tank (Bencan and Levin 2008). Other assays that evaluate anxiety rely on identifying the spatial preference of fish in the horizontal axis and they are inspired by the assays previously described in rodents (Champagne et al. 2010). One of the most documented of these methodologies is the light-dark spatial preference assay (Blaser and Penalosa 2011). Commonly, stressed zebrafish tend to prefer being in a dark or shaded area and avoid an illuminated area. Additionally, the evaluation of spatial preference in fish can also be varied by presenting different aversive stimuli on one side of the tank. Examples of these stimuli are animations of an expanding dot on an LCD monitor (Luca and Gerlai 2012), images of predators (Bass and Gerlai 2008), or even robotic models of a predator fish or bird (Cianca et al. 2013). A comparison of rodent versus zebrafish assays of emotional reactivity is presented in Fig. 8.4.

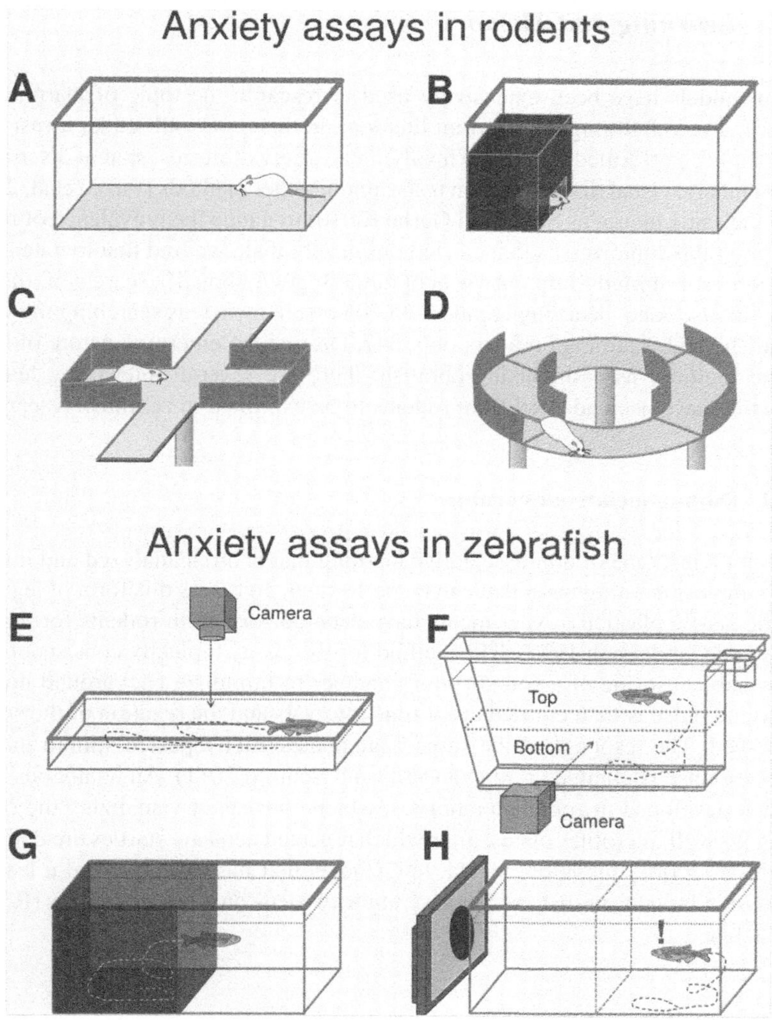

Fig. 8.4 Many assays designed to assess emotional reactivity (e.g., anxiety) in zebrafish were adapted from rodent assays. Some examples of anxiety tests in rodents are the (**a**) open-field test, (**b**) the light/dark exploration test, and different elevated mazes such as the (**c**) plus maze and the (**d**) zero maze. In zebrafish, the open field tests can be performed in different tank configurations, such as a (**e**) shallow tank or a (**f**) deep and narrow tank where the "diving test" can be performed, and the (**g**) light/dark exploration test is performed in a manner very similar to the (**b**) rodent counterpart. (**h**) Assessment of fear and anxiety in zebrafish can also be performed by exposing the fish to aversive visual stimuli on a computer screen, such as pictures of predatory fish or an expanding dot mimicking a looming predator

8.4.7 *Learning and Memory*

Rodent models have been extensively used to research the topic of learning and memory; a search through the current literature in this topic will render a vast number of studies performed in rodents involving a variety of mazes, spatial discrimination chambers, visual discrimination tests, among other methods (Puzzo et al. 2014). Using the same metric as Sison and Gerlai (2010) to gauge the prevalence of rodent research in this topic versus that carried out in zebrafish, we find that to date (April 14, 2016) a PubMed (http://www.ncbi.nlm.nih.gov/pubmed/) search of the key words "mouse" and "learning" renders 25,705 results, whereas searching the words "zebrafish" and "learning" renders only 347. Despite the emerging nature of learning and memory assessment in zebrafish, there are several noteworthy learning assays that have been adapted from rodents to be exploited in zebrafish research.

8.4.7.1 Non-associative Learning

Habituation is a type of non-associative learning that is often analyzed and it is one of the simplest learning tasks that can be performed. To assess this form of learning, acoustic startle plasticity experiments have been performed in rodents for decades (Davis and Gendelman 1977). The method for this assay typically consists of placing the subject inside of a sound-proof chamber to minimize background noise, a loud sound pulse is then emitted at set time intervals and the reaction of the subject is observed. The response of the subject attenuates after repeated stimuli until no startle response is elicited at all (Valsamis and Schmid 2011). An analogous assay has been developed in zebrafish embryos, which consists of visualizing the organisms in 96-well microtiter plates, after which repeated acoustic startles are delivered (Best et al. 2008), this is one of the few studies that has characterized a learning paradigm in larval zebrafish, as the vast majority focus on adult organisms (Roberts et al. 2014).

8.4.7.2 Associative Learning

Associative learning involves learning through linking different stimuli to one another. The topic of behavioral assays to assess associative learning in zebrafish has been previously covered in detail. In zebrafish, associative learning has been assessed using the plus maze (Sison and Gerlai 2010), the T-maze (Colwill et al. 2005), spatial alternation tests (Williams et al. 2002; Smith and Weis 1997) and classical conditioning (Valente et al. 2012). Associative learning experiments have been utilized to assess learning and memory in zebrafish exposed to arsenic (de Castro et al. 2009) and to methylmercury (Smith et al. 2010).

8.4.8 Predator Avoidance and Prey Capture

In the context of AOP models for environmental risk assessment, two of the most useful indicators of the success or decline of a population are the capacity of individuals to avoid predators and to capture prey. Many models rely on the assumption that the effects on these two endpoints directly affect population dynamics. However, there are far fewer studies that measure predator avoidance and prey capture than those that assess more fundamental behaviors, such as spontaneous locomotion or startle response. One reason for this may be the inherent technical challenges of properly performing predator-prey interaction experiments, in addition to the widely accepted notion that the much simpler spontaneous locomotion assays render acceptable indicators of predator avoidance and prey capture success. Weis and collaborators (Weis and Candelmo 2012) have been at the forefront of documenting the study of pollutants on predator-prey interactions in fish. In a recent review by this research group (Weis and Candelmo 2012), the authors made reference to several studies that measured predator-prey interactions in fish, curiously, a substantial number of the current published assessments of predator-prey interactions in response to pollutants have been carried out by Weis and collaborators themselves (Weis and Candelmo 2012; Weis et al. 2003; Weis and Weis 1995a, b; Zhou et al. 2001; Zhou and Weis 1998). Other published works cited in this study were dated (20 years, often more), and large time periods existed between studies. This illustrates the fact that not many research laboratories focus on the analysis of these endpoints for ecotoxicology risk assessment despite the fact that predator-prey interaction assays render data that can readily be placed in an ecological context.

In zebrafish larvae, prey capture has been long considered among the fundamental constituents of their behavioral repertoire. One of the best documented methods to measure this endpoint is to supply the larvae with paramecia, then count the number of paramecia captured by the larvae in a fixed amount of time under a microscope (Budick and O'Malley 2000). However, for toxicology purposes, similar approaches have been utilized in two studies that assessed the effects of methylmercury in prey capture success (Mora-Zamorano et al. 2016a; Samson et al. 2001). Feeding assays in adult zebrafish are also extremely scarce. Notable examples are the use of brine shrimp nauplii as positive reinforcement in an associative learning assay in adult zebrafish exposed to methylmercury (Smith et al. 2010).

Predator avoidance assays in zebrafish have been mostly carried out utilizing simulated predators. In zebrafish larvae, it has been reported that the fish swim away from an aversive visual stimulus (e.g., an animated moving circle or bar), presumed to be interpreted by the larvae as a threat (Colwill and Creton 2011a). Additionally, an aversive response can be elicited in larval Atlantic croaker when presented with the image of a black oval on a white card, which is swung towards the larvae by a remotely operated pendulum. This approach is believed to simulate the cross section of a predatory fish approaching the larvae, and it has been used to assess the effects of methylmercury on predator evasion (Alvarez Mdel et al. 2006). In adult zebrafish, the approaches to simulate predators have ranged range from painting a black bar on a white rotating drum to elicit an escape response every time the fish encounters the bar (Li and Dowling 1997), to animated sympatric predators and expanding dots on a computer monitor (Luca and Gerlai 2012; Gerlai et al. 2009) and robotic models of predatory fish and birds (Cianca et al. 2013). Out of these approaches, the rotating drum and bar has been successfully utilized to assess vision impairment in adult zebrafish exposed to methylmercury, and to determine whether selenomethionine mitigates said vision impairments (Weber et al. 2008).

8.5 Challenges for Neurobehavioral AOP Development

Currently, the integration of behavioral endpoints into AOP models for risk assessment faces a number of challenges. Neurobehavioral toxicology studies often focus on the identification of abnormal phenotypes upon exposure to a chemical of interest and do not necessarily anchor the observed phenotype to a specific molecular and cellular insult. These actions hinder the discovery of molecular initiating events that serve as the foundation of an AOP model. However, recent studies have started to elucidate the molecular and cellular consequences associated with abnormal neurobehavioral outcomes as a result of emerging toxicology and pharmacology research. For example, it has been shown that waterborne exposure of embryonic zebrafish to 2, 2′,4,4-tetrabromodiphenyl ether (BDE-47; a predominant congener of polybrominated diphenyl ethers [PBDEs] in the environment) significantly disrupted spontaneous activity, decreased touch response (see Sect. 8.4.4) and free swimming speed, and perturbed larval behavioral responses to illuminated versus dark periods (i.e., visual motor response, see Sect. 8.4.2.1) (Chen et al. 2012). Interestingly, these abnormal neurobehavioral phenotypes were associated with significantly reduced axonal growth of primary and secondary motor neurons. Similarly, Wang et al. have revealed that developmental exposure of zebrafish to the endocrine-disrupting plasticizer, bisphenol A (BPA; 15 μM), produced a decrease in the spontaneous movement, swimming speed, and touch response (compared to control) associated with hindered axonal growth of spinal neurons and the abnormal development of the axial musculature (Wang et al. 2013). Conversely, low-dose developmental exposure (0.0068 μM; 1000-fold lower that the accepted human exposure) of BPA and bisphenol S (i.e., commonly used replacement analog of BPA) has been shown to induce hyperactivity in zebrafish via precocious hypothalamic neurogenesis (Kinch et al. 2015). Neurochemical studies have begun to elucidate the role of neurotransmitter signaling in altering neurobehavior (Basu 2015).

Raftery and Volz have demonstrated that exposure of zebrafish embryos to abamectin (an avermectin insecticide) eliminated spontaneous tail contractions via modulation of the γ-aminobutyric (GABA) receptor (Raftery and Volz 2015). Jin et al. have also demonstrated that exposure (until 96 hpf) of zebrafish embryos to imazalil (300 μg/L), a fungicide that is extensively used in agriculture, significantly reduced the average swimming speed and distance upon exposure via a concomitant reduction in acetylcholinesterase (AChE) gene expression and enzymatic activity compared to control (Jin et al. 2016). Moreover, exposures (until 5 dpf) of zebrafish to the organophosphate flame retardants, tri-n-butyl phosphate (3125 μg/L) and tris (2-butoxyethyl) phosphate (6250 μg/L), yielded a reduced swimming speed (in both locomotor and visual motor response assays) associated with a significant decrease in the gene expression of AChE without concurrent impact on AChE activity compared to control (Sun et al. 2016a). However, in the same study, a reduction in *both* AChE activity and gene expression was associated with reduced swimming speed upon exposure to chorpyrifos (300 μg/L), an organophosphate pesticide (OPP). Other research further supports the possible relationship between OPP-induced alterations in neurobehavior and the inhibition of AChE (Yen et al. 2011), while other studies suggest the contrary (Richendrfer and Creton 2015). Collectively, this evidence suggests that AChE may be a common locus involved in the molecular initiating events of neurotoxicant-induced alteration of neuromotor control that necessitates further research.

Environmental contaminants also mediate neurotoxicity via targeting the signaling pathways of the neurotransmitters dopamine and serotonin. Ek et al. (2016) have reported that zebrafish exhibit similar behavioral phenotypes to those of rats and humans, upon activation of the dopaminergic system, thus exemplifying the zebrafish as a predictive translational model of neurobehavioral pharmacology and toxicology. Disruption of the dopaminergic signaling has been shown to impact memory and associative learning (Naderi et al. 2016), the consolidation of latent learning of spatial information (Naderi et al. 2016), social and anxiety-related behavior (Wang et al. 2016b), and locomotion (Tran et al. 2015) in zebrafish. Subchronic exposure (45 days) of adult zebrafish to titanium oxide nanoparticles (TiO_2 NP, 5–40 μg/L) impaired spatial recognition memory (as assessed via a Y-maze assay) and locomotion compared to control (Sheng et al. 2014). This altered behavior was associated with reduced concentrations of norepinephrine, dopamine, and serotonin, neuronal apoptosis, and dysregulated expression of memory-related genes in TiO_2NP-exposed zebrafish brains compared to control. Wang et al. have reported decreased locomotor behavior in larval zebrafish exposed (2–120 hpf) to DE-71 (i.e., a mixture of polybrominated diphenyl ethers) compared to control. The observed reduction in locomotion paralleled significant reductions in whole-body concentrations of dopamine, down-regulation of genes related to the development of dopaminergic neurons, and decreased expression of tyrosine hydroxylase and dopamine transporter proteins in dopaminergic neurons (Wang et al. 2016a). Using the same experimental paradigm, Wang et al. have also shown that DE-71 exposure disrupts neurogenesis and inhibits serotonin synthesis (Wang et al. 2015). Insult of the serotonergic pathway has been linked to abnormal social behavior and anxiety in zebrafish (Herculano and Maximino 2014).

In addition to understanding the molecular mechanisms that govern changes in neurobehavior, an effort must be made to classify behaviors in order of complexity and establish links between the fundamental and complex behaviors. Kalueff and collaborators (2013) have compiled a document describing 190 different behavioral outputs in zebrafish larvae and adults, then proceeded to create a conceptual diagram of the relationship between behaviors and other biological phenomena, including interaction among organisms (i.e., an ecologically relevant scenario) and psychiatric disorders (i.e., human-health relevant scenario). With such a large number of behavioral endpoints that can potentially be analyzed, it is imperative to establish the relationship of said behaviors with putative adverse outcomes. This challenging task will require analyzing an assortment of different behavioral endpoints in parallel under standardized experimental conditions, preferably choosing a variety of behaviors of different levels of complexity, rather than conducting studies where only one behavioral endpoint is analyzed. There is also a need to perform cross-species validation of AOP models. The baseline behaviors of different species can vary greatly, regardless of chemical exposure, therefore cross-species comparisons should be performed whenever possible, especially if behaviors observed in fish models are to be translated to behaviors in rodents or humans. Furthermore, it is vital to ensure analogous experimental conditions when assessing behavior in different species, thus it is crucial to know the characteristics of the organisms to be used as models and use this information to match experimental conditions as accurately as possible (e.g., developmental stage of chemical exposure, developmental stage at which behavioral assays are performed, etc.).

8.6 Conclusions

AOP models based on behavioral data are still a relatively new approach for environmental and human health risk assessment. Few studies have made use of behavioral data to make predictions of adverse outcomes. More neurobehavioral toxicology studies that complement observed abnormal phenotypes with transcriptomic, proteomic, and/or metabolomic data in effort to identify key molecular intersections (i.e., common molecular initiating events among multiple classes of chemicals) are a necessity. This will serve to link neurotoxicant-induced molecular and/or cellular responses to adverse apical outcomes at the population and/or ecosystem level(s). In turn, the identification of such intersections will improve the ability to predict system-level impacts and, ultimately, expedite human and ecological risk assessment. The increasing use and simplification of NBSA methodologies, paired with systems and molecular/biochemical biology-based approaches and the promising features of the AOP model, will strengthen the use of behavior as a predictor of adverse outcomes in the future.

References

Abdeljalil J, Hamid M, Abdel-Mouttalib O, Stéphane R, Raymond R, Johan A, José S, Pierre C, Serge P (2005) The optomotor response: a robust first-line visual screening method for mice. Vis Res 45(11):1439–1446

Ahmad F, Noldus LPJJ, Tegelenbosch RAJ, Richardson MK (2012) Zebrafish embryos and larvae in behavioural assays. Behaviour 149(10–12):1241–1281

Ahmed TS, Fernandes Y, Gerlai R (2012) Effects of animated images of sympatric predators and abstract shapes on fear responses in zebrafish. Behaviour 149(10–12):1125–1153
Alvarez Mdel C, Murphy CA, Rose KA, McCarthy ID, Fuiman LA (2006) Maternal body burdens of methylmercury impair survival skills of offspring in Atlantic croaker (*Micropogonias undulatus*). Aquat Toxicol 80(4):329–337. doi: S0166-445X(06)00369-9
Ankley GT, Bennett RS, Erickson RJ, Hoff DJ, Hornung MW, Johnson RD, Mount DR, Nichols JW, Russom CL, Schmieder PK (2010) Adverse outcome pathways: a conceptual framework to support ecotoxicology research and risk assessment. Environ Toxicol Chem 29(3):730–741
Arini A, Head JA, Murphy CA, Carvan MJ, Goetz R, Klingler RH, Nam D-H, Basu N (2016) Neuroendocrine biochemical effects in methylmercury-exposed yellow perch. Comparative Biochemistry and Physiology Part C: Toxicology & Pharmacology 187:10–18
Bailey J, Oliveri A, Levin ED (2013) Zebrafish model systems for developmental neurobehavioral toxicology. Birth Defects Res C Embryo Today Rev 99(1):14–23
Bal-Price A, Crofton KM, Leist M, Allen S, Arand M, Buetler T, Delrue N, FitzGerald RE, Hartung T, Heinonen T (2015) International STakeholder NETwork (ISTNET): creating a developmental neurotoxicity (DNT) testing road map for regulatory purposes. Arch Toxicol 89(2):269–287
Bang PI, Yelick PC, Malicki JJ, Sewell WF (2002) High-throughput behavioral screening method for detecting auditory response defects in zebrafish. J Neurosci Methods 118(2):177–187
Bass SLS, Gerlai R (2008) Zebrafish (*Danio rerio*) responds differentially to stimulus fish: the effects of sympatric and allopatric predators and harmless fish. Behav Brain Res 186(1):107–117
Basu N (2015) Applications and implications of neurochemical biomarkers in environmental toxicology. Environ Toxicol Chem 34(1):22–29
Bencan Z, Levin ED (2008) The role of α7 and α4β2 nicotinic receptors in the nicotine-induced anxiolytic effect in zebrafish. Physiol Behav 95(3):408–412
Benzer S (1967) Behavioral mutants of drosophila isolated by countercurrent distribution. Proc Natl Acad Sci U S A 58(3):1112–1119
Best JD, Berghmans S, Hunt JJFG, Clarke SC, Fleming A, Goldsmith P, Roach AG (2008) Non-associative learning in larval zebrafish. Neuropsychopharmacology 33(5):1206–1215
Bhandiwad AA, Sisneros JA (2016) Revisiting psychoacoustic methods for the assessment of fish hearing. In: Fish hearing and bioacoustics. Springer, Cham, pp 157–184
Bhandiwad AA, Zeddies DG, Raible DW, Rubel EW, Sisneros JA (2013) Auditory sensitivity of larval zebrafish (*Danio rerio*) measured using a behavioral prepulse inhibition assay. J Exp Biol 216(Pt 18):3504–3513. https://doi.org/10.1242/jeb.087635
Bhinder G, Tierney KB (2012) Olfactory-evoked activity assay for larval zebrafish. Zebrafish Protoc Neurobehav Res 66:71–84
Blaser RE, Penalosa YM (2011) Stimuli affecting zebrafish (*Danio rerio*) behavior in the light/dark preference test. Physiol Behav 104(5):831–837
Blechinger SR, Kusch RC, Haugo K, Matz C, Chivers DP, Krone PH (2007) Brief embryonic cadmium exposure induces a stress response and cell death in the developing olfactory system followed by long-term olfactory deficits in juvenile zebrafish. Toxicol Appl Pharmacol 224(1):72–80
Branson K, Robie AA, Bender J, Perona P, Dickinson MH (2009) High-throughput ethomics in large groups of drosophila. Nat Methods 6(6):451–457
Braubach OR, Wood H-D, Gadbois S, Fine A, Croll RP (2009) Olfactory conditioning in the zebrafish (*Danio rerio*). Behav Brain Res 198(1):190–198
Buck LMJ, Winter MJ, Redfern WS, Whitfield TT (2012) Ototoxin-induced cellular damage in neuromasts disrupts lateral line function in larval zebrafish. Hear Res 284(1):67–81
Budick SA, O'Malley DM (2000) Locomotor repertoire of the larval zebrafish: swimming, turning and prey capture. J Exp Biol 203(Pt 17):2565–2579
Burgess HA, Granato M (2007) Sensorimotor gating in larval zebrafish. J Neurosci 27(18):4984–4994. https://doi.org/10.1523/JNEUROSCI.0615-07.2007
Burgess HA, Johnson SL, Granato M (2009) Unidirectional startle responses and disrupted left–right co-ordination of motor behaviors in robo3 mutant zebrafish. Genes Brain Behav 8(5):500–511

Champagne DL, Hoefnagels CCM, de Kloet RE, Richardson MK (2010) Translating rodent behavioral repertoire to zebrafish (*Danio rerio*): relevance for stress research. Behav Brain Res 214(2):332–342
Chen J, Huang C, Zheng L, Simonich M, Bai C, Tanguay R, Dong Q (2011) Trimethyltin chloride (TMT) neurobehavioral toxicity in embryonic zebrafish. Neurotoxicol Teratol 33(6):721–726
Chen X, Huang C, Wang X, Chen J, Bai C, Chen Y, Chen X, Dong Q, Yang D (2012) BDE-47 disrupts axonal growth and motor behavior in developing zebrafish. Aquat Toxicol 120:35–44
Cianca V, Bartolini T, Porfiri M, Macrì S (2013) A robotics-based behavioral paradigm to measure anxiety-related responses in zebrafish. PLoS One 8(7):e69661
Clift D, Richendrfer H, Thorn RJ, Colwill RM, Creton R (2014) High-throughput analysis of behavior in zebrafish larvae: effects of feeding. Zebrafish 11(5):455–461
Coffin AB, Ramcharitar J (2016) Chemical ototoxicity of the fish inner ear and lateral line. In: Fish hearing and bioacoustics. Springer, Cham, pp 419–437
Colwill RM, Creton R (2011a) Imaging escape and avoidance behavior in zebrafish larvae. Rev Neurosci 22(1):63–73
Colwill RM, Creton R (2011b) Locomotor behaviors in zebrafish (Danio Rerio) larvae. Behav Process 86(2):222–229
Colwill RM, Raymond MP, Ferreira L, Escudero H (2005) Visual discrimination learning in zebrafish (Danio Rerio). Behav Process 70(1):19–31
Creton R (2009) Automated analysis of behavior in zebrafish larvae. Behav Brain Res 203(1):127–136
Cronin CJ, Feng Z, Schafer WR (2006) Automated imaging of C. Elegans behavior. Methods Mol Biol 351:241
Cryan JF, Holmes A (2005) The ascent of mouse: advances in modelling human depression and anxiety. Nat Rev Drug Discov 4(9):775–790
Davis M, Gendelman PM (1977) Plasticity of the acoustic startle response in the acutely decerebrate rat. J Comp Physiol Psychol 91(3):549
de Castro MR, Lima JV, de Freitas DPS, de Souza VR, Dummer NS, de Aguiar RB, dos Santos LC, Marins LF, Geracitano LA, Monserrat JM (2009) Behavioral and neurotoxic effects of arsenic exposure in zebrafish (*Danio rerio*, Teleostei: Cyprinidae). Comparative Biochemistry and Physiology Part C: Toxicology & Pharmacology 150(3):337–342
DeAngelis DL, Grimm V (2014) Individual-based models in ecology after four decades. F1000Prime Rep 6(39):6
Deeti S, O'Farrell S, Kennedy BN (2014) Early safety assessment of human oculotoxic drugs using the zebrafish visualmotor response. J Pharmacol Toxicol Methods 69(1):1–8
Detrich HW III, Zon LI, Westerfield M (2009) Essential zebrafish methods: genetics and genomics. Academic Press, Amsterdam
Dubovický M, Kovačovský P, Ujházy E, Navarová J, Brucknerová I, Mach M (2008) Evaluation of developmental neurotoxicity: some important issues focused on neurobehavioral development. Interdiscip Toxicol 1(3–4):206–210
Easter SS Jr, Nicola GN (1996) The development of vision in the zebrafish (Danio Rerio). Dev Biol 180(2):646–663
Eddins D, Cerutti D, Williams P, Linney E, Levin ED (2010) Zebrafish provide a sensitive model of persisting neurobehavioral effects of developmental chlorpyrifos exposure: comparison with nicotine and pilocarpine effects and relationship to dopamine deficits. Neurotoxicol Teratol 32(1):99–108
Egan RJ, Bergner CL, Hart PC, Cachat JM, Canavello PR, Elegante MF, Elkhayat SI, Bartels BK, Tien AK, Tien DH (2009) Understanding behavioral and physiological phenotypes of stress and anxiety in zebrafish. Behav Brain Res 205(1):38–44
Ek F, Malo M, Åberg Andersson M, Wedding C, Kronborg J, Svensson P, Waters S, Petersson P, Olsson R (2016) Behavioral analysis of dopaminergic activation in zebrafish and rats reveals similar phenotypes. ACS Chem Neurosci 7(5):633–646

Emran F, Rihel J, Dowling JE (2008) A Behavioral assay to measure responsiveness of zebrafish to changes in light intensities. J Vis Exp 20:923. https://doi.org/10.3791/923

Eriksson P, Fredriksson A (1991) Neurotoxic effects of two different pyrethroids, bioallethrin and deltamethrin, on immature and adult mice: changes in behavioral and muscarinic receptor variables. Toxicol Appl Pharmacol 108(1):78–85

Geisler HC, Westerga J, Gramsbergen A (1993) Development of posture in the rat. Acta Neurobiol Exp (Wars) 53(4):517–524

Gerlai R, Fernandes Y, Pereira T (2009) Zebrafish (Danio Rerio) responds to the animated image of a predator: towards the development of an automated aversive task. Behav Brain Res 201(2):318–324

Goldey ES, Kehn LS, Lau C, Rehnberg GL, Crofton KM (1995) Developmental exposure to polychlorinated biphenyls (Aroclor 1254) reduces circulating thyroid hormone concentrations and causes hearing deficits in rats. Toxicol Appl Pharmacol 135(1):77–88

Gonzalez ST, Remick D, Creton R, Colwill RM (2016) Effects of embryonic exposure to polychlorinated biphenyls (PCBs) on anxiety-related behaviors in larval zebrafish. Neurotoxicology 53:93–101

Groh KJ, Carvalho RN, Chipman JK, Denslow ND, Halder M, Murphy CA, Roelofs D, Rolaki A, Schirmer K, Watanabe KH (2015) Development and application of the adverse outcome pathway framework for understanding and predicting chronic toxicity: II. A focus on growth impairment in fish. Chemosphere 120:778–792

Heiden TK, Hutz RJ, Carvan MJ 3rd (2005) Accumulation, tissue distribution, and maternal transfer of dietary 2,3,7,8,-tetrachlorodibenzo-p-dioxin: impacts on reproductive success of zebrafish. Toxicol Sci 87(2):497–507. https://doi.org/10.1093/toxsci/kfi201

Herculano AM, Maximino C (2014) Serotonergic modulation of zebrafish behavior: towards a paradox. Prog Neuro-Psychopharmacol Biol Psychiatry 55:50–66

Hirata H, Takahashi M, Yamada K, Ogino K (2011) The biological role of the glycinergic synapse in early zebrafish motility. Neurosci Res 71(1):1–11

Howard IP, Simpson WA (1989) Human optokinetic nystagmus is linked to the stereoscopic system. Exp Brain Res 78(2):309–314

Howe K, Clark MD, Torroja CF, Torrance J, Berthelot C, Muffato M, Collins JE, Humphray S, McLaren K, Matthews L (2013) The zebrafish reference genome sequence and its relationship to the human genome. Nature 496(7446):498–503

Hurd MW, Debruyne J, Straume M, Cahill GM (1998) Circadian rhythms of locomotor activity in zebrafish. Physiol Behav 65(3):465–472

Isbister GK, Bowe SJ, Dawson A, Whyte IM (2004) Relative toxicity of selective serotonin reuptake inhibitors (SSRIs) in overdose. J Toxicol Clin Toxicol 42(3):277–285

Jin Y, Zhu Z, Wang Y, Yang E, Feng X, Fu Z (2016) The fungicide imazalil induces developmental abnormalities and alters locomotor activity during early developmental stages in zebrafish. Chemosphere 153:455–461

Kalueff AV, Gebhardt M, Stewart AM, Cachat JM, Brimmer M, Chawla JS, Craddock C, Kyzar EJ, Roth A, Landsman S (2013) Towards a comprehensive catalog of zebrafish behavior 1.0 and beyond. Zebrafish 10(1):70–86

Kinch CD, Ibhazehiebo K, Jeong JH, Habibi HR, Kurrasch DM (2015) Low-dose exposure to bisphenol a and replacement bisphenol S induces precocious hypothalamic neurogenesis in embryonic zebrafish. Proc Natl Acad Sci U S A 112(5):1475–1480. https://doi.org/10.1073/pnas.1417731112

Kokel D, Peterson RT (2011) Using the zebrafish photomotor response for psychotropic drug screening. Methods Cell Biol 105:517–524. https://doi.org/10.1016/B978-0-12-381320-6.00022-9

Kokel D, Bryan J, Laggner C, White R, Cheung CYJ, Mateus R, Healey D, Kim S, Werdich AA, Haggarty SJ (2010) Rapid behavior-based identification of neuroactive small molecules in the zebrafish. Nat Chem Biol 6(3):231–237

Lambert AM, Bonkowsky JL, Masino MA (2012) The conserved dopaminergic diencephalospinal tract mediates vertebrate locomotor development in zebrafish larvae. J Neurosci 32(39):13488–13500. https://doi.org/10.1523/JNEUROSCI.1638-12.2012
Landesmann B, Mennecozzi M, Berggren E, Whelan M (2013) Adverse outcome pathway-based screening strategies for an animal-free safety assessment of chemicals. Altern Lab Anim 41(6):461–471
Li L, Dowling JE (1997) A dominant form of inherited retinal degeneration caused by a non-photoreceptor cell-specific mutation. Proc Natl Acad Sci U S A 94(21):11645–11650
Lin SY, Vollrath MA, Mangosing S, Shen J, Cardenas E, Corey DP (2015) The zebrafish pinball wizard gene encodes WRB, a tail-anchored-protein receptor essential for inner-ear hair cells and retinal photoreceptors. J Physiol 594(4):895–914
Lindsay SM, Vogt RG (2004) Behavioral responses of newly hatched zebrafish (Danio Rerio) to amino acid chemostimulants. Chem Senses 29(2):93–100
Luca RM, Gerlai R (2012) In search of optimal fear inducing stimuli: differential behavioral responses to computer animated images in zebrafish. Behav Brain Res 226(1):66–76
MacPhail RC, Brooks J, Hunter DL, Padnos B, Irons TD, Padilla S (2009) Locomotion in larval zebrafish: influence of time of day, lighting and ethanol. Neurotoxicology 30(1):52–58
Maximino C, Araujo J, Leão LKR, Abaj G, Krm O, Lima MG, EdJO B, Crespo-López ME, Gouveia A Jr, Herculano AM (2011) Possible role of serotoninergic system in the neurobehavioral impairment induced by acute methylmercury exposure in zebrafish (Danio rerio). Neurotoxicol Teratol 33(6):727–734
Miklósi Á, Andrew RJ (2006) The zebrafish as a model for behavioral studies. Zebrafish 3(2):227–234
Miles FA (1993) The sensing of rotational and translational optic flow by the primate optokinetic system. Rev Oculomot Res 5:393–403
Mora-Zamorano FX, Klingler R, Murphy CA, Basu N, Head J, Carvan MJ III (2016a) Parental whole life cycle exposure to dietary methylmercury in zebrafish (Danio Rerio) affects the behavior of offspring. Environ Sci Technol 50(9):4808–4816
Mora-Zamorano FX, Svoboda KR, Carvan MJ III (2016b) The nicotine-evoked locomotor response: a Behavioral paradigm for toxicity screening in zebrafish (Danio Rerio) embryos and Eleutheroembryos exposed to methylmercury. PLoS One 11(4):e0154570
Naderi M, Jamwal A, Chivers DP, Niyogi S (2016) Modulatory effects of dopamine receptors on associative learning performance in zebrafish (*Danio rerio*). Behav Brain Res 303:109–119
Nathan FM, Ogawa S, Parhar IS (2015) Kisspeptin1 modulates odorant-evoked fear response via two serotonin receptor subtypes (5-HT1A and 5-HT2) in zebrafish. J Neurochem 133(6):870–878
Neuhauss SCF (2003) Behavioral genetic approaches to visual system development and function in zebrafish. J Neurobiol 54(1):148–160
Obholzer N, Wolfson S, Trapani JG, Mo W, Nechiporuk A, Busch-Nentwich E, Seiler C, Sidi S, Sollner C, Duncan RN, Boehland A, Nicolson T (2008) Vesicular glutamate transporter 3 is required for synaptic transmission in zebrafish hair cells. J Neurosci 28(9):2110–2118. https://doi.org/10.1523/JNEUROSCI.5230-07.2008
Okada S (2015) The taste system of small fish species. Biosci Biotechnol Biochem 79(7):1039–1043
Orger MB, Gahtan E, Muto A, Page-McCaw P, Smear MC, Baier H (2009) Behavioral screening assays in zebrafish. In: Essential zebrafish methods: genetics and genomics. Academic Press, Amsterdam, p 113
Patton EE, Zon LI (2001) The art and design of genetic screens: zebrafish. Nat Rev Genet 2(12):956–966
Perkins EJ, Ankley GT, Crofton KM, Garcia-Reyero N, LaLone CA, Johnson MS, Tietge JE, Villeneuve DL (2013) Current perspectives on the use of alternative species in human health and ecological hazard assessments. Environ Health Perspect 121(9):1002

Petzold AM, Balciunas D, Sivasubbu S, Clark KJ, Bedell VM, Westcot SE, Myers SR, Moulder GL, Thomas MJ, Ekker SC (2009) Nicotine response genetics in the zebrafish. Proc Natl Acad Sci U S A 106(44):18662–18667. https://doi.org/10.1073/pnas.0908247106

Pilz PKD, Schnitzler H-U (1996) Habituation and sensitization of the acoustic startle response in rats: amplitude, threshold, and latency measures. Neurobiol Learn Mem 66(1):67–79

Portugues R, Engert F (2009) The neural basis of visual behaviors in the larval zebrafish. Curr Opin Neurobiol 19(6):644–647

Prusky GT, West PWR, Douglas RM (2000) Behavioral assessment of visual acuity in mice and rats. Vis Res 40(16):2201–2209

Puzzo D, Lee L, Palmeri A, Calabrese G, Arancio O (2014) Behavioral assays with mouse models of Alzheimer's disease: practical considerations and guidelines. Biochem Pharmacol 88(4):450–467

Raftery TD, Volz DC (2015) Abamectin induces rapid and reversible hypoactivity within early zebrafish embryos. Neurotoxicol Teratol 49:10–18

Reif DM, Truong L, Mandrell D, Marvel S, Zhang G, Tanguay RL (2015) High-throughput characterization of chemical-associated embryonic behavioral changes predicts teratogenic outcomes. Arch Toxicol 90(6):1459–1470

Richendrfer H, Créton R (2013) Automated high-throughput behavioral analyses in zebrafish larvae, e50622. J Vis Exp (77). https://doi.org/10.3791/50622

Richendrfer H, Creton R (2015) Chlorpyrifos and malathion have opposite effects on behaviors and brain size that are not correlated to changes in AChE activity. Neurotoxicology 49:50–58

Roberts AC, Bill BR, Glanzman DL (2014) Learning and memory in zebrafish larvae. Frontiers in Neural Circuits 7(126):225–235. https://doi.org/10.3389/fncir.2013.00131

Saint-Amant L, Drapeau P (1998) Time course of the development of motor behaviors in the zebrafish embryo. J Neurobiol 37(4):622–632

Samson JC, Goodridge R, Olobatuyi F, Weis JS (2001) Delayed effects of embryonic exposure of zebrafish (*Danio rerio*) to methylmercury (MeHg). Aquat Toxicol 51(4):369–376

Schneider CA, Rasband WS, Eliceiri KW (2012a) NIH image to ImageJ: 25 years of image analysis. Nat Methods 9(7):671–675

Schneider H, Klee EW, Clark KJ, Petzold AM, Mock VL, Abarr JM, Behrens JL, Edelen RE, Edwards BA, Hobgood JS (2012b) Zebrafish and drug development: a Behavioral assay system for probing nicotine function in larval zebrafish. Zebrafish Protocols Neurobehav Res 6:53–70

Schnörr SJ, Steenbergen PJ, Richardson MK, Champagne DL (2012) Measuring thigmotaxis in larval zebrafish. Behav Brain Res 228(2):367–374

Scott GR, Sloman KA (2004) The effects of environmental pollutants on complex fish behaviour: integrating behavioural and physiological indicators of toxicity. Aquat Toxicol 68(4):369–392

Scudder CA (2009) Optokinetic response adaptation. In: Encyclopedia of neuroscience. Springer, London, pp 3042–3044

Selderslaghs IWT, Hooyberghs J, De Coen W, Witters HE (2010) Locomotor activity in zebrafish embryos: a new method to assess developmental neurotoxicity. Neurotoxicol Teratol 32(4):460–471

Sheng L, Wang L, Su M, Zhao X, Hu R, Yu X, Hong J, Liu D, Xu B, Zhu Y (2014) Mechanism of TiO2 nanoparticle-induced neurotoxicity in zebrafish (*Danio rerio*). Environ Toxicol 31(2):163–175. https://doi.org/10.1002/tox.22031

Sison M, Gerlai R (2010) Associative learning in zebrafish (*Danio rerio*) in the plus maze. Behav Brain Res 207(1):99–104

Smith GM, Weis JS (1997) Predator-prey relationships in mummichogs (*Fundulus heteroclitus* (L.)): effects of living in a polluted environment. J Exp Mar Bio Ecol 209(1):75–87

Smith LE, Carvan MJ, Dellinger JA, Ghorai JK, White DB, Williams FE, Weber DN (2010) Developmental selenomethionine and methylmercury exposures affect zebrafish learning. Neurotoxicol Teratol 32(2):246–255

Spulber S, Kilian P, Ibrahim WNW, Onishchenko N, Ulhaq M, Norrgren L, Negri S, Di Tuccio M, Ceccatelli S (2014) PFOS induces behavioral alterations, including spontaneous hyperactivity that is corrected by dexamfetamine in zebrafish larvae. PLoS One 9(4):e94227

Stewart AM, Gaikwad S, Kyzar E, Kalueff AV (2012) Understanding spatio-temporal strategies of adult zebrafish exploration in the open field test. Brain Res 1451:44–52

Sun L, Xu W, Peng T, Chen H, Ren L, Tan H, Xiao D, Qian H, Fu Z (2016a) Developmental exposure of zebrafish larvae to organophosphate flame retardants causes neurotoxicity. Neurotoxicol Teratol 55:16–22

Sun Y, Zhang G, He Z, Wang Y, Cui J, Li Y (2016b) Effects of copper oxide nanoparticles on developing zebrafish embryos and larvae. Int J Nanomedicine 11:905–918. https://doi.org/10.2147/IJN.S100350

Tanimoto M, Ota Y, Horikawa K, Oda Y (2009) Auditory input to CNS is acquired coincidentally with development of inner ear after formation of functional afferent pathway in zebrafish. J Neurosci 29(9):2762–2767. https://doi.org/10.1523/JNEUROSCI.5530-08.2009

Thomas LT, Welsh L, Galvez F, Svoboda KR (2009) Acute nicotine exposure and modulation of a spinal motor circuit in embryonic zebrafish. Toxicol Appl Pharmacol 239(1):1–12

Tiedeken JA, Ramsdell JS, Ramsdell AF (2005) Developmental toxicity of domoic acid in zebrafish (Danio Rerio). Neurotoxicol Teratol 27(5):711–717

Tierney KB (2011) Behavioural assessments of neurotoxic effects and neurodegeneration in zebrafish. Biochimica et Biophysica Acta (BBA)-Molecular Basis of Disease 1812(3):381–389

Tilson HA, MacPhail RC, Crofton KM (1995) Defining neurotoxicity in a decision-making context. Neurotoxicology 16(2):363–375

Tran S, Nowicki M, Muraleetharan A, Gerlai R (2015) Differential effects of dopamine D1 and D2/3 receptor antagonism on motor responses. Psychopharmacology 232(4):795–806

Truong L, Harper SL, Tanguay RL (2011) Evaluation of embryotoxicity using the zebrafish model. In: Drug safety evaluation: methods and protocols. Springer, USA pp 271–279

Valente A, Huang KH, Portugues R, Engert F (2012) Ontogeny of classical and operant learning behaviors in zebrafish. Learn Memory 19(4):170–177. https://doi.org/10.1101/lm.025668.112

Valsamis B, Schmid S (2011) Habituation and prepulse inhibition of acoustic startle in rodents. J Vis Exp 55:e3446–e3446

Vendrell-Llopis N, Yaksi E (2015) Evolutionary conserved brainstem circuits encode category, concentration and mixtures of taste. Sci Rep 5:17825. https://doi.org/10.1038/srep17825

Wang B, Du Y (2013) Cadmium and its neurotoxic effects. Oxidative Med Cell Longev 2013:898034

Wang X, Dong Q, Chen Y, Jiang H, Xiao Q, Wang Y, Li W, Bai C, Huang C, Yang D (2013) Bisphenol a affects axonal growth, musculature and motor behavior in developing zebrafish. Aquat Toxicol 142:104–113

Wang X, Yang L, Wu Y, Huang C, Wang Q, Han J, Guo Y, Shi X, Zhou B (2015) The developmental neurotoxicity of polybrominated diphenyl ethers: effect of DE-71 on dopamine in zebrafish larvae. Environ Toxicol Chem 34(5):1119–1126

Wang X, Yang L, Wang Q, Guo Y, Li N, Ma M, Zhou B (2016a) The neurotoxicity of DE-71: effects on neural development and impairment of serotonergic signaling in zebrafish larvae. J Appl Toxicol. https://doi.org/10.1002/jat.3322

Wang Y, Li S, Liu W, Wang F, Hu L-F, Z-m Z, Wang H, Liu C-F (2016b) Vesicular monoamine transporter 2 (Vmat2) knockdown elicits anxiety-like behavior in zebrafish. Biochem Biophys Res Commun 470(4):792–797. https://doi.org/10.1016/j.bbrc.2016.01.079

Weber DN, Connaughton VP, Dellinger JA, Klemer D, Udvadia A, Carvan MJ (2008) Selenomethionine reduces visual deficits due to developmental methylmercury exposures. Physiol Behav 93(1):250–260

Weis JS, Candelmo A (2012) Pollutants and fish predator/prey behavior: a review of laboratory and field approaches. Current Zoology 58(1):9–20

Weis JS, Weis P (1995a) Effects of embryonic exposure to methylmercury on larval prey-capture ability in the mummichog, *Fundulus heteroclitus*. Environ Toxicol Chem 14(1):153–156

Weis JS, Weis P (1995b) Swimming performance and predator avoidance by mummichog (*Fundulus heteroclitus*) larvae after embryonic or larval exposure to methylmercury. Can J Fish Aquat Sci 52(10):2168–2173

Weis JS, Samson J, Zhou T, Skurnick J, Weis P (2003) Evaluating prey capture by larval mummichogs (*Fundulus heteroclitus*) as a potential biomarker for contaminants. Mar Environ Res 55(1):27–38

Williams FE, White D, Messer WS (2002) A simple spatial alternation task for assessing memory function in zebrafish. Behav Process 58(3):125–132

Wolman MA, Jain RA, Liss L, Granato M (2011) Chemical modulation of memory formation in larval zebrafish. Proc Natl Acad Sci U S A 108(37):15468–15473. https://doi.org/10.1073/pnas.1107156108

Yen J, Donerly S, Levin ED, Linney EA (2011) Differential acetylcholinesterase inhibition of chlorpyrifos, diazinon and parathion in larval zebrafish. Neurotoxicol Teratol 33(6):735–741

Zeddies DG, Fay RR (2005) Development of the acoustically evoked behavioral response in zebrafish to pure tones. J Exp Biol 208(Pt 7):1363–1372. https://doi.org/10.1371/journal.pone.0082182

Zhang X, Hong Q, Yang L, Zhang M, Guo X, Chi X, Tong M (2015) PCB 1254 exposure contributes to the abnormalities of optomotor responses and influence of the photoreceptor cell development in zebrafish larvae. Ecotoxicol Environ Saf 118:133–138. https://doi.org/10.1016/j.ecoenv.2015.04.026

Zhou T, Weis JS (1998) Swimming behavior and predator avoidance in three populations of *Fundulus heteroclitus* larvae after embryonic and/or larval exposure to methylmercury. Aquat Toxicol 43(2):131–148

Zhou T, Scali R, Weis JS (2001) Effects of methylmercury on ontogeny of prey capture ability and growth in three populations of larval *Fundulus heteroclitus*. Arch Environ Contam Toxicol 41(1):47–54

Chapter 9
The Application of Omics Data to the Development of AOPs

Mary T. McBride

Abstract Omics approaches offer potential for use in chemical hazard and risk assessments when applied as part of a systems toxicology or integrative approach, and when considered in the context of the adverse outcome pathway (AOP) framework. Omics data provide individual snapshots of gene expression, protein expression and metabolite activity. When integrated, these individual snapshots yield deep biological insights. Omics can provide mechanistic information about the effects of chemicals and can help decipher toxicity mechanisms and modes of action. Omics data have the potential to increase confidence in species extrapolation, and can be used to identify biomarkers of exposure and toxicity. Although omics have been used for more than a decade, acceptance of omics data in regulated applications has been slow. The toxicology community is grappling with how to make use of omics data in a regulatory framework, and how to use AOPs to drive regulatory decision-making processes. In this chapter, an overview of major omics is provided that includes recent advances and describes the potential application of omics data to the development of AOPS while defining some of the challenges associated with the broader acceptance of omics within a regulatory toxicology framework.

9.1 Introduction: Omics and Adverse Outcome Pathways

Traditional toxicity testing of industrial chemicals, pesticides, and pharmaceuticals involves exposing animals to high doses of toxicants, observing the effects, and trying to set safe exposure levels in humans by extrapolating to expected human responses at lower doses. Whole animal testing based on well-established endpoints provides a means to directly measure and quantify adverse effects at the tissue, organ, and organism level (Suter et al. 2004). However, these tests typically do not yield insights about crucial cellular or molecular responses from which we can

M.T. McBride (✉)
Global Market Regulations and Standards Strategy, Agilent Technologies, Inc.,
5301 Stevens Creek Blvd, Santa Clara, CA 95050, USA
e-mail: Mary_mcbride@agilent.com

N. Garcia-Reyero, C.A. Murphy (eds.), *A Systems Biology Approach to Advancing Adverse Outcome Pathways for Risk Assessment*,
DOI 10.1007/978-3-319-66084-4_9

begin to understand and identify fundamental mechanisms or mode-of-action underlying toxicity. Mechanistic insights are critical to understand the causal linkages between exposures and adverse outcomes, and to formulate quantitative linkages between molecular and cellular events that go beyond simple correlations.

In an effort to overcome the shortcoming and limitations of traditional animal-based toxicity testing, many groups are turning to approaches that integrate biochemical and cell-based assays with high-content omics technologies, bioinformatics and computational tools – approaches that offer potential for development of predictive toxicology (Bhattacharya et al. 2011; Cote et al. 2012; Hege Harrill and Rusyn 2008; NRC 2007). For example, the Human Toxome project (http://human-toxome.com/) aims to map human toxicity pathways (i.e., the human toxome). This consortium has begun by mapping estrogenic pathways in human breast cancer cells using a combination of transcriptomics and metabolomics (Bouhifd et al. 2014; Hartung and McBride 2011). In another approach, scientists at the Hamner Institutes have used well-characterized compounds to map and model a small number of well-studied "prototype" pathways, including ligand activated pathways (estrogen receptor, PPARα, and AhR) and stress activated pathways (DNA damage and oxidative stress) that respond to environmental toxicants (Andersen et al. 2011). Case studies are being developed to demonstrate how to use the new types of information for human safety assessments. Detailed mechanistic studies reveal the differences in nuclear receptor biology (a) between rats and humans, and (b) in vitro and in vivo in the rat. These results have informed development of a novel in vitro assay for receptor-mediated cell proliferation that have been extensively validated using prototypical CAR agonists (McMullen et al. 2014, McMullen et al. 2016). In additional to individual investigations and smaller-scale consortium, several large-scale programs designed to develop, capture, catalog, and utilize mechanistic data obtained from biological testing systems (including omics) have also been established in recent years. These program include EPA's ToxCast (http://epa.gov/ncct/Tox21/) (Dix et al. 2007; Judson et al. 2010; Martin et al. 2010), the Tox-21 Consortium (Tox21C), a collaborative research effort between the EPA, NIH and FDA (http://tox21.org/) (Attene-Ramos et al. 2013), SEURAT (http://www.seurat-1.eu/) and the OECD Adverse Outcome Pathway (AOP) initiative (OECD 2013a, b; Tralau and Luch 2015).

Omics are methods for the comprehensive study and analysis of complex biological samples. Although there are numerous omics, toxicology applications focus primarily on genes (genomics), mRNA (transcriptomics), proteins (proteomics), and metabolites (metabolomics). Traditional toxicology evaluates end points (e.g., phenotypic changes, disease, death) while omics measurements made at molecular and cellular levels provide information that, when combined, reveal the relationships between genes, proteins, and metabolites, and facilitate understanding of molecular and cellular processes as an integrated system rather than as a collection of disparate measurements or individual endpoints. Omics data can form the basis of computational models that can be used to quantify the degree of molecular or cellular perturbations and may also accelerate the development of dynamic multi-scale biological models that will extend our ability to link exposures beyond the

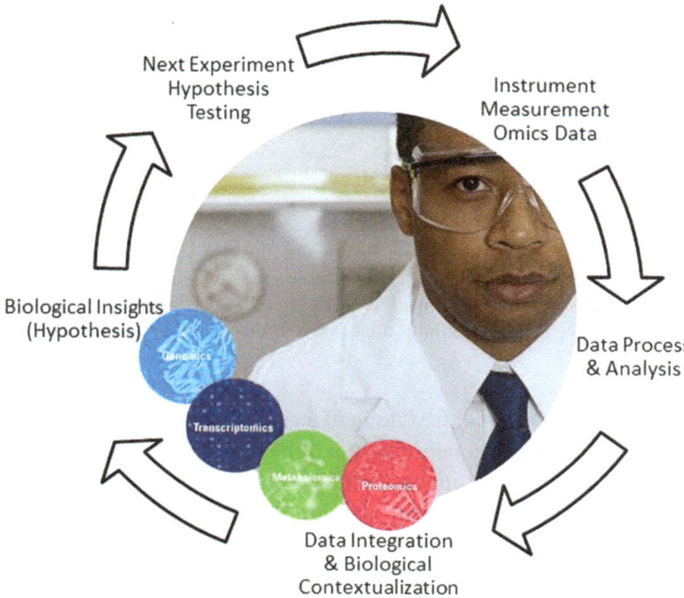

Fig. 9.1 A schematic depiction of an integrated omics experiment. Omics experiments may utilize one or more types of experimental data sets (e.g., genomics, proteomics, metabolomics) measured using different instrument platforms. Data sets can then be individually process and then integrated, and using appropriate bioinformatics tools, the relationships between genes, proteins and metabolites can be visualized, garnering new insights into fundamental disease mechanisms. Placing data into biological context promotes deeper insights and better understanding, and facilitates hypothesis testing while helping to formulate future experiments

molecular and cellular levels to tissue or organ-level responses. Omics have been used for more than 10 years to identify, classify, characterize, screen and prioritize chemical compounds. For toxicology, most omics studies extend well beyond simple interpretations of gene expression or creating lists of metabolites; most studies integrate endpoints at higher levels of biological organization (Connon et al. 2012). Some examples include associating mechanistic responses to changes in reproduction, growth rates, viability of offspring, or other or physiological functions (Garcia-Reyero et al. 2011; Van Aggelen et al. 2010). Omics are used to evaluate the effects of compounds across doses, exposure times, and species; to identify novel signatures or biomarkers of toxicity; to study toxicity pathways and to elucidate mode-of-action (Waters and Fostel 2004; Fig. 9.1).

Moving towards more mechanistically-based risk assessments has increased the complexity of data that toxicologists must now consider, including in vitro assays, high-throughput screening results, computational models, and omics experiments which generate very large data sets. The concept of adverse outcome pathways (AOPs) was proposed to provide a framework for collecting and organizing all of the existing knowledge associated with toxicological processes, from exposure to

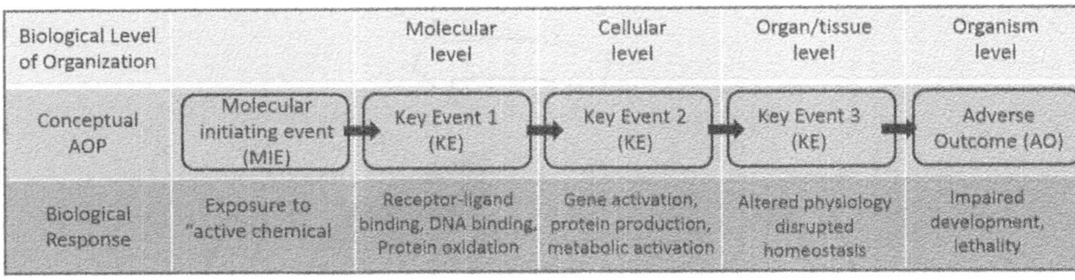

Fig. 9.2 Conceptual diagram of an adverse outcome pathway (AOP). Each AOP begins with a molecular initiating event (MIE) in which a chemical interacts with a biological target leading to a series of key events (KEs) that result in an adverse outcome (AO). Biological levels of organization are shown in the *top row*, while potentially measurable typical biological responses are shown in the *bottom row*

adverse outcomes. Ankley et al. first defined an AOP as "a conceptual construct that portrays existing knowledge concerning the linkage between a direct molecular initiating event (e.g., a molecular interaction between a xenobiotic and a specific biomolecule) and an adverse outcome at a biological level of organization relevant to risk assessment" (Ankley et al. 2010). Although AOPs are often depicted as a series of linear events (Fig. 9.2) they involve multiple independent, interacting response networks where linked events span biological levels and involve many biological entities (EFSA 2014).

The linkages between the molecular initiating event (MIE), the series of intermediate key events (KEs), and the adverse outcome (AO) may be causal, mechanistic, inferred, or correlative and the information about these linkages may come from various sources including in vivo tests, molecular and cell-based screening assays, omics measurements, and computational methods (Groh et al. 2015). The KEs individually correspond to empirically observable precursor steps that form parts of toxicity pathways and mode-of-action (MOA); as such, they should be definable and make sense from physiological and biochemical perspectives (OECD 2013a, b). The AOP framework provides structure – a way to collect, organize, and display knowledge across multiple levels of biological organization (i.e., molecular, cellular, tissue, etc.) that also helps to identify key gaps and facilitates prioritization of research needed to fill those gaps (Villeneuve et al. 2014).

Well-developed AOPs are intended to serve as the central element of a toxicological knowledge framework and are expected to guide experimental testing approaches to support risk assessments and regulatory decision-making. While there are currently more than 100 AOPs under development, only a few could be considered complete or nearly complete. However, even incomplete AOPs can inform chemical grouping, and "read-across" (predicting the toxicity of a chemical based on structural similarities to other chemicals). (OECD 2013a, b), AOPs that are more complete, with linkages that include quantitative information could form the basis for an integrated approach to testing and assessment (IATA) and guide development of integrated testing strategies (ITS) (Tollefsen et al. 2014).

Omics technologies present both opportunities and challenges for toxicology, and for human health risk assessments. Although omics have been used for more than a decade, acceptance of omics data in regulated applications has been slow. The toxicology community is grappling with how to make use of omics data in a regulatory framework, and how to use AOPs to drive regulatory decision-making processes. The goal of this chapter is to illustrate the application of omics to the development of AOPs. Here, an overview of each the four major omics – genomics, transcriptomics, proteomics and metabolomics – is presented, including recent development and advances in each of these areas that are applicable to their use in AOPs, followed by a summary of how omics data has been or could potentially be applied to the development of AOPs, and concludes with a discussion of some of the challenges associated with multi-omics data and the issues facing the broader acceptance of omics within a regulatory toxicology framework.

9.2 Overview of OMICS

9.2.1 Genomics

The completion in 2003 of the Human Genome Project catalyzed the application of genomics to understand the effects of drugs, industrial chemicals and other environmental stressors on biological systems (Collins et al. 2003). Genomics is a scientific discipline that studies genome structure and function. Genome sequencing provides the specific order and identity of DNA nucleotide bases. Sequence information can be used to identify functional regions of the genome (e.g., protein-coding genes, regulatory sequences, non-coding regions), and genomes can be compared to look for differences between genomes as well as for structural variations within a single genome (e.g., single nucleotide polymorphisms, insertions, deletions, duplications, copy number variations, methylation).

Recent Advances in Genomics The field of genomics is progressing so rapidly that the National Human Genome Research Institute (NHGRI), one of the 27 Institutes at the National Institutes of Health (NIH) produces a monthly highlight of what it considers to be the "coolest genomic advances, broadly defined". http://www.genome.gov/27543594. Genomics technologies have developed and evolved at an amazing pace in recent years, transforming our ability to catalogue and study the information stored in genomes. Conventional capillary electrophoresis (CE) sequencing, also known as "Sanger" or "first generation" sequencing methods have been used for several decades to determine the order of the DNA base-pairs one by one, and the technology played a central role in the Human Genome Project. While CE sequencers are still used, and in fact still provide the "gold standard" for sequencing accuracy, these applications have all but been replaced by next-generation sequencing (NGS) approaches that enable massively parallel sequencing of billions of DNA molecules simultaneously. NGS (also referred to as "second generation" sequencing) has substantially reduced the time and costs of sequencing and dramatically increased sequence output. The human genome, comprising 3.1 billion nucleotide bases required nearly $22 M and several years to sequence just a decade ago; NGS platforms today can sequence that same genome for less than $10 K in a matter of a few days and continuing innovations in NGS benchtop sequencers are rapidly closing in on the goal of achieving the $1 K genome. A number of excellent reviews describe various NGS platforms in detail and compare their performance characteristics (Mehinto et al. 2012; Loman et al. 2012; Metzker, 2010; Su et al. 2011).

Ultrafast DNA sequencing represents the third generation in DNA sequencing and many strategies are under development; these include sequencing-by-hybridization, nanopore sequencing, and sequencing-by-synthesis. These technologies sequence single DNA molecules in real time, in contrast to next-generation instruments that sequence millions of very short DNA fragments in parallel. Third generation strategies and platforms have been reviewed and compared but all of these approaches provide improvements over current methods including higher-throughput, faster turn-around times, longer read lengths, and reduced costs (Pareek et al. 2011).

Advances in NGS have enabled a host of new applications such as the 1000 Genomes Project (a population-based whole genome sequencing effort to identify common genetic variants), the Cancer Genome Atlas (TCGA), an effort to accelerate our understanding of the molecular basis of cancer, and many other large-scale research efforts (Green et al. 2011). The lower costs and higher throughput of genome sequence information enabled by NGS have fueled tremendous growth in disease research that aims to associate or correlate structural variations in DNA with diseases. The most common structural variation is the single nucleotide polymorphism (SNP). SNPs are mutations in single nucleotides found throughout the genome that have a phenotypic consequence that can often be associated with a disease. Consequently, considerable effort has been focused to identify these SNP "biomarkers". Another structural variation is called the copy number variation (CNV) which is an alteration where there is a gain or loss of large amounts of DNA sequence; a number of studies have established correlations between CNVs and disease (Girirajan et al. 2011).

NGS, coupled to other technologies such as oligonucleotide microarrays have also enabled significant advances. Genome-wide association studies (GWAS) were one of the most commonly used approaches to compare genomes: a typical GWAS experiment might involve comparisons of large numbers of genomes from well-phenotyped individuals to look for structural variants including SNPs. GWAS studies are focused largely on finding small differences between genomes; these discoveries direct research towards targeted therapeutics for diseases, and have given rise to entirely new disciplines such as epigenetics (looking for ways in which the DNA itself rather than the nucleotide bases or the sequence gets modified) which in turn effects gene expression and gene regulation. Hybrid technologies, like chromatin immune-precipitation coupled to DNA microarray (ChIP-chip) or sequencing

(ChIP-seq) have been used to probe the genome-wide location and function of DNA binding proteins and facilitate studies of DNA-protein interactions to unravel how various transcription factors and other proteins interact with DNA to regulate gene expression, while RNA sequencing (RNA-seq) enables sequencing of RNA transcripts, a technique that vastly expands upon, and compliments, microarray-based gene expression studies.

CRISPR/Cas9 is a new genomics advance that enables genome editing, allowing researchers to make precise, targeted changes to the genome. This tool allows scientists to study how changes in a gene sequence can affect gene function (Konermann et al. 2015; Swiech et al. 2015). While the system has been studied since the 1980s (Ishino et al. 1987) it has garnered significant attention since a 2012 publication appeared, in which a team a scientists reported the use of the system as a gene-editing tool (Jinek et al. 2012). That paper launched a flurry of genetic engineering activity, with more than 20 new papers published each week on the topic today (Cong et al. 2013).

Challenges for Genomics Demand for NGS systems, driven by lower costs and higher throughput, is robust and does not show any signs of abating. The biggest challenge for researchers using NGS approaches is managing the huge amounts of data that are generated. Tools are required for data collection, storage, tracking, and processing; adequate tools are not really readily available although these issues constitute active areas of research and development. Enabling better data flow between data producers and data consumers or end-users will require specialized data architecture and better integration with information systems, while interpretation of data for clinical/diagnostics use will require the development of specialized tools and the development and sharing of genome knowledge-bases.

9.2.2 Transcriptomics

The human genome is estimated to contain about 21,000 genes, but at any given time, only a small fraction of genes are active. Assessing global gene expression in response to environmental stress, genetic perturbations, or cell lifecycle is an essential step toward unraveling and understanding toxicological mechanisms or mode-of-action. Gene activity can be inferred by identifying proteins but protein studies are often very complex and challenging. It is far simpler to study gene expression by examining the RNA message or "transcript". The transcriptome comprises ~100,000 mRNA molecules, and also includes non-coding RNAs such as rRNA, tRNA, and micro-RNAs.

Transcriptomics studies are usually carried out using oligonucleotide microarrays, NGS approaches, or real-time polymerase chain reaction (RT-PCR), although RT-PCR is not used nearly as often for transcriptomics studies as either microarrays or sequencing. Sequencing methods, collectively referred to as RNA-seq, include methods for determining sequence content, as well as the abundance of mRNAs, non-coding RNAs and miRNAs, as well as ChIP-seq (chromatin immunoprecipitation methods for measuring DNA-protein complexes) and methyl-seq (used to study methylation sites). Microarrays and RNA-seq both capture the characteristic and specific patterns of gene expression (i.e., "signatures") for thousands of genes simultaneously that result from exposures to a given toxicant under a given set of experimental conditions and provide quantitative measurements of the dynamic expression of mRNA molecules. This is in contrast to the static measure of DNA provided by gene sequencing.

Microarray-based transcriptomics have been used for many years and is by far the most widely-used of the omic approaches in toxicology. Cellular response to toxicant exposures for the entire genome can be probed in a single microarray experiment. Gene expression profiling enables the identification of specific genes that are differentially expressed as a result of changes in environmental conditions. Linking these gene changes to a chemically-induced phenotype (i.e., "phenotypic anchoring") facilitates predictive toxicity and elucidation of mode-of-action (Cui and Paules 2010). Gene expression profiles obtained on separate arrays can be compared to evaluate the effects of different compounds, doses and exposure times across species, or between/within populations (Gerecke et al. 2009). Genes from different samples that exhibit the same or similar expression profiles can be identified using statistical methods (e.g., clustering techniques), leading to potential insights regarding common pathways, or mode-of-action, assuming the clustered genes are functionally related (Afshari et al. 2011). Gene function and gene relationships within networks can be established and verified using gene knockout or silencing techniques. Gene expression signatures also enable toxicants to be grouped or classified into different toxicity classes, usually based on potency or mode-of-action, and facilitate the prediction of toxicity of chemically-related compounds (Fielden et al. 2007).

Gene expression profiles can guide the identification of biomarkers of toxicity, even at very low exposure doses when no phenotypic changes have been observed. For example, Heinloth et al. demonstrated how the analysis of gene expression profiles from liver samples obtained from rats exposed to sub-toxic doses of acetaminophen indicated subtle cellular injury that was not detectable by histopathology or clinical chemistry methods (Heinloth et al. 2004). Such biomarkers of toxicity could identify potentially toxic drug candidates even when there are no indicators of toxicity in preclinical studies (McBurney et al. 2009; McBurney et al. 2012). These biomarkers could serve as the basis for suites of in vitro assays to assist in compound screening, to group chemicals by toxicity class or mode-of-action, to monitor drug therapies for safety and efficacy, and to monitor for exposures to environmental toxicants, even at sub-critical exposure levels.

Recent Advances in Transcriptomics Advances in NGS are driving advances in transcriptomics. RNA transcripts can be sequenced in a cell and used to study RNA expression patterns, point mutations, alternative gene spliced transcripts, post-transcriptional changes, gene fusion, SNPs and other mutations and changes in gene expression. RNA-seq is increasingly being used to discover and study different types of RNA (miRNA, siRNA, lincRNA, tRNA, etc); results of these studies have the potential to identify new biomarkers as well. Compared to microarrays, RNA-seq offers improved sensitivity, better precision, a much greater dynamic range (microarrays lack sensitivity for genes expressed at either very low or very high levels), and better reproducibility for both technical and biological replicates (Chen et al. 2012). Progress in the application of transcriptomics to toxicology has been impeded by a lack of discovery and data mining tools. Two of the largest toxicogenomics databases, the Japanese Toxicogenomics

Project (TGP or TG-GATEs) and DrugMatrix were made publically available in 2011. The two databases were described and compared in a recent review (Chen et al. 2012). Access to these large data repositories is expected to accelerate the development of new bioinformatics and data mining tools and to provide new opportunities for knowledge discovery.

Challenges for Transcriptomics Because gene expression profiling is largely global in nature, such experiments generate massive amounts of data. Analysis of this data requires a combination of statistical tools, bioinformatics, and databases, and usually requires expertise in the biological system under study. While many of the bioinformatics software tools and databases have become standardized, interpretation of the data remains a significant challenge. Linking observed changes in gene expression profiles to conventional toxicological endpoints (i.e., phenotypic anchoring) remains a central challenge. Gene expression analysis results often do not directly correlate to results from proteomics or metabolomics; although all proteins are based on mRNA precursors, the expression level of a given gene that codes for production of a protein does not correspond to the amount of protein produced, as the expression level alone does not account for post-translational modifications or other ways in which proteins are regulated. This example underscores the need to fully utilize all available biological information like that obtained from integrated omics studies and to combine that information with computational modeling. No single biomarker or set of signatures yet effectively serves as a disease or disease-state indicator and expression-based diagnostics are not yet at the point where they can reliably predict disease or disease outcomes. This may be due to a variety of factors, including cellular heterogeneity (lack of pure cell populations that yield distinct profiles) or genetic heterogeneity (individuals in a population will not have the same expression profiles, even when they all have the same disease) (Chuang et al. 2010). RNA-seq transcriptomics presents many of the same challenges that were discussed for genomics (above), namely dealing with large amounts of data, and developing tools for the management, processing, analysis, and interpretation of this data.

9.2.3 Proteomics

Proteomics is the comprehensive study of the entire compliment of proteins and their modifications (i.e., the proteome) of an organism. The human proteome, estimated to comprise between 250,000 and 1 million proteins (along with their post-transcriptional, translational, and post-translational modifications) is highly dynamic – it varies over time and even varies from cell to cell. Proteins exist in concentrations that can span nine orders of magnitude, making low abundance proteins extremely difficult to detect and characterize. Thus, proteomics measurements are far more complex and challenging compared to the relatively straightforward and somewhat static human genome and the smaller, more tractable human transcriptome. Researchers perform a variety of proteomics studies that include the global identification of all proteins in a sample (protein profiling) using discovery or "shotgun" proteomics, the quantitative measurement of protein expression (i.e., abundance), the study of protein structures, including protein variations and modifications, and the interactions of proteins and other molecules (e.g., protein-protein, protein-DNA, etc.).

Recent Advances in Proteomics The main technologies used in proteomics are two-dimensional gel electrophoresis (2-DE) and liquid chromatography tandem mass spectrometry (LC–MS/MS). Advances in mass spectrometry in the past few years now enable the routine identification and quantification of thousands of protein components in samples and consequently, most proteomics studies are now performed using liquid chromatography/mass spectrometry (LC/MS) because of its sensitivity, selectivity, accuracy, speed and throughput (Chen and Pramanik 2009).

To date, most proteomics research has been done in an untargeted or discovery mode. This approach has been used primarily to identify all proteins in a given sample (protein profiling) without any prior knowledge of what proteins might be present in a sample. More recently it has been used for differential quantification of the identified proteins. A typical proteomics workflow is shown in Fig. 9.3. In this approach, proteins extracted from cells, tissues or other complex sample matrices are prepared in a series of steps (determined by experimental objectives) that may include cell lysis, pre-fractionation, or other separation, purification and concentration techniques. Proteins are enzymatically digested into their smaller constituent peptide fragments. Samples containing multiple proteins will generate many thousands or hundreds of thousands of peptide fragments. To simplify analysis, peptides are separated using liquid chromatography; peptides within LC fractions are ionized and passed to the mass spectrometer. The mass analyzer filters the ions and records their mass-to-charge (m/z) ratio along with their relative abundance as peaks that populate a mass spectrum. Ions comprising specific peaks (precursor ions) are selected and further analyzed by tandem MS (MS/MS) to generate characteristic fragment ions. The combinations of precursor m/z and their associated fragment ions are then compared to sequences of known peptide fragments and identified. Fragments are further assembled to enable identification of the protein sequence. While this approach has enabled significant advances to whole proteome

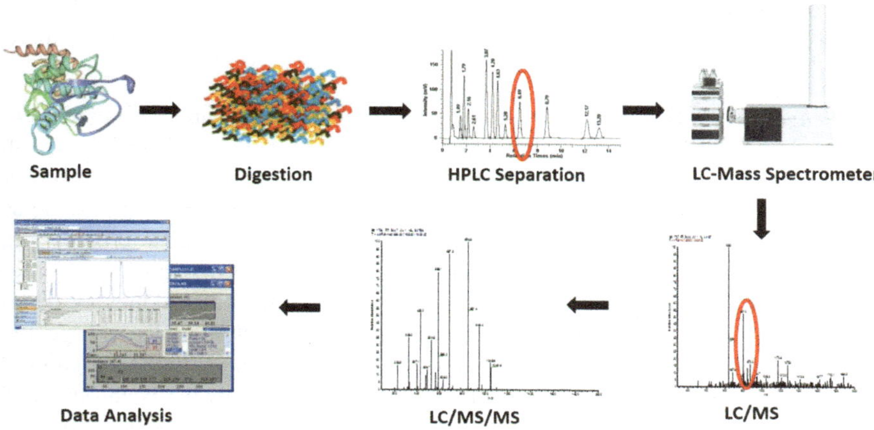

Fig. 9.3 A typical LC-MS/MS proteomics workflow, for either discovery or targeted proteomics. Extracted proteins are digested to peptide fragments; peptide fractions are then further separated and identified using HPLC. HPLC peptide fractions (a single fraction is indicated in the red circle in the HPLC chromatogram) are ionized and passed to the mass spectrometer. The mass analyzer filters the ions and records their mass-to-charge (m/z) ratio along with their relative abundance as peaks that populate a mass spectrum. Ions comprising specific peaks (precursor ions, indicated by the red circle in the LC/MS spectrum) are selected and further analyzed by tandem LC (MS/MS) to generate characteristic fragment ions. The combinations of precursor m/z and their associated fragment ions are then compared to sequences of known peptide fragments and identified. Fragments are then quantified and may be further assembled to enable identification of the protein sequence

identification and mapping, it suffers from significant shortcomings: (1) the analysis of a complete proteome remains challenging, expensive and time-consuming and only a few labs have become truly expert in this approach; (2) results often cannot be reproduced because of the way in which precursor ions are selected- even within the same lab using the same sample; (3) the approach does not enable identification of low-abundance proteins; and (4) in any experiment designed to address a specific scientific question, a large numbers of "irrelevant" proteins will be identified, while some number of relevant proteins will be missed (Domon and Aebersold 2010).

The emerging strategy of targeted proteomics enables researchers to detect, identify and quantify specific aspects of the proteome. In a targeted approach, the proteins of interest are known in advance and the MS is programmed to select only those certain signature peptides using a technique known as selected reaction monitoring (SRM) or sometimes referred to as multiple reaction monitoring (MRM). This approach enables much greater sensitivity over discovery-based approaches, and enables detection of low-abundance proteins. Absolute and relative quantification is possible. SRM approaches are inherently multiplexed; tens to hundreds of proteins can be monitored during the same experiment. It also provides vastly improved reproducibility such that multiple labs can produce identical results (Marx 2013). One challenges to the targeted proteomics approach is that despite mass spectrometry sensitivity at the attomolar level, not all proteins will be detected and only a limited number of measureable proteins can be included in the same experiment (Titz et al. 2014).

Challenges for Proteomics Within toxicology, proteomics research efforts (toxicoproteomics) have been largely directed towards identification of biomarkers with prognostic or diagnostic value, reflecting the fact that discovery (untargeted) proteomics has been the dominant strategy for the past decade. Biomolecules serve as early indicators of disease and they can be used to monitor disease progression, pharmacologic therapeutic response, and adverse responses to toxicants. Biomarker discovery and identification has been largely focused on liver and kidney as a consequence of studies driven by the pharmaceutical sector, although disease-specific markers have also been identified (van Vliet 2011; Altelaar et al. 2013). Progress in biomarker discovery, identification and validation for toxicology has been very slow and many early supporters in the field have become disillusioned. The slow progress does not reflect a lack of suitable biomarkers; rather, it reflects the inherent challenges of using an untargeted approach to discovery. Targeted proteomics is enabling rapid advances within in vitro toxicology, for both biomarker discovery as well as for expanding and developing our understanding of pathway-based molecular mechanisms of toxicity. For example, identification and quantitation of proteins in a sample can reveal that a signaling pathway is active; conversely, knowledge of signaling pathways can be used to map and model human responses to chemical exposures or to pharmaceuticals (Collings and Vaidya 2008).

Like other omic approaches, proteomics experiments generate very large data sets that present significant data management, storage, transfer, analysis, and interpretation challenges. Analysis is complex, and requires specific tools for data processing, including statistical methods, databases, and bioinformatics tools. Although significant progress has been made, much more needs to be done. Another major challenge that will confront regulators is the lack of standardization across proteomics technologies. For example, standardized sample preparation, handling, and processing protocols should be implemented because molecular profiles obtained from omics studies may be very sensitive – results can vary widely as a consequence of differences in specimen type, collection/isolation/storage/processing methods, the volume of sample used versus volume sample required for accurate result, the number of replicate samples run vs. the number of replicates needed for statistical analysis, and so on.

9.2.4 Metabolomics

Metabolites are small molecules, such as amino acids, lipids, organic acids and sugars that are intermediate or end products of metabolism. Unlike genes and protein that can be altered and are subject to regulatory processes, metabolites are the downstream products of gene expression (and also the end product of a toxic insult) and directly reflect biochemical end products that are closer to the phenotype (van Ravenzwaay et al. 2007). Metabolomics is the study of metabolites and is used to identify all of the metabolites present in a given cell or organism at a specific time (global metabolite profiling) or to characterize specific metabolites with respect to concentration or other parameters. In 2007, scientists completed the first draft of the human metabolome, cataloging approximately 2500 metabolites, 1200 drugs and 3500 food components; this information is available in the Human Metabolome Database (www.hmdb.ca), although it is still incomplete (Wishart et al. 2007).

Recent Advances in Metabolomics Modern metabolomics research had its origins in nuclear magnetic resonance spectroscopy (NMR) but over the past two decades mass spectrometry-based studies of metabolomics have become much more prevalent than NMR, due to the high sensitivity, specificity, and ability of MS to detect and identify large numbers of metabolites. Gas chromatography/mass spectrometry (GC/MS) was used to study complex samples and later researchers expanded into liquid chromatography/mass spectrometry (LC/MS), driven by the advent of affordable, accurate mass, time-of-flight (TOF) instruments. The advantages and limitations of each technology have been the subject of numerous reviews (Bouhifd et al. 2013; Dunn and Ellis 2005).

Metabolomics experiments are conducted using either targeted or untargeted strategies (Fig. 9.4). Targeted metabolomics is a method used to determine the relative abundances and concentrations of a specific set of pre-selected metabolites, usually related to a specific metabolic pathway. Targeted applications typically employ triple quadruple LC/MS or GC/MS because the QQQ provides reliable, sensitive and reproducible quantitative analysis. The method requires that the exact structure of metabolites are known; therefore the instrument is first optimized against standard compounds in selected reaction monitoring. While the method is quantitative and enables direct comparisons of metabolites between samples, it also requires that the exact structures of the metabolites under study are known and usually requires the use of analytical standards. Therefore, targeted metabolite studies are limited to those metabolites catalogued in searchable mass spectra libraries; available metabolomics databases, along with bioinformatics tools to facilitate data analysis and interpretation have been described (Baker 2011; Go 2010; Patti et al. 2012; Wishart et al. 2013).

Untargeted (discovery) metabolomics methods are used to establish the metabolite profile of a given sample. Untargeted metabolomics approaches usually employ TOF or QTOF mass analyzers, as the instrument enables high resolution and accurate mass measurements for identification and characterization, particularly with unknown compounds. Discovery metabolomics experiments involve examining an untargeted and unbiased suite of metabolites, finding the ones with statistically significant variations in abundance within a set of experimental versus control samples, and determining their chemical structure. An interpretation step allows the researcher to connect the metabolite with the biological process or condition.

Metabolomics has been expanding rapidly and applications are now routine in the areas of system biology, drug discovery, pharmaceutical research, early disease

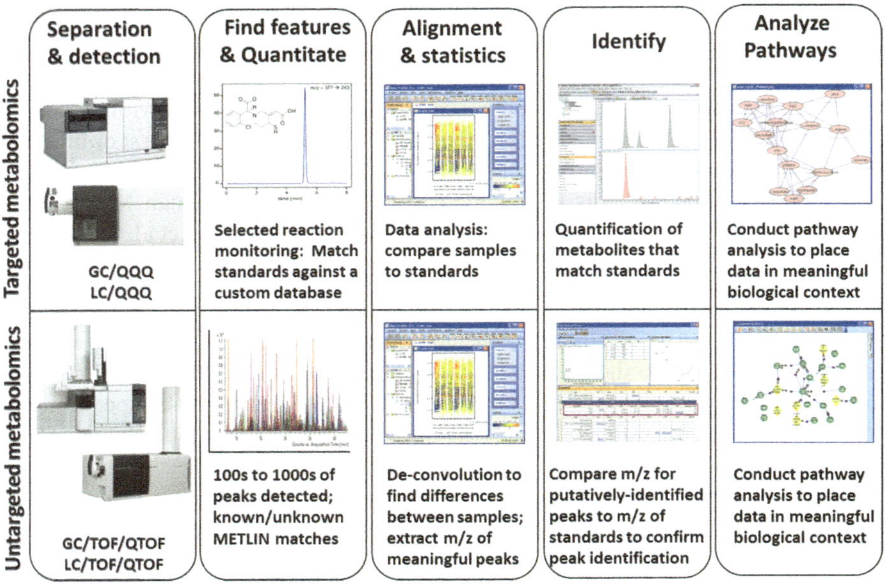

Fig. 9.4 Mass-spectrometry-based metabolomics workflows, for targeted (*upper*) and untargeted (*lower*) applications. Targeted metabolomics is used to determine the relative abundances and concentrations of a specific set of pre-selected metabolites, usually related to a specific metabolic pathway. Sample metabolites are compared to standards and exact matches quantified. Untargeted (discovery) metabolomics experiments involve examining an untargeted and unbiased suite of metabolites, finding the ones with statistically significant variations in abundance within a set of experimental versus control samples, and determining their chemical structure. An interpretation step allows the researcher to connect the metabolite with the biological process or condition

detection, toxicology, newborn screening, food safety and nutrition science and others. Metabolomics is finding broad acceptance and ready adoption in toxicology. Even as early as 2000, metabolomics was explored as a technique for rapid in vivo screening. The Consortium for Metabonomic Toxicology (COMET) performed NMR-based studies to predict liver and kidney toxicity using serum and urine samples from rodents; that data is still used today (Lindon et al. 2005). The same approach has been extended more broadly and now in vivo metabolomics are routinely used in drug development to screen for potential toxic effects of drug candidates, as well as for mode-of-action studies (van Ravenzwaay et al. 2012). Metabolomics is also being applied to in vitro toxicology. Rameriz et al. have provided a long list of suggested in vitro metabolomics applications for toxicology and connected these suggestions to their actual implementation through active research efforts (Ramirez et al. 2013). Just a few of the application areas they identified are: (1) development of prediction models, where metabolite profiles obtained from training compounds of known toxicities could be compared to unknown compounds to predict their potential toxicity; (2) to rank/prioritize compounds and to sort or classify molecules with respect to their mode-of-action or predicted toxicities; (3) to use pathway-based knowledge to pinpoint potential drug/compound molecular targets and predict their mode-of-action and to map and model pathways of toxicity; and (4) biomarker discovery.

Challenges for Metabolomics In order to gain acceptance by the regulatory community, metabolomics will have to overcome a number of technical challenges. Quality control methods must be developed that ensure that in the process of preparing and analyzing samples, potential artifacts are not introduced. Advances are needed for sample throughput that enable faster, more robust, reliable and repeatable sample preparation, measurement and analysis especially since mass spectrometry experiments require that a large number of biological replicates are analyzed. Metabolite identification is limited by existing databases and additional effort is needed to expand and continuously update these databases. Similar to proteomics, metabolomics experiments suffer from a lack of standardization that spans technology platforms, analytical methods, statistical methods, data analysis and interpretation, etc. Metabolomics also generates very large data sets that present significant data management, storage, transfer, analysis, and interpretation challenges. Metabolomics analysis is complex, and requires specific tools for data processing, including statistical methods, databases, and bioinformatics tools.

9.3 Application of OMICS Data to AOPs

Omics approaches hold considerable promise for the future of toxicity testing and for the development of hazard assessments tools including AOPs. Their utility has been demonstrated over the past decade, where omics approaches have been used to identify, classify, characterize, screen and prioritize chemical compounds; to evaluate the effects of compounds across doses, exposure times, and species; to identify novel signatures or biomarkers of toxicity; to study toxicity pathways and to elucidate mode-of-action (Waters et al. 2004). Omics studies can be used to bridge in vitro and in vivo data as well; for example, a hypothesis for MOA in a system can be evaluated using appropriate in vitro and/or in vivo omics studies to verify/corroborate a hypothesis. Omics studies can also be used to verify postulated links between the upstream events and those that occur at the cellular and subcellular levels, and omics data can also help fill in information gaps for poorly-defined AOPs. Additionally, since potentially many AOPS may share a common MOA, omics data may play a central role in fully mapping the early stage key events to better define the points of divergence between AOPS that begin from a common MOA.

Omics also provides endpoints of chemically-induced adverse effects for key events associated with AOPs (EFSA 2014). Integrated omics data sets can be used to develop comprehensive molecular, cellular, and organ – level profiles of key events in AOPs, setting a foundation for species comparison studies as well as for studies that consider the range of chemical responses attributable to human variability. Conversely, AOPs can be used to inform development of omic-based predictive assays that could potentially be applied for hazard or risk assessments of previously-uncharacterized chemical compounds or for use with complex mixtures of chemicals. Taken together, the application of omics data to the development of AOPs holds considerable promise for hazard and risk assessments, and may help to validate integrated testing strategies and reduce reliance on animal testing.

Application of Genomics Data to AOPs As costs for NGS technologies continue to decline while volume of sequence information generated continues to increase, genomics-based research will also continue to accelerate and fuel tremendous growth in research that aims to associate or correlate structural variations in DNA with diseases. Genomics applications will continue to inform mechanism-of-action studies, and new discoveries will direct research towards targeted therapeutics for diseases. These studies will yield new insights into mechanisms underlying toxicity and guide approaches towards eliminating, minimizing, or by-passing normal cell responses to reduce adverse outcomes.

Application of Transcriptomics Data to AOPs The utility of microarrays has been demonstrated in countless applications, but the three most common applications are: grouping/classification of compounds, elucidation of mode-of-action, and biomarker identification. Others have reviewed and reported the application of transcriptomics data for hazard assessments (OECD 2013a, b; Thomas et al. 2013; US-EPA 2013). RNA-seq holds promise with improved sensitivity, better precision, a much greater dynamic range (microarrays lack sensitivity for genes expressed at either very low or very high levels), and better reproducibility.

Application of Proteomics Data to AOPs Proteomics are increasingly being used in toxicology and hazard assessment (Van Summeren et al. 2012). Some typical applications of proteomics to human hazard assessment of chemicals include: (1) the identification of toxicant protein targets to understand MOA; (2) biomarker discovery and validation in major initiatives like the FDA's Critical Path Initiative (Woodcock and Woolsley 2008) or the EU-based InnoMed PredTox project.

Application of Metabolomics Data to AOPs Metabolites can be created in response to chemicals that originate endogenously (inside the body) or exogenously (outside of the body). Small changes in the genome or proteome can be easily detected in the metabolome; the metabolome also reflects an organism's response to changes in the environment. For these reasons, the metabolome is often referred to as the "ome" closest to the phenotype. Biomarker discovery and drug safety screens are two examples where metabolomics has already enabled informed decision making.

9.4 Challenges to the Application of OMICs Data to AOPS

General Challenges A fundamental challenge confronting the use of omics approaches across all scientific disciplines is that the omics experiments are technically very challenging, requiring complex molecular and analytical techniques, highly specialized training, and sophisticated bioinformatics tools to analyze very large data sets. In addition, researchers must have deep fundamental understanding of biology in order to interpret the data and place it into a meaningful biological context. Another key issue relates to the sensitivity of the methodologies which may lead to the detection of changes that may not be biologically or toxicologically relevant (EFSA 2014).

The lack of standardization and validation across the omics (especially proteomics and metabolomics) has been previously described; this is indeed a challenge that must be addressed and overcome if omics data are to be accepted by the regulatory community. Quality control, sample preparation, sample processing, data processing, data analysis, and data interpretation are all areas that are ripe for improved method development, standardization and harmonization.

Data Integration: Bioinformatics and Visualization Tools Experimental omics approaches are high-throughput, data-driven, top-down approaches that generate large amounts of data (Zhang et al. 2010). Combining data from different platforms and assays across multiple experiments into a coherent approach that appropriately weighs and evaluates the different data sources is quite challenging and represents the next generation of pathway identification tools. The two main challenges for integration of omics data sets are the limitations of bioinformatics and visualization tools to enable researchers to analyze and interpret their data within a meaningful biological context, and the overall processing, storage, and curation of data into databases such that data can be easily accessed, retrieved, shared, and archived. Bioinformatics tools will need to be built on novel, flexible architectures, to provide a broad foundation for joint analysis and visualization of orthogonal data. Several key processes, including transfer of different kinds of data between different software applications, facilitating new custom visualizations, enabling statistical analyses involving pathway databases, and providing workflow and help facilities in order to ensure that the software is accessible to users with different levels of experience, are critical to pathway-based orthogonal analysis and must be considered. The development and refinement of methods and tools for integration of omics data constitutes an active area of research and several recent reviews provide excellent summaries of available tools (Wanichthanarak et al. 2015; Fukishima et al. 2014).

Turning omics data sets into results that advance our understanding of the fundamental biology underlying the data requires considerable analysis and interpretation. Heterogeneous, multi-dimensional data must first be processed and analyzed to extract the features of interest (e.g., genes, metabolites, etc.). Next, application-specific steps, such as feature extraction (for metabolomics data, for example) and identification or expression analysis (for transcriptomics data, for example) are applied. The processed data sets are then ready for integration. This requires that databases are available that allow correlations between features, including genes, proteins, metabolites, chemicals and other compounds of interest, along with consideration of an array of experimentally-determined meta data (e.g., cell-based imaging, pathology, chemical analysis, etc.). Any omics integration approach will employ pathway-based analysis tools. Such analysis is limited by access to curated biochemical pathways such as WikiPathways, KEGG, or BioCyc. These databases represent an excellent starting point for data integration, but they are limited in that they only provide a static view of biochemical pathways.

Network-based analyses extends the pathway approach by representing complex interactions between genes, metabolites, and proteins without relying on predefined or pre-determined biochemical pathways. These networks are useful to map multiple omics experimental results and help identify altered regions of the networks (Wanichthanarak et al. 2015). For example, GeneSpring software can be used to combine heterogeneous data, such as genomic sequencing, gene expression, and metabolomics abundance into one project, allowing investigators to analyze and view results from different experiments in a single user interface.

The GeneSpring Pathway Architect module enables visualization and analysis of curated pathway content using a variety of publically-available pathway databases for building, annotating and querying biological pathways. GeneSpring incorporates Gene Ontology (GO) analysis, Gene Set Enrichment Analysis (GSEA), Gene Set Analysis (GSA) and network analysis tools (Fasani et al. 2016). Correlation-based tools (based on methods such as principal component analysis, canonical correlation, analysis, or discriminant analysis) can be used to look at associations between entities from a single type of omics data set, or between entities from different types of omics data (Rajasundaram and Selbig 2016). For example, identification of co-regulated entities, such as genes and metabolites enables identification or mapping to networks or pathways, which enables potential identification of mechanisms. Correlation analysis, combined with predictive statistical approaches, such as sparse partial least squares regression may reveal correlations with known biological functions as well as correlations for which biological relevance remains to be verified (Rajasundaram and Selbig 2016). Intuitive graphical displays that employ a variety of plots, graphs and diagrams help users conceptualize and interpret the information in their data, and other interactive visualization tools make it easy to import/export graphical images and to compare results from different experiments.

The pathway, network, and correlation tools discussed here are relatively simple to use and are standard in most commercially-available data integration packages. Other types of correlation-based tools (e.g., Bayesian networks, partial correlations) that are far more sophisticated are also available, but they are also more difficult to use, may be computationally challenging, and generally require a fairly detailed biological understanding of the system under study.

Publically-Accessible Databases Given the complexity and sheer volume of data generated in omics studies, there is an emerging need for comprehensive, publically-accessible databases. Databases such as CEBS, ACToR, PubChem, GO, Gene Map

Annotator and Pathway Profiler, Science Signaling Connections Map, BioCarta, Reactome and KEGG are useful in this regard; of the more than 1000 biologically-relevant databases are already publically available, several hundred are specifically relevant to toxicology but many of those contain data that is not necessarily in a format that is directly useable (Judson 2010). EPA's ACToR, the Aggregated Computational Toxicology Resource, is an example of a knowledgebase that brings together diverse types of information into a system where interrelationships of individual database elements (e.g., traditional toxicology, chemical structure information, high throughput screening data, molecular pathway analysis, chemical data repositories, peer reviewed published literature, and internal Agency databases) can be explored and utilized (Judson et al. 2008). The ACToR database links information from more than 400 source databases and data sets on chemical identity. All published data associated with the ToxCast, ToxRefDB (a mineable, searchable database of pesticide toxicity data) and Tox21C programs are consolidated within ACToR and the knowledgebase is publically accessible. Given the existing utility and advanced stage of development of ACToR, it could serve as the foundation upon which to build out a complete knowledgebase for all twenty-first century toxicology testing data and metadata.

9.5 Conclusion

Advances on omics approaches, combined with molecular toxicology provide both opportunities and challenges for regulatory agencies and others who must consider the use of these new tools and technologies when conducting hazard and risk assessments. A variety of issues will need to be considered and resolved before omics data can contribute significantly to risk assessments, but applying omics data today to AOPs should help the toxicology community establish better linkages a-between key events, which should in turn, lead to more quantitative AOPs.

References

Afshari CA, Hamadeh HK, Bushel PR (2011) The evolution of bioinformatics in toxicology: advancing toxicogenomics. Toxicol Sci 120:S225–S237

Altelaar AFM, Munoz J, Heck AJR (2013) Next-generation proteomics: towards an integrative view of proteome dynamics. Nat Rev Genet 14:35–48

Andersen ME, Clewell HJ (III), Carmichael PL et al (2011) Toxicity testing in the 21st century: a vision and a strategy. Food for thought… can case study approaches speed implementation of the NRC report, ALTEX, 28, pp 175–183

Ankley GT, Bennett RS, Erickson RJ et al (2010) Adverse outcome pathways: a conceptual framework to support ecotoxicology research and risk assessment. Environ Toxicol Chem 29:730–741

Attene-Ramos MS, Miller N, Huang R et al (2013) The Tox21 robotic platform for the assessment of environmental chemicals – from vision to reality. Drug Discov Today 18:716–723

Baker M (2011) Metabolomics: from small molecules to big ideas. Nat Methods 8:117–121
Bhattacharya S, Zhang Q, Carmichael PL et al (2011) Toxicity testing in the 21st century: defining new risk assessment approaches based on perturbation of intracellular toxicity pathways. PLoS One 6(6):e20887
Bouhifd M, Hartung T, Hogberg HT et al (2013) Review: toxicometabolomics. J Appl Toxicol 33:1365–1383
Bouhifd M, Hogberg H, Kleensang A et al (2014) Mapping the human toxome by systems toxicology. BCPT 115:24–31
Chen G, Pramanik BN (2009) Application of LC/MS to proteomics studies: current status and future prospects. Drug Discov Today 14:465–471
Chen M, Zhang M, Borlak J et al (2012) A decade of Toxicogenomics research and its contribution to toxicological science. Toxicol Sci 130:217–228
Chuang H-Y, Hofree M, Ideker T (2010) A decade of systems biology. Annu Rev Cell Dev Biol 26:721–744
Collings FB, Vaidya VS (2008) Novel technologies for the discovery and quantitation of biomarkers of toxicity. Toxicology 245:167–174
Collins FS, Morgan M, Patrinos A (2003) The human genome project: lessons from large-scale biology. Science 300:286–290
Cong L, Ran FA, Cox D et al (2013) Multiplex genome engineering using CRISPR. Cas systems. Science 339:819–823
Connon RE, Geist J, Werner I (2012) Effect-based tools for monitoring and predicting the ecotoxicological effects of chemicals in the aquatic environment. Sensors 12:12741–12771
Cote I, Anastas PT, Birnbaum LS et al (2012) Advancing the next generation of health risk assessment. Environ Health Perspect 120:1499–1502
Cui Y, Paules RS (2010) Use of transcriptomics in understanding mechanisms of drug-induced toxicity. Pharmacogenomics 11:573–585
Dix DJ, Houck KA, Martin MT et al (2007) The ToxCast program for prioritizing toxicity testing of environmental chemicals. Toxicol Sci 95:5–12
Domon B, Aebersold R (2010) Options and considerations when selecting a quantitative proteomics strategy. Nat Biotech 28:710–721
Dunn WB, Ellis DI (2005) Metabolomics: current analytical platforms and methodologies. TrAC Trends Anal Chem 24:285–294
European Food Safety Authority (2014) Modern methodologies and tools for human hazard assessment of chemicals. EFSA J 12:3638, 87 pp. doi:10.2903/j.efsa.2014.3638
Fasani RA, Livi CB, Choudhury DR et al (2016) The human toxome collaboratorium: a shared environment for multi-omic computational collaboration within a consortium. Technology Report ARTICLE. Front Pharmacol 6:322. https://doi.org/10.3389/fphar.2015.00322
Fielden MR, Brennan R, Gollub J (2007) A gene expression biomarker provides early prediction and mechanistic assessment of hepatic tumor induction by nongenotoxic chemicals. Toxicol Sci 99:90–100
Fukishima A, Kanaya S, Nishida K (2014) Integrated network analysis and effective tools in plant systems biology. Plant Syst Synth Biol 5:598
Garcia-Reyero N, Lavelle CM, Escalon BL et al (2011) Behavioral and genomic impacts of a wastewater effluent on the fathead minnow. Aquat Toxicol 101:38–48
Gerecke DR, Chen M, Isukapalli SS et al (2009) Differential gene expression profiling of mouse skin after sulfur mustard exposure: extended time response and inhibitor effect. Toxicol Appl Pharmacol 234:156–165
Girirajan S, Campbell CD, Eichler EE (2011) Human copy number variation and complex disease annual rev. Genetics 45:203–226
Go EP (2010) Database resources in metabolomics: an overview. J Neuroimmune Pharmacol 5:18–30
Green ED, Guyer MS, The National Human Genome Research Institute (2011) Charting a course for genomic medicine from base pairs to bedside. Nature 470:203–213

Groh KJ, Carvalho RN, Chipman JK et al (2015) Development and application of the adverse outcome pathway framework for understanding and predicting chronic toxicity: I. Challenges and research needs in ecotoxicology. Chemosphere 120:764–777

Hartung T, McBride M (2011) Food for thought...on mapping the human toxome. ALTEX 28:83–93

Hege Harrill A, Rusyn I (2008) Systems biology and functional genomics approaches for the identification of cellular responses to drug toxicity. Expert Opin Drug Metab Toxicol 4:1379–1389

Heinloth AN, Irwin RD, Boorman GA et al (2004) Gene expression profiling of rat livers reveals indicators of potential adverse effects. Toxicol Sci 80:193–202

Ishino Y, Shinagawa H, Makino K et al (1987) Nucleotide sequence of the *iap* gene, responsible for alkaline phosphatase isozyme conversion in *Escherichia coli*, and identification of the gene product. J Bacteriol 169:5429–5433

Jinek M, Chylinski K, Fonfara I et al (2012) A programmable dual-RNA-guided DNA endonuclease in adaptive bacterial immunity. Science 337:816–821

Judson R (2010) Public databases supporting computational toxicology. J Toxicol Environ Health B Crit Rev 13:218–231

Judson R, Richard A, Dix D et al (2008) ACToR–aggregated computational toxicology resource. Toxicol Appl Pharmacol 233:7–13

Judson RS, Houck KA, Kavlock RJ et al (2010) In vitro screening of environmental chemicals for targeted testing prioritization: the ToxCast project. Env Health Perspect 118:485–492

Konermann S, Brigham MD, Trevino AE et al (2015) Genome-scale transcriptional activation by an engineered CRISPR-Cas9 complex. Nature 517:583–588

Lindon JC, Keun HC, Ebbels TM et al (2005) The consortium for metabonomic toxicology (COMET): aims, activities and achievements. Pharmacogenomics 6:691–699

Loman NJ, Misra RV, Dallman TJ et al (2012) Performance comparison of benchtop high-throughput sequencing platforms. Nat Biotechnol 30:434–439

Martin MT, Dix DJ, Judson RS et al (2010) Impact of environmental chemicals on key transcription regulators and correlation to toxicity end points within EPA's ToxCast program. Chem Res Toxicol 23:578–590

Marx V (2013) Targeted proteomics. Nat Methods 10:19–22

McBurney RN, Hines WM, Von Tungeln LS et al (2009) The liver toxicity biomarker study: phase 1 design and preliminary results. Toxicol Pathol 37:52–64

McBurney RN, Hines WM, Von Tungeln LS et al (2012) The liver toxicity biomarker study: phase 1 markers for the effects of tolcapone or entacapone. Toxicol Pathol 40:951–964

McMullen PD, Bhattacharya S, Woods CG et al (2014) A map of the PPAR-α transcription regulatory network for primary human hepatocytes. Chem Biol Interact 209:14–24

McMullen PD, Bhattacharya S, Woods CG et al (2016) PPARα signaling in liver: exploring the biological basis of human-rat differences in phenotypic responses to hepatic nuclear receptor activation. Toxicol Sci (in press)

Mehinto AC, Martyniuk CJ, Spade DJ et al (2012) Applications of next-generation sequencing in fish ecotoxicogenomics. Front Genet 3:1–10

Metzker ML (2010) Sequencing technologies – the next generation. Nat Rev Genet 11:31–46

National Research Council (2007) Toxicity testing in the 21st century: a vision and a strategy. National Academies Press, Washington, DC

OECD, Organisation for Economic Co-operation and Development (2013a) Guidance document on developing and assessing adverse outcome pathways. Series on testing and assessment, No. 184. Paris, ENV/JM/MONO(2013)6

OECD, Organisation for Economic Co-operation and Development (2013b) Users' handbook supplement to the guidance document for developing and assessing AOPs. Available at https://aopkb.org/common/AOP_Handbook.pdf

Pareek CS, Smoczynski R, Tretyn A (2011) Sequencing technologies and genome sequencing. J Appl Genet 52:413–435

Patti GJ, Yanes O, Siuzdak G (2012) Metabolomics: the apogee of the omics trilogy. Nat Rev Mol Cell Biol 13:263–270
Rajasundaram D, Selbig J (2016) More effect-more results: recent advances in integrative "omics" data analysis. Curr Opin Plant Biol 30:57–61
Ramirez T, Daneshian M, Kamp H et al (2013) t4 report: metabolomics in toxicology an preclinical research. ALTEX 30:209–225
Su Z, Ning B, Fang H, Hong H, Perkins R, Tong W, Shi L (2011) Next-generation sequencing: a revolutionary tool for toxicogenomics. Gen Appl Syst Toxicol. Published Online: 15 Sep 2011. doi: https://doi.org/10.1002/9780470744307.gat232
Suter L, Babiss LE, Wheeldon EB (2004) Toxicogenomics in predictive toxicology in drug development. Chem Biol 11:161–171
Swiech L, Heidenreich M, Banerjee A et al (2015) In vivo interrogation of gene function in the mammalian brain using CRISPR-Cas9. Nat Biotechnol 33:99–103
Thomas RS, Philbert MA, Auerbach SS et al (2013) Incorporating new technologies into toxicity testing and risk assessment: moving from 21st century vision to a data-driven framework. Toxicol Sci 136:4–18
Titz B, Elamin A, Martin F et al (2014) Proteomics for systems toxicology. Comput Struct Biotechnol J 11:73–90
Tollefsen KE, Scholz S, Cronin MT et al (2014) Applying adverse outcome pathways (AOPs) to support integrated approaches to testing and assessment (IATA). Regul Toxicol Pharmacol 70:629–640
Tralau T, Luch A (2015) Moving from rats to cellular omics in regulatory toxicology: great challenges toward sustainability or "up shit creek without a paddle"? Arch Toxicol 89:819–821
US-EPA (U.S. Environmental Protection Agency) (2013) Next generation risk assessment: incorporation of recent advances in molecular, computational, and systems biology. Available online http://cfpub.epa.gov/ncea/risk/recordisplay.cfm?deid=259936#Download
Van Aggelen G, Anklet GT, Baldwin WS et al (2010) Integrating omic technologies into aquatic ecological risk assessment and environmental monitoring: hurdles, achievements, and future outlook. Environ Health Perspect 118:1–5
van Ravenzwaay B, Cunha GC, Leibold E et al (2007) The use of metabolomics for the discovery of new biomarkers of effect. Toxicol Lett 172:21–28
van Ravenzwaay B, Herold M, Kamp H et al (2012) Metabolomics: a tool for early detection of toxicological effects and an opportunity for biology based grouping of chemicals-from QSAR to QBAR. Mutat Res 746:144–150
Van Summeren A, Renes J, van Delft JH et al (2012) Proteomics in the search for mechanisms and biomarkers of drug-induced hepatotoxicity. Toxicol In Vitro 26:373–385
van Vliet E (2011) Current standing and future prospects for the technologies proposed to transform toxicity testing in the 21st century. ALTEX 28:17–44
Villeneuve DL, Crump D, Garcia-Reyero N et al (2014) Adverse outcome pathway development II: best practices. Toxicol Sci 142:321–330
Wanichthanarak K, Fahrmann JF, Grapov D (2015) Genomic, proteomic, and Metabolomic data integration strategies. Biomark Insights 10(S4):1–6
Waters MD, Fostel JM (2004) Toxicogenomics and systems toxicology: aims and prospects. Nat Rev Genet 5:936–948
Wishart DS, Tzur D, Knox C et al (2007) HMDB: the human metabolome database. Nucleic Acids Res 35(Database issue):D521–D526
Wishart DS, Jewison T, Guo AC et al (2013) HMDB 3.0—the human metabolomics database in 2013. Nucleic Acids Res 41:D801–D807
Woodcock J, Woolsley R (2008) The FDA critical path initiative and its influence on new drug development. Annu Rev Med 59:1–12
Zhang W, Li F, Nie L (2010) Integrating multiple 'omics' analysis for microbial biology: application and methodologies. Microbiology 156:287–230

Chapter 10
Use of Genetic Manipulation for Evaluating and Understanding Adverse Outcome Pathways

Christopher Warner, Natàlia Garcia-Reyero, and Edward Perkins

Abstract Innovations in biology have brought forth a new era of genetic manipulation ranging from the creation of molecular scissors for targeted single-nucleotide alterations, to a simultaneous inactivation of 62 genes in pig embryos to "humanize" transplant tissue. Genetic engineering advances allow for novel testing paradigms to understand chemical interactions and information flow in biological systems. Emerging platforms may provide mechanistic knowledge of chemical stressor interactions in biological systems to facilitate the development of alternative testing methods, as well as prioritize higher tier toxicity testing for risk assessment. This chapter will discuss recent advances in genetic manipulation and describe how these techniques improve our understanding of toxicity across multiple biological scales. These efforts will ultimately aid in validation of Adverse Outcome Pathway (AOP) key event relationships for ecological risk assessment.

10.1 The Need for Validation Systems in Chemical Hazard Assessments

Chemical hazard assessment has long relied on apical data generated in animal toxicity tests and the application of both uncertainty factors and conservative assumptions for decision making. However, due to the cost and time limitations, it is not practical nor feasible to test all chemicals that could adversely affect ecosystems using animal models with phenotypic end points (NRC 2007). Chemical assessment approaches and regulatory efforts in the US and in Europe are moving towards computational chemistry, high-throughput screening (HTS), in vitro assays and biological pathway based measures to more effectively assess the potential for

C. Warner • N. Garcia-Reyero (✉) • E. Perkins
US Army Engineer Research and Development Center, Environmental Laboratory,
Vicksburg, MS 39180, USA
e-mail: natalia@icnanotox.org

N. Garcia-Reyero, C.A. Murphy (eds.), *A Systems Biology Approach to Advancing Adverse Outcome Pathways for Risk Assessment*,
DOI 10.1007/978-3-319-66084-4_10

chemical to cause toxicity (Dix et al. 2007; Worth et al. 2014). These programs highlight the need for developing, improving, and validating new laboratory tools based on recent scientific advances to understand the hazards and risks posed by chemicals. There have been a number of advances in tool development, including biomarker discovery, model system development, systems biology, bioinformatic analytics and computational toxicology (Mathieu 2013). Integration of these efforts relies on an overarching framework that can extrapolate initiation of a toxicological pathway to an adverse outcome, The Adverse Outcome Pathway (AOP) framework provides a structure for organizing knowledge about the progression of toxicity events across scales of biological organization that ultimately lead to adverse outcomes at the organism or population level (Ankley et al. 2010). AOPs consist of a sequence of key events from a molecular-level initiating event, where a chemical binds to a receptor and an ensuing cascade of cellular, organ, and organism level effects culminating in an adverse outcome of regulatory significance (Fig. 10.1; Villeneuve et al. 2014). AOPs have been developed for a number of important systems, including the endocrine (Russom et al. 2014), acetyl cholinesterase (Watanabe et al. 2011), oxidative-phosphorylation (Wilbanks et al. 2014), among many others (Perkins et al. 2015; Tollefsen et al. 2014; Willett 2014).

The AOP framework relies on detailed knowledge on how key events within a pathway interact to create response-response relationships across at all levels of biological organization. Efforts to understand these relationships rely on mechanistic understandings of biochemical and genetic interactions (Patlewicz et al. 2013). However, conventional toxicological methods often fail to identify or describe causal relationships across biological scales that are needed to confidently link changes at the molecular initiating event (MIE) through a cascade of key events (KE) to the adverse outcome (AO). Alternative approaches, such as tools recently developed for genetic engineering of cells and organisms, enable an unprecedented ability to understand the mechanistic underpinnings of toxicological pathways and adverse outcomes. In this chapter, we identify commonly used genetic manipulation tools, then apply these tools to establish mechanistic relationships across multiple levels of biological organization in the AOP framework.

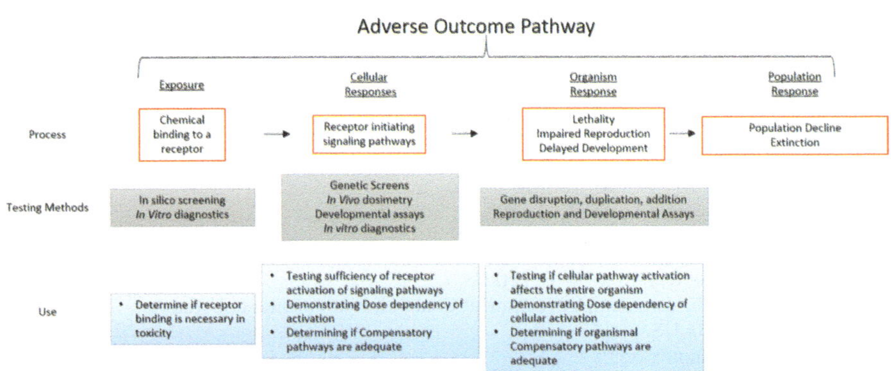

Fig. 10.1 Genetic engineering efforts integrated into an adverse outcome pathway

10.2 Genetic Manipulation Tools

Genetic engineering is the modification of an organism's genetic composition by artificial means, often involving the transfer or modification of specific genes, from one organism into entirely different species. Genetically modified organisms include collections of permanent, conditional knock-outs (deletions), or knock-ins (gene addition or duplication). Within the last 10 years, there have been numerous tools developed for genetic modification of many species (Baltimore et al. 2015). These also include transcriptional level modifications that can alter levels of protein within a modified cell. Recent advancements in methods for genome modification, such as Clustered Regularly Interspaced Short Palindromic Repeats (CRISPR) technologies and CRISPR-associated (CRISPR/Cas9), enable experimentation that is cheaper and faster than previous methods while permitting research that was not previously possible (Yang et al. 2015).

These methods hold a promise in becoming standard experimental strategies for testing the functional role of genes in both model and non-model organisms. The recent generation of engineered nuclease-mediated mutants in rat, zebrafish, maize and tobacco testifies to the significance of the methods and the list is expanding rapidly (McMahon et al. 2012). Both of these methods can disrupt genes either permanently or temporarily and allow for gain-of-function (overexpression of a gene or dominant active), loss-of-function (gene expression knock down or dominant negative), mosaic analysis, lineage-restricted studies and cell tracing experiments. Transgenesis, the process of creating stable mutants, allows for targeted changes of specific genes while transient reverse genetic approaches are temporary, typically quicker, cheaper, and require little animal facility space (Hogan et al. 2008).

10.2.1 Engineered Nucleases and Homologous Recombination: Examining the Role of Specific Genes in an AOP

A direct way to demonstrate causality in a pathway and linkage to an AO is to block a key event from activating in the presence of conditions that would otherwise activate the event and downstream events. This can be done genetically using engineered nucleases and homologous recombination to make genes overexpress (always on or expressed at much higher levels than normal) or remove genes to observe their role in biological pathways and AOs. Often referred to as "molecular scissors", nuclease proteins allow for precision edits and additions to an organism's genome. A nuclease creates specific double-strand breaks along DNA at desired locations in the genome, often guided using specific sequences. Nucleases edit DNA through covalent bond alterations on DNA nucleotides, akin to a copy-paste function (Zhao et al. 2014). Subsequent DNA breakage recruits the cell's native repair machinery. This results in a genome alteration by replacing a region of DNA with the desired insert (Esvelt and Wang 2013). There are currently four families of engineered nucleases being used, including Zinc Finger Nucleases (ZFNs), Transcription Activator-Like Effector Nucleases (TALENs), the CRISPR/Cas9 system, and engineered meganuclease re-engineered homing endonucleases. Each of these proteins have distinct advantages in terms of availability, sequence specificity, ease of use, targeting efficiency and off target mutations (Tan et al. 2012).

10.2.2 Altering Gene Expression Using RNA Interference

One limitation of methods that directly add to, or change, genes in genomes is that these changes, once induced, are generally permanent, continuously affect the cell or animal, and require lengthy selection to get transgenic lines for analysis. RNA interference, or RNAi, is an approach that can be used to temporarily reduce or eliminate expression of genes without the need for lengthy generation of transgenic lines. RNAi is a gene regulation pathway that is controlled by small regulatory molecules of RNA. In the pathway, short double-stranded RNA molecules are bound by the protein Dicer in a cell's cytoplasm and are cleaved to produce a passenger strand, which is degraded, and a guide strand. The guide strand directs the RNA-induced silencing complex (RISC) to selectively destroy specific mRNAs that are complementary to the guide strand RNA (Bagasra and Prilliman 2004). In the laboratory, double-stranded RNA can be synthesized with a sequence that is complementary to a gene of interest and introduced into a cell or organism, where it activates the RNAi pathway, degrading the mRNA of the complementary gene, and consequently decreasing expression of the targeted gene. Since RNA targeting is introduced directly into the cells that are targeted, application of RNAi approaches generally provide only temporary inhibition of gene function. RNAi can also result in unpredictable off-target effects due to partial matching to unintended targets in the genome (Alic et al. 2012).

10.2.3 Genetic Manipulation in Developing In Vitro and In Vivo Assays to Assess Key Events

MIE and KE are pragmatically defined as specific, measurable biological events. In vivo and in vitro assays provide methods to quantify these observations. In vitro methods utilize biological molecules (e.g. receptors, enzymes, or ion channels), or biological tissue (cell lines, xenografts, or *ex vivo* organs) outside of whole animal to determine if a chemical will bind to a biological target, activate a specific pathway, or affect a cellular process (Fig. 10.2). Whereas in vivo methods include use of whole animals for toxicity testing. Bacteria, yeast, worms, and zebrafish embryos

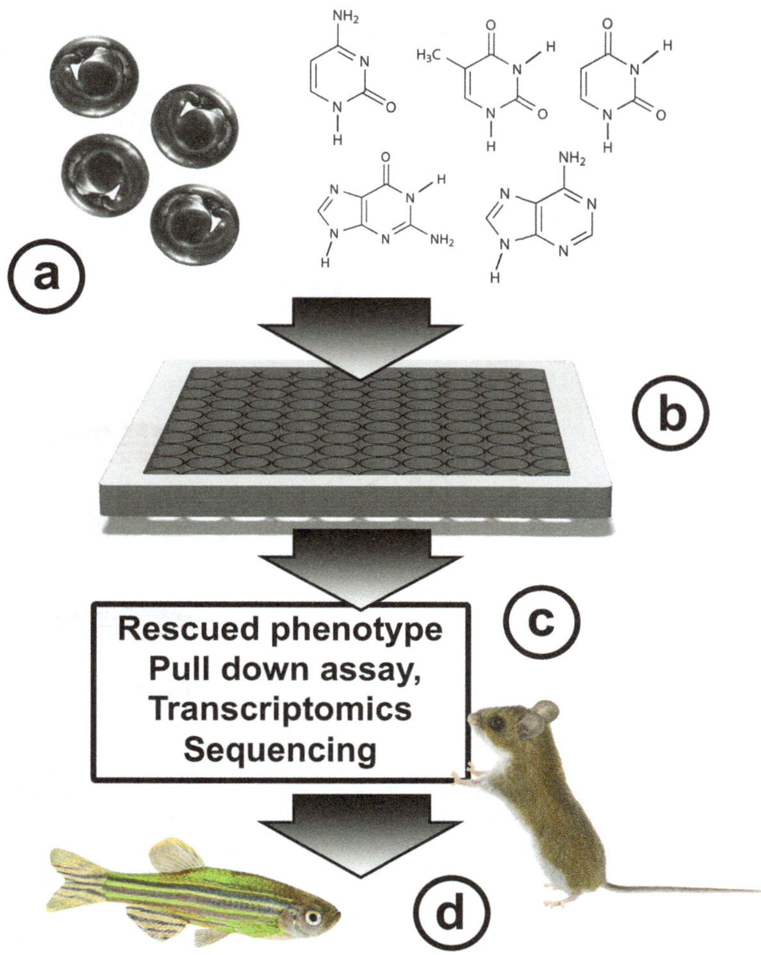

Fig. 10.2 Chemical screening in zebrafish. Zebrafish embryos are assayed using large-scale, high-throughput manipulation and analysis. (**a**) Identification of an embryonic phenotype that is a relevant toxicity model is a key step in this process. Organic dyes, developmental abnormalities, delayed development or expression of a genetically modified reporter can be used as metrics to sort out chemical hits. (**b**) Once a relevant embryo phenotype is found or genetically modified, embryos can be distributed into 96 well plates. Each well will receive a distinct small molecule, either manually or with the aid of a liquid-handling robot. This method has been applied to screens ranging from 1000 to 26,000 molecules. (**c**) Identifying the mechanism of hits depends upon the type of screening employed. (**d**) While mechanistic evaluation is ongoing, chemical toxicity can be validated in multiple higher tier testing methods, including transgenic disease models or full animal studies

are preferred due to their small size, low cost, ease of genetic manipulation, and short generation time. Figure 10.3 demonstrates one form of in vivo testing, where a reporter gene is controlled by a response element, such as a developmental or stress response promotor. Many assay systems exist that utilize genetic

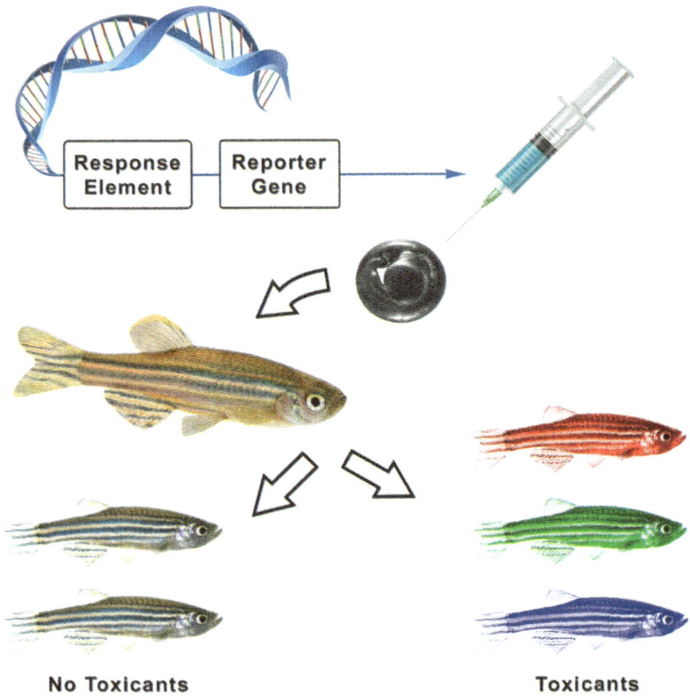

Fig. 10.3 In Vivo genetic circuit testing in zebrafish. DNA is synthesized with a response elements to drive the expression of different fluorescent reporter proteins. DNA is transfected into zebrafish embryos and maintained throughout the fish's life cycle, often transmitting into offspring. Stable transgenic zebrafish can be exposed to different environmental pollutants and toxicants. Cellular responses to chemical stimulation can be visualized with fluorescent microscopy. Moreover, tissue specific responses can be quantified by comparing against non-induced controls

manipulation (Table 10.1), each one targets a specific biomolecule and provides various levels of information. We will describe in detail the Yeast 2 hybrid system as it provides an elegant example of how genetic manipulation can be utilized to assess the physiological relevance of signaling dynamics within a cell.

10.2.3.1 Yeast 2 Hybrid Systems to Identify Important Protein Interactions in Key Events

The Yeast 2 Hybrid system (Y2H) screens for interacting proteins based on transcriptional activation. The protein of interest, or "bait" is fused to a DNA binding domain. Proteins that bind, or "fish" are fused to a transcriptional activation domain. Any protein that binds to the "bait" will activate the transcription of a *his* reporter gene. Y2H screens begin by constructing the bait plasmid, and a library of cDNAs in the fish library. Each plasmid contains a selectable marker, such as an antibiotic,

Table 10.1 Inducible systems for in vivo/in vitro genetic studies

System	Description	Tissue specific localization	Animal models	Level of inducible control	Advantages
Tetracycline-controlled transcriptional activation	Transcription at a *tetO* promoter is reversibly turned on or off by the presence of a tetracyline antibiotic	Yes	Mammalian cell culture, mice, xenographs, bacteria	Intracellular concentration of antibiotic	Tight, reversible control of transcription
Gal4/UAS	Native promoters or other DNA elements (upstream activation sequence; UAS) are tested by cloning the transcription activator GAL4 gene downstream. An active UAS activates GAL4 to activate a reporter for easy readout	Yes	Drosophila, Xenpus, zebrafish	Genetic	Can test the performance of promoter sequences
Optogenetics	Light sensitive ion channel proteins are expressed in neurons. Specific wavelengths of light will excite or inhibit biochemical activity of the protein in real time	No	Worms, fruit flies, mice, rats, and monkeys	Light exposure	Real-time spatiotemporal control of nervous system components

(continued)

Table 10.1 (continued)

System	Description	Tissue specific localization	Animal models	Level of inducible control	Advantages
Engineered promoters	Temperature, Osmolarity, nutrient, etc.. Also includes a wide variety of promoter strengths, to control expression level relative to other genes	Yes	Mostly *E. coli* and mammalian cells	Varies	Tunable control of expression through a variety of physico-chemical gradients
Lac operon	The lac repressor inhibits transcription of a gene until lactose binds and frees the site for RNAP to transcribe	No	Bacteria	Genetic	Simple system for transcriptional activation
Cre-lox recombination	LoxP promoter sites can be inserted before genes of interest. When Cre proteins are expressed, LoxP sites are cut an inverted, causing transcriptional silencing	Yes	Plants, zebrafish, drosophila	Genetic	Validated system for genetic manipulation
Fusion proteins	Adding an observable tag to the target protein, such as fluorescent, luminescent, or chromogenic	Yes	Many	Genetic	Simple system for visualizing protein expression with a moderate level of spatial and temporal resolution
Morpholinos	RNA analogue for competitive inhibition of transcripts and decrease expression	No	Mice, zebrafish, drosophila, Xenpus, sea urchins	Delivery (Micoinjection) into embryo	Disrupts pre-mRNA of protein expression

or essential amino acid. Plasmids are transformed into yeast cells. Cells are grown in selective media so that cells containing both plasmids will grow. Cells are plated onto media lacking histidine, so that colonies with interacting plasmids will grow. Proteins from colonies are identified through DNA sequencing (Brückner et al. 2009; Lodish et al. 1995). There are many variations of this method, including the type of reporter and cell lines used, as well as inducible systems and multi-component systems that provide insights into more complex protein-protein interactions (Kelly and Stumpf 2008; Charbonnier et al. 2008). This screen is relatively simple to perform and can identify interacting proteins in a scalable system. This screen has been used to for both hypothesis testing as well as exploratory work. For example, Y2H have been used to confirm critical targets of the F-box protein ubiquitin ligase involved in methylmercury toxicity (Lee et al. 2015). Y2H has also been used to query chemical activity on estrogen signaling pathways for high throughput endocrine disruption testing (Nishihara et al. 2000). This system, like many testing platforms, is far from perfect. Both false positives and negatives plague the system, however, where validation approaches are required. Moreover, expression bias among cDNA libraries is well documented, and some protein partners may require post-translational modifications, chaperone proteins, or other multiunit complexes before interactions are observed (Chen et al. 2010). Ultimately, Y2H screens provide a functional assay where exposure of genetic manipulated cells provides evidence for toxicity pathways and ultimately AOPs.

10.3 Application of Genetic Manipulation Approaches to Understand and Define Key Event Relationships

Genetic manipulation can be used by toxicologists at all biological levels of organization within the AOP framework. Genetically manipulated organisms and tissues provide a tool kit for both direct hypothesis testing as well as elucidating toxicity pathways in an exploratory manner (Table 10.2)

Table 10.2 Overview of genetic engineering efforts to aid validating AOP relationships

Biological level	Description	Examples
Receptor binding (MIE)	Chemical-protein interactions	Protein mutants for enhanced NMR and crystallography GPCR receptors, PTM proteins in bacteria and yeast
Cellular response	Genetic screens Reporter circuits	Functional genomics Genetic devices, biomarker discovery, pathway specific assays
Organism response	Simplified model systems Reproduction and developmental assays Disease models	Reverse engineering of biological circuitry, disease models Reproducanimal models, iPSC, cell line immortalization

10.3.1 Use of Genetic Manipulation to Define MIEs

Testing whether a chemical interacts with a biological receptor, an MIE, is complicated. Protein or nucleic acid receptor binding is typically tested through a variety of in vivo and in vitro methods. Below we discuss how genetic engineering has advanced our ability to understand MIEs.

10.3.1.1 Genetic Manipulation of Proteins for Greater Understanding of Chemical-Protein Interactions

Genetic manipulation of peptides provides toxicologists a toolbox to better understand protein structure and function. Functional relationships embody an underpinning for the chemical-protein interactions that define receptor docking. For instance, amino acid substitutions, where every version of an enzyme is made by switching each amino acid for glycine is a routine practice to determine regions of interest. A biologist can compare the activity or specificity of the mutated protein to the wild-type, and map the enzymes active, allosteric and inhibitory binding sites. Structure-activity relationships can be established based off of these peptide maps to understand the primary structure of an enzyme. Protein maps provide working knowledge that can be used for either specific hypothesis testing concerning chemical affinity, or for targets in high throughput screening for exploratory work. For example, genetically engineered forms of native receptors allow enhanced structural analysis using NMR and X-ray crystallography (Ellison et al. 2011), which is a fundamental requirement to *ab initio* analysis of receptor docking (Ritchie 2008).

Genetic manipulation also provides tools to understand higher order levels of protein assembly. Protein folding and post translational modifications can be tested by cloning peptides from one species into another one. With contrasting protein assembly machinery, cells of one organism can express a protein distinctive from cells of another organism. Species to species comparison is possible through these assays that will inform MIE specificity.

Genetic manipulation can also be used to modify the epigenetic landscape of a cell's DNA. Engineered nucleases are able to target DNA in a sequence specific manner and add or remove methylation to explore the role of methylation in modifying chemical responses. Moreover, nucleic acid architecture, which is orchestrated by a legion of proteins, can be modified through protein engineering. These epigenetic tools can be used to inform toxicologists how toxicity progresses during exposure through critical non-genetic pathways.

In vitro testing of receptor docking provides direct evidence for an MIE. Comparing organisms that express the receptor to ones that do not can definitively link an MIE to subsequent key events and adverse outcomes. Inducible expression and overexpression of targeted receptors allows one to examine response-response relationships between key events leading to adverse outcomes. Knocking out and overexpression of receptors are possible in many cell types: prokaryotes, fungi, plants, invertebrates, fish, bird, and mammalian cell lines. Moreover, genetic engineering techniques have advanced the ability for complex proteins to be expressed in simpler model systems. For example complex membrane and nuclear receptors, such as G-protein coupled receptors (Skretas and Georgiou 2008), post-translational modified proteins (Kaminoka 2011), and steroid receptors (Wooge et al. 1992) can be expressed in bacteria (Mattanovich et al. 2012) and yeast (Zoonens and Miroux 2010). These genetically altered cells are more convenient to work with because of lower costs, faster life cycles and greater control of genetic backgrounds.

10.3.2 Genetic Approaches to Assess or Modify KE at the Cellular Level

Chemical interactions with a receptor or other MIE is only biologically relevant when downstream cellular pathways are activated and lead to a physiological change at higher levels. To demonstrate causality, dose-response relationships between chemicals and different events or response-response relationships between Key Events are required. Moreover, threshold responses need to be established where pathway activation is above a critical level that compensatory pathways are not equipped to buffer the cell from a toxic response. Below are a number of the methods to determine cellular responses from perturbation of receptors or other MIEs. While genetic screens can confirm if a gene or pathway is necessary to induce a toxicity pathway, more nuanced approaches are required to determine how much of a toxicant is required to induce the toxic response. In vivo dosimetry provides such a platform. Typical experiments include dosing an animal or model system with an increasing amount of compound. Phenotypic end points are measured and compared against the amount of chemical present. Exposure-response curves provide direct evidence for a threshold of perturbation and linkage between events according to the modified Bradford hill criteria, i.e. temporal relationships, strength of response, coherence of response etc.

10.3.2.1 Genetic Screen Based Determination of Biological Pathway/ Processes

Genetic manipulation provides materials that can directly link receptor activation to downstream effects. When specific genes or pathways are involved, targeted knock outs provide a clear link to the response by abolishing toxicity when the corresponding gene is removed or knocked out of the organism (Gaytán and Vulpe 2014). For exploratory work, genetic manipulation has incredibly expanded the options available to perform genetic screens where a phenotype of concern is selected from a mutagenized population. Engineered nucleases and RNAi (Cullen and Arndt 2005) or CRISPR/Cas9 knock-out libraries (Shalem et al. 2014) can identify specific genes responsible for the toxicity phenotype using high throughput assays. Genetic screens using transgenic techniques have been applied to a number of ecological models, including fruit flies (Danielsen et al. 2016), zebrafish (Holtzman et al. 2016), worms (Zugasti et al. 2016), and mice (Mohr et al. 2016).

10.3.2.2 Reporter Circuits

Genetic manipulation provides tools to generate biomarkers that simplify testing methods. Many of the end points used in developing response curves can be either complicated or expensive, for example, cell viability, cell proliferation, cytochrome activity, kinase induction, DNA mutation, and hormone signaling use expensive reagents and equipment. Engineered genetic devices, on the other hand, are simplified biological systems that can act as a pathway dipstick for toxicity testing. Composed of DNA circuits, DNA devices can be either embedded in a cell or in a standalone cell-free system. DNA circuitry enables chemical and genetic interactions to be precisely queried in a system stripped of all other components, including protein machinery, metabolic biproducts, genetic variability etc. Genetic devices have been shown to report on a number of physiological cues, including intracellular nutrient levels, oxidative-reduction environment, cell-cycle advancement and others (Haynes and Silver 2009). In these systems, a signal-responsive transcription factor is fused to a DNA binding domain (e.g., Gal4) that binds to synthetic regulatory elements upstream of a minimal promoter and target gene. Sensors can also be built by assembling minimal promoters with natural regulatory DNA elements that are induced by endogenous transcription factors that respond to various stimuli. For instance, the cellular metabolite thiamine pyrophosphate (TPP) is sensed by synthetic riboswitches (Yamauchi et al. 2008). Using these engineered systems, monitoring biomarkers for stress is simplified. KEs relating to cellular toxicity can be established based off of standalone testing systems that are free from the complicated network of repair mechanisms, compensatory pathways and environmental factors.

Biomarkers with spatial-temporal resolution are possible through genetic manipulation. Visualizing proteins and nucleic acids through genetic fusions to reporter proteins has become a rapidly growing field. Research laboratories working with rats (Ma et al. 2014), mice (Yang et al. 2013), flies (Yu et al. 2014) and fish (Peterson and MacRae 2011) have used epitope tags and fluorescent proteins to label endogenous proteins and generated gene expression reporters. In flies, a histone acetyltransferase protein encoded by the gene *chameau* was C-terminally tagged with GFP, and myc was used to tag an uncharacterized gene, *CG4221* (Yu et al. 2013). In mice, the *Sox2* gene was tagged with the V5 epitope (Yang et al. 2013). Additionally, two different fluorescent reporters were generated for the genes *nanog* and *Oct4* (Yang et al. 2013). These reporters used either the viral 2A peptide or an internal ribosome entry site (IRES) to express fluorescent proteins with the same expression pattern as the endogenous gene but not fused to the protein product. While these groups used standard fluorescent proteins, a spectrum of fluorescent proteins of different colors and with diverse functions are available (Dean and Palmer 2014; Harrison et al. 2014). Moreover, because the CRISPR–Cas9 system is amenable to multiplexing, tags could be added simultaneously to multiple genes or different splice isoforms of a single gene. There is an ever-growing number of genetically encoded molecular tags that can be used for functional analysis, protein purification, or protein and RNA localization studies. Recruitment of cellular toxicity pathways can be confirmed through visualization of genetic fusions.

10.3.3 Understanding Events at the Organism Level with Genetic Manipulation

Understanding how the concert of signaling pathways give rise to a toxic response in any organism is a significant challenge. Moreover, deriving toxicity information across organisms is even more complicated. Genetic manipulation provides genetic test beds for hypothesis testing as well as model systems to explore how accurate KERs describe or predict a toxic response.

10.3.3.1 Simplified Model Systems for Deriving Causality in Complicated Biological Networks

Genetic manipulation provides systems to test ways in which biological components interact. KEs are difficult to identify in complex biological systems as linear models for dose-response relationships are hard to find in non-engineered systems. Genetically engineered systems offer a biologically based model with reduced complexity for teasing out mechanisms of toxicity. For example, genetically engineered tumor models offer tissue specific phenotypes that can be probed to validate stressor activity. Genetic modifications to cells provide increase sensitization to a particular stressor, which can act to highlight relevant biological machinery of interest. For example tumorigenecity of dioxin can be traced to sustained activation of the aryl-hydrocarbon receptor (AhR). Overexpression of AhR in mouse models demonstrates both the receptor is required for a tumor formation, but the complex interaction of cell signaling pathways determine the severity of tumor formation in a species specific manner. Tumor models with modified expression of AhR provided direct evidence of dioxin's effect on cell proliferation as the rate of growth was proportional to expression levels (Shimba et al. 2002; Becker et al. 2015; Behnisch et al. 2001).

Genetically modified model systems can be used to establish species specific responses. Differences in dioxin sensitivity have been observed across a number of species; vertebrates are more sensitive than other organisms. Amino acid substitutions on the AhR explain some of the response differences (Karchner et al. 2006; Head et al. 2008). However, AhR structure and expression of the aryl hydrocarbon hydroxylase enzyme (AHH) does not account for all differences (Moriguchi et al.

2003). Internal signaling, such as cytochrome mediated xenobiotic metabolism provides resistance to dioxin stressors. Through genetic manipulation, differential expression of these signaling pathways, along with the differences in AhR structure and AHH activity clarify the species specific toxicity of dioxin (Uno et al. 2009) and provide evidence for the dioxin AOP.

Organisms of higher life forms are also expressed into lower life forms through genetic engineering, to provide understanding of biological information across species in an efficient manner. For example, human immune systems have been cloned into rat and mice for test beds in pharmaceutical sciences to establish KEs critical for drug development. Mammalian proteins have been engineered into zebrafish for hypothesis testing and high throughput screening of potential pollutants. Last, engineered organisms provide test beds for understanding gene regulation and protein expression. These methods demonstrate the power genetic manipulation has for validating AOP relationships.

10.3.3.2 Reproduction and Developmental Assays

Genetic manipulation provides an assortment of tools to assess KE relationships involved with reproduction and development, which contribute to population level effects. There have been a number of tools aiding hypothesis testing of healthy germline tissue. Plant development and reproduction, especially in flowering plants has been challenged by long generation times and low transformation efficiency with loss-of function assays. Instead, using a transcriptional gene silencing mechanism to repress expression of specific MADS-box genes, Lu et al. (2007) demonstrated flower morphology could be used as an indicator of reproductive success. This method provided an efficient way to study genes involved in reproductive stages of organisms with long life cycles and simplified exposure testing for appraising AOPs.

Genetic manipulation has advanced access to key biomarkers involved in reproductive pathways. Fusion proteins of vertebrate hormones have provided simplified systems for observing reproductive outcomes during exposure. Zhang et al. (2014) deleted the hormone-specific β-genes of both FSH and LH in zebrafish using TALENs and showed clear genetic relationships for key reproductive events, including gonadal differentiation, puberty onset, gametogenesis, final maturation, and fertility. Xu et al. 2014 investigated silkworm (*Bombyx mori*) reproduction by combining transgenesis with TALEN technologies. The authors showed transgenic animals co-expressing TALEN left and right arms targeting the female-specific *Bmdsx* exon resulted in somatic mutations and female mutants lost fecundity because of lack of egg storage and abnormal external genitalia, again demonstrating simplified methods for elucidating mechanisms of reproductive toxicity.

Genetic manipulation has also provided cells throughout the differentiation process that can be used to understand KER in critical steps in the developmental process. Immortalization of cells through genetic manipulation provides biologic material for in vitro as discussed above, however, stem cells offer the unique advantage of testing progression through the critical phases of growth and development. Stem cells are capable of renewing themselves; that is, they can be continuously cultured in an undifferentiated state, giving an indefinite pool of cells to work with for large scale analysis (Yu et al. 2007). Moreover, stem cells have the ability to differentiation into any cell type and form self-organizing organ like structures, providing a platform to recreate genomic, cellular and organismal tissues for precise disease modeling and functional physiological studies (González 2016). While human stem cells have been a major focus with recent research and development efforts, animal model stem cells have also recently been developed, such as mouse (Okita et al. 2007), rat (Kim et al. 2009), and zebrafish (Grandel et al. 2006; Chen and Zon 2009) stem cells which can be used extensively for (eco)toxicology *in vitro* testing. For instance, Davila et al. (2004) demonstrated how genetically manipulated stem cells provide better in vitro models for screening genotoxic, epigenetic toxicants and reproductive toxicology than conventional methods. Wang et al. (2014) used mouse stem cells at various stages of development and differentiation to show a susceptibility window to bisphenol A that leads to tumorigenesis through cell compartment and self-renewal functions in the developing organisms. Stem cells offer unprecedented advantages for toxicity assays, however, methods must be standardized and cell lines need to be validated before experimentation is implemented.

10.3.3.3 Disease Models

Disease models offer toxicologists biological material with an observable pathology expressed with known etiology. Disease models can be useful to confirm AOP linkages at higher levels of biological organization. Creation of these models requires precise genetic modifications and environmental exposure conditions. For example, hepatoxicity models have been generated in zebrafish (Hill et al. 2005) using genetic screens. Zebrafish mutants exhibiting hepatic pathology have been used to validate functional roles in lipid metabolism and confirm toxic responses with specific gene anomalies (Carten and Farber 2009). Metabolic disorders involving lipid metabolism have been tested in murine models as well.

Transgenic mouse models have been made for various metabolic disease states. Mouse models for lipotoxicity have been created in which excess lipid uptake, driven by overexpression of fatty acid transport proteins in the heart leads to cardiac dysfunction. In MHC-ACS mice, excess unmetabolized lipid is associated with cardiomyocyte apoptosis, systolic heart failure and premature death (Chiu et al. 2001). This novel mouse model uses a tissue specific tag to induce expression of a protein. Using this targeting specificity, local perturbations in myocardial lipid metabolism in the pathogenesis of inherited and acquired forms of heart failure are able to be investigated. Transgenic overexpression of other enzymes involved in metabolic pathways have been expressed in specific tissues using a similar approach, including: liver-type phosphofructokinase (Elson et al. 1994), Tissue-Nonspecific Alkaline Phosphatase (Savinov et al. 2015) in Vascular Endothelium, Cardiomyocyte-Specific PPAR β/δ Overexpression (Kim et al. 2013) and many others (Masuzaki et al. 2001). These disease models provide means to capture relevant biomolecular interaction that give rise to a toxic response.

Disease models have been generated for neurodegenerative diseases caused by gain of function mechanisms, where malformed proteins accumulate to a toxic level. Spinocerebellar ataxia type 1 (SCA1) is one such disease, characterized by loss of motor coordination due to the degeneration of cerebellar Purkinje cells and brain stem neurons. Transgenic mice with a mutated SCA1 allele were generated using conventional cloning and microinjection techniques (Burright et al. 1995). These mice were crossbred with mice conditionally expressing heat shock proteins (HSP70) (Marber et al. 1995). The resulting transgenic mouse has been used to identify a myriad of neurological pathways and provide dose-response toxicity evidence for a variety of stress response pathways. (Adachi et al. 2003; Muchowski and Wacker 2005; Kim et al. 2016).

10.4 Future Considerations

With the gold rush of genetic advances, it is important to understand how these will manifest into tools and platforms to advance understanding of biological systems. Many of the complex interactions in biological systems can be identified, investigated, and validated using these tools. Toxicologists can use these tools to confirm key event relationships in AOPs, develop biomarkers for impact assessment, and create high throughput test systems for efficient screening of chemicals. Careful consideration should be employed when extrapolating evidence from engineered systems, however. Genetic manipulation interferes with the inner workings of natural systems in ways that are not fully understood. Weight of evidence, and plausibility criteria are important to include when evaluating AOPs. Genetic manipulation can simplify the process to understand chemical toxicity, however, rigorous studies are needed before a final determination.

References

Adachi H, Katsuno M, Minamiyama M et al (2003) Heat shock protein 70 chaperone overexpression ameliorates phenotypes of the spinal and bulbar muscular atrophy transgenic mouse model by reducing nuclear-localized mutant androgen receptor protein. J Neurosci 23(6):2203–2211

Alic N, Hoddinott MP, Foley A et al (2012) Detrimental effects of RNAi: a cautionary note on its use in drosophila ageing studies. PLoS One 7(9):e45367. doi:10.1371/journal.pone

Ankley GT, Bennett RS, Erickson RJ et al (2010) Adverse outcome pathways: a conceptual framework to support ecotoxicology research and risk assessment. Environ Toxicol Chem 29(3):730–741. doi:10.1002/etc.34

Bagasra O, Prilliman KR (2004) RNA interference: the molecular immune system. J Mol Histol 35(6):545–553. doi:10.1007/s10735-004-2192-8

Baltimore D, Berg P, Botchan D et al (2015) A prudent path forward for genomic engineering and germline gene modification. Science 348:36–38. doi:10.1126/science.aab1028

Becker RA, Patlewicz G, Simon TW et al (2015) The adverse outcome pathway for rodent liver tumor promotion by sustained activation of the aryl hydrocarbon receptor. Regul Toxicol Pharmacol 73(1):172–190. doi:10.1016/j.yrtph.2015.06.015

Behnisch PA, Hosoe K, Sakai S (2001) Bioanalytical screening methods for dioxins and dioxin-like compounds – a review of bioassay/biomarker technology. Environ Int 27(5):413–439. doi:10.1016/S0160-4120(01)00028-9

Brückner A, Polge C, Lentze N et al (2009) Yeast Two-Hybrid, a Powerful Tool for Systems Biology. Int J Mol Sci 10(6): 2763–2788. doi:10.3390/ijms10062763

Burright EN, Clark HB, Servadio A et al (1995) SCA1 transgenic mice: a model for neurodegeneration caused by an expanded CAG trinucleotide repeat. Cell 82(6):937–948. doi:10.1016/0092-8674(95)90273-2. ISSN 0092-8674

Carten JD, Farber SA (2009) A new model system swims into focus: using the zebrafish to visualize intestinal lipid metabolism in vivo. Clin Lipidol 4(4):501–515. doi:10.2217/clp.09.40

Charbonnier S, Gallego O, Gavin AC (2008) The social network of a cell: recent advances in interactome mapping. Biotechnol Annu Rev 14:1–28

Chen YC, Rajagopala SV, Stellberger T et al (2010) Exhaustive benchmarking of the yeast two-hybrid system. Nat Method 7(9): 667–668. doi:10.1038/nmeth0910-667

Chen AT, Zon LI (2009) Zebrafish blood stem cells. Cell Biochem 108(1):35–42. doi:10.1002/jcb.22251

Chiu HC, Kovacs A, Ford DA et al (2001) A novel mouse model of lipotoxic cardiomyopathy. J Clin Invest 107(7):813–822. doi:10.1172/JCI10947

Cullen LM, Arndt GM (2005) Genome-wide screening for gene function using RNAi in mammalian cells. Immunol Cell Biol 83:217–223. doi:10.1111/j.1440-1711.2005.01332.x

Danielsen ET, Moeller ME, Yamanaka N et al (2016) A drosophila genome-wide screen identifies regulators of steroid hormone production and developmental timing. Dev Cell 37(6):558–570

Davila JC, Cezar GG, Thiede M et al (2004) Use and application of stem cells in toxicology. Toxicol Sci 79(2):214–223. doi:10.1093/toxsci/kfh100

Dean KM, Palmer AE (2014) Advances in fluorescence labeling strategies for dynamic cellular imaging. Nat Chem Biol 10(7):512–523. doi:10.1038/nchembio.1556

Dix DJ, Houck KA, Martin MT et al (2007) The ToxCast program for prioritizing toxicity testing of environmental chemicals. Toxicol Sci 95(1):5–12

Ellison CM, Enoch SJ, Cronin MT (2011) A review of the use of in silico methods to predict the chemistry of molecular initiating events related to drug toxicity. Expert Opin Drug Metab Toxicol 7(12):1481–1495. doi:10.1517/17425255.2011.629186

Elson A, Levanon D, Weiss Y et al (1994) Overexpression of liver-type phosphofructokinase (PFKL) in transgenic-PFKL mice: implication for gene dosage in trisomy 21. Biochemist 299(2):409–415

Esvelt KM, Wang HH (2013) Genome-scale engineering for systems and synthetic biology. Mol Syst Biol 9(1):641. doi:10.1038/msb.2012.66

Gaytán BD, Vulpe CD (2014) Functional toxicology: tools to advance the future of toxicity testing. Front Genet 5:110. doi:10.3389/fgene.2014.00110

González F (2016) CRISPR/Cas9 genome editing in human pluripotent stem cells: harnessing human genetics in a dish. Dev Dyn 245(7):788–806. doi:10.1002/dvdy.24414

Grandel H, Kaslin J, Ganz J et al (2006) Neural stem cells and neurogenesis in the adult zebrafish brain: origin, proliferation dynamics, migration and cell fate. Dev Biol 295(1):263–277. doi:10.1016/j.ydbio.2006.03.040

Harrison MM, Jenkins BV, O'Connor-Giles KM et al (2014) A CRISPR view of development. Genes Dev 28(17):1859–1872. doi:10.1101/gad.248252.114

Haynes KA, Silver PA (2009) Eukaryotic systems broaden the scope of synthetic biology. J Cell Biol 187:589–596. doi:10.1083/jcb.200908138

Head JA, Hahn ME, Kennedy SW (2008) Key amino acids in the aryl hydrocarbon receptor predict dioxin sensitivity in avian species. Environ Sci Technol 42(19):7535–7541

Hill AJ, Teraoka H, Heideman W et al (2005) Zebrafish as a model vertebrate for investigating chemical toxicity. Toxicol Sci 86(1):6–19. doi:10.1093/toxsci/kfi110

Hogan BM, Verkade H, Lieschke GJ et al (2008) Manipulation of gene expression during zebrafish embryonic development using transient approaches. Methods Mol Bio 469:273–300

Holtzman NG, Iovine MK, Liang JO et al (2016) Learning to fish with genetics: a primer on the vertebrate model *Danio rerio*. Genetics 203(3):1069–1089

Kaminoka M (2011) Engineering of therapeutic proteins production in Escherichia coli. Curr Pharm Biotechnol 12(2):268–274. doi:10.2174/138920111794295693

Karchner SI, Franks DG, Kennedy SW et al (2006) The molecular basis for differential dioxin sensitivity in birds: role of the aryl hydrocarbon receptor. PNAS 103(16):6252–6257. doi:10.1073/pnas.0509950103

Kelly W, Stumpf M (2008) Protein-protein interactions: from global to local analyses. Curr Opin Biotechnol 19: 396–403.

Kim D, Kim CH, Moon JI et al (2009) Generation of human induced pluripotent stem cells by direct delivery of reprogramming proteins. Cell Stem Cell 4(6):472–476. doi:10.1016/j.stem.2009.05.005

Kim T, Zhelyabovska O, Liu J et al (2013) Generation of an inducible, cardiomyocyte-specific transgenic mouse model with PPAR β/δ overexpression. Methods Mol Biol 952:57–65. doi:10.1007/978-1-62703-155-4_4

Kim JY, Kim N, Zheng Z et al (2016) 70-kDa heat shock protein downregulates dynamin in experimental stroke a new therapeutic target? Stroke 47:2103–2111. doi:10.1161/Strokeha.116.012763

Lee JY, Ishida Y, Kuge S et al (2015) Identification of substrates of F-box protein involved in methylmercury toxicity in yeast cells. FEBS Lett 14 (589; 19 Pt B): 2720–5. doi:10.1016/j.febslet.2015.08.016

Lodish H, Baltimore D, Berk A et al (1995) Molecular cell biology (Vol. 3). New York: Scientific American Books

Lu H-C, Chen H-H, Tsai W-C et al (2007) Strategies for functional validation of genes involved in reproductive stages of orchids. Plant Physiol 143(2):558–569. doi:10.1104/pp.106.092742

Ma Y, Zhang L, Huang X (2014) Genome modification by CRISPR/Cas9. FEBS J 281(23):5186–5193. doi:10.1111/febs.13110

Marber MS, Mestril R, Chi SH et al (1995) Overexpression of the rat inducible 70-kD heat stress protein in a transgenic mouse increases the resistance of the heart to ischemic injury. J Clin Invest 95(4):1446–1456. doi:10.1172/JCI117815

Masuzaki H, Paterson J, Shinyama NM (2001) A transgenic model of visceral obesity and the metabolic syndrome. Science 294(5549):2166–2170

Mathieu V (2013) The adverse outcome pathway concept: a pragmatic tool in toxicology. Toxicology 312:158–165. ISSN 0300-483X

Mattanovich D, Branduardi P, Dato L et al (2012) Recombinant protein production in yeasts. Methods Mol Biol 824:329–358. doi:10.1007/978-1-61779-433-9

McMahon MA, Rahdar M, Porteus M (2012) Gene editing: not just for translation anymore. Nat Methods 9(1):28–31

Mohr SE, Hu Y, Ewen-Campen B et al (2016) CRISPR guide RNA design for research applications. FEBS J 283(17):3232–3238. doi:10.1111/febs.13777

Moriguchi T, Motohashi H, Hosoya T et al (2003) Distinct response to dioxin in an arylhydrocarbon receptor (AHR)-humanized mouse. PNAS 100(10):5652–5657. doi:10.1073/pnas.1037886100

Muchowski PJ, Wacker JL (2005) Modulation of neurodegeneration by molecular chaperones. Nat Rev Neurosci 6(1):11–22

Nishihara T, Nishikawa J, Kanayama T et al (2000) Estrogenic Activities of 517 Chemicals by Yeast Two-Hybrid Assay. J Health Sci 46(4): 282-29. doi:10.1248/jhs.46.282

(NRC) National Research Council (2007) Toxicity testing in the 21st century: a vision and a strategy. Committee on Toxicity Testing and Assessment of Environmental Agents; Board on Environmental Studies and Toxicology; Institute for Laboratory Animal Research; Division on Earth and Life Studies, Washington, DC. doi:10.17226/11970

Okita K, Ichisaka T, Yamanaka S (2007) Generation of germline-competent induced pluripotent stem cells. Nature 448:313–317. doi:10.1038/nature05934

Patlewicz G, Simon T, Goyak K et al (2013) Use and validation of HT/HC assays to support 21st century toxicity evaluations. Regul Toxicol Pharmacol 65(2):259–268. ISSN 0273-2300

Perkins EJ, Antczak P, Burgoon L et al (2015) Adverse outcome pathways for regulatory applications: examination of four case studies with different degrees of completeness and scientific confidence. Toxicol Sci 148(1):14–25. doi:10.1093/toxsci/kfv181

Peterson RT, MacRae CT (2011) Systematic approaches to toxicology in the zebrafish. Annu Rev Pharmacol Toxicol 52:433–453. doi:10.1146/annurev-pharmtox-010611-134751

Ritchie DW (2008) Recent progress and future directions in protein-protein docking. Curr Protein Pept Sci 9(1):1–15

Russom CL, LaLone CA, Villeneuve DL et al (2014) Development of an adverse outcome pathway for acetylcholinesterase inhibition leading to acute mortality. Environ Toxicol Chem 33:2157–2169. doi:10.1002/etc.2662

Savinov AY, Salehi M, Yadav MC et al (2015) Transgenic overexpression of tissue-nonspecific alkaline phosphatase (TNAP) in vascular endothelium results in generalized arterial calcification. J Am Heart Assoc 4(12):e002499. doi:10.1161/JAHA.115.002499

Shalem ONE, Sanjana E, Hartenian X et al (2014) Genome-scale CRISPR-Cas9 knockout screening in human cells. Science 343(6166):84–87. doi:10.1126/science.1247005

Shimba S, Kazuo K, Moro I et al (2002) Overexpression of the aryl hydrocarbon receptor (AhR) accelerates the cell proliferation of A549 cells. Biochemist 132(5):795–802

Skretas G, Georgiou G (2008) Engineering G protein-coupled receptor expression in bacteria. PNAS 105(39):14747–14748. doi:10.1073/pnas.0807741105

Tan WS, Carlson DF, Walton MW et al (2012) Precision editing of large animal genomes. Adv Genet 80:37–97. doi:10.1016/B978-0-12-404742-6.00002-8

Tollefsen KE, Scholz S, Cronin MT et al (2014) Applying adverse outcome pathways (AOPs) to support integrated approaches to testing and assessment (IATA). Regul Toxicol Pharmacol 70(3):629–640. doi:10.1016/j.yrtph.2014.09.009

Uno S, Endo K, Ishida Y et al (2009) CYP1A1 and CYP1A2 expression: comparing 'humanized' mouse lines and wild-type mice; comparing human and mouse hepatoma-derived cell lines. Toxicol Appl Pharmacol 237(1):119–126. doi:10.1016/j.taap.2009.03.001

Villeneuve DL, Crump D, Garcia-Reyero N et al (2014) Adverse outcome pathway (AOP) development I: strategies and principles. Toxicol Sci 142(2):312–320. doi:10.1093/toxsci/kfu199

Wang D, Gao H, Bandyopadhyay A et al (2014) Pubertal bisphenol a exposure alters murine mammary stem cell function leading to early neoplasia in regenerated glands. Cancer Prev Res 7(4):445–455. doi:10.1158/1940-6207.CAPR-13-0260

Watanabe KH, Andersen ME, Basu N et al (2011) Defining and modeling known adverse outcome pathways: domoic acid and neuronal signaling as a case study. Environ Toxicol Chem 30:9–21. doi:10.1002/etc.373

Wilkbanks MS, Gust KA, Atwa S et al (2014) Validation of a genomics-based hypothetical adverse outcome pathway: 2,4-Dinitrotoluene perturbs PPAR Signaling thus impairing energy metabolism and exercise endurance. Toxicol Sci 141(1):44–58. doi:10.1093/toxsci/kfu104

Willett C (2014) Adverse outcome pathways: development and use in toxicology. In: Wexler P (ed) Encyclopedia of toxicology, 3rd edn. Academic, Oxford, pp 95–99. doi:10.1016/B978-0-12-386454-3.01244-6. isbn:9780123864550

Wooge CH, Nilsson GM, Heierson A et al (1992) Structural requirements for high affinity ligand binding by estrogen receptors: a comparative analysis of truncated and full length estrogen receptors expressed in bacteria, yeast, and mammalian cells. Mol Endocrinol 6(6):861–869. doi:10.1210/mend.6.6.1495491

Worth A, Barroso J, Bremer S et al (2014) Alternative methods for regulatory toxicology – a state of the art review. JRC Sci Policy Rep EUR 26797:1–475

Xu J, Wang Y, Li Z et al (2014) Transcription activator-like effector nuclease (TALEN)-mediated female-specific sterility in the silkworm, *Bombyx mori*. Insect Mol Biol 23:800–807. doi:10.1111/imb.12125

Yamauchi T, Miyoshi D, Kubodera T et al (2008) Riboswitches for enhancing target gene expression in eukaryotes. Chembiochem 9:1040–1043. doi:10.1002/cbic.200700782

Yang H, Wang H, Shivalila CS et al (2013) One-step generation of mice carrying reporter and conditional alleles by CRISPR/Cas-mediated genome engineering. Cell 154:1370–1379

Yang L, Güell M, Niu D et al (2015) Genome-wide inactivation of porcine endogenous retroviruses (PERVs). Science 350:1101–1104

Yu J, Vodyanik MA, Smuga-Otto K et al (2007) Induced pluripotent stem cell lines derived from human somatic cells. Science 318(585):1917–1920. doi:10.1126/science.1151526

Yu Z, Chen H, Liu J et al (2014) Various applications of TALEN- and CRISPR/Cas9-mediated homologous recombination to modify the drosophila genome. Biol Open 3:271–280

Yu Z, Ren M, Wang Z et al (2013). Highly efficient genome modifications mediated by CRISPR/Cas9 in Drosophila. Genetics 195: 289–291. doi:10.1534/genetics.113.153825

Zhang Z, Zhu B, Ge W (2014) Genetic analysis of zebrafish gonadotropin (FSH and LH) functions by TALEN-mediated gene disruption. Mol Endocrinol 29(1):76–98

Zhao P, Zhang Z, Ke H et al (2014) Oligonucleotide-based targeted gene editing in *C. elegans* via the CRISPR/Cas9 system. Cell Res 24:247–250. doi:10.1038/cr.2014.9

Zoonens M, Miroux B (2010) Expression of membrane proteins at the Escherichia coli membrane for structural studies. Methods Mol Biol 601:49–66. doi:10.1007/978-1-60761-344-2

Zugasti O, Thakur N, Belougne J et al (2016) A quantitative genome-wide RNAi screen in *C. elegans* for antifungal innate immunity genes. BMC Biol 14(35). doi:10.1186/s12915-016-0256-3

Chapter 11
Considering Epigenetics in Adverse Outcome Pathways

Kristine L. Willett

Abstract While the concept of epigenetics was first recognized in the 1940s, appreciation for the potential of epigenetic change to be either the molecular initiating event or a key event underlying a phenotypic adverse outcome is much more recent. Now it is well established that epigenetic transcriptional regulatory processes are critical both during normal development and disease progression. Environmental factors that act epigenetically during key developmental stages can cause irreversible changes in gene expression, tissue structure or function and increase the risk of developing adult disease. Furthermore, certain epigenetic consequences (e.g. DNA methylation status) can be passed between generations impacting offspring that were not ever exposed to the stressor. To date, the incorporation of epigenetic events into adverse outcome pathways is limited by incomplete understanding of the basic mechanisms underlying epigenetic regulation of gene transcription and how that is conserved, however advances are being made very quickly in the field.

11.1 Epigenetics Definition

Working in the 1940s and 1950s, Conrad Waddington (Waddington 1942, 1956) is attributed with first describing the concept of epigenetics (e.g. "above genetics") in his experiments wherein environmental stress combined with developmental plasticity resulted in new phenotypes that were inheritable (Noble 2015). Today the term epigenetics is interpreted more widely to be both heritable processes independent of DNA sequence (e.g. X-chromosome inactivation and genome imprinting) but also transcriptional regulatory processes that influence many different cellular properties (Greally and Jacobs 2013). In biomedical research, comprehensive understanding of the role of epigenetic mechanisms in health and disease is both a research priority and rapidly expanding field (Portela and Esteller 2010). Epigenetic

K.L. Willett (✉)
University of Mississippi, University, MS 38677, USA
e-mail: kwillett@olemiss.edu

N. Garcia-Reyero, C.A. Murphy (eds.), *A Systems Biology Approach to Advancing Adverse Outcome Pathways for Risk Assessment*,
DOI 10.1007/978-3-319-66084-4_11

inheritance is well recognized and relatively common in plants (Heard and Martienssen 2014). In higher organisms, certain associations between epigenetic changes and disease phenotype, even if the exposure was during early development, are becoming accepted, but true transgenerational epigenetic inheritance particularly in humans is less clear (Heard and Martienssen 2014) despite high profile publicity of the phenomena in the lay literature (e.g. *Time* cover story (Cloud 2010); *Science News* (Saey 2013)).

When considering incorporating epigenetics into adverse outcome pathways, the key events may be changes in: DNA methylation, histone modification or non-coding RNA (Fig. 11.1). Each of these is a measurable response and associates, as described in more detail below, with altered gene expression (e.g. key event relationship). While there is strong evidence of the key event relationship between these particular epigenetic changes and altered gene expression, the molecular initiating event, or how a chemical directly causes the epigenetic perturbation is still largely unknown. Furthermore, while there is growing evidence in animal models that some stressors cause epigenetic change, altered gene expression and adverse phenotypic outcomes, much research is still needed to fill in subsequent event relationships to quantitatively link all the key events.

DNA methylation is the most understood of the mechanisms of epigenetic control and has been extensively reviewed (Attwood et al. 2002; Ko et al. 2015; Szyf 2012). DNA methylation is an important mechanism regulating chromatin structure, transcriptional control, and normal cellular function (Doerfler 1983). DNA methylation is sequence specific; methylated cytosines are mostly found in the dinucleotide sequence ‘CG’. CpG islands are regions with a high density of ‘CG’ dinucleotides associated with the promoter regions of genes and are typically unmethylated in active genes. Methylation of normally unmethylated CpG islands, located in the 5′ promoter region of genes, is associated with transcriptional inactivation of chromosomes, transgenes, disease genes and certain developmentally regulated genes (Kass et al. 2002). However, methylation of cytosine can also occur in non-CpG sites including CHG and CHH sequence contexts, where H is an A, C, or T (Feng et al. 2010) and can serve unknown functions. More recently, thanks to new analytical technologies that facilitate whole genome methylation analysis (Bock et al. 2010; Harris et al. 2010), studies have expanded from the focus on CpG islands near transcriptional start sites and as a result the associations with gene expression have become more complicated. CpG islands located in gene bodies tend to be methylated in tissue specific patterns while methylation of CpGs not located in islands is more dynamic (Jones 2012). In fact, in human embryonic stem cells, methylation of CpG islands at the 3′ end of genes conferred tissue- and cell-type specific gene activation whereas promoter CpG island methylation repressed activation (Yu et al. 2013). CpGs methylation may also have a regulatory role in gene splicing (Laurent et al. 2010). So while DNA methylation remains the most studied aspect of epigenetic regulation, clearly it is a rapidly progressing field with many unanswered questions.

CpG methylation is maintained by the enzymes DNA (cytosine-5)-methyltransferases (DNMTs). During mammalian embryonic development DNMT3a and

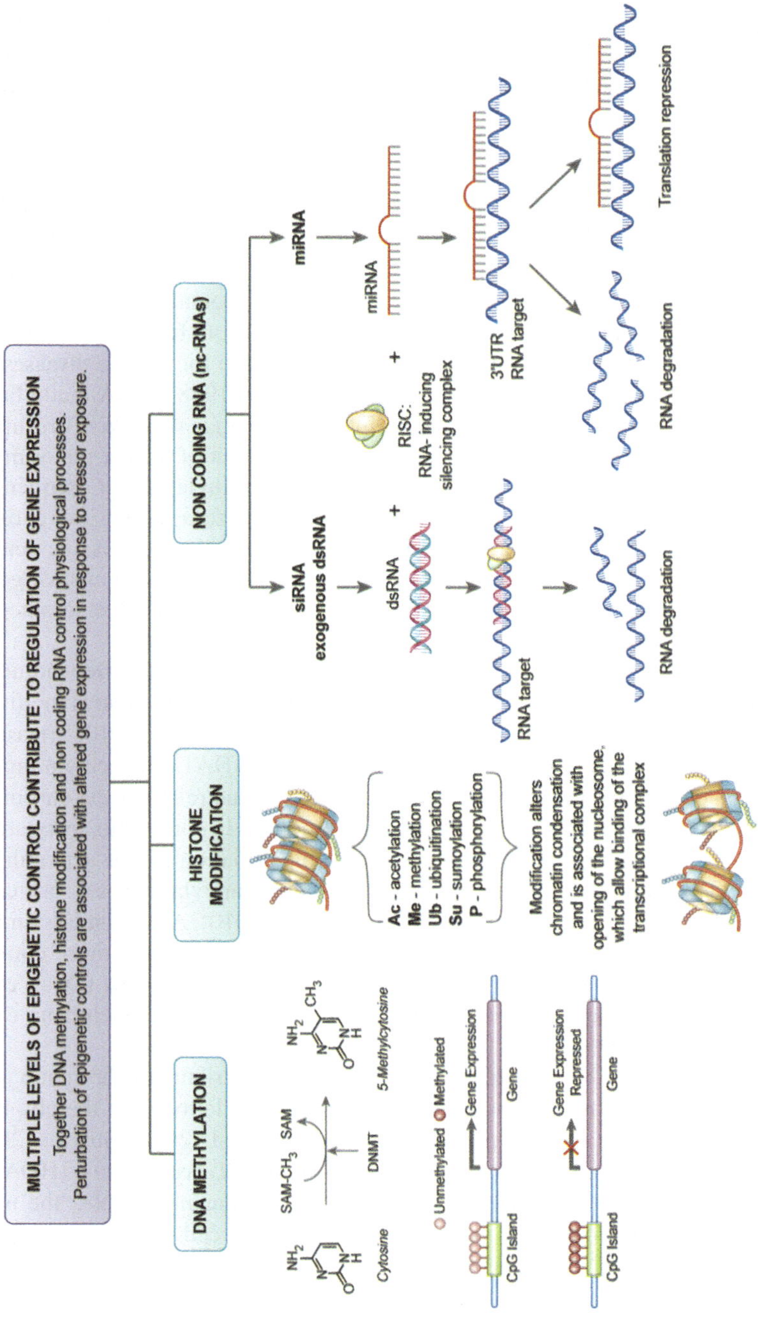

Fig. 11.1 Molecular mechanisms of epigenetic control of gene expression

DNMT3b establish *de novo* methylation while DNMT1 maintains methylation patterns during DNA replication (Subramaniam et al. 2014). *S*-adenosyl-methionine (SAM), a product of the folate/methionine cycles, serves as the methyl donor for DNMT reactions, and dysregulation of SAM homeostasis is implicated in various diseases (Martinez-Lopez et al. 2008; Mato and Lu 2007; Padmanabhan et al. 2013). Demethylation activity is provided by TET1, TET2 and TET3 enzymes which cause the ten-eleven translocation (TET)-mediated methylcytosine hydroxylation (5mC to 5hmC) (Shen et al. 2013). Proteins that selectively bind to 5hmCs (hydroxymethylcytosines) may also contribute to transcriptional regulation and thus play epigenetic roles (Iurlaro et al. 2013).

DNA methylation patterns can be altered by environmental factors that induce epigenetic changes in DNA, in turn altering gene expression. It has been estimated that in humans 37% of all germ-line mutations responsible for genetic diseases are localized to CpG dinucleotides (Cooper and Youssoufian 1988). DNA methylation may increase the mutation rate of an imprinted allele (that is inactive) resulting in loci that have a high mutation rate. For example, five out of the six major mutational hot spots in the p53 tumor suppressor gene are methylated CpG sequences, and these sites correlate with enhanced reactivity with benzo(a)pyrene diol epoxide (Weisenbeger and Romano 2002). For this reason, particularly for stressors that are mutagenic, there may be interdependence between epigenetic mechanisms, genotoxicity and/or DNA repair in the key event relationships of an adverse outcome pathway (Heard and Martienssen 2014).

Histone modifications are another mediator of gene expression. Depending on the covalent modification and which amino acid residue of the histone is modified, transcription is predictably activated or repressed (Choudhuri et al. 2010; Srivastava and Ahn 2015). For example, histone acetyltransferases (HATs) by acetylating the histone lysine residues, decrease the affinity of the histone relaxing the chromatin so that transcriptional activators can initiate transcription. In contrast to acetylations which are primarily activating, methylation and ubiquitination can be activating or repressing depending on the residue modified, and sumoylations are primarily repressing (Choudhuri et al. 2010; Portela and Esteller 2010). As the understanding of histone modification advances, it like methylation, is becoming more complicated in that cross-talk between histones with various modifications (Duan et al. 2008) as well as interactions with DNMTs have been reported (Tachibana et al. 2008; Wang et al. 2009).

Small non-coding RNAs (or microRNAs) can also contribute to chromatin state maintenance by targeting loci for histone or DNA methylation, and thereby help mediate epigenetically gene expression (Choudhuri et al. 2010; Sharma 2014; Stuwe et al. 2014; Szyf 2015). For example, histone deacetylase enzyme (HDAC1) expression can be repressed in cancer cells by a particular microRNA, the miR-449a, highlighting the role of the microRNAs in regulating cell growth and viability (Noonan et al. 2009). RNA interference occurs when microRNAs bind to 3′ untranslated regions in target mRNAs and destabilize, degrade, and/or translationally repress their targets. Six nucleotides of the 5′ region of the microRNA are the seed sequence that interacts with the target RNA. Computational analysis has predicted that more than 60% of human protein coding genes could pair with microRNAs (Friedman et al. 2009), but the biological relevance of all the predicted interactions is still being discovered.

11.2 Epigenetic Change as a Key Event in AOPs

Clearly, the molecular mechanisms underlying epigenetic control of gene expression are complicated and interrelated. Fundamental new knowledge about how DNA methylation and histone modification is controlled and the roles of microRNAs in either process is being rapidly discovered. Undoubtedly, epigenetic change will play a role in stressor adverse outcomes and will needed to be incorporated into AOP approaches (Fig. 11.2). That said, there are several time-scales wherein epigenetic changes can influence adverse outcomes including: the developmental basis for adult diseases, multigenerational impacts or even transgenerational impacts. Examples of each have been reported in various animal models and/or humans as highlighted below.

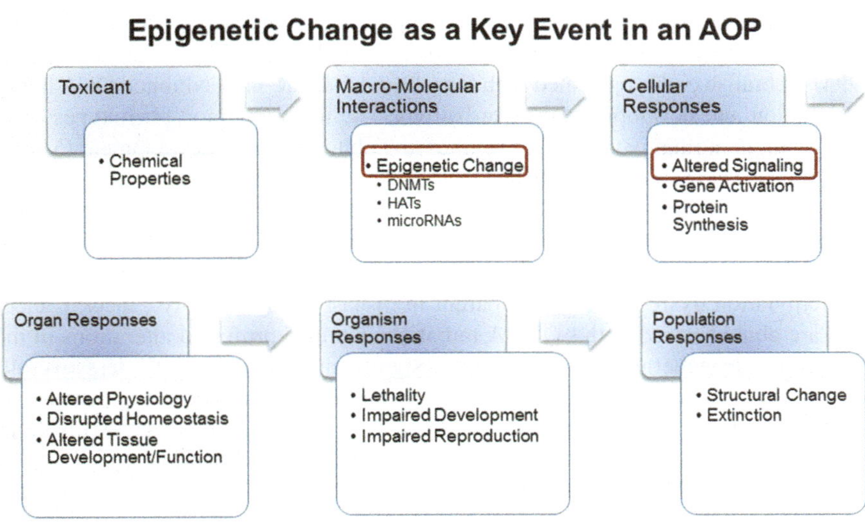

Fig. 11.2 Epigenetic change can be incorporated as a key macro-molecular event responsible for altered signaling in stressor adverse outcome pathways

11.2.1 Epigenetics and Developmental Origins of Health and Disease (DOHaD)

The toxicology community has begun to recognize that susceptibility to disease can be set *in utero* as a result of exposure to contaminants or nutritional deficits (Choudhuri et al. 2010; Heindel 2008; Szyf 2009). Similarly, differences early life experiences have been correlated with epigenetic changes. For example, using rodent models numerous studies have shown postnatal stress and/or lack of maternal care results in adverse behaviors in adult offspring (reviewed in (Maccari et al. 2014)). One of the first studies linked maternal care, persistent changes in DNA methylation and histone acetylation in hippocampal glucocorticoid receptor, and heightened stress response in adult offspring (Weaver et al. 2004). The relationships between adult behavior and epigenetic mechanisms were further validated by treatments including methionine to stimulate DNA methylation (Weaver et al. 2005) and trichostain A to inhibit histone deacetylase (Weaver et al. 2004). With the appreciation that during brain development, maturation, and learning epigenetic mechanisms play a key role (Bender and Weber 2013; Gabel and Greenberg 2013; Lister et al. 2013), it is not surprising that there is a plethora of new studies into the potential role epigenetics plays in mental diseases such as autism and psychotic disorders (Abdolmaleky et al. 2005, 2015) and drug addiction (Kenny 2014).

The developmental importance of methylation homeostasis is provided by disruptions caused by 5-Azacytidine (5azaC), an established inhibitor of DNA methylation. In the chick embryo model 5azaC caused hypomethylation and activation of several genes resulting in developmental abnormalities and arrest (Zagris and Podimatas 1994). Zebrafish embryos treated with 5-azaC and 5-azadC also exhibited DNA hypomethylation and developmental perturbations. The most common phenotypes were loss of tail, abnormal patterning of somites and abnormal head development (Martin et al. 1999). These studies established that DNA methylation was required for normal gastrulation and subsequent patterning of the dorsal mesoderm in fish. Administration of 5azaC to pregnant mice also resulted in perturbation of embryonic DNA synthesis, low fetal weight and death of rapidly proliferating cells (Rogers et al. 1994).

Methylation dysregulation is apparent in diseases most notably cancer. Cancer cells are characterized both by DNA mutations but also profound alterations in the epigenome. In general, cancer cells have significantly less (20–60% less overall) 5-methylcytosine than normal cells, (Portela and Esteller 2010) but hypermethylation of certain promoters (e.g. tumor suppressor and DNA repair genes) is common. Additionally, miRNA downregulation and histone modifications (e.g. reduced monoacetylated H4K16) are typical in human tumors, reviewed in (Portela and Esteller 2010). Despite that numerous epigenetic changes that are constantly being identified in all of the various cancer subtypes, it remains a challenge to distinguish if the epigenetic changes are the key event or more of a bystander or consequence effect. That said, for certain cancers new drug development is targeting epigenetic mechanisms including DNMT inhibitors (Decitabine [Dacogen], 5-Azacytidine [Vidaza]) and HDAC inhibitors (Vorinostat [Zolinza], Romidepsin [Istodax], Belinostat [Beleodaq]) (Hamm and Costa 2015; Yoo and Jones 2006).

11.2.2 Epigenetics and Multigenerational AOs

One of the seminal studies highlighting the potential for multigenerational epigenetic conservation of phenotypes used the Agouti model (Dolinoy et al. 2007a; Morgan et al. 1999). In this mouse model a retrotransposon is inserted upstream of the agouti gene. Agouti protein expression is related to yellow fur, obesity, and diabetic phenotypes. Because the transposon is controlled epigenetically, offspring will express a mosaic of phenotypic fur color ranging from yellow to brown. Using this model, bisphenol A (BPA), a well-recognized endocrine disrupting compound, decreased retrotransposon methylation and offspring fur color was shifted to yellow. Furthermore, when maternal diet was supplemented with methyl donors including folic acid, vitamin B12 and betaine the BPA-mediated hypomethylation was reversed (Dolinoy et al. 2007b).

Similarly studies in sheep wherein ewes during the periconceptional period were fed diets depleted in vitamin B12, folate, or methionine led in the adult offspring to both methylation status changes in 4% of 1400 CpG islands investigated as well as adverse outcomes. Adverse phenotypes predominated in the male offspring and included increased blood pressure, adiposity, insulin resistance and altered immune response (Sinclair et al. 2007).

Epidemiology cohorts also support the idea that nutritional deficits cause multigenerational epigenetic impacts in humans. Many studies have tested the "thrifty phenotype hypothesis" proposed by Hales and Barker (Hales and Barker 2001). They proposed the now widely supported relationship between poor fetal growth (as a consequence of poor fetal nutrition) and subsequent permanently compromised glucose-insulin metabolism resulting in type 2 diabetes and metabolic syndrome. The Dutch "Hunger Winter" cohort, adults that experienced famine during their periconceptional period, had persistent hypermethylation in the insulin-like growth factor 2 (IGF2) gene (Heijmans et al. 2008). Follow-up methylation studies of 15 additional loci in this cohort indicated that changes in DNA methylation were dependent not only on nutritional stress but the timing and sex of the exposed offspring (Tobi et al. 2009). Likewise, human offspring born to diabetic (preexisting or gestational) mothers are more likely to have higher birth weights and ultimately develop obesity and diabetes (Fraser and Lawlor 2014; Patti 2013). Furthermore many epigenetic changes have been identified in relevant growth and metabolic genes (Quilter et al. 2014) (e.g. leptin, adiponectin, ABCA1 etc. reviewed in (Ma et al. 2015)).

The foundations of the epigenetics field were in nutritional impacts on the epigenome, but now the potential for multigenerational impacts of environmental toxicants with an epigenetic key event is also emerging. For example, polycyclic aromatic hydrocarbons (PAHs) are a ubiquitous class of combustion-associated contaminants that are known carcinogens and reproductive toxicants. Epidemiologic studies have shown prenatal exposure to PAHs (from maternal inhalation) is related to neural tube defects (Ren et al. 2011), a lower mental development index at age 3 (Perera et al. 2006) and decrements in full-scale IQ and verbal IQ at age 5 after adjustment for other confounding factors (Perera et al. 2009). PAHs are also linked to preterm deliveries and small size for gestational age (Choi et al. 2008; Langlois et al. 2014), low birth weights (Siddiqui et al. 2008) and cleft lip ± palate (Langlois et al. 2013). Disproportionate fetal growth is, in turn, related to coronary heart disease (Barker et al. 1993; Barker 1997). Multigenerational adverse outcomes of paternal smoking are also beginning to be appreciated (Northstone et al. 2014). Furthermore, recent studies have identified differences in global methylation and gene specific promoter CpG island methylation in cord blood, placenta, newborns and children exposed *in utero* to tobacco smoke or PAHs (Breton et al. 2009; Herbstman et al. 2012; Joubert et al. 2012; Suter et al. 2011).

A clear challenge when considering either nutritional deficits or complex mixtures of environmental contaminant exposures is that multiple adverse developmental outcomes can manifest in offspring, and it is likely that epigenetic dysregulation of an individual gene/protein or even pathway will not be responsible for all the phenotypes. While single generational epigenetic transmission is well supported, the human health impacts of truly transgenerational epigenetic inheritance is more controversial.

11.2.3 Epigenetics and Transgenerational AOs

Similar to what was described for single generation nongenetic inheritance, in humans the basis for transgenerational impacts is related to epidemiological studies of dietary distress. Dietary distress in cohorts from Northern Sweden was sex-dependently linked to mortality and diabetic deaths in grandchildren (Bygren et al. 2014; Pembrey 2010). As depicted in Fig. 11.3, in mammals, transgenerational effects of a stressor cannot be reached until the F3 generation because both F1 and the F2 germ cells are potentially exposed during the female Fo exposure (Skinner 2008). Studies of human transgenerational epigenetic inheritance are obviously complicated by the length of the studies to capture three or more generations and the retrospective nature of the exposure. Additionally, a key question remains related to how epigenetic changes escape the reprogramming that occurs twice both after fertilization and during primordial germ cell differentiation (Szyf 2015). Better understanding of the molecular mechanisms that allow for epigenetic marks to be transferred across generations will fundamentally strengthen the case for transgenerational epigenetic inheritance (Heard and Martienssen 2014). Some of this understanding is being provided from various laboratory models that include rodents and fish.

A good example to illustrate the state of the knowledge on transgenerational epigenetic inheritance is found in studies of vinclozolin, a dicarboximide fungicide used on fruits, vegetables and turf grasses. The toxicity of vinclozolin is associated with its anti-androgenic mechanisms particularly of its two metabolites M1 and M2. Human reproductive hazards associated with exposure appear low (Zober et al. 1995) but incompletely assessed (James 1997). In rodents, perinatal vinclozolin

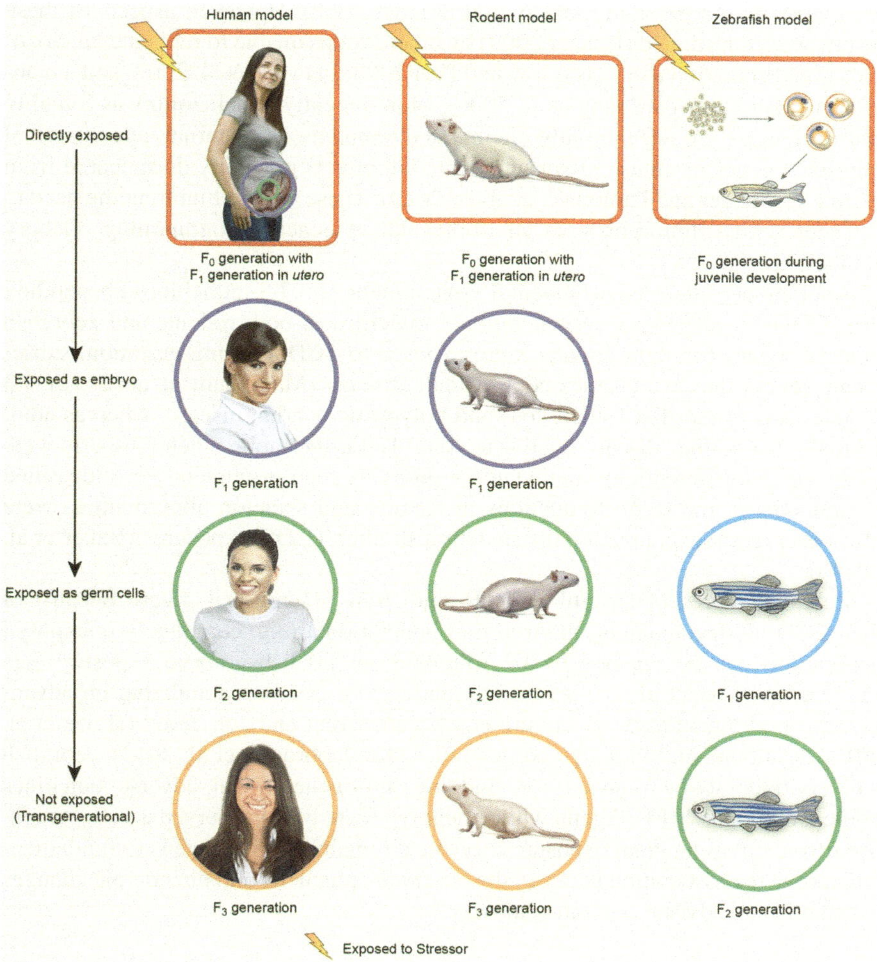

Fig. 11.3 Multigenerational adverse outcomes can result from parental exposures to stressors including chemical or nutrient deficiencies (depicted by lightning bolt). In pregnant mammals, a F_0 exposure can also directly expose F_1 and F_2 generations in the embryonic and germ cell stages, respectively. Potential transgenerational effects are not realized until the F_3 generation. In contrast, because early development is external in fish, transgenerational effects can be measured in the F_2 generation

exposure causes demasculization of the male reproductive system including decreased anogenital distance, retained nipples, hypospadias, reduced sperm numbers (Gray Jr. et al. 1999) and altered sexual behaviors (Colbert et al. 2005) in offspring. The Skinner laboratory has published a series of papers reporting that perinatal exposure to vinclozolin causes transgenerational (to F4) sperm defects (Anway et al. 2005, 2008a), epigenetic altered gene expression in the testis (Anway et al. 2008b) and prostate (Anway and Skinner 2008). However, aspects of these results were questioned (Renner 2009) by other groups unable to replicate vinclozolin's transgenerational effects (Gray and Furr 2008; Inawaka et al. 2009) and a manuscript was retracted (Chang et al. 2006). More recently, vinclozolin was found to cause changes in differentially methylated domains of paternal and maternal imprinted genes in mouse offspring sperm, but effects gradually disappeared from F1 to F3 (Stouder and Paoloni-Giacobino 2010). These studies highlight the need to consider sexual dimorphism in environmental epigenetic programming (Gabory et al. 2009).

Another persistent environmental contaminant, 2,3,7,8-tetrachlorodibenzodiox in (TCDD), has shown transgenerational toxicities in both rodents and zebrafish models. After F0 rodent females were exposed to TCDD during gestation, subsequent generations were assessed for adult disease (Manikkam et al. 2012). F3 females had primordial follicle loss and polycystic ovarian disease, whereas adult F3 males had kidney disease relative to controls. Furthermore, when F3 sperm were analyzed, 50 differentially methylated regions in gene promoters were identified (Manikkam et al. 2012). Reductions in fertility and skeleton abnormalities were also observed transgenerationally in zebrafish after TCDD exposures (Baker et al. 2014b).

Zebrafish represent a useful model to study transgenerational effects. Because of their external development, the true transgenerational (e.g. completely unexposed generation) can be reached by F2 following an F0 fish embryo exposure (see Fig. 11.3) (Baker et al. 2014a; Greally and Jacobs 2013). Capitalizing on advantages of high fecundity, low culture costs, transparent and conserved (Howe et al. 2013) developmental biology and genomics (Kettleborough et al. 2013), zebrafish will continue to be relevant for studying multigenerational adverse outcomes (Villeneuve et al. 2014). Again, while phenotypes are being observed across generations from environmental exposures, a critical remaining challenge is to fundamentally prove the association between the observed epigenetic and phenotypic changes to establish an adverse outcome pathway.

11.3 Experimental Challenges and Future Perspectives

Going forward, the major goal of AOP development is to improve regulatory decision making (Edwards et al. 2016). Given the extensive evidence that both epigenetic mechanisms are fundamental in development, health, and disease and dietary and environmental stressors significantly impact the epigenetic homeostasis, there is no doubt that ultimately epigenetics will need to be incorporated in key event relationships of various AOPs. That said, there is a significant lack of fundamental understanding of epigenetic mechanisms and nongenetic inheritance. Several reviews have highlighted the issues associated with incorporating epigenetic impacts into safety assessments and human health risk assessments (Alyea et al. 2014;

Goodman et al. 2010; LeBaron et al. 2010). In a case study of vinclozolin epigenetic human risk, the need for causal relationships between toxic endpoints and epigenetic alterations and the dose-dependence of epigenetic changes were specifically highlighted (Alyea et al. 2014). Teasing out whether epigenetic change is a cause or a result of a particular toxicity will remain a significant challenge. Furthermore, researchers will need to be diligent in choosing appropriate research animal models, developmental stages, tissue subtypes and epigenomic assays for subsequent applicability (Greally and Jacobs 2013). Finally, adverse outcomes mediated by epigenetic change are not only relevant to human risk assessments and must also be broadly considered in ecotoxicology as epigenetics may play a role in wildlife fitness and resilience (Lee et al. 2015; Mirbahai and Chipman 2014; Schwindt 2015; Wang et al. 2014). In sum, epigenetic change and its role in adverse outcomes within and across generations is an exciting area of research that is undergoing exceedingly rapid new discovery and has significant implications for the study of health and disease.

References

Abdolmaleky HM, Thiagalingam S, Wilcox M (2005) Genetics and epigenetics in major psychiatric disorders: dilemmas, achievements, applications, and future scope. Am J Pharmacogenomics 5:149–160

Abdolmaleky HM, Zhou JR, Thiagalingam S (2015) An update on the epigenetics of psychotic diseases and autism. Epigenomics 7:427–449

Alyea RA, Gollapudi BB, Rasoulpour RJ (2014) Are we ready to consider transgenerational epigenetic effects in human health risk assessment? Environ Mol Mutagen 55:292–298

Anway MD, Skinner MK (2008) Transgenerational effects of the endocrine disruptor vinclozolin on the prostate transcriptome and adult onset disease. Prostate 68:517–529

Anway MD, Cupp AS, Uzumcu M, Skinner MK (2005) Epigenetic transgenerational actions of endocrine disruptors and male fertility. Science 308:1466–1469

Anway MD, Rekow SS, Skinner MK (2008a) Comparative anti-androgenic actions of vinclozolin and flutamide on transgenerational adult onset disease and spermatogenesis. Reprod Toxicol 26:100–106

Anway MD, Rekow SS, Skinner MK (2008b) Transgenerational epigenetic programming of the embryonic testis transcriptome. Genomics 91:30–40

Attwood JT, Yung RL, Richardson BC (2002) DNA methylation and the regulation of gene transcription. Cell Mol Life Sci 59:241–257

Baker TR, King-Heiden TC, Peterson RE, Heideman W (2014a) Dioxin induction of transgenerational inheritance of disease in zebrafish. Mol Cell Endocrinol 398:36–41

Baker TR, Peterson RE, Heideman W (2014b) Using zebrafish as a model system for studying the transgenerational effects of dioxin. Toxicol Sci 138:403–411

Barker DJ (1997) The fetal origins of coronary heart disease. Acta Paediatr Suppl 422:78–82

Barker DJ, Osmond C, Simmonds SJ, Wield GA (1993) The relation of small head circumference and thinness at birth to death from cardiovascular disease in adult life. BMJ 306:422–426

Bender A, Weber M (2013) DNA methylation: an identity card for brain cells. Genome Biol 14:131

Bock C, Tomazou EM, Brinkman AB, Muller F, Simmer F, Gu H, Jager N, Gnirke A, Stunnenberg HG, Meissner A (2010) Quantitative comparison of genome-wide DNA methylation mapping technologies. Nat Biotechnol 28:1106–1114

Breton CV, Byun HM, Wenten M, Pan F, Yang A, Gilliland FD (2009) Prenatal tobacco smoke exposure affects global and gene-specific DNA methylation. Am J Respir Crit Care Med 180:462–467

Bygren LO, Tinghog P, Carstensen J, Edvinsson S, Kaati G, Pembrey ME, Sjostrom M (2014) Change in paternal grandmothers' early food supply influenced cardiovascular mortality of the female grandchildren. BMC Genet 15:12

Chang HS, Anway MD, Rekow SS, Skinner MK (2006) Transgenerational epigenetic imprinting of the male germline by endocrine disruptor exposure during gonadal sex determination. Endocrinology 147:5524–5541

Choi H, Rauh V, Garfinkel R, Tu Y, Perera FP (2008) Prenatal exposure to airborne polycyclic aromatic hydrocarbons and risk of intrauterine growth restriction. Environ Health Perspect 116:658–665

Choudhuri S, Cui Y, Klaassen CD (2010) Molecular targets of epigenetic regulation and effectors of environmental influences. Toxicol Appl Pharmacol 245:378–393

Cloud (2010) Why your DNA isn't your destiny, 175th edn. Time, pp 48–53

Colbert NK, Pelletier NC, Cote JM, Concannon JB, Jurdak NA, Minott SB, Markowski VP (2005) Perinatal exposure to low levels of the environmental antiandrogen vinclozolin alters sex-differentiated social play and sexual behaviors in the rat. Environ Health Perspect 113:700–707

Cooper DN, Youssoufian H (1988) The CpG dinucleotide and human genetic disease. Hum Genet 78:151–155

Doerfler W (1983) DNA methylation and gene activity. Annu Rev Biochem 52:93–124

Dolinoy DC, Das R, Weidman JR, Jirtle RL (2007a) Metastable epialleles, imprinting, and the fetal origins of adult diseases. Pediatr Res 61:30R–37R

Dolinoy DC, Huang D, Jirtle RL (2007b) Maternal nutrient supplementation counteracts bisphenol A-induced DNA hypomethylation in early development. Proc Natl Acad Sci U S A 104:13056–13061

Duan Q, Chen H, Costa M, Dai W (2008) Phosphorylation of H3S10 blocks the access of H3K9 by specific antibodies and histone methyltransferase. Implication in regulating chromatin dynamics and epigenetic inheritance during mitosis. J Biol Chem 283:33585–33590

Edwards SW, Tan YM, Villeneuve DL, Meek ME, McQueen CA (2016) Adverse outcome pathways-organizing toxicological information to improve decision making. J Pharmacol Exp Ther 356:170–181

Feng S, Cokus SJ, Zhang X, Chen PY, Bostick M, Goll MG, Hetzel J, Jain J, Strauss SH, Halpern ME, Ukomadu C, Sadler KC, Pradhan S, Pellegrini M, Jacobsen SE (2010) Conservation and divergence of methylation patterning in plants and animals. Proc Natl Acad Sci U S A 107:8689–8694

Fraser A, Lawlor DA (2014) Long-term health outcomes in offspring born to women with diabetes in pregnancy. Curr Diab Rep 14:489

Friedman RC, Farh KK, Burge CB, Bartel DP (2009) Most mammalian mRNAs are conserved targets of microRNAs. Genome Res 19:92–105

Gabel HW, Greenberg ME (2013) Genetics. The maturing brain methylome. Science 341:626–627

Gabory A, Attig L, Junien C (2009) Sexual dimorphism in environmental epigenetic programming. Mol Cell Endocrinol 304:8–18

Goodman JI, Augustine KA, Cunnningham ML, Dixon D, Dragan YP, Falls JG, Rasoulpour RJ, Sills RC, Storer RD, Wolf DC, Pettit SD (2010) What do we need to know prior to thinking about incorporating an epigenetic evaluation into safety assessments? Toxicol Sci 116:375–381

Gray LE, Furr J (2008) Vinclozolin treatment induces reproductive malformations and infertility in male rats when administered during sexual but not gonadal differentiation; however, the effects are not transmitted to the subsequent generations. Biol Reprod 78:228

Gray LE Jr, Ostby J, Monosson E, Kelce WR (1999) Environmental antiandrogens: low doses of the fungicide vinclozolin alter sexual differentiation of the male rat. Toxicol Ind Health 15:48–64

Greally JM, Jacobs MN (2013) In vitro and in vivo testing methods of epigenomic endpoints for evaluating endocrine disruptors. ALTEX 30:445–471

Hales CN, Barker DJ (2001) The thrifty phenotype hypothesis. Br Med Bull 60:5–20

Hamm CA, Costa FF (2015) Epigenomes as therapeutic targets. Pharmacol Ther 151:72–86

Harris RA, Wang T, Coarfa C, Nagarajan RP, Hong C, Downey SL, Johnson BE, Fouse SD, Delaney A, Zhao Y, Olshen A, Ballinger T, Zhou X, Forsberg KJ, Gu J, Echipare L, O'Geen H, Lister R, Pelizzola M, Xi Y, Epstein CB, Bernstein BE, Hawkins RD, Ren B, Chung WY, Gu H, Bock C, Gnirke A, Zhang MQ, Haussler D, Ecker JR, Li W, Farnham PJ, Waterland RA, Meissner A, Marra MA, Hirst M, Milosavljevic A, Costello JF (2010) Comparison of sequencing-based methods to profile DNA methylation and identification of monoallelic epigenetic modifications. Nat Biotechnol 28:1097–1105

Heard E, Martienssen RA (2014) Transgenerational epigenetic inheritance: myths and mechanisms. Cell 157:95–109

Heijmans BT, Tobi EW, Stein AD, Putter H, Blauw GJ, Susser ES, Slagboom PE, Lumey LH (2008) Persistent epigenetic differences associated with prenatal exposure to famine in humans. Proc Natl Acad Sci U S A 105:17046–17049

Heindel JJ (2008) Animal models for probing the developmental basis of disease and dysfunction paradigm. Basic Clin Pharmacol Toxicol 102:76–81

Herbstman JB, Tang D, Zhu D, Qu L, Sjodin A, Li Z, Camann D, Perera FP (2012) Prenatal exposure to polycyclic aromatic hydrocarbons, benzo[a]pyrene-DNA adducts, and genomic DNA methylation in cord blood. Environ Health Perspect 120:733–738

Howe K, Clark MD, Torroja CF, Torrance J, Berthelot C, Muffato M, Collins JE, Humphray S, McLaren K, Matthews L, McLaren S, Sealy I, Caccamo M, Churcher C, Scott C, Barrett JC, Koch R, Rauch GJ, White S, Chow W, Kilian B, Quintais LT, Guerra-Assuncao JA, Zhou Y, Gu Y, Yen J, Vogel JH, Eyre T, Redmond S, Banerjee R, Chi J, Fu B, Langley E, Maguire SF, Laird GK, Lloyd D, Kenyon E, Donaldson S, Sehra H, Meida-King J, Loveland J, Trevanion S, Jones M, Quail M, Willey D, Hunt A, Burton J, Sims S, McLay K, Plumb B, Davis J, Clee C, Oliver K, Clark R, Riddle C, Eliott D, Threadgold G, Harden G, Ware D, Mortimer B, Kerry G, Heath P, Phillimore B, Tracey A, Corby N, Dunn M, Johnson C, Wood J, Clark S, Pelan S, Griffiths G, Smith M, Glithero R, Howden P, Barker N, Stevens C, Harley J, Holt K, Panagiotidis G, Lovell J, Beasley H, Henderson C, Gordon D, Auger K, Wright D, Collins J, Raisen C, Dyer L, Leung K, Robertson L, Ambridge K, Leongamornlert D, McGuire S, Gilderthorp R, Griffiths C, Manthravadi D, Nichol S, Barker G, Whitehead S, Kay M, Brown J, Murnane C, Gray E, Humphries M, Sycamore N, Barker D, Saunders D, Wallis J, Babbage A, Hammond S, Mashreghi-Mohammadi M, Barr L, Martin S, Wray P, Ellington A, Matthews N, Ellwood M, Woodmansey R, Clark G, Cooper J, Tromans A, Grafham D, Skuce C, Pandian R, Andrews R, Harrison E, Kimberley A, Garnett J, Fosker N, Hall R, Garner P, Kelly D, Bird C, Palmer S, Gehring I, Berger A, Dooley CM, Ersan-Urun Z, Eser C, Geiger H, Geisler M, Karotki L, Kirn A, Konantz J, Konantz M, Oberlander M, Rudolph-Geiger S, Teucke M, Osoegawa K, Zhu B, Rapp A, Widaa S, Langford C, Yang F, Carter NP, Harrow J, Ning Z, Herrero J, Searle SM, Enright A, Geisler R, Plasterk RH, Lee C, Westerfield M, de Jong PJ, Zon LI, Postlethwait JH, Nusslein-Volhard C, Hubbard TJ, Roest CH, Rogers J, Stemple DL (2013) The zebrafish reference genome sequence and its relationship to the human genome. Nature 496:498–503

Inawaka K, Kawabe M, Takahashi S, Doi Y, Tomigahara Y, Tarui H, Abe J, Kawamura S, Shirai T (2009) Maternal exposure to anti-androgenic compounds, vinclozolin, flutamide and procymidone, has no effects on spermatogenesis and DNA methylation in male rats of subsequent generations. Toxicol Appl Pharmacol 237:178–187

Iurlaro M, Ficz G, Oxley D, Raiber EA, Bachman M, Booth MJ, Andrews S, Balasubramanian S, Reik W (2013) A screen for hydroxymethylcytosine and formylcytosine binding proteins suggests functions in transcription and chromatin regulation. Genome Biol 14:R119

James WH (1997) Is vinclozolin a reproductive hazard to men? Occup Environ Med 54:285

Jones PA (2012) Functions of DNA methylation: islands, start sites, gene bodies and beyond. Nat Rev Genet 13:484–492

Joubert BR, Haberg SE, Nilsen RM, Wang X, Vollset SE, Murphy SK, Huang Z, Hoyo C, Midttun O, Cupul-Uicab LA, Ueland PM, Wu MC, Nystad W, Bell DA, Peddada SD, London SJ (2012) 450K epigenome-wide scan identifies differential DNA methylation in newborns related to maternal smoking during pregnancy. Environ Health Perspect 120:1425–1431
Kass SU, Pruss D, Wolffe AP (2002) How does DNA methylation repress transcription. Trends Genet 13:444–449
Kenny PJ (2014) Epigenetics, microRNA, and addiction. Dialogues Clin Neurosci 16:335–344
Kettleborough RN, Busch-Nentwich EM, Harvey SA, Dooley CM, van de Be EF, Sealy I, White RJ, Herd C, Nijman IJ, Fenyes F, Mehroke S, Scahill C, Gibbons R, Wali N, Carruthers S, Hall A, Yen J, Cuppen E, Stemple DL (2013) A systematic genome-wide analysis of zebrafish protein-coding gene function. Nature 496:494–497
Ko M, An J, Rao A (2015) DNA methylation and hydroxymethylation in hematologic differentiation and transformation. Curr Opin Cell Biol 37:91–101
Langlois PH, Hoyt AT, Lupo PJ, Lawson CC, Waters MA, Desrosiers TA, Shaw GM, Romitti PA, Lammer EJ (2013) Maternal occupational exposure to polycyclic aromatic hydrocarbons and risk of oral cleft-affected pregnancies. Cleft Palate Craniofac J 50:337–346
Langlois PH, Hoyt AT, Desrosiers TA, Lupo PJ, Lawson CC, Waters MA, Rocheleau CM, Shaw GM, Romitti PA, Gilboa SM, Malik S (2014) Maternal occupational exposure to polycyclic aromatic hydrocarbons and small for gestational age offspring. Occup Environ Med 71:529–535
Laurent L, Wong E, Li G, Huynh T, Tsirigos A, Ong CT, Low HM, Kin Sung KW, Rigoutsos I, Loring J, Wei CL (2010) Dynamic changes in the human methylome during differentiation. Genome Res 20:320–331
LeBaron MJ, Rasoulpour RJ, Klapacz J, Ellis-Hutchings RG, Hollnagel HM, Gollapudi BB (2010) Epigenetics and chemical safety assessment. Mutat Res 705:83–95
Lee JW, Won EJ, Raisuddin S, Lee JS (2015) Significance of adverse outcome pathways in biomarker-based environmental risk assessment in aquatic organisms. J Environ Sci 35:115–127
Lister R, Mukamel EA, Nery JR, Urich M, Puddifoot CA, Johnson ND, Lucero J, Huang Y, Dwork AJ, Schultz MD, Yu M, Tonti-Filippini J, Heyn H, Hu S, Wu JC, Rao A, Esteller M, He C, Haghighi FG, Sejnowski TJ, Behrens MM, Ecker JR (2013) Global epigenomic reconfiguration during mammalian brain development. Science 341:1237905
Ma RC, Tutino GE, Lillycrop KA, Hanson MA, Tam WH (2015) Maternal diabetes, gestational diabetes and the role of epigenetics in their long term effects on offspring. Prog Biophys Mol Biol 118:55–68
Maccari S, Krugers HJ, Morley-Fletcher S, Szyf M, Brunton PJ (2014) The consequences of early-life adversity: neurobiological, behavioural and epigenetic adaptations. J Neuroendocrinol 26:707–723
Manikkam M, Tracey R, Guerrero-Bosagna C, Skinner MK (2012) Dioxin (TCDD) induces epigenetic transgenerational inheritance of adult onset disease and sperm epimutations. PLoS One 7:e46249
Martin CC, Laforest L, Akimenko M-A, Ekker M (1999) A role for DNA methylation in gastrulation and somite patterning. Dev Bio 206:189–205
Martinez-Lopez N, Varela-Rey M, Ariz U, Embade N, Vazquez-Chantada M, Fernandez-Ramos D, Gomez-Santos L, Lu SC, Mato JM, Martinez-Chantar ML (2008) S-adenosylmethionine and proliferation: new pathways, new targets. Biochem Soc Trans 36:848–852
Mato JM, Lu SC (2007) Role of S-adenosyl-L-methionine in liver health and injury. Hepatology 45:1306–1312
Mirbahai L, Chipman JK (2014) Epigenetic memory of environmental organisms: a reflection of lifetime stressor exposures. Mutat Res Genet Toxicol Environ Mutagen 764–765:10–17
Morgan HD, Sutherland HG, Martin DI, Whitelaw E (1999) Epigenetic inheritance at the agouti locus in the mouse. Nat Genet 23:314–318
Noble D (2015) Conrad Waddington and the origin of epigenetics. J Exp Biol 218:816–818
Noonan EJ, Place RF, Pookot D, Basak S, Whitson JM, Hirata H, Giardina C, Dahiya R (2009) miR-449a targets HDAC-1 and induces growth arrest in prostate cancer. Oncogene 28:1714–1724

Northstone K, Golding J, Davey SG, Miller LL, Pembrey M (2014) Prepubertal start of father's smoking and increased body fat in his sons: further characterisation of paternal transgenerational responses. Eur J Hum Genet 22:1382–1386

Padmanabhan N, Jia D, Geary-Joo C, Wu X, Ferguson-Smith AC, Fung E, Bieda MC, Snyder FF, Gravel RA, Cross JC, Watson ED (2013) Mutation in folate metabolism causes epigenetic instability and transgenerational effects on development. Cell 155:81–93

Patti ME (2013) Intergenerational programming of metabolic disease: evidence from human populations and experimental animal models. Cell Mol Life Sci 70:1597–1608

Pembrey ME (2010) Male-line transgenerational responses in humans. Hum Fertil (Camb) 13:268–271

Perera FP, Rauh V, Whyatt RM, Tsai WY, Tang D, Diaz D, Hoepner L, Barr D, Tu YH, Camann D, Kinney P (2006) Effect of prenatal exposure to airborne polycyclic aromatic hydrocarbons on neurodevelopment in the first 3 years of life among inner-city children. Environ Health Perspect 114:1287–1292

Perera FP, Li Z, Whyatt R, Hoepner L, Wang S, Camann D, Rauh V (2009) Prenatal airborne polycyclic aromatic hydrocarbon exposure and child IQ at age 5 years. Pediatrics 124:e195–e202

Portela A, Esteller M (2010) Epigenetic modifications and human disease. Nat Biotechnol 28:1057–1068

Quilter CR, Cooper WN, Cliffe KM, Skinner BM, Prentice PM, Nelson L, Bauer J, Ong KK, Constancia M, Lowe WL, Affara NA, Dunger DB (2014) Impact on offspring methylation patterns of maternal gestational diabetes mellitus and intrauterine growth restraint suggest common genes and pathways linked to subsequent type 2 diabetes risk. FASEB J 28:4868–4879

Ren A, Qiu X, Jin L, Ma J, Li Z, Zhang L, Zhu H, Finnell RH, Zhu T (2011) Association of selected persistent organic pollutants in the placenta with the risk of neural tube defects. Proc Natl Acad Sci U S A 108:12770–12775

Renner R (2009) Key environmental epigenetics paper challenged. Environ Sci Technol 43:8009–8010

Rogers JM, Francis BM, Sulik KK, Alles AJ, Massaro EJ, Zucker RM, Elstein KH, Rosen MB, Chernoff N (1994) Cell death and cell cycle perturbation in the developmental toxicity of the demethylating agent, 5-aza-2′-deoxycytidine. Teratology 50:332–339

Saey (2013) From great grandma to you, 183th edn. pp 18–21

Schwindt AR (2015) Parental effects of endocrine disrupting compounds in aquatic wildlife: is there evidence of transgenerational inheritance? Gen Comp Endocrinol 219:152–164

Sharma A (2014) Novel transcriptome data analysis implicates circulating microRNAs in epigenetic inheritance in mammals. Gene 538:366–372

Shen L, Wu H, Diep D, Yamaguchi S, D'Alessio AC, Fung HL, Zhang K, Zhang Y (2013) Genome-wide analysis reveals TET- and TDG-dependent 5-methylcytosine oxidation dynamics. Cell 153:692–706

Siddiqui AR, Gold EB, Yang X, Lee K, Brown KH, Bhutta ZA (2008) Prenatal exposure to wood fuel smoke and low birth weight. Environ Health Perspect 116:543–549

Sinclair KD, Allegrucci C, Singh R, Gardner DS, Sebastian S, Bispham J, Thurston A, Huntley JF, Rees WD, Maloney CA, Lea RG, Craigon J, McEvoy TG, Young LE (2007) DNA methylation, insulin resistance, and blood pressure in offspring determined by maternal periconceptional B vitamin and methionine status. Proc Natl Acad Sci U S A 104:19351–19356

Skinner MK (2008) What is an epigenetic transgenerational phenotype? F3 or F2. Reprod Toxicol 25:2–6

Srivastava R, Ahn SH (2015) Modifications of RNA polymerase II CTD: connections to the histone code and cellular function. Biotechnol Adv 33:856–872

Stouder C, Paoloni-Giacobino A (2010) Transgenerational effects of the endocrine disruptor vinclozolin on the methylation pattern of imprinted genes in the mouse sperm. Reproduction 139:373–379

Stuwe E, Toth KF, Aravin AA (2014) Small but sturdy: small RNAs in cellular memory and epigenetics. Genes Dev 28:423–431

Subramaniam D, Thombre R, Dhar A, Anant S (2014) DNA methyltransferases: a novel target for prevention and therapy. Front Oncol 4:80

Suter M, Ma J, Harris A, Patterson L, Brown KA, Shope C, Showalter L, Abramovici A, Agaard-Tillery KM (2011) Maternal tobacco use modestly alters correlated epigenome-wide placental DNA methylation and gene expression. Epigenetics 6:1284–1294

Szyf M (2009) The early life environment and the epigenome. Biochim Biophys Acta 1790:878–885

Szyf M (2012) The early-life social environment and DNA methylation. Clin Genet 81:341–349

Szyf M (2015) Nongenetic inheritance and transgenerational epigenetics. Trends Mol Med 21:134–144

Tachibana M, Matsumura Y, Fukuda M, Kimura H, Shinkai Y (2008) G9a/GLP complexes independently mediate H3K9 and DNA methylation to silence transcription. EMBO J 27:2681–2690

Tobi EW, Lumey LH, Talens RP, Kremer D, Putter H, Stein AD, Slagboom PE, Heijmans BT (2009) DNA methylation differences after exposure to prenatal famine are common and timing- and sex-specific. Hum Mol Genet 18:4046–4053

Villeneuve D, Volz DC, Embry MR, Ankley GT, Belanger SE, Leonard M, Schirmer K, Tanguay R, Truong L, Wehmas L (2014) Investigating alternatives to the fish early-life stage test: a strategy for discovering and annotating adverse outcome pathways for early fish development. Environ Toxicol Chem 33:158–169

Waddington CH (1942) The epigenotype. Endeavour 1:18–20

Waddington CH (1956) The genetic assimilation of the biothorax phenotype. Evolution 10:1–13

Wang J, Hevi S, Kurash JK, Lei H, Gay F, Bajko J, Su H, Sun W, Chang H, Xu G, Gaudet F, Li E, Chen T (2009) The lysine demethylase LSD1 (KDM1) is required for maintenance of global DNA methylation. Nat Genet 41:125–129

Wang X, Li Q, Lian J, Li L, Jin L, Cai H, Xu F, Qi H, Zhang L, Wu F, Meng J, Que H, Fang X, Guo X, Zhang G (2014) Genome-wide and single-base resolution DNA methylomes of the Pacific oyster *Crassostrea gigas* provide insight into the evolution of invertebrate CpG methylation. BMC Genomics 15:1119

Weaver IC, Cervoni N, Champagne FA, D'Alessio AC, Sharma S, Seckl JR, Dymov S, Szyf M, Meaney MJ (2004) Epigenetic programming by maternal behavior. Nat Neurosci 7:847–854

Weaver IC, Champagne FA, Brown SE, Dymov S, Sharma S, Meaney MJ, Szyf M (2005) Reversal of maternal programming of stress responses in adult offspring through methyl supplementation: altering epigenetic marking later in life. J Neurosci 25:11045–11054

Weisenbeger DJ, Romano LJ (2002) Cytosine methylation in a CpG sequence leads to enhanced reactivitiy with benzo[*a*]pyrene diol epoxide that correlates with a comformational change. J Biol Chem 274:23948–23955

Yoo CB, Jones PA (2006) Epigenetic therapy of cancer: past, present and future. Nat Rev Drug Discov 5:37–50

Yu DH, Ware C, Waterland RA, Zhang J, Chen MH, Gadkari M, Kunde-Ramamoorthy G, Nosavanh LM, Shen L (2013) Developmentally programmed 3′ CpG island methylation confers tissue- and cell-type-specific transcriptional activation. Mol Cell Biol 33:1845–1858

Zagris N, Podimatas T (1994) 5-Azacytidine changes gene expression and causes developmental arrest of early chick embryo. Int J Dev Biol 38:741–744

Zober A, Hoffmann G, Ott MG, Will W, Germann C, van Ravenzwaay B (1995) Study of morbidity of personnel with potential exposure to vinclozolin. Occup Environ Med 52:233–241

Chapter 12
Tiered Approaches to Incorporate the Adverse Outcome Pathway Framework into Chemical-Specific Risk-Based Decision Making

Jeremy Leonard, Shannon Bell, Noffisat Oki, Mark Nelms, Yu-Mei Tan, and Stephen Edwards

Abstract The concept of Adverse Outcome Pathways (AOPs) arose as a means of addressing the challenges associated with establishing relationships between high-throughout (HT) in vitro dose response data and in vivo biological outcomes. However, AOP development has also been met with challenges of its own, such as the time, effort, and expertise necessary to achieve a scientifically sound construct able to support ecotoxicology and human health risk assessment. Thus, a staged development process has been developed to match the information content of an AOP with the decision context in which it will be used. This approach allows effort to be spent on detailed evidence evaluation and quantitative assessment of the dose-response characteristics for those AOPs where this level of confidence and

---

Disclaimer The information in this document has been funded wholly (or in part) by the U. S. Environmental Protection Agency. It has been subjected to review by the National Health and Environmental Effects Research Laboratory and approved for publication. Approval does not signify that the contents reflect the views of the Agency, nor does mention of trade names or commercial products constitute endorsement or recommendation for use.

J. Leonard • N. Oki • M. Nelms
Oak Ridge Institute for Science and Education, Oak Ridge, TN 37831, USA
e-mail: leonard.jeremy@epa.gov; oki.noffisat@epa.gov; nelms.mark@epa.gov

S. Bell
National Institute of Environmental Health Sciences, Research Triangle Park, NC 27709, USA
e-mail: shannon.bell2@nih.gov

Y.-M. Tan
National Exposure Research Laboratory, United States Environmental Protection Agency, Research Triangle Park, NC 27709, USA
e-mail: tan.cecilia@epa.gov

S. Edwards (✉)
National Health and Environmental Effects Research Laboratory, United States Environmental Protection Agency, 109 TW Alexander Dr. Mail Code B105-03, Research Triangle Park, NC 27709, USA
e-mail: Edwards.steve@epa.gov

N. Garcia-Reyero, C.A. Murphy (eds.), *A Systems Biology Approach to Advancing Adverse Outcome Pathways for Risk Assessment*,
DOI 10.1007/978-3-319-66084-4_12

precision is needed. In addition, through advances in computational analytical methodologies that integrate HT data (e.g., transcriptomic data) with traditional toxicology information spanning a broad chemical and biological space, computationally predicted AOPs can be rapidly generated to help accelerate the curation of AOPs. AOPs are chemical agnostic thereby allowing a single AOP to be coupled with in vitro dose-response information from a variety of chemicals. To predict an in vivo outcome, however, exposure and pharmacokinetic characteristics (i.e., absorption, metabolism, distribution, and elimination) must be considered. As with the staged development process for AOPs, it is possible to develop ADME predictions in a tiered manner such that lower tiers provide qualitative or semi-quantitative predictions when data is lacking, and higher tiers provide quantitative predictions with increasing confidence when data is abundant. Tiered approaches to AOP development and ADME predictions provide a mechanism for using AOPs, with chemical-specific exposure and pharmacokinetic considerations, for risk assessment both in data poor and data rich scenarios. They also provide a natural mechanism for identifying areas of research that would have the highest impact on risk-based decision making by highlighting AOPs and/or ADME predictions that are insufficient to address the decision context in which they could be used.

12.1 Integration of Adverse Outcome Pathways into the Twenty-First Century Toxicity Testing Paradigm

Currently, tens of thousands of chemicals exist in commerce (Egeghy et al. 2012), and there is a 12–16% increase in the rate of chemicals that are newly registered by the Chemical Abstracts Service on a yearly basis (Binetti et al. 2008). Traditionally, toxicity testing for these chemicals has involved a complete array of in vivo animal studies that provide apical endpoints associated with toxic outcomes arising from exposures to environmental chemicals within whole biological systems (Krewski et al. 2009). Testing using animal models offers several advantages, in that chemicals can be studied in detail, and experimental conditions are well controlled. In addition, pharmacokinetic (PK) properties (i.e., absorption, distribution, metabolism, and elimination [ADME]) are included in the test itself, though extrapolation from the test species to the target species is still required. Animal models hold several disadvantages as well, such as large investments in time and cost, use of inbred subjects in some cases, and challenges associated with extrapolating results from the animal model to relevant human-health outcomes (Zurlo et al. 2001; Phillips et al. 2009; Soldatow et al. 2013). More importantly, the limited resources available for traditional toxicity testing renders it difficult to determine toxicity-related information for the majority of the chemicals currently in commerce or to keep pace with new registrations. Given these disadvantages, alternative toxicity testing protocols are needed to provide toxicity information in order to keep pace with the more rapid production of new chemicals.

In 2004, the National Toxicology Program (NTP) proposed a new "roadmap" for toxicity testing in the twenty-first century that focused on the refinement, replacement, and reduction of animal studies with rapid screening protocols able to maintain scientific quality, promote animal welfare, and protect human health (Shukla et al. 2010). Priority was placed on identification of primary pathways and molecular mechanisms that can be linked to disease (Andersen and Krewski 2009). One strategy designed to meet this goal involved the development of high-throughput (HT) in vitro assays related to computational modeling and toxicity (NRC 2007). Such assays provide a rapid, cost-efficient means of evaluating thousands of understudied chemicals across hundreds of pathway-based toxicity endpoints at concentrations that are relevant to both environmental and human health (Sun et al. 2012), and can aid in chemical prioritization for more extensive in vivo testing (Austin et al. 2008; Kavlock et al. 2009).

In vitro assays are designed to determine the responses of technological targets, which often act as surrogates for in vivo biological targets at selected chemical concentrations, and these data are used to assess hazard. Establishing relevance of these in vitro perturbations to in vivo responses is often difficult, as these assays lack the biological context of an in vivo system. For example, an in vitro assay can reveal whether or not a certain chemical has the ability to bind with and perturb a technological target, but there remains some uncertainty as to what the resulting outcome might be in a living organism. Interpreting the relationships between toxicant perturbations on an in vitro molecular target and adverse outcomes observed in a living organism, as well as identifying the complex toxicity pathways leading to in vivo adverse outcomes, gives rise to several challenges that should be addressed before HT testing can be optimally applied in twenty-first century toxicity testing.

In response to these challenges, the concept of Adverse Outcome Pathways (AOPs) was developed to aid in understanding the mechanistic basis of in vivo toxicity by establishing linkages between HT testing results and adverse biological outcomes that are of concern in risk assessment and chemical management practices. An AOP is a linear construct whereby existing knowledge regarding the mechanistic basis for chemical toxicity is described by a series of key events that connect an upstream molecular initiating event (MIE) to a downstream adverse outcome (AO). This framework takes the toxicity pathway concept set forth by the National Research Council (NRC 2007) and extends it to higher levels of biological organization, up to the population level (Villeneuve et al. 2014b). The original aim of the AOP framework involved the use of pathway-based data to support ecotoxicology risk assessment and research through non-direct measures of apical toxicological outcomes (Ankley et al. 2010). The AOP framework has further been extended to support human health risk assessment (Tollefsen et al. 2014) due to its ability to act as a scaffold onto which various pathway-based data can be arranged, thereby allowing a mechanistic connection between levels of biological organization to be established (Vinken 2013).

Two fundamental components are utilized when describing an AOP: key events (KEs) and key event relationships (KERs) (Villeneuve et al. 2014b). The KEs that comprise an AOP describe measurable and requisite changes in the biological state at each level of organization. The inaugural KE within an AOP is the MIE, which is defined as the interaction of a xenobiotic stressor with a molecular target (Ankley et al. 2010) and is often analogous to a technological target used in an in vitro assay. Upon sufficient perturbation, the MIE is followed by a series of intermediate KEs, ideally at least one for each level of biological organization between the MIE and the final endpoint in the AOP. The AOP is anchored on the other end by one or more special KEs, the AOs that represent apical endpoints sufficient to support a chemical management decision. Such outcomes can be captured at the individual level (e.g., organ dysfunction, cancer, abnormalities) for human health risk assessment or at the population level (e.g., reduced species recruitment) in the case of an ecological risk assessment (Kramer et al. 2011; Villeneuve et al. 2014b). The KERs link adjacent KEs and provide scientific evidence supporting such a linkage. Additionally, KERs introduce directionality to the AOP framework by identifying that KE in the relationship which can be found upstream and which can be found downstream.

An AOP is often comprised of one MIE, one AO, and at least one KE at each intermediate level of biological organization (Ankley et al. 2010; Villeneuve et al. 2014a). Each AOP is considered to be separate from other AOPs during its development. This is not intended to disregard the complexity contained within biological systems and the ability of multiple signaling pathways to exert significant influence upon one another; it is a strategy that more easily enables AOPs to be identified and elucidated. Additionally, it is acknowledged that a single specific KE is likely to be a component within multiple AOPs. These shared KEs allow AOPs to be joined together to form AOP networks, in order to inform integrated approaches to testing assessment (Tollefsen et al. 2014; Villeneuve et al. 2014b). The AOP framework provides a biological context to the interpretation of in vitro hazard data (Tollefsen et al. 2014), and thus, it enhances the applicability of new toxicity testing approaches by providing information regarding relevant in vitro concentrations capable of perturbing an MIE. Due to its chemical-agnostic nature, the AOP framework enables the evaluation of data from HT in vitro assays that are designed to simultaneously measure the activity of large numbers of chemicals for a given molecular target in only a fraction of the time required for traditional toxicity tests (Becker et al. 2015). While the importance of chemical-specific characteristics (e.g., exposure and ADME properties) is recognized by developers of the AOP framework, by nature AOPs are most relevant to the hazard component of risk assessment.

Since the 1983 publication of the Red Book (NRC 1983), the core of risk assessment has been defined by hazard identification, followed by the combination of dose-response analysis and exposure assessment. This was refined in 2009 with the publication of Science and Decisions (NRC 2009), which recommended improvements to risk assessment by (1) introducing a decision context into the assessment process; (2) developing more explicit decisions regarding the use of default safety factors; and (3) handling uncertainty and variability in an appropriate manner when establishing a reference concentration or dose (Abt et al. 2010). Hazard identification and dose-response assessment have traditionally relied on overt toxicity data generated from laboratory animal studies. Such studies have the advantage of including both the toxicokinetics and toxicodynamics of the chemical being tested, but with the disadvantages described above. Exposure assessment has traditionally relied upon chemical-specific monitoring data or predictions to identify source, fate and transport processes, along with final receptors for chemical exposure (MacIntosh and Spengler 2000). Together these pieces of information can be used for risk characterization of chemicals, often with little or no knowledge of the underlying mechanisms driving either exposure or toxicity.

Risk assessment based on the new paradigms of HT toxicity and exposure evaluation seeks to integrate in vitro screening assays able to identify hazard (Judson et al. 2011; Vinken 2013), in silico models able to estimate exposure (Wambaugh et al. 2014; Isaacs et al. 2014), and quantitative structure activity relationship (QSAR)-based models, for prioritizing thousands of data-poor environmental chemicals. Though the approaches used in traditional and HT risk assessment have changed significantly over the years, both involve the necessary inputs of hazard AND exposure. For AOPs to be applied in an optimal manner during risk assessment, in vitro concentrations able to perturb the molecular target and induce an MIE can be extrapolated to biologically-effective target tissue doses. These in turn can be converted to external exposure levels through reverse dosimetry (Lin and Lu 1997; Simmons et al. 2005; Stadnicka-Michalak et al. 2014; Groh et al. 2015) to support risk-based decision-making processes (Benford et al. 2010). The ADME behaviors of a chemical, which are lacking in in vitro assays, are the driving factors mediating this biologically-effective dose.

The use of the AOP framework facilitates interpretation of HTT because an AOP can be developed using a limited number of reference chemicals and then used to interpret the HTT results for potentially hundreds of chemicals exhibiting positive results for a given assay or assay battery. While the process for fully describing and evaluating the AOP is laborious and time-consuming, AOPs at all stages of development are available for use depending on the decision-making context. Similarly, full characterization of the ADME properties for a chemical and accurate prediction of its toxicokinetic behaviors requires a considerable effort, thus limiting the number of chemicals for which such characterization can be performed. Fortunately, both the AOP development process and ADME modeling are amenable to tiered approaches that allow predictions to be made across the continuum from data-poor to data-rich situations, albeit with tradeoffs in both confidence and precision for those estimates (Fig. 12.1). These tiered approaches not only enable the use of AOPs and ADME predictions from lower tiers when the decision-making context allows for a higher degree of uncertainty and/or lower precision in the estimate, but they also provide a mechanism for prioritizing AOP development to match the description of the AOP with the needs of decision makers. This avoids having a handful of fully characterized AOPs for cases where that level of precision isn't needed, while also avoiding cases in which other less developed AOPs are not sufficient to support the decisions for which they would be used.

At one extreme, a paucity of data or need for expedited processing may require best estimates of KEs and their relationships, or may necessitate the qualitative assessment of whether or not a stressor that binds with a technological target in vitro might be capable of reaching the molecular analogue of that target in vivo. Such lower-tier approaches likely do not provide a high level of confidence for making informed decisions, but they can act as the groundwork for higher-tier approaches.

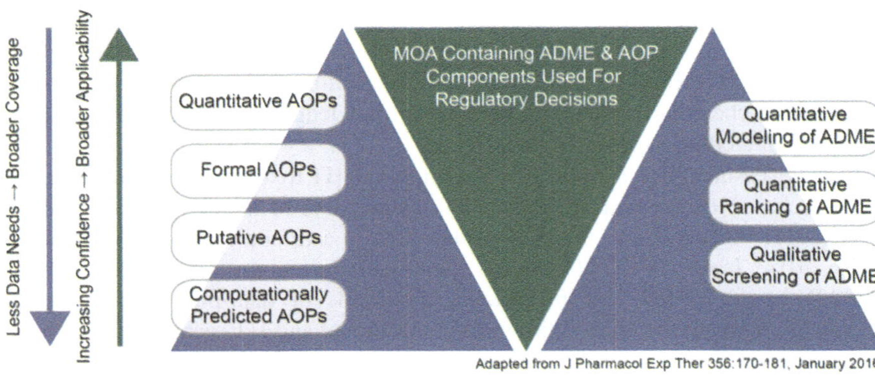

Fig. 12.1 Graphical representation of the tiered structure of the Adverse Outcome Pathway (AOP) and pharmacokinetic (e.g., absorption, distribution, metabolism, and elimination [ADME]) components needed to determine mode of action (MOA) for supporting risk-based decisions. *Blue triangles* represent the relative number of AOPs or ADME predictions that can be completed with a fixed amount of effort. Lower tiers in each case require less input data and time expended evaluating the available information. For certain decisions such as screening and prioritization, this level of confidence may be sufficient. Use of higher tiers increases the utility for a broader range of risk-based decisions (illustrated by the *green triangle* in the *center*), but it also limits the number of cases that can be considered due to the requirements of more input data and time for experimentation and computational modeling that are necessary to achieve this higher level of confidence

Researchers seeking to conduct higher-tier analyses but lacking the necessary resources or information can use results from lower-tier analyses as a means to identify data gaps. In doing so, the appropriate level of time and effort might be expended in obtaining such critical information. At the other extreme, available data may be sufficient to allow for development of AOPs using researcher expertise and knowledge, as well as the development of computational models with the ability to predict external chemical doses expected to result in an MIE. As the amount of empirical data necessary for evaluation of relationships between KEs and AOs or parameterization of models increases, so does confidence in the mechanistic processes leading to an AO, along with an exposure or ADME model's predictive capabilities. Thus, risk assessors are provided with the ability to make more informed decisions. The tiered approaches involved in AOP development and its application to risk assessment will be discussed in further detail throughout the remainder of this chapter.

12.2 Accelerating Adverse Outcome Pathway Development Via Systems Approaches

In 2012, the AOP development program was initiated by the Organization for Economic Co-operation and Development (OECD) to promote the development and use of AOPs. As an accompaniment to the development program, the OECD released a handbook that acts to advise AOP developers in regards to the key information that should be included within the AOP, and also provides developers with a template that can be used to aid in assembling and organizing such information in support of the AOP (OECD 2013a; Villeneuve et al. 2014b). Villeneuve et al. (Villeneuve et al. 2014b) described five fundamental principles for developing AOPs: (1) AOPs are not chemical specific, i.e. any chemical able to sufficiently perturb the MIE and the intermediate KEs may induce the AO; (2) AOPs are modular and composed of reusable components - notably key events (KEs) and key event relationships (KERs); (3) an individual AOP, composed of a single sequence of KEs and KERs, is a pragmatic unit of AOP development and evaluation; (4) networks composed of multiple AOPs that share common KEs and KERs are likely to be the functional unit of prediction for most real-world scenarios; and (5) AOPs are living documents that will evolve over time as new knowledge is generated. By following these principles during development, the individual KEs and KERs enable AOPs to mimic the modularity held by biological systems, whereby different processes may be conserved across multiple biological pathways.

The first step in the development process is to identify the KEs involved in the progression of the AOP and the relationship(s) among them; thus, providing a scaffold onto which the supporting evidence and information can be arranged. A number of strategies have been proposed that may be utilized when developing an AOP: (1) bottom-up, i.e. start with MIE data and work to identify the mechanistic information that links the MIE to the downstream KEs; (2) middle-out, i.e. start with data for an intermediate KE and work to identify the mechanistic information that anchors this KE to both an MIE and an AO; and (3) top-down, i.e. start with data for an observable AO and work to identify the mechanistic information that links the AO to the upstream KEs within the AOP (Villeneuve et al. 2014b; Groh et al. 2015). The mechanistic information used to link each of the KEs present in an AOP may be derived from a variety of sources: the available literature, in silico techniques, in vitro assays, or in vivo tests. It is envisioned that each of these sources of information will be used in concert during AOP development.

Typically, evaluations regarding the essentiality of each KE within an AOP are aided by data from knock-out, knock-down, or reversibility studies. For example, knocking out a specific gene associated with a given upstream KE could demonstrate the essentiality of that KE by showing that perturbations upstream do not result in an AO when that earlier step is blocked. These types of studies provide researchers with the ability to determine whether preventing a perturbation of an upstream KE will lead to a concomitant reduction in the observation of the downstream KE(s). Data generated from experiments that measure the difference in observing the downstream KE(s) after the upstream KE has been impeded could provide a high degree of weight of evidence that the upstream KE is essential. In contrast, if no such evidence exists, or the results from certain experiments can be disputed, the weight of evidence in essentiality would be weak (OECD 2013a; Becker et al. 2015).

When sufficient information has been assembled, the evidence for each KE and KER can be systematically evaluated, thereby providing support during the assessment of the entire AOP. Conducting the evaluation process in this manner enables identification of any data gaps that may be present, whilst also acting as a guide for the most appropriate use of the AOP within a risk management setting. Typically, modified Bradford-Hill considerations are utilized when assessing the evidence for each component within a given AOP as they can help determine the relevance of the identified supporting information (Hill 1965; Meek et al. 2014; Becker et al. 2015). When taken together, the Bradford-Hill considerations can enable assessments regarding the essentiality of each KE and the empirical support and biological plausibility for each KER that comprise the AOP (OECD 2013a).

Supporting data for each KE within the AOP should include the following: (1) describing the role the KE plays under normal biological (homeostatic) conditions and how the KE might be perturbed during the course of the AOP; and (2) describing the assay(s) that may be conducted to test for the impact of perturbation upon the KE. The KERs in the AOP capture the evidence supporting the causal relationships among the KEs, which is essential for the use of the AOP. This evidence consists of the following components: (1) the biological plausibility of the relationship between the two KEs, based upon current knowledge of how they interact under homeostatic conditions; (2) the specific evidence establishing an association between the upstream and downstream KEs, i.e. does the evidence that is present support the proposal that a perturbation in the upstream KE induces a change in the downstream KE; and (3) the uncertainty (if any) pertaining to the relationship between the two KEs, i.e. is there evidence within the literature that contests the relationship between the KEs. Additionally, if possible, the KER descriptions should provide quantitative information regarding the relationship between the upstream and downstream KEs, i.e. what level of response in the upstream KE elicits a response in the downstream KE.

The most significant element within a weight of evidence determination for an AOP over its entirety is assessment of the biological plausibility of each KER present (OECD 2013a; Meek et al. 2014). Well-established biological knowledge and related information is used to identify the supporting mechanistic evidence regarding the presence of the KER between upstream and downstream KEs. A high degree of confidence would be derived from well-established and well- documented mechanistic information that is accepted as true by the broader scientific community. Meanwhile, a low degree of confidence would be derived from two KEs being statistically associated with one another without mechanistic understanding supporting the KER (OECD 2013a; Meek et al. 2014).

Three main factors should be addressed when assessing the level of empirical support for each KER, namely: (1) response-response concordance, i.e. is the change in a downstream KE preceded by an appropriate change in a related upstream KE; (2) temporal concordance, i.e. is a downstream KE observed to a greater degree and at later time points than a related upstream KE; and (3) incidence concordance, i.e. is a downstream KE observed with a lower incidence than that of a related upstream KE. As the KER delineates the causative relationship between the upstream and downstream KEs, it would be expected that, experimentally, the upstream KE should be observed at lower chemical doses, earlier time points, and with increased incidence than the downstream KE. However, due to the technical limitations inherent in the methods used to measure the perturbations of the KEs, these assumptions might not necessarily hold true. Therefore, one must be careful to consider these limitations when assembling the information for use within a weight of evidence assessment for empirical support. As such, when conducting weight of evidence, a high degree of confidence would be derived from extensive evidence describing dose-response, temporal, and incidence concordance between two KEs when assessing exposure to a variety of stressors. A low level of weight of evidence would be achieved if studies were to show significant inconsistencies in dose-response, temporal, and/or incidence concordance, or if data from different species/taxa did not align when expected to due to conserved biological processes (OECD 2013a; Becker et al. 2015).

Additionally, supporting evidence can be utilized to define the domain of applicability for the AOP and its respective components, i.e. the species, life-stages, and sexes of the organism(s) for which the AOP is relevant. Generally, the most restrictive KE present in the AOP is used to define the applicability domain of the entire AOP. However, information for alternative species may be used for AOPs that have been developed to assess a human health endpoint when the underlying biological process is conserved. Working through each of these steps in turn, from the identification of the relevant KEs and KERs within an AOP, to assembling the supporting evidence for each of those KE/KERs, to the culminating evaluation of the collected evidence, is critical.

The process of AOP development represents a continuum in which supporting evidence expands over time, and which is consistent with the fifth principle of development previously described (Fig. 12.2). However, AOPs can be roughly classified according to their respective stage of development. Putative AOPs (pAOPs) are manually curated by domain experts and may be the result of a literature review or presented as part of a research publication. These pAOPs can include hypothetical linkages based solely on correlative evidence among KEs, with biological plausibility based primarily on the judgment of domain experts and without an exhaustive review of the literature. Formal AOPs (fAOPs) have undergone a formal evaluation of the evidence supporting the KE relationships and essentiality of the KEs as defined in the OECD AOP development handbook (OECD 2013b). AOPs at this stage may be submitted to the OECD for review and possible endorsement. Quantitative AOPs (qAOPs) incorporate dose-response data from reference chemicals to quantitatively define the response-response relationships between each pair of KEs, and allow predictions regarding the level of activation required for early KEs to elicit a meaningful response at the AO. These efforts can use pAOPs or fAOPs as the scaffold and will most often provide the evidence needed for formal evaluation of a pAOP as a by-product of the quantitative data collected during the process of defining the qAOP.

Recognizing the need for tools with the ability to facilitate the expert-driven AOP development process, an AOP Knowledge base (AOP-KB) was developed via an international collaboration under the auspices of the OECD AOP development program. The AOP-KB is comprised of four modules: AOPXplorer, AOP-Wiki,

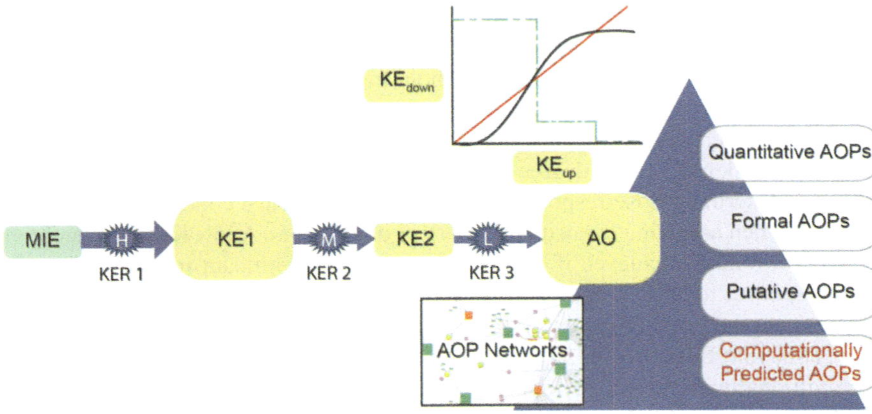

Fig. 12.2 Illustration of the phases of Adverse Outcome Pathway (AOP) development along the continuum highlighted in Fig. 12.1. Computationally-predicted AOPs (cpAOPs) are created via network inference algorithms and provide hypothetical AOPs in a network context for expert evaluation. AOPs that have undergone a preliminary evaluation by an expert are classified as putative AOPs. Formal AOPs have undergone a weight of evidence evaluation as outlined by the Organization for Economic Cooperation and Development handbook and have a weight of evidence call for each key event relationship (KER) and essentiality calls for certain key events (KEs) (*represented by node size in figure*). Quantitative AOPs define response-response relationships that allow for the prediction of dose-response behavior at downstream KEs based on the dose-response behavior of an upstream KE for a specific chemical. In most cases this will be based on a computational model that describes the mechanisms underlying the AOP

Effectopedia, and the Intermediate Effects DataBase. The first three modules enable community-led efforts to develop and evaluate AOPs across all stages of AOP development as described below. The last module provides a mechanism for submitting AOP-related information for regulatory consideration in Europe and provides tools linking AOPs to more chemical-specific toxicity data.

The pAOP development process can be facilitated via the use of the AOPXplorer module. The AOPXplorer is a computational tool developed by the United States Army Engineer Research and Development Center that provides users with a graphical representation of the networks present within AOPs. This tool incorporates data from multiple sources and provides bioinformatics capabilities that allow for data integration. It fully incorporates existing AOPs from the other AOP-KB modules, thus placing novel discoveries in the context of existing knowledge. It can also act as a platform to incorporate computationally-predicted AOPs (cpAOPs) into the development process, as discussed below.

After its development, the pAOP can be used as a scaffold onto which scientific evidence supporting the inclusion of each KE or KER can be assembled; it is at this stage that the AOP-Wiki can be utilized. The AOP-Wiki (https://aopwiki.org) is a tool created by the European Commission's Joint Research Center (JRC) and the United States Environmental Protection Agency (USEPA) for organizing available knowledge and published data, via crowd-sourcing, by providing a set of web-based forms tied specifically to the information requirements laid out in the OECD handbook (OECD 2013b). These structured forms can be used for depositing pAOPs and fAOPs (and associated supporting data) that can be shared with the larger scientific community. Using this peer-based, data sharing approach, one group can enter a pAOP of interest into the AOP-Wiki and include a basic rationale for the assembly of this pAOP. Experts in an area of biology specifically related to that pAOP could then adopt it and provide the evidence for or against the AOP based on their more extensive knowledge within that particular field. Scientists are thereby provided with a mechanism to solicit expert feedback from other researchers working in the areas of toxicology, public health, or biology to increase the impact of their own studies when supporting risk assessors and decision makers.

Two distinct types of qAOPs require consideration: probabilistic and mechanistic (Perkins et al. 2015). The qAOPs developed in a probabilistic manner can be explored using bayesian network analysis to, for example, identify minimally sufficient nodes or indicate whether inclusion of additional assays or KEs might increase confidence in the AO. The AOPXplorer module can incorporate the evidence evaluated during the fAOP development process with additional data, when creating these probabilistic qAOPs. As the number of AOPs increase, AOP networks should emerge based upon KEs that are common to multiple AOPs within the AOP-Wiki and Effectopedia. The AOPXplorer can then provide estimates of probabilities of triggering AOs based on the interrelated AOPs collected from these sources.

In cases where more quantitative precision is required to predict the dose-response of an AO for a chemical based on dose-response information from early KEs, the development of a mechanistic qAOP is necessary. Mechanistic qAOPs integrate dose or concentration-dependent quantitative information in order to examine mathematical relationships along the pathway, and, thus, more closely represent the underlying biology. Their development requires quantitative response-response information for each pair of KEs so that the level of change in the upstream KE required to induce the downstream KE can be elucidated (Villeneuve et al. 2014a). Effectopedia (www.effectopedia.org) is designed to be an open-knowledge aggregation and collaboration tool and was developed by the OECD to provide details about the development of structured and cpAOPs in an encyclopedic manner. It also enables development and analysis of mechanistic qAOPs.

It is expected that reliable information of the highest quality will be derived from expert-driven AOP development, especially when peer review from organizations such as OECD is included in the development process. However, obtaining this high level of reliability comes at a cost, namely the time that is required for experts to invest in generation and review the AOPs. This problem especially arises when generating probabilistic qAOPs and when using resources such as the AOPXplorer, which are best suited for use with multiple AOPs for a given AO or with larger AOP networks. Use of the AOP framework in context of a HT testing strategy, and as a means of driving the development of testing batteries, requires sufficient AOP coverage of the biological assay targets and information on how such targets are related in vivo. In addition to the knowledge required regarding the relationships between a given KE to other KEs within a given AOP, there is also a need to identify possible connections (if any) of that KE with other AOPs, in order to develop more informed testing strategies. Given that biological systems are quite flexible and maintain an inherent redundancy, it is not unreasonable to assume that multiple AOPs can exist for a given AO and that the same KEs have the ability to span both multiple and distinct AOPs. The labor-intensive curation of fAOPs and pAOPs implies the inability of these AOP classes to meet the current needs for coverage of biological space. Furthermore, because fAOPs can be considered as distinct units rather than more integrated systems (e.g., networks or series of AOPs), they do not lend themselves as-is to the designing of larger integrated testing strategies.

System biology-based computational methods have developed out of the need to harness the more abundant coverage present in HT data (e.g., transcriptomic data and other high content data types), and to meet the challenges associated with the requirement for efficiently covering a vast chemical and biological space (Fig. 12.2 – *lowest tier*). These HT datasets, when combined with computational analytical approaches, can facilitate and accelerate the AOP development process while simultaneously increasing knowledge of the biological space covered by these AOPs (Bell et al. 2016; Oki et al. 2016). By leveraging large amounts of existing publically available data, computational approaches can be applied in the integration of the various levels of biological organization in order to generate a network with the ability to relate changes in biological pathways to measured phenotypes and AOs (Perkins et al. 2011; Kleinstreuer et al. 2011; AbdulHameed et al. 2014; Oki et al. 2016). These cpAOPs not only can serve as scaffolds to help accelerate the curation of pAOPs and fAOPs, but they can also aid in providing guidance during formulation of testing strategies.

Oki et al. (2016) describes publically available datasets that provide information at the various levels of biological organization and that can be used alongside experimental data to develop cpAOP networks (Fig. 12.3). Experimental reference data can be integrated by identifying direct linkages across experimental results or through an identifier, or they may be described by identifying the co-occurrence of certain items (frequent itemset mining) with the ability to span multiple experimental datasets possessing unclear associations (Bell et al. 2016). These connections form a network, which is defined as a set of nodes (e.g., chemicals, phenotypes, pathways, AOs, or assays) connected by edges (presence of a relationship). By exploring the topology of the networks, the identity of AO-specific sub-networks and cpAOPs can be determined.

These sub-networks offer a computational approach (via their structure) that allows identification of nodes with high connectivity (e.g., KEs shared by multiple cpAOPs), in order to determine sufficient KEs predictive of an AO, and to provide insight on assay coverage. Overlaying this assay space onto a cpAOP network significantly facilitates the identification of KEs that lack sufficient coverage (Bell et al. 2016). Such identification can prove especially useful when searching for shared and sufficient KEs that are optimal in the design of a minimal testing strategy, such as determining the minimum number of biological assay targets required to predict an AO. Knowledge derived from pAOPs and higher classes (e.g., fAOPs and qAOPs) can be used to increase confidence in the cpAOP networks.

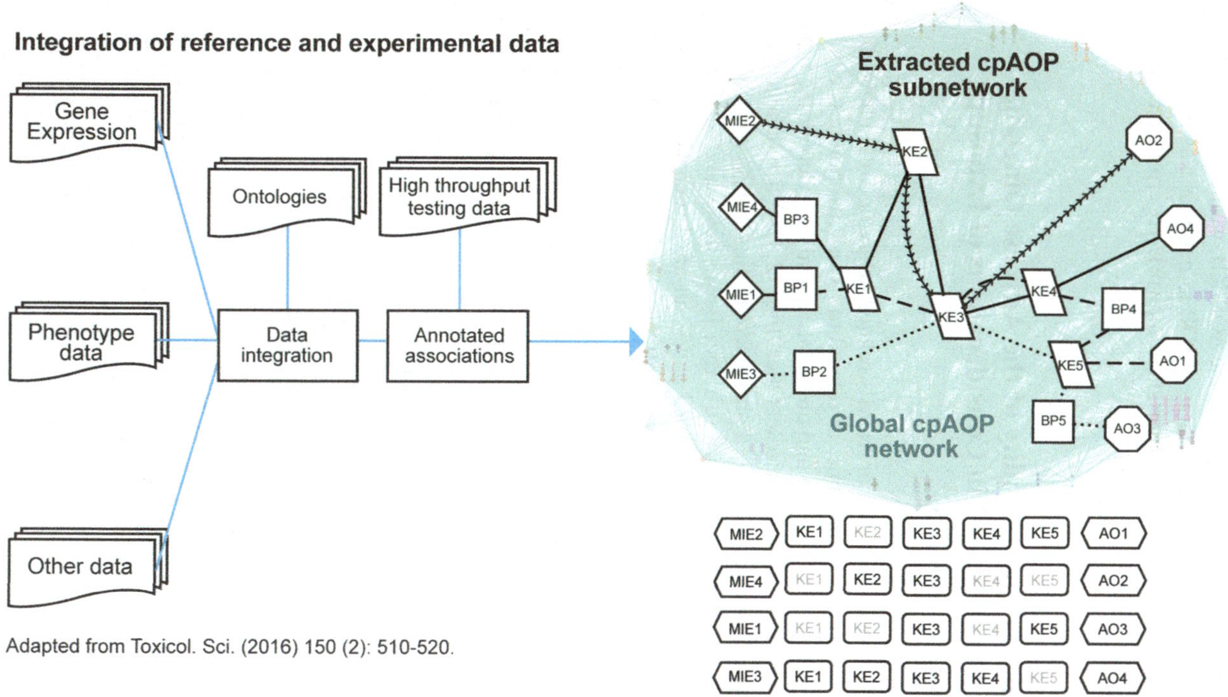

Fig. 12.3 Overview of the computationally-predicted AOP (cpAOP) development process. The integration of various data sources (*left*) facilitate creation of global cpAOP networks (*top right*). These global networks can then be queried for specific diseases or molecular targets of interest to extract sub-networks for expert evaluation. During the expert evaluation phase, the individual AOPs can be defined while maintaining the network context (*bottom right*). The network context can then be used to identify key events (KEs) within the individual AOPs for assay development. For example, assays developed for KE3 would prove to be excellent early screening assays, as they cover a broad range of molecular initiating events (MIEs) and adverse outcomes (AOs). That is, KE3 is the only KE found to be in each of the individual AOPs. Additional assays could then be developed/used to better define the precise mechanism for a specific chemical. The figure shows four example profiles based on the network connectivity

The development of cpAOPs would benefit tremendously from inclusion of ontologies and controlled vocabularies, as use of these standardized lexicons can ease data integration by facilitating interoperability across databases. Such ontologies, which are currently under development for the AOP-KB, would serve a critical role in computationally relating a human-curated AOP to one that was built in an automated fashion, and can provide the appropriate frameworks for cross-referencing of terms.

12.3 Expanding the Applicability of the Adverse Outcome Pathway Framework Via Considerations of Exposure and Pharmacokinetics

Evaluating the influence that exposure or ADME-related behaviors might have on a chemical's in vivo toxicological outcome, as related to its in vitro potency, can be achieved through a series of qualitative and quantitative analyses, depending upon data availability and specific goals of the researcher (Fig. 12.4). When data is abundant, development of physiologically based pharmacokinetic (PBPK) and pharmacodynamic (PD) models linking exposure to target tissue doses and subsequent target tissue responses may be easily achieved and is an ideal scenario (Caldwell et al. 2012). However, the sheer numbers of chemicals circulating throughout the environment and in production (Egeghy et al. 2012), along with their wide spectrum of ADME properties, render the effort required to develop such chemical-specific models time- and resource-prohibitive. In these data-poor situations, alternative in silico methods and cheminformatics tools can be used to explore the large chemical space more effectively by identifying molecular properties of

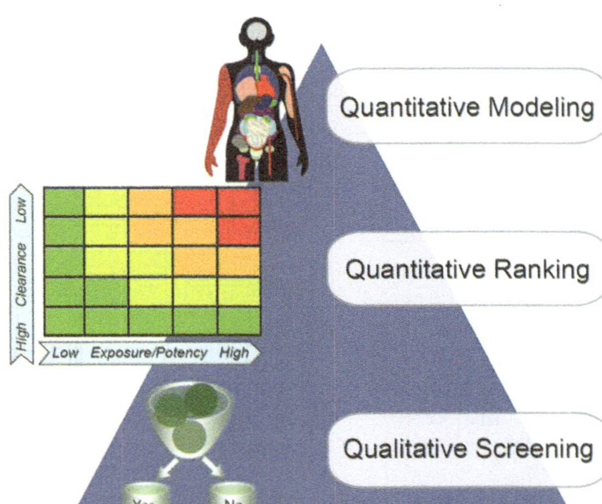

Fig. 12.4 Tiers for incorporating information related to chemical exposure and pharmacokinetic behaviors when applying Adverse Outcome Pathways into risk-based decision making

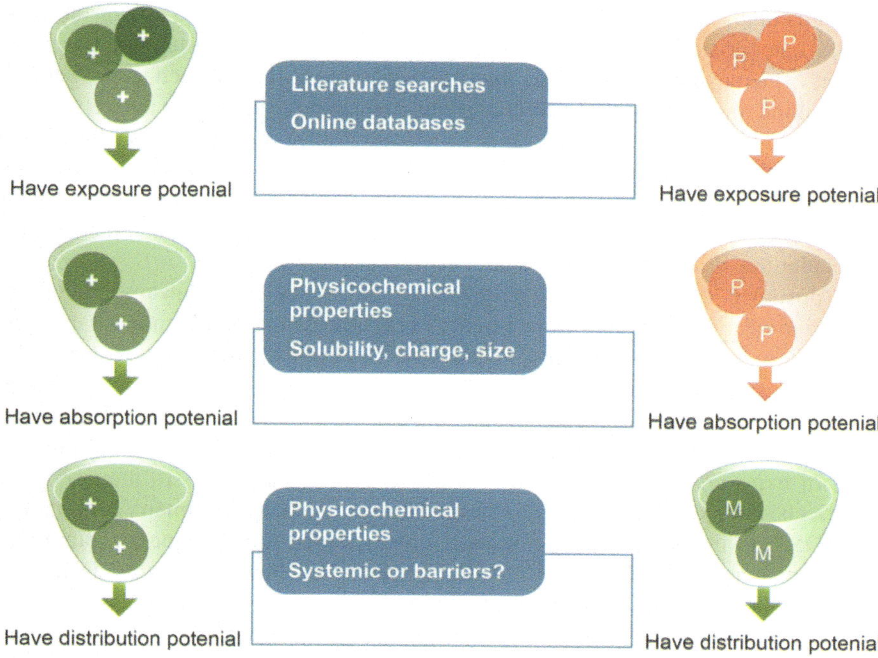

Fig. 12.5 Queries of potential factors that allow investigators to qualitatively evaluate the ability of a chemical to reach its intended internal (e.g., blood, tissue, cell, enzyme, membrane receptor) target. For parent compounds, the chemical's ability to reach its intended target should be tracked from external exposure to distribution to that target. Unless metabolites are manufactured or are generated through environmental processes (e.g., photolysis), the exposure and absorption potential of their parent compounds should be determined, and the ability to reach the intended target for the metabolite itself should then be evaluated

chemicals that influence exposure and ADME-related behaviors and that aid in their prediction (Sun 2005; Wegner et al. 2006). Using these alternative computational approaches to refine HT in vitro results is preferable to building a detailed model with predictions that cannot be validated in the absence of available PK data.

A qualitative workflow was developed recently to assess the potential for in vitro active chemicals to reach an in vivo molecular target and trigger an MIE (Phillips et al. 2016). The workflow begins with the selection of an AOP of interest, which allows the associated molecular target induced by the MIE to be determined. Chemicals found to be active in in vitro assays can be present in the environment, or they might be active metabolic moieties of parent compounds that may or may not exhibit in vitro activity themselves. Both the exposure and absorption potential of the parent compound of an active metabolite require evaluation, as these two processes are precursors to metabolite generation within the body. Assessing the ability of the metabolite itself to access the molecular in vivo target should follow this evaluation. If the metabolite is also known to be present in the environment, its own exposure and absorption potential should also be queried (Fig. 12.5).

In this qualitative workflow, exposure potentials are assigned to one of the four general categories that are often intuitively accounted for: (1) exposure to the general public; (2) exposure to individuals under special circumstances, such as workers or those taking medication; (3) uncertain or unknown exposure; and (4) unlikely exposure. As this first, qualitative, tier aims to be conservative in nature, those chemicals with uncertain or unknown exposure are included with those chemicals exposed to individuals and the general population, and advanced to the next step of the workflow. Those chemicals known to lack exposure potential will not reach the in vivo molecular target, and, thus, are considered "low priority".

Exposure potential can be determined through a variety of means, including extraction of published data from research articles, technological monographs, or registration eligibility documents. While there is a large degree of time invested in compiling such empirical data, confidence in conclusions regarding exposure potential is also much higher, and such data can be used in more computationally-intense higher tiers. Alternatively, HT exposure models can provide rapid predictions of exposure at several population percentiles for hundreds to thousands of chemicals. For example, the HT Stochastic Human Exposure Dosimetry Simulation (SHEDS-HT) model integrates population use patterns (e.g., frequency, duration, time, and magnitude) derived from the Consolidated Human Activity Database (McCurdy et al. 2000), chemical weight fractions (Goldsmith et al. 2014), and known age-specific physiological parameters (e.g., skin surface area, inhalation rate) to simulate exposure of chemicals comprising 200 consumer product types to 100,000 individuals across all age groups (Isaacs et al. 2014). It should be noted that such HT exposure models, by design, take limited amounts of data as input to provide estimates with a large amount of uncertainty and error. Thus, investigators are left with the decision regarding their desire to balance time and effort with precision and accuracy.

Absorption and distribution potential can also be determined for chemicals using empirical data extracted from resources such as those listed above. Alternatively, physicochemical descriptors and chemical-specific properties can be examined to assess absorption potential, i.e. can a chemical be absorbed via skin, and distribution potential, i.e. can a chemical cross the placenta. These descriptors can be predicted using cheminformatics tools and molecular chemistry models, such as pharmacophore modeling, geometric optimization, and conformational analysis (Goldsmith et al. 2012). There are several open-source and commercial platforms available that provide a variety of QSAR models and algorithms based on two-dimensional and three-dimensional structures to predict chemical-specific ADME properties capable of mediating absorption and distribution. Chemicals that can be systematically distributed may require further assessment regarding their potential to reach a specific molecular target, i.e. is the chemical able to penetrate the blood-brain barrier to access brain acetylcholinesterase (AChE). Those chemicals that are capable of absorption and distribution can be considered of high priority and advanced in order to undergo additional quantitative screening. Although elimination is considered a vital component of the ADME process, evaluating it in a qualitative manner is challenging, as it can be assumed that all chemicals should eventually leave the body at some time in one form or another. Rather, the quantitative relationship between intake rates and elimination rates is much more critical in influencing chemical toxicity.

One of the pitfalls involved with the use of in vitro assays is the erroneous omission of chemicals that might appear to be inactive under assay conditions but that may resolve into active metabolites or that may be active under in vivo conditions (Eisenbrand et al. 2002; Kirkland et al. 2014), or inclusion of chemicals that are rapidly bio-inactivated and eliminated quickly. If metabolites are included in in vitro tests, it is because investigators recognize that these metabolites are known to cause adverse health effects, and the parent compounds of these metabolites are generally also known. Unfortunately, this scenario is more often the exception rather than the norm, especially with regards to chemicals that are newly developed and distributed to market. Identifying the multitudes of potential metabolites generated from a parent compound can prove arduous (Shlomi et al. 2008). Ideally, in vivo testing can aid in this identification, and the different metabolites can then be subjected to in vitro testing to verify whether they have the ability to induce a molecular response (NRC 2007). However, this testing approach requires a great deal of time and effort. Alternatively, QSAR-based modeling approaches can aid in evaluating the potential for parent compounds to become metabolized, as well as in predicting possible metabolites based on enzymatic activity and chemical structure (Dimelow et al. 2011; Andrade et al. 2014; Kirchmair et al. 2015). While exogenous metabolites are often detoxification products, some may be biologically active moieties for particular molecular targets. Using fragment-based and molecular fingerprint analyses (Willett et al. 1998; Myint and Xie 2010) or chemotyping (Yang et al. 2015), the functional groups that mediate in vitro activity for known active compounds can be compared to those of predicted metabolites in order to determine which metabolites are most likely to also be active. These computational approaches can be accomplished in only a fraction of the time of that required for in vivo testing, and with lower cost than that associated with in vitro testing.

The utility of the qualitative screening workflow was demonstrated through a case study involving prioritization of chemicals tested in the ToxCast™ in vitro human AChE inhibition assay (Phillips et al. 2016). Of the 146 chemicals tested within the assay, 30 were found to be active, and only 20 were retained as being high priority after querying for exposure, absorption, and distribution potential. In addition to identifying false positives, which are referred to as those active chemicals with the inability to reach an in vivo molecular target, similarity searching using molecular fingerprints and a similarity threshold score of 75% identified 22 false negatives. False negatives are referred to as those inactive chemicals that may be parent compounds of active metabolites, or that exhibit activity in vivo but not in vitro; in this case such false negatives were represented primarily by organophosphates and carbamates with weak in vivo AChE inhibition activity. Consideration of ADME behaviors and exposure can aid in the refinement of in vitro results through elimination of those chemicals that might otherwise have undergone additional testing. In addition, consideration of the presence of false negatives in an assay can aid in identification of possible inactive progenitors of active metabolites, as well as increase confidence in the resulting "active" or "inactive" hits in in vitro assays. For example, if fewer false negatives are identified, this may suggest that an assay performs reasonably well.

This lower tier qualitative approach is meant to be conservative due to large uncertainties and data gaps regarding exposure and ADME information. As a result, the number of chemicals designated as "high priority and of possible concern" may remain larger than that required for sufficient allocation of funds and resources for more extensive testing. When some amount of exposure and ADME data are available, either through measurements or reliable predictions, these data can be integrated using higher tier quantitative analyses to estimate an in vivo concentration that may result in induction or inhibition of a molecular target as a basis for chemical prioritization. There are multiple options available when conducting quantitative analyses, and the integrity of data is likely to be the most critical driving factor in determining an appropriate approach. For example, when there is a lack of empirical data concerning exposure levels and ADME properties, in silico methods may be used to estimate these quantitative inputs. However, it should be recognized that large amounts of uncertainty might be present in such predictions. In such a case, it may be more appropriate to place chemicals into prioritization bins to allow room for error, and investigators are left with the decision to subject all, or only some, chemicals falling within the highest priority bin(s) to further testing.

In cases where data are abundant or sufficient to allow for parameterization and evaluation of PK/PBPK and PD models, chemicals can then be investigated on an individual basis to identify a point of departure capable of inducing in vivo toxicity (Filipsson et al. 2003; Davis et al. 2011) and margin of exposure (MOE). Such an approach, while being more low-throughput in nature when compared to relative ranking of chemicals in priority bins, offers a more objective comparison for decision makers. Chemical-specific data, even when not lacking, may still hold many inconsistencies due to the large variety in methodological procedures used in individual research studies. In instances where data gaps and uncertainties exist, identification of parameters for which errors might hold a significant impact on the predictability of a quantitative model is critical in gauging whether model outputs are appropriate for higher tier risk assessment. Two possible higher tier approaches involving quantitative analyses were demonstrated in case examples that are further described below.

The first case study utilized a PK/PD model to prioritize 25 AChE inhibiting chemicals based on their in vitro potency levels, daily absorbed rates, and clearance rates (Leonard et al. 2016a). Specifically, a PK model consisting of (1) inputs of daily absorbed doses of active chemicals or inactive parents of active metabolites; (2) parameters that describe stoichiometric yield and metabolic rate of parental biotransformation to active metabolites; and (3) clearance rates of both parents and active metabolites, was used to estimate blood concentrations of active moieties at average plasma concentrations integrated over time (C_{Avg}). The C_{Avg}, together with in vitro potency data describing the concentration necessary to inhibit/induce molecular target activity by 50% (EC_{50}) and maximum inhibitive/inductive activity (E_{Max}), was then incorporated into a PD model to derive the in vivo toxicological activity of a chemical at a given absorbed dose and to allow chemicals to be placed into discretized bins of increasing concern based on rankings of their activity. It was found that those chemicals with moderate to high potency and exposure were placed into higher priority bins, and that those chemicals that exhibited a very rapid clearance were placed in a lower priority bin, even if potency was high (Leonard et al. 2016a).

An additional component of this case study involved replacement of empirical data describing model parameters with predicted values to examine how the uncertainty in these variables might affect the priority ranking. Compared to other variables, a higher number of bins were misassigned when daily absorbed doses were predicted (Leonard et al. 2016a). This finding was unsurprising, as human exposure can vary widely due to the plethora of activities that may allow an individual to be exposed to a specific chemical, and to what degree, leading to a large range of potential exposure levels for the population as a whole. In this case study, SHEDS-HT was used to estimate daily absorbed doses for the population, and the resulting distributions reflect the wide variety of consumer use patterns, including age-specific, gender-specific, and occupation-specific activities (Isaacs et al. 2014). Differences in individual exposure levels are also influenced by the number of products certain chemicals are found in, as well as the weight fractions of those chemicals in these products (Goldsmith et al. 2014).

While there is great value in coupling an AOP with lower-tier qualitative or quantitative approaches for priority ranking, further expansion of the utility of the AOP framework in HT risk assessment can be achieved using higher-tier computational models to compare external exposure concentrations and internal target doses capable of triggering an MIE (Groh et al. 2015). When chemical-specific and physiological data is abundant, PBPK models can be developed to follow chemicals throughout a biological system based on their ADME properties (Meibohm and Derendorf 1997). In this manner, in vivo chemical target tissue concentrations able to elicit an adverse biological response, as determined through in vitro testing, can be linked to external exposure levels that will result in such biologically-effective internal chemical concentrations. Derivation of a MOE that is readily interpretable by risk assessors can be accomplished by comparing these biologically-effective external exposure levels to those levels likely to be encountered by the population (Fig. 12.6). This quantitative approach has been used to estimate external conazole fungicide concentrations with the ability to alter the xenobiotic CAR/PXR signaling pathway in vivo (Judson et al. 2011), though in this example, only the two ADME properties of intrinsic clearance and fraction of the chemical unbound to plasma proteins were considered.

In the second quantitative case study, an integrated framework was developed to utilize a PBPK/PD model, a HT exposure model, and in vitro potency data to derive MOEs for chemicals of varying ability to inhibit the thyroid peroxidase (TPO) enzyme (Leonard et al. 2016b). TPO plays a significant role in the synthesis of the thyroid hormones thyroxine and triiodothyronine through its conversion of iodide to iodine and by hydrogen peroxide-mediated oxidation (Jameson and Wheetman 2001). Specifically in this case study, a PBPK model was used to estimate thyroid

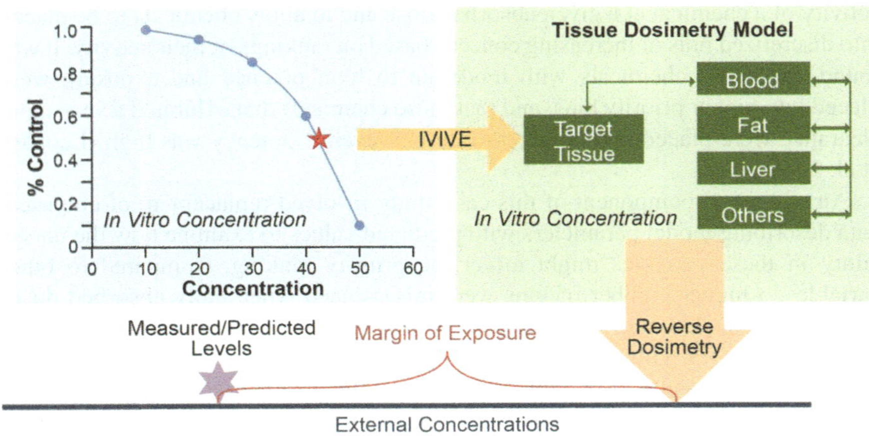

Fig. 12.6 Applying in vitro to in vivo extrapolation (IVIVE) and reverse dosimetry approaches during quantitative assessment of chemical risk. Such a workflow includes determining a biological point of departure (POD), and external exposure concentration that will lead to that POD, by integrating in vitro dose-response data and physiological mechanisms (e.g., tissue volume and blood flow) into a pharmacokinetic (PK) or physiologically based pharmacokinetic (PBPK) and pharmacodynamic (PD) model. A prioritization of chemicals can then be achieved by comparing their PODs to known or predicted population exposure concentrations to estimate a margin of exposure

chemical concentration, and this concentration was entered into a PD model, along with EC_{50} and E_{Max} potency data, to determine the oral equivalent dose necessary to reduce the production and release of thyroid hormones by 10%, followed by reverse dosimetry to derive the external exposure level necessary to result in the interal biologically-effective tissue dose (ED_{10}). The ED_{10} was then compared to external exposure levels predicted at the 50th and 95th population percentiles, using the SHEDS-HT model, to derive a MOE. This integrated framework can be applied to other AOPs using the most appropriate PBPK and PD models. For example, the PBPK model can be expanded to investigate AOs across multiple life stages, such as fetal exposure through placental transfer or infant exposure through contaminated breast milk. This case study also highlights the value of chemical-specific exposure and ADME considerations in regards to increasing confidence in HT risk assessment. For example, it was found that the ultraviolet absorber benzophenone-2 (Downs et al. 2013) exhibited the slowest estimated clearance and the third highest potency of the six tested TPO-inhibiting chemicals (Paul et al. 2014), leading to it having the lowest MOE. In addition, although the antimicrobial chemical triclosan exhibited only the 2nd lowest potency, its widespread abundance in a number of consumer and industrial products (Thompson et al. 2005) resulted in the highest predicted exposure level, leading to it having the 2nd lowest MOE (Leonard et al. 2016b).

One of the core objectives of the new twenty-first century toxicity testing paradigm is to utilize HT in vitro data to investigate toxicological pathways resulting from chemical exposure. However, interaction of a chemical with a molecular target does not necessarily lead to an MIE and subsequent AO. Rather, a chemical that is active in vitro must reach the in vivo target at the correct time and concentration necessary to induce the MIE, especially in regards to AOPs involved with developmental toxicity. Evaluating the risk of a chemical based on its in vitro potency alone can certainly lead to erroneous conclusions in the risk assessment process, and considerations of both exposure and ADME-related properties should enable managers to make better-informed and more confident decisions.

12.4 From AOPs to Mode of Action (MOA): Considerations for the Use of AOPs

While AOPs were originally developed in the context of ecological risk assessment (Ankley et al. 2010), the mode of action (MOA) framework was developed to meet the needs of researchers and regulators seeking to elucidate the mechanistic processes leading to adverse biological outcomes for human health risk assessment (Meek et al. 2003, 2014; Seed et al. 2005; Boobis et al. 2008). Many parallels exist between the AOP and MOA frameworks, especially reliance on the identification of KEs to describe the mechanistic basis for chemical toxicity when a complete description of the mechanism of action is lacking. Just as with AOPs, MOA analysis requires causal linkages to be established between upstream and downstream KEs. More recently, there has been a concerted effort to synchronize the AOP and MOA frameworks. As an example, the evaluation of weight of evidence and description of the quantitative understanding for AOPs is derived from the corresponding literature for MOA analysis (Meek et al. 2014).

The key difference between the AOP and MOA frameworks is the first principle of AOP development; that is, AOPs are chemical-agnostic. This means that while the perturbation will be from a chemical in most cases, no information specific to a single chemical is included in the definition of an AOP. For example, an AOP in which the MIE includes protein alkylation will necessarily be initiated by the binding of a chemical to the protein, but the AOP will be general for any chemical that can serve as an alkylating agent for that protein. AOPs are intended to inform HT toxicity testing where a large number of chemicals are screened across a battery of assays. In this case, the AOP must be designed such that positive results in a particular assay for any chemical can be interpreted in light of that AOP. MOA analysis, in contrast, is intended to directly inform decision makers regarding the risk of a specific chemical. In the latter case, some direct measure of toxicity for the chemical in question is required along with other chemical-specific information, such as metabolism, that might influence the activity of the chemical.

Accounting for these subtle nuances, the MOA framework can be thought of as an extension of the AOP framework, through consideration of chemical-specific toxicity and ADME information. Integration of the AOP and MOA frameworks allows for the expansion of hazard assessment to dose-response assessment for specific chemicals (Mackay et al. 2014) through three critical components: (1) toxicokinetics of the chemical based on exposure and ADME information; (2) chemical-agnostic toxicodynamics as determined through the AOP; and (3) chemical-specific dose-response information derived from HT in vitro assays. Chemical-specific dose-response relationships in vivo can then be obtained using chemical-specific ADME information to map external exposure to target tissue dose, followed by investigating the interaction of the chemical with the molecular target and the ability of this interaction to lead to an adverse biological response. As an example, a model that integrates data from 18 estrogen receptor-related in vitro assays linked to an estrogen receptor signaling AOP has shown predictability that is comparable to previous in vivo assays for identifying estrogenic compounds (Browne et al. 2015; Judson et al. 2015). These dose-response estimates for specific chemicals from the in vitro assays can be combined with exposure predictions, along with PBPK/PD model-based predictions of human equivalent doses, to assess whether those chemicals can attain external exposure levels that might lead to adverse reproductive responses (Judson et al. 2014; Wetmore et al. 2015).

When assessing the utility of a particular AOP, it is important to consider its relevance to species, life stage, and sex (Villeneuve et al. 2014a; Groh et al. 2015). It should be noted that should such differences exist within or among species, only PD differences can be addressed using AOPs. Identification of KEs related to ubiquitous AOs across species (e.g., aromatase inhibition leading to reproductive difficulties in species with estrogen receptors) provides an opportunity for scientists to investigate species extrapolation for those KEs, such that toxicological studies including the downstream KEs can be performed with a limited set of representative species and extrapolated to the species of concern. In addition, human relevance/species concordance may be determined based on available information for tested species using the MOA framework, as it allows for estimation of quantitative differences in PK properties and behaviors that might vary across species or life-stages (Meek et al. 2014). It should be recognized that establishing relevance among tested taxa with other species or humans requires identification of conserved compartments and KEs across all biological levels of organization, including the initial molecular target, cellular mechanisms, and organ similarities.

If toxicity is implied for a specific chemical, in accordance with both a HT test and whole organism test, that chemical can likely be listed as being of concern. The HT assay evaluating, in essence, the same MOA as the whole organism test should not be held to a higher standard in terms of its linkage to the AO. For example, the concordance between an HT test and an in vivo test that relate to the same KE should be evaluated in light of the concordance among the results from the in vivo assays for that KE. If two tests address different KEs within an AOP, the overall confidence in the portion of the AOP that lies downstream of each KE must be considered. If one KE is upstream of a weak KER and the other is not, the confidence in the assay connected to the downstream KE will be higher for accurately predicting that AO. However, due consideration should be placed on the ability of the in vivo assay to include ADME characteristics in addition to the AOP-related concerns.

Combining AOPs with existing chemical toxicity data can provide a structured framework for communicating toxicological outcomes to risk managers. For example, QSAR-based in silico approaches can be coupled with existing in vitro toxicity data for structurally similar chemicals in order to predict the toxicity of chemicals that have not yet been tested, that lack of sufficient information for a decision, or that are under development (Patlewicz et al. 2015; Alves et al. 2015a, b). Such QSAR-based approaches can also allow in vivo toxicity to be predicted for chemicals that are structurally similar to other chemicals that have undergone more extensive in vivo toxicity testing. To determine whether additional chemical testing is needed, AOPs have been identified as a key component of Integrated Approaches to Testing and Assessment (IATA) strategies (Tollefsen et al. 2014). Confidence in using read-across methods to design alternative testing strategies based on decision context, mechanistic information, structural similarities, and data availability for chemical groups is based on the understanding of toxicokinetics and toxicodynamics for specific chemicals and can be increased with supporting in vitro data (Patlewicz et al. 2015). When investigators are able to confidently predict in vivo behaviors of specific chemicals tested through HT in vitro means, using sufficient data obtained from MOA analysis, the synergy between the AOP and MOA frameworks becomes obvious.

12.5 Conclusions

A broad coverage of toxicological space is required for AOPs to support the risk assessment of chemicals. Investing effort on defining many AOPs and forgoing the full evaluation of each will provide this broad coverage. Cases in which the AOPs are determined to lack sufficient evidence to support a given decision can then be used to prioritize those AOPs for further evaluation. Fortunately, as AOPs continue to evolve, they take advantage of the incorporation of emerging technologies and computational resources that allow for rapid development and that aid in better understanding of the pathways involved in AOs of interest.

Though AOPs are chemical-independent by nature, when integrating information related to characteristics that influence chemical toxicity, such as ADME-related properties and exposure, results from HT in vitro assays can be refined to provide greater confidence for decision makers. Incorporation of ADME behaviors also enables the AOP framework to be extended to instances in which data availability allows for establishing linkages between chemical-specific dose-responses and external exposure levels, or for determination of in silico predicted toxicity. The contents presented in this chapter have only just begun to touch upon the challenges faced during AOP development, along with the opportunities and advancements the AOP framework can provide to the emerging field of toxicity testing in the twenty-first century.

References

AbdulHameed MDM, Tawa GJ, Kumar K et al (2014) Systems level analysis and identification of pathways and networks associated with liver fibrosis. PLoS One 9:e112193. doi:10.1371/journal.pone.0112193

Abt E, Rodricks JV, Levy JI et al (2010) Science and decisions: advancing risk assessment. Risk Anal 30:1028–1036. doi:10.1111/j.1539-6924.2010.01426.x

Alves VM, Muratov E, Fourches D et al (2015a) Predicting chemically-induced skin reactions. Part II: QSAR models of skin permeability and the relationships between skin permeability and skin sensitization. Toxicol Appl Pharmacol 284:273–280. doi:10.1016/j.taap.2014.12.013

Alves VM, Muratov E, Fourches D et al (2015b) Predicting chemically-induced skin reactions. Part I: QSAR models of skin sensitization and their application to identify potentially hazardous compounds. Toxicol Appl Pharmacol 284:262–272. doi:10.1016/j.taap.2014.12.014

Andersen ME, Krewski D (2009) Toxicity testing in the 21st century: bringing the vision to life. Toxicol Sci 107:324–330. doi:10.1093/toxsci/kfn255

Andrade CH, Silva DC, Braga RC (2014) *In silico* prediction of drug metabolism by P450. Curr Drug Metab 15:514–525

Ankley GT, Bennett RS, Erickson RJ et al (2010) Adverse outcome pathways: a conceptual framework to support ecotoxicology research and risk assessment. Environ Toxicol Chem 29:730–741. doi:10.1002/etc.34

Austin C, Kavlock RJ, Tice R (2008) Tox21: putting a lens on the vision of toxicity testing in the 21st century. Available via AltTox.org. http://alttox.org/tox21-putting-a-lens-on-the-vision-of-toxicity-testing-in-the-21st-century/. Accessed 30 May 2016

Becker RA, Ankley GT, Edwards SW et al (2015) Increasing scientific confidence in adverse outcome pathways: application of tailored Bradford-Hill considerations for evaluating weight of evidence. Regul Toxicol Pharmacol 72:514–537. doi:10.1016/j.yrtph.2015.04.004

Bell SM, Angrish MM, Wood CE, Edwards SW (2016) Integrating publicly available data to generate computationally predicted Adverse Outcome Pathways for fatty liver. Toxicol Sci 150:510–520. doi:10.1093/toxsci/kfw017

Benford D, Bolger PM, Carthew P et al (2010) Application of the margin of exposure (MOE) approach to substances in food that are genotoxic and carcinogenic. Food Chem Toxicol 48(Suppl 1):S2–24. doi:10.1016/j.fct.2009.11.003

Binetti R, Costamagna FM, Marcello I (2008) Exponential growth of new chemicals and evolution of information relevant to risk control. Ann Ist Super Sanita 44:13–15

Boobis AR, Cohen SM, Dellarco V et al (2008) IPCS framework for analyzing the relevance of a cancer mode of action for humans. Crit Rev Toxicol 36:781–792. doi:10.1080/10408440600977677

Browne P, Judson RS, Casey WM et al (2015) Screening chemicals for estrogen receptor bioactivity using a computational model. Environ Sci Technol 49:8804–8814. doi:10.1021/acs.est.5b02641

Caldwell JC, Evans MV, Krishnan K (2012) Cutting edge PBPK models and analyses: providing the basis for future modeling efforts and bridges to emerging toxicology paradigms. J Toxicol. doi:10.1155/2012/852384

Davis JA, Gift JS, Zhao QJ (2011) Introduction to benchmark dose methods and U.S. EPA's benchmark dose software (BMDS) version 2.1.1. Toxicol Appl Pharmacol 254:181–191. doi:10.1016/j.taap.2010.10.016

Dimelow RJ, Metcalfe PD, Thomas S, Lyubimov AV (2011) *In silico* models of drug metabolism and drug interactions. In: Lyubimov A (ed) Encyclopedia of drug metabolism and interactions, vol V1. Wiley, Hoboken, pp 1–55

Downs CA, Kramarsky-Winter E, Fauth JE et al (2013) Toxicological effects of the sunscreen UV filter, benzophenone-2, on planulae and *in vitro* cells of the coral, *Stylophora pistillata*. Ecotoxicology 23:175–191. doi:10.1007/s10646-013-1161-y

Egeghy PP, Judson R, Gangwal S et al (2012) The exposure data landscape for manufactured chemicals. Sci Total Environ 414:159–166. doi:10.1016/j.scitotenv.2011.10.046

Eisenbrand G, Pool-Zobel B, Baker V et al (2002) Methods of *in vitro* toxicology. Food Chem Toxicol 40:193–236. doi:10.1016/S0278-6915(01)00118-1

Filipsson AF, Sand S, Nilsson J, Victorin K (2003) The benchmark dose method–review of available models, and recommendations for application in health risk assessment. Crit Rev Toxicol 33:505–542. doi:10.1080/10408440390242360

Goldsmith MR, Peterson SD, Chang DT et al (2012) Informing mechanistic toxicology with computational molecular models. Methods Mol Biol 929:139–165. doi:10.1007/978-1-62703-050-2_7

Goldsmith M-R, Grulke CM, Brooks RD et al (2014) Development of a consumer product ingredient database for chemical exposure screening and prioritization. Food Chem Toxicol 65:269–279. doi:10.1016/j.fct.2013.12.029

Groh KJ, Carvalho RN, Chipman JK et al (2015) Development and application of the adverse outcome pathway framework for understanding and predicting chronic toxicity: I. Challenges and research needs in ecotoxicology. Chemosphere 120:764–777. doi:10.1016/j.chemosphere.2014.09.068

Hill AB (1965) The environment and disease: association or causation? Proc R Soc Med 58:295–300

Isaacs KK, Glen WG, Egeghy P et al (2014) SHEDS-HT: an integrated probabilistic exposure model for prioritizing exposures to chemicals with near-field and dietary sources. Environ Sci Technol 48:12750–12759. doi:10.1021/es502513w

Jameson JL, Wheetman AP (2001) Disorders of the thyroid gland. In: Braunwald E, Fauci AS, Kasper DL et al (eds) Harrsison's principles of internal medicine; chapter 405, 11th edn. McGraw-Hill, New York, pp 2048–2060

Judson RS, Kavlock RJ, Setzer RW et al (2011) Estimating toxicity-related biological pathway altering doses for high-throughput chemical risk assessment. Chem Res Toxicol 24:451–462. doi:10.1021/tx100428e

Judson R, Houck K, Martin M et al (2014) *In vitro* and modelling approaches to risk assessment from the U.S. Environmental Protection Agency ToxCast programme. Basic Clin Pharmacol Toxicol 115:69–76. doi:10.1111/bcpt.12239

Judson RS, Magpantay FM, Chickarmane V et al (2015) Integrated model of chemical perturbations of a biological pathway using 18 *in vitro* high-throughput screening assays for the estrogen receptor. Toxicol Sci 148:137–154. doi:10.1093/toxsci/kfv168

Kavlock RJ, Austin CP, Tice R (2009) Toxicity testing in the 21st century: implications for human health risk assessment. Risk Anal 29:485–497. doi:10.1111/j.1539-6924.2008.01168.x

Kirchmair J, Göller AH, Lang D et al (2015) Predicting drug metabolism: experiment and/or computation? Nat Rev Drug Discov 14:387–404. doi:10.1038/nrd4581

Kirkland D, Zeiger E, Madia F et al (2014) Can *in vitro* mammalian cell genotoxicity test results be used to complement positive results in the Ames test and help predict carcinogenic or *in vivo* genotoxic activity? I. Reports of individual databases presented at an EURL ECVAM workshop. Mutat Res Genet Toxicol Environ Mutagen 775–776:55–68. doi:10.1016/j.mrgentox.2014.10.005

Kleinstreuer NC, Judson RS, Reif DM et al (2011) Environmental impact on vascular development predicted by high-throughput screening. Environ Health Perspect 119:1596–1603. doi:10.1289/ehp.1103412

Kramer VJ, Etterson MA, Hecker M et al (2011) Adverse outcome pathways and ecological risk assessment: bridging to population-level effects. Environ Toxicol Chem 30:64–76. doi:10.1002/etc.375

Krewski D, Andersen ME, Mantus E, Zeise L (2009) Toxicity testing in the 21st century: implications for human health risk assessment. Risk Anal 29:474–479. doi:10.1111/j.1539-6924.2008.01150.x

Leonard JA, Sobel Leonard A, Chang DT et al (2016a) Evaluating the impact of uncertainties in clearance and exposure when prioritizing chemicals screened in high-throughput assays. Environ Sci Technol. doi:10.1021/acs.est.6b00374. Epub ahead of print

Leonard JA, Tan Y-M, Gilbert M et al (2016b) Estimating margin of exposure to thyroid peroxidase inhibitors using high-throughput *in vitro* data, high-throughput exposure modeling, and physiologically based pharmacokinetic/pharmacodynamic modeling. Toxicol Sci 151:57–70. doi:10.1093/toxsci/kfw022

Lin JH, Lu AYH (1997) Role of pharmacokinetics and metabolism in drug discovery and development. Pharmacol Rev 49:403–449

MacIntosh DL, Spengler JD (2000) Human exposure assessment. World Health Organization, Geneva, p 375. Accessed 31 Aug 2016. http://www.inchem.org/documents/ehc/ehc/ehc214.htm

Mackay D, Giesy JP, Solomon KR (2014) Fate in the environment and long-range atmospheric transport of the organophosphorus insecticide, chlorpyrifos and its oxon. Rev Environ Contam Toxicol 231:35–76. doi:10.1007/978-3-319-03865-0_3

McCurdy T, Glen G, Smith L, Lakkadi Y (2000) The national exposure research laboratory's consolidated human activity database. J Expo Anal Environ Epidemiol 10:566–578

Meek MEB, Bucher JR, Cohen SM et al (2003) A framework for human relevance analysis of information on carcinogenic modes of action. Crit Rev Toxicol 33:591–653. doi:10.1080/713608373

Meek ME, Boobis A, Cote I et al (2014) New developments in the evolution and application of the WHO/IPCS framework on mode of action/species concordance analysis. J Appl Toxicol 34:1–18. doi:10.1002/jat.2949

Meibohm B, Derendorf H (1997) Basic concepts of pharmacokinetic/pharmacodynamic (PK/PD) modelling. Int J Clin Pharmacol Ther 35:401–413

Myint KZ, Xie X-Q (2010) Recent advances in fragment-based QSAR and multi-dimensional QSAR methods. Int J Mol Sci 11:3846–3866. doi:10.3390/ijms11103846

National Research Council (NRC) (1983) Risk assessment in the Federal Government: managing the process. National Academies Press, Washington, DC

National Research Council (NRC) (2007) Toxicity testing in the 21st century: a vision and a strategy. National Academies Press, Washington, DC

National Research Council (NRC) (2009) Science and decisions. National Academies Press, Washington, DC

Oki NO, Nelms MD, Bell SM et al (2016) Accelerating adverse outcome pathway development using publicly available data sources. Curr Environ Health Rep 3:53–63. doi:10.1007/s40572-016-0079-y

Organization for Economic Cooperation and Development (OECD) (2013a) Guidance document on developing and assessing adverse outcome pathways. Available via OECD. http://www.oecd.org/officialdocuments/publicdisplaydocumentpdf/?cote=env/jm/mono(2013)6&doclanguage=en. Accessed 30 May 2016

Organization for Economic Cooperation and Development (OECD) (2013b) Users' handbook supplement to the guidance document for developing and assessing AOPs. Available via OECD. http://aopkb.org/common/AOP_Handbook.pdf. Accessed 14 Apr 2016

Patlewicz G, Simon TW, Rowlands JC et al (2015) Proposing a scientific confidence framework to help support the application of adverse outcome pathways for regulatory purposes. Regul Toxicol Pharmacol 71:463–477. doi:10.1016/j.yrtph.2015.02.011

Paul KB, Hedge JM, Rotroff DM et al (2014) Development of a thyroperoxidase inhibition assay for high-throughput screening. Chem Res Toxicol 27:387–399. doi:10.1021/tx400310w

Perkins EJ, Chipman JK, Edwards S et al (2011) Reverse engineering adverse outcome pathways. Environ Toxicol Chem 30:22–38. doi:10.1002/etc.374

Perkins EJ, Antczak P, Burgoon L et al (2015) Adverse outcome pathways for regulatory applications: examination of four case studies with different degrees of completeness and scientific confidence. Toxicol Sci 148:14–25. doi:10.1093/toxsci/kfv181

Phillips RD, Bahadori T, Barry BE et al (2009) Twenty-first century approaches to toxicity testing, biomonitoring, and risk assessment: perspectives from the global chemical industry. J Expo Sci Environ Epidemiol 19:536–543. doi:10.1038/jes.2009.38

Phillips MB, Leonard JA, Grulke CM et al (2016) A workflow to investigate exposure and pharmacokinetics influences on high-throughput *in vitro* chemical screening based on adverse outcome pathways. Environ Health Perspect 124:53–60. doi:10.1289/ehp.140945

Seed J, Carney EW, Corley RA et al (2005) Overview: using mode of action and life stage information to evaluate the human relevance of animal toxicity data. Crit Rev Toxicol 35:664–672
Shlomi T, Cabili MN, Herrgard MJ et al (2008) Network-based prediction of human tissue-specific metabolism. Nat Biotechnol 26:1003–1010. doi:10.1038/nbt.1487. [pii]10.1038/nbt.1487
Shukla SJ, Huang R, Austin CP, Xia M (2010) The future of toxicity testing: a focus on *in vitro* methods using a quantitative high throughput screening platform. Drug Discov Today 15:997–1007. doi:10.1016/j.drudis.2010.07.007
Simmons JE, Evans MV, Boyes WK (2005) Moving from external exposure concentration to internal dose: duration extrapolation based on physiologically based pharmacokinetic derived estimates of internal dose. J Toxicol Environ Health A 68:927–950. doi:10.1080/15287390590912586
Soldatow VY, LeCluyse EL, Griffith LG, Rusyn I (2013) *In vitro* models for liver toxicity testing. Toxicol Res (Camb) 2:23–39. doi:10.1039/C2TX20051A
Stadnicka-Michalak J, Tanneberger K, Schirmer K, Ashauer R (2014) Measured and modeled toxicokinetics in cultured fish cells and application to *in vitro – in vivo* toxicity extrapolation. PLoS One 9:e92303. doi:10.1371/journal.pone.0092303
Sun H (2005) Predicting ADMET properties by projecting onto chemical space: benefits and pitfalls. Curr Comput – Aided Drug Des 1:179–193. doi:10.2174/1573409053585708
Sun H, Xia M, Austin CP, Huang R (2012) Paradigm shift in toxicity testing and modeling. AAPS J 14:473–480. doi:10.1208/s12248-012-9358-1
Thompson A, Griffin P, Stuetz R, Cartmell E (2005) The fate and removal of triclosan during wastewater treatment. Water Environ Res 77:63–67. doi:10.2175/106143005X41636
Tollefsen KE, Scholz S, Cronin MT et al (2014) Applying adverse outcome pathways (AOPs) to support integrated approaches to testing and assessment (IATA). Regul Toxicol Pharmacol 70:629–640. doi:10.1016/j.yrtph.2014.09.009
Villeneuve DL, Crump D, Garcia-Reyero N et al (2014a) Adverse outcome pathway development II: best practices. Toxicol Sci 142:321–330. doi:10.1093/toxsci/kfu200
Villeneuve DL, Crump D, Garcia-Reyero N et al (2014b) Adverse outcome pathway (AOP) development I: strategies and principles. Toxicol Sci 142:312–320. doi:10.1093/toxsci/kfu199
Vinken M (2013) The adverse outcome pathway concept: a pragmatic tool in toxicology. Toxicology 312:158–165. doi:10.1016/j.tox.2013.08.011
Wambaugh JF, Wang A, Dionisio KL et al (2014) High throughput heuristics for prioritizing human exposure to environmental chemicals. Environ Sci Technol 48:12760–12767. doi:10.1021/es503583j
Wegner JK, Fröhlich H, Mielenz HM, Zell A (2006) Data and graph mining in chemical space for ADME and activity data sets. QSAR Comb Sci 25:205–220. doi:10.1002/qsar.200510009
Wetmore BA, Wambaugh JF, Allen B et al (2015) Incorporating high-throughput exposure predictions with dosimetry-adjusted *in vitro* bioactivity to inform chemical toxicity testing. Toxicol Sci 148:121–136. doi:10.1093/toxsci/kfv171
Willett P, Barnard JM, Downs GM (1998) Chemical similarity searching. J Chem Inf Comput Sci 38:983–996. doi:10.1021/ci9800211
Yang C, Tarkhov A, Marusczyk J et al (2015) New publicly available chemical query language, CSRML, to support chemotype representations for application to data mining and modeling. J Chem Inf Model 55:510–528. doi:10.1021/ci500667v
Zurlo J, Rudacille D, Goldberg AM (2001) Animals and alternatives in testing: history, science, and ethics. Mary Ann Liebert, Larchmont

Chapter 13
The Development of Quantitative AOPs

Irvin R. Schultz and Karen H. Watanabe

Abstract A quantitative adverse outcome pathway (qAOP) is a mathematical/computational model that represents the dynamic processes linking a molecular initiating event with an adverse outcome. A unique feature that distinguishes a qAOP from other biologically based mathematical models is the prediction of key events that are part of the qualitative adverse outcome pathway and are measurable experimentally. This chapter reviews the evolution of qAOPs, describes methods to develop qAOPs, and provides two case study examples focused on reproduction in fish.

13.1 Introduction

The concept of an Adverse Outcome Pathway (AOP) gained attention in 2009 and was the focus of a 2009 SETAC Pellston Workshop entitled, "A Vision and Strategy for Predictive Ecotoxicology in the 21st Century: Defining Adverse Outcome Pathways Associated with Ecological Risk." As described by Ankley et al. (2010), AOPs provide a framework for organizing information about a molecular initiating event (MIE) and the key events (KEs) that lead to an outcome of interest for risk assessment. Based upon available scientific data, AOPs provide a qualitative/conceptual description of the sequence of events leading to an adverse outcome and guides toxicity testing strategies, particularly in the development of in vitro assays. However, a large component of risk assessment relies on quantitative analyses that must address a wide variety of toxicants, exposure concentrations, and their effects upon wildlife species in a predictive fashion. Thus, an essential extension of an AOP for quantitative risk assessment is the development of mathematical/computational

I.R. Schultz (✉)
National Marine Fisheries Service, Lynker Tech & NW Fisheries Science Center, NOAA, Seattle, Washington 98112, United States
e-mail: irvin.schultz@noaa.gov

K.H. Watanabe
School of Mathematical and Natural Sciences, Arizona State University, Tempe, AZ, USA
e-mail: Karen.Watanabe@asu.edu

N. Garcia-Reyero, C.A. Murphy (eds.), *A Systems Biology Approach to Advancing Adverse Outcome Pathways for Risk Assessment*,
DOI 10.1007/978-3-319-66084-4_13

model(s) that represent the dynamic processes in an AOP, which we refer to as a quantitative AOP (qAOP).

There are now many examples in the toxicological literature of how quantitative models can aid the risk assessment process that include prediction of (i) toxicant exposure and disposition within the body (pharmaco−/toxicokinetics), (ii) specific toxic effects (pharmaco−/toxicodynamics) or interactions associated with the MIE, and (iii) adverse responses at higher biological scales (Nichols et al. 1991; Krauss et al. 2012; Sturla et al. 2014). Quantitative models provide specific, unambiguous predictions of toxicant effect(s) and/or changes in model variables that are linked to apical endpoints (e.g. 17β-estradiol, vitellogenin (VTG), and reproduction). By doing so, quantitative models also permit critical evaluation of hypotheses and assumptions associated with an AOP. This may identify knowledge gaps that exist in the system being modeled and guide future experimentation. Quantitative models have been shown to be valuable because they can organize experimental facts and assumptions in a logical manner, explore the implications of change on model parameters through simulations, estimate parameter values that are difficult or impossible to determine experimentally and prioritize research needs (Andersen et al. 1995). While valuable in their own right, existing models typically represent one or two scales of biological organization and thus cannot represent an entire AOP without modification.

In this chapter, we describe methods and a few case studies of qAOPs that have been developed since our first examination of how one could model a known AOP at the Pellston Workshop in 2009 (Watanabe et al. 2011). Our working definition of a qAOP is that it mathematically describes the processes of an AOP from the MIE to the adverse outcome. Prior to qAOP model development an established qualitative AOP is needed to identify the MIE, KEs, and the outcome of interest. By definition, KEs are measureable experimentally and thus are natural endpoints for qAOP model prediction. In addition, a qAOP model will be able to predict many other endpoints for which model evaluation data may be available, but these endpoints may not necessarily be identified as KEs. To date, two approaches have been used to develop qAOP models: one that links existing biologically based models by taking output from one model as input into another model; and *ab initio* model development. However, the development of qAOP models is just beginning and as our understanding of biological processes improves, computer technology advances, and AOP networks are identified, methods of model development will also change to keep pace with technology.

13.2 Developing Quantitative Adverse Outcome Pathways

An AOP is by definition a multiscale process that seeks to causally link changes at lower levels of biological organization with adverse effects at higher levels such as individuals and populations. This creates a challenge for qAOP model development for a number of reasons, but perhaps most notably are the temporal differences among biological scales. For example, many sub-cellular processes such as receptor binding, signal transduction and enzyme kinetics operate at time scales of seconds (or less) to minutes, while higher-level processes such as pharmacokinetic properties of hormones occur over hours-days or weeks. Tissue growth and development can occur over a period of months while other organismal processes associated with maturation and aging may take years. Further increasing biological scale, population-level processes operate at generational time scales, which can be many tens-to-hundreds of years. Thus, linking models of different scales requires careful thought regarding the type of output from one scale and how it will subsequently be used as input for a higher scale. A simple way to link biological scales and account for time differences is to take a hierarchical approach, with output from lower scales sequentially passed on to the next scale (Cilfone et al. 2015). For example, a simple multi-scale model may be composed of a cell or tissue-based compartment (timescale in minutes or hours) with a single predicted output that is used as an input parameter for a model that predicts a whole-animal level process. This approach has been used to relate a biochemical biomarker response with an adverse effect such as lethality. A good example of this approach is the model developed by Wu et al. (2015) to describe the relationship of fish gill Na+/K+ ATPase activity with metal induced toxicity. In their model, predicted gill Na+/K+ ATPase activity is used to estimate a "damage" parameter that is linked to a model of acute toxicity. The hierarchical approach is initially attractive as it follows the organization of most qualitative AOPs. However, strict hierarchical organization is limiting because the exchange of information typically needs to occur in both directions across scales (Cilfone et al. 2015). From an AOP perspective, this reflects the potential for time-dependent changes at higher scales to alter the behavior or output of lower scales. Examples of where this may occur would be hormonal feedback mechanisms and adaptation or acquired tolerance to a toxicant during long-term exposures.

Another consideration in multiscale AOP modeling is the level of understanding or detail in knowledge regarding a specific biological scale. For example, the ability to interrogate whole genome responses to toxicant exposure has provided an increasingly thorough understanding of cellular processes that may be altered. This has permitted development of highly complex cellular models of gene networks that can describe processes or key events at a relatively fine degree of resolution (Le Novere 2015). Quantitative models at higher levels of biological scale are typically "coarser" and do not describe processes with the same level of resolution. These differences in resolution may cause problems with parameter estimation, especially in situations where significant data gaps exist about processes contributing to a key event. In these situations the modeler may be forced to use data collected at higher biological scales (e.g. whole organism, apical endpoints) to guide parameter estimation at lower biological scales. This is sometimes referred to as disaggregation, because values measured at higher scales typically reflect the interaction of many factors occurring at lower scales and therefore need to be teased apart, which can introduce greater error in their estimates (Yang 2013). From a qAOP perspective, the ideal situation is to have the flow of information (output from one scale) proceed from finer to coarser scales (Yang 2013). This tends to promote a bottom-up approach in

AOP development that would begin with a detailed quantitative description of the molecular initiating event and its effect at the cellular level. However, both middle-out (e.g. tissue, organ system level response data) (Groh et al. 2015) and top-down approaches (e.g. apical endpoint data) to AOP development are to be expected because for many toxicants, knowledge about the MIE may be limited and toxicity data may only be available from higher biological scales. In these latter cases, the modeler may be forced to simplify quantitative descriptions at lower scales due to difficulties in accurately estimating model parameters.

The approaches used for linking existing or newly developed models require the modeler to first answer several important questions. What data are available to inform model development and parameterization? How many levels of organization (scales) are needed for the AOP? Will it be necessary to include large transitions in scale (e.g. gene expression to apical endpoint)? How much detail (number of equations and parameters) can be included within a scale? Answers to these questions will emerge from the available toxicity data and knowledge about the biology of the organism. It is helpful to keep in mind that all models reflect deliberate choices made to include or exclude reactions or processes. There will always be a need to manage model complexity by seeking to limit the number of equations and parameters to those considered essential for model performance, and it is common to lump discrete processes into a single or few parameters to manage model complexity. The challenge to the qAOP modeler is to identify where and how best to lump parameters. There is no standardized or approved method for doing this. There are various statistical tools that can help guide when to increase or decrease model complexity (Yamaoka et al. 1978) along with parameter sensitivity analysis methods that can help determine whether a model parameter or groups of parameters actually improves model performance. However, decisions on how many parameters to include or exclude in a model is primarily based on available data, biological knowledge, anticipated use of the model and prior modeling experience. In the next sections, we discuss approaches used to create a qAOP that incorporates currently available mathematical models and an AOP that requires new models to be developed.

13.2.1 Incorporating Existing Models at Different Biological Scales to Create a qAOP

One approach to develop a qAOP is to maximize existing resources by utilizing quantitative (e.g., mathematical, statistical, and/or computational) models developed for different processes within an AOP. These models may represent one KE relationship or span multiple KEs and relationships. Though an AOP is chemical agnostic, a qAOP requires specification of the chemical(s) of interest and the species to which the qAOP applies. Thus, to adapt an existing model for use in a qAOP, model parameters must be known for the chemical and species of interest, which includes fixed parameters in the model that may need adjustment because of interspecies or chemical differences, and user-defined input parameters that are set by the user for a specific application. The following summarizes the main steps in formulating a qAOP model.

1. First obtain a qualitative AOP description from the AOP Wiki (https://aopkb.org/aopwiki/index.php/Main_Page). The AOP Wiki is part of the OECD Adverse Outcome Pathway Knowledge Base and serves as a repository for AOPs and their supporting evidence. If a new (q)AOP is being created, consult the AOP Wiki to see how these are formulated and organized.
2. Once an AOP is selected, a literature review should be performed to find any quantitative models that have been developed for different parts of the AOP.
3. The existing models should be evaluated for applicability in the qAOP model. For example, a model may exist for a KE relationship in a related species, but required input data or parameter values may not be known for the species of interest. In addition, model predictions need to be compatible or usable as input into the following model that represents a higher level of biological organization. Adjustments may be needed to translate units of measure, or time intervals at which predictions are produced. Finally, differences in computer platforms and software licensing requirements may need to be addressed. Thus, when utilizing existing models, collaboration between model developers is helpful.
   (a) When no model exists for one or more KE relationships, the qAOP developer will need to develop the KE relationship(s). KEs, by definition, are measurable experimentally. Thus data, if not already available, should be obtained in order to develop the KE relationship(s) that is/are needed.
4. Once a set of models representing the AOP are linked in a qAOP model, evaluation of the qAOP model predictions with independent data sets should be performed. If each existing model was developed for the species of interest, and was independently evaluated, then it may not be necessary to perform an evaluation of the output from the qAOP, though performing spot checks to ensure that the links between models are working properly is always advisable.

13.2.2 Recent Examples of qAOP Model Development

In this section, three examples from the recent scientific literature of qAOP development are described. These three examples illustrate some of the challenges involved in developing qAOP's for eco-toxicologically relevant species, which require linking data and output across several biological scales. For many species of interest, the available biological and toxicological data may be limited to one or two scales. Incorporating complex mathematical descriptions for each scale may be inappropriate at present and simpler, more empirical approaches may be necessary to link output between biological scales. A good example of this approach is presented in a study by Miller et al. (2015) who focused on the effects of pulp mill effluents on reproduction in a population of white sucker (*Catostomus commersoni*). Many years of field monitoring had indicated an association between exposure to effluents and delays in time to maturation and decreased fecundity. An AOP was developed that considered inhibition of ovarian steroidogenesis as the initiating event, which was assumed to be associated with circulating levels of testosterone (Miller et al. 2015). The authors then developed an empirically-derived equation that was used to relate circulating levels of testosterone with fecundity and the relative proportion of breeding females in the population (Miller et al. 2015). The predicted effects on reproduction were then used as inputs into a density-dependent population model of white suckers to predict population level impacts of effluent exposure. In this example, a tissue level measurement (circulating sex hormone) was used as the starting point for the qAOP. This decision was based on the availability of testosterone measurements, which were routinely made during monitoring. It also reflects a compromise made by the authors in that more specific information on the effects of effluent exposure on sex steroid synthesis was not available. Thus, extending the model to a lower biological scale, such as the ovarian follicle and a more explicit description of sex steroid synthesis would have required additional equations and parameters that would have been difficult to estimate.

Another example of qAOP development is described by Ananthasubramaniam et al. (2015) who worked with the freshwater invertebrate *Daphnia magna*. In this study, a highly detailed bioenergetic model of *Daphnia* populations was developed from laboratory data and included a large number of parameters describing many physiological processes associated with feeding, growth, development and reproduction. A sensitivity analysis of these parameters was performed to determine how changes in their values affected the predicted lifetime reproduction and long-term growth rate of the population. Other experiments focused on gene expression changes in *Daphnia* exposed to a suite of model toxicants. Toxicant associated gene expression changes were then mapped to various physiological processes, which were linked to parameters describing these processes in the population model. The authors then suggest that relative changes in gene expression can be used to adjust the corresponding model parameters to predict exposure impacts on *Daphnia* populations (Ananthasubramaniam et al. 2015). Here, gene expression changes are directly being used to guide parameterization of a population-level model. However the large transition in model scales (primarily sub-cellular to population level) reduces the diagnostic power of the model as changes in fecundity were typically associated with a generalized pattern of altered gene expression. The authors acknowledged that a more mechanistic understanding of how subsets of genes influence daphnia physiology would likely improve specificity of model predictions for diverse toxicants.

In a final example, a qAOP was developed for an aquatic plant *Myriophyllum spicatum* exposed to a photosynthesis inhibitor (Riedl et al. 2015). Detailed metabolomic analysis was performed on leaf extracts to derive a dose-specific "metabolic effect level index", which condenses the observed changes in concentration of various small molecules (Riedl et al. 2015). Other endpoints measured were traditional plant apical endpoints including main shoot length, dry weight change and photosynthetic efficiency. A statistical model was then developed to relate changes in the metabolic effect level index with the apical endpoints (Riedl et al. 2015). Thus, tissue level changes (in the metabolome) were used to estimate whole organism level effects (growth). One challenge with this approach noted by the authors and associated with expressing metabolomic responses as an index, was that it tends to average all changes and assume that all discrete responses contribute equally to the adverse outcome (Riedl et al. 2015). The authors suggest future improvements in their qAOP would be to incorporate more specificity of the metabolomic response, recognizing that some metabolites are more closely associated with a specific biochemical pathway that may have more or less contribution to apical endpoint responses.

The recent interest in developing qAOP models has tended to focus on new model development. This will likely change over time as more models are developed, providing more options for the aspiring qAOP modeler. Clearly, there are tradeoffs to any qAOP model development, whether the modeler relies on linking existing models or building a new model. The decision process for deciding whether the advantages of one approach outweigh the disadvantages depends in part upon the intended use of the qAOP model. Indeed, maximizing all available resources and data should help to reduce the resources needed to develop the qAOP. If the existing models have been evaluated independently, a certain amount of confidence can be imparted upon the model predictions. However, translation of model output to be used as input into a higher-level model requires care and diligence if manually done, or slight modifications to the model code may be needed to automate the process. These steps require time and effort. Having one quantitative model that spans the entire AOP eliminates any concerns about compatible units of measure between models or computer platform differences. Examples of both approaches are described in the next section that focuses on two case studies.

13.3 Case Studies: Fish Reproduction

Reproduction is one of several core apical endpoints that a qAOP model may need to consider or predict because it can be translated to population outcomes. Over the past 25 years, public concerns over endocrine disruption in wild fish populations have spurred extensive research on the endocrinology of the fish reproductive system. Endocrine control of reproduction in fishes is functionally similar to other vertebrates and consists of the hypothalamus, the pituitary, the gonads and for fishes and other egg-laying vertebrates the liver, and is often referred to as the HPGL axis. Communication between these tissues occurs via the blood and specialized neurosecretory fibers from the hypothalamus to the pituitary. A minimum of five hormones plays an essential role in sexual maturation starting with production of gonadotropin-releasing hormone (GnRH) in the hypothalamus. The GnRH is sent to the pituitary where it stimulates the synthesis of the gonadotropins (GTHs): follicle-stimulating hormone (FSH) and luteinizing hormone (LH). In females, FSH stimulates growth of the ovarian follicles, each of which contains a single oocyte (immature egg). The LH is involved in maturation and triggering ovulation. Estradiol-17β (E2) in females, 11-ketotestosterone in males, along with progesterones such as 17,20β-dihydroxy-4-pregnen-3-one (DHP), promote gonad growth and induce gamete maturation, respectively. These hormones interact with each other through positive and negative feedback creating an intricate network of pathways that ultimately synchronize processes culminating in the timely production of mature eggs or sperm. In females, the liver is important because it synthesizes VTG, which is induced by E2 and is an essential component of the growing oocytes. Recent efforts to develop mathematical descriptions of fish reproduction have been encouraged by similar efforts made for farm animals and humans (Pring et al. 2012; Roblitz et al. 2013) and illustrate several of the challenges associated with qAOP model development such as incorporating time delays associated with different biological scales, the need to incorporate multiple sources of in vivo and in vitro derived data and to extend organism-level models to the population level. This section will present two case studies describing HPGL model development in different types of fishes to highlight some of the approaches used to overcome these modeling challenges.

13.3.1 **Ab initio** *qAOP Model: HPGL Model Development and Use with qAOP in Rainbow Trout (***Oncorhynchus mykiss***)*

Reproduction in most fish species including those important to regulatory testing and environmental health research, can be divided into two large groups: group synchronous spawners (a single large clutch of oocytes develop synchronously for one spawning event) or asynchronous spawners (several small clutches are spawned at different times during a reproductive season; fathead minnow, zebrafish, anchovy). Among synchronous spawning fishes such as the salmonids (trout, salmon, chars), some exhibit semelparity where death occurs after a single reproductive event, while others exhibit iteroparity, where repeated spawning events can occur over many years. Rainbow trout (*Oncorhynchus mykiss*) and other salmonid species have a long history of experimental use including many toxicological studies that provide a rich background of biological knowledge (Thorgaard et al. 2002). In addition, trout and salmon are important aquaculture species with 2013 estimates of global production exceeding 3,000,000 tons (FAO, http://www.fao.org/fishery/cultured-species/search/en). Thus, there is strong interest in developing a mathematical model of the salmonid reproductive axis to form a core of qAOP models for reproductive effects of chemical and physical stressors.

An early mathematical model of female salmon reproduction focused on the semelparous coho salmon (*Oncorhynchus kisutch*; (Kim et al. 2006)). One advantage of working with semelparous species is that gamete maturation is highly synchronous (e.g. a much larger proportion of oocytes and spermatozoa exist in identical states of development throughout the cycle; (Campbell et al. 2003; Luckenbach et al. 2008)). This facilitates a simpler description of gamete growth, which in the Kim et al. (2006) study was treated as a sequential series of five steps. All oocytes were assumed to be in each step or stage, with an empirically derived duration (days) before all oocytes "jumped" to the next stage. This approach also permits segregation of GTH activity, which was used to arbitrarily restrict LH release until E2 declines during a narrowly defined time period preceding final oocyte maturation (FOM) in the model. Another simplification was to treat GnRH as a type of "on" or "off" switch; GnRH was only on (at a fixed level) between days 75 and 250 of the reproductive cycle where it could stimulate FSH and LH synthesis. These model features appeared biologically appropriate for salmon where LH is released from the pituitary as a single massive surge prior to ovulation and FSH release is greatly reduced during the few months preceding spawning. This model also did not include an explicit description of vitellogenesis and assumed that impacts associated with E2 synthesis directly reflected effects on oocyte maturation and spawning. However, as efforts were begun to adapt this model to the iteroparous rainbow trout, it became clear that a more detailed description of these processes including vitellogenesis, was necessary.

Rainbow trout exhibit group synchronous reproduction with spawning occurring annually once sexual maturity is attained, although spawning sometimes occurs every 2 years depending on environmental conditions (Crim et al. 1992; Seamons and Quinn 2010). The major environmental cue for the seasonality of the trout reproductive cycle is annual photoperiod changes. Although trout are synchronous, it is known that growth and development among individual oocytes can vary substantially during much of the reproductive cycle (Tyler et al. 1990). Thus, it was felt that the HPGL model needed to include a capacity to allow sub-populations of oocytes to exist within the ovary, particularly during the time period of active vitellogenesis when size variance among oocytes is most pronounced (Tyler et al. 1990). Also, as a consequence of spawning multiple times in their lifetime, model assumptions about GnRH and FSH production needed to be updated along with the mechanism of releasing LH from the pituitary.

To make the model development process more tractable, it was decided to initially separate the description of vitellogenesis from the rest of the model. This can easily be done because the primary external input to VTG production by the liver is blood plasma E2 while the output is VTG. This permits the vitellogenesis model to be viewed as a sub-model and incorporate as much biological detail as considered necessary to describe many of the individual steps involved in VTG synthesis and secretion. This approach was taken in the VTG model described by Sundling et al. (2014). Their VTG model exemplifies several of the approaches used to describe gene expression and protein synthesis with the added feature that the protein (VTG) is actively secreted into the bloodstream. A common modeling approach to describe the synthesis of a gene product (e.g. hormone or VTG) is to assume production rate is the product of a biosignal (e.g. such as mRNA) and a proportionality constant that relates protein production rate to the corresponding biosignal. This approach is often appropriate because for example, the synthesis of a protein such as FSH is proportional to the amount of its mRNA in the pituitary. Time delays can be incorporated to account for the lag time between biosignal appearance and the initial appearance of hormone or protein in the bloodstream. More complex relationships between biosignal and the rate of hormone production, have been described and can be incorporated as needed (Jusko and Ko 1994). With regard to VTG synthesis, the Sundling et al. (2014) model assumes synthesis is regulated by estrogen receptors (ER), which when bound by E2 activate transcription of genes associated with VTG causing formation of VTG mRNA and protein. The model assumes E2 reversibly binds to its receptor with rate constants used to characterize the binding of E2 to ER. The E2-ER complex triggers synthesis VTG mRNA, which actively produces VTG inside the liver. An amplification factor is included in the equation describing VTG synthesis, which is needed to account for observations that one mRNA molecule may be translated many times. From the liver, VTG is transferred into plasma, where the plasma kinetics of secreted VTG is described using a clearance-volume pharmacokinetic model similar to that described in Schultz et al. (Schultz et al. 2001).

The trout VTG model has recently been incorporated into a second generation HPG model to continuously predict annual spawning in female rainbow trout (Gillies et al. 2016). In addition to continuously describing GnRH production, the revised model explicitly describes oocyte size, which is used to define growth and to differentiate the developmental stages of maturation. The model also permits multiple subpopulations of oocytes to exist in the ovary to make it more biologically consistent. A conceptual view of the second-generation model is shown in Fig. 13.1 (Left).

One application for the female trout HPGL model is for qAOP evaluation of toxicants using in vitro derived data. The HPGL model is not exclusively linked with one qAOP model or toxic mode of action, rather it is a quantitative tool to convert experimentally measured effects on hormone synthesis or action in target tissues such as the pituitary, ovary or liver into predicted effects on oocyte growth and ovulation. Reverse toxicokinetic modeling approaches would then be used to estimate environmental exposures needed to cause predicted adverse target organ concentrations. Predicted effects on oocyte growth or maturation would also need to be used as input into a population model. This overall approach offers advantages particularly when only in vitro data are available, as many model parameters such as those associated with gonadotropin synthesis and secretion (parameters mFSH, mLH, FSH, LH), ovarian synthesis of estrogen and 17,20 DHP (parameters for basal E2 and FSH stimulated synthesis and DHP) and estrogen induced synthesis of VTG (e.g. *vtg* mRNA, VTG) can be estimated from cell culture methods. Thus, multiple toxic modes of action and target tissues (direct effects on both the pituitary and ovary for example) can be accommodated. An example of model simulations is shown in Fig. 13.1 (right) and a summary of the overall qAOP process described is shown in Fig. 13.2.

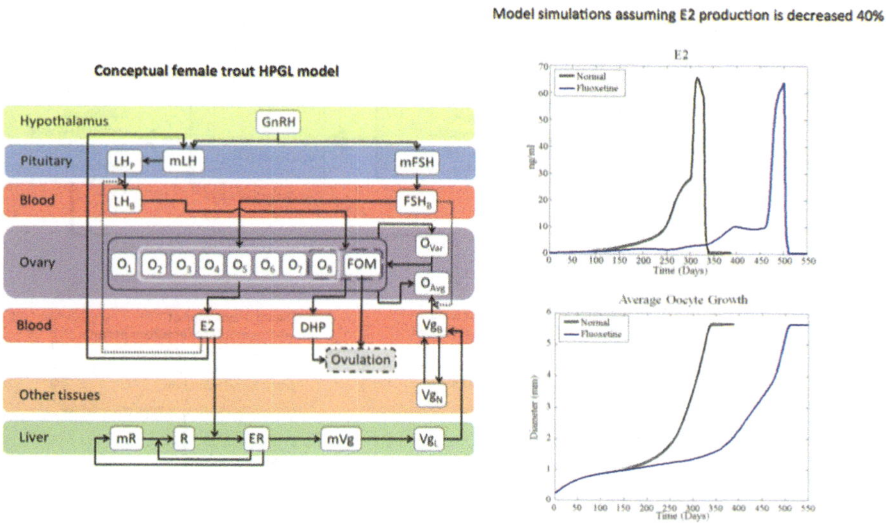

Fig. 13.1 (*Left*) Conceptual description of a second-generation HPGL model for female trout. The model is used to extrapolate results from select in vitro studies of tissues comprising the reproductive axis. HPGL model simulations provide estimates of target tissue levels that would be considered adverse (e.g. LOEC, AC50 or other internal dose metric) towards successful reproduction. The corresponding environmental exposure levels needed to achieve the unwanted internal dose metric is estimated using toxicokinetic models. Sources of input data for model parametrization can come from in vivo and in vitro studies. The (*right*) panel shows an example of changes in E2 production and oocyte growth during a hypothetical exposure to the pharmaceutical fluoxetine. Model simulations were performed assuming the rate of E2 synthesis decreased by 40%

13.3.2 *Linking Existing Models to Create a qAOP Model: Aromatase Inhibition in Fathead Minnow (*Pimephales promelas*)*

An AOP for aromatase inhibition was described by Ankley et al. (2010) and is available in the AOP Wiki (https://aopwiki.org/; AOP:25). Figure 13.3 depicts the AOP and Watanabe et al. (2014) provided an overview of qAOP development for aromatase inhibition in fathead minnow, which is described in greater detail here. Independently, models that simulate different KEs in the AOP were developed for the hypothalamic-pituitary-gonadal axis (HPG) (Mayo et al. 2012; Cheng et al. 2016); oocyte growth dynamics (Li et al. 2011; Watanabe et al. 2016); and population dynamics (Miller and Ankley 2004). The HPG axis model includes the MIE, inhibition of the enzyme aromatase, and predicts changes in steroid hormones (e.g., 17β-estradiol, testosterone) and plasma VTG. The oocyte growth dynamics model uses plasma VTG concentration as input, and predicts oocyte growth and daily spawning. Predictions of spawning as a function of time were then input into the population dynamics model to predict future trajectories. Through a

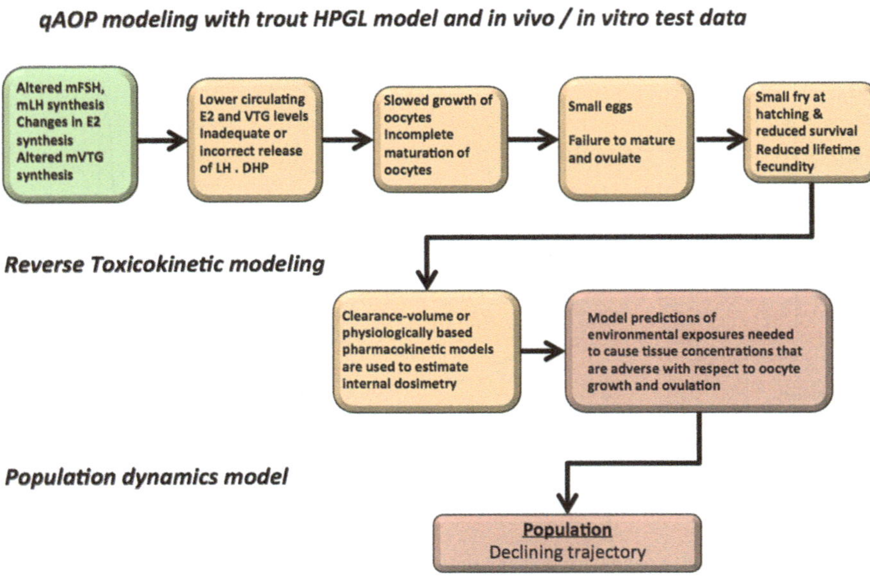

Fig. 13.2 Generalized overview of quantitative AOP using the trout HPGL model

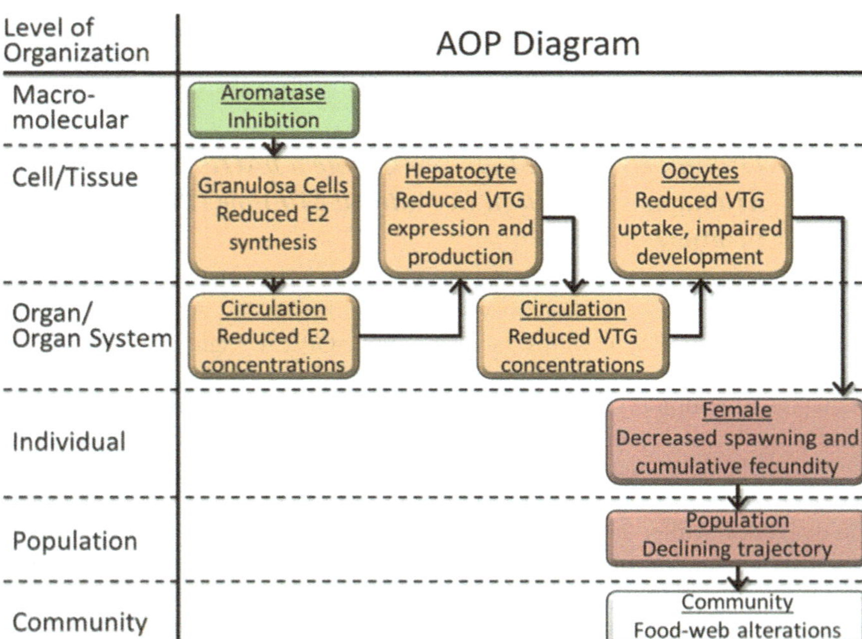

Fig. 13.3 Overview of qAOP development for aromatase inhibition in fathead minnow

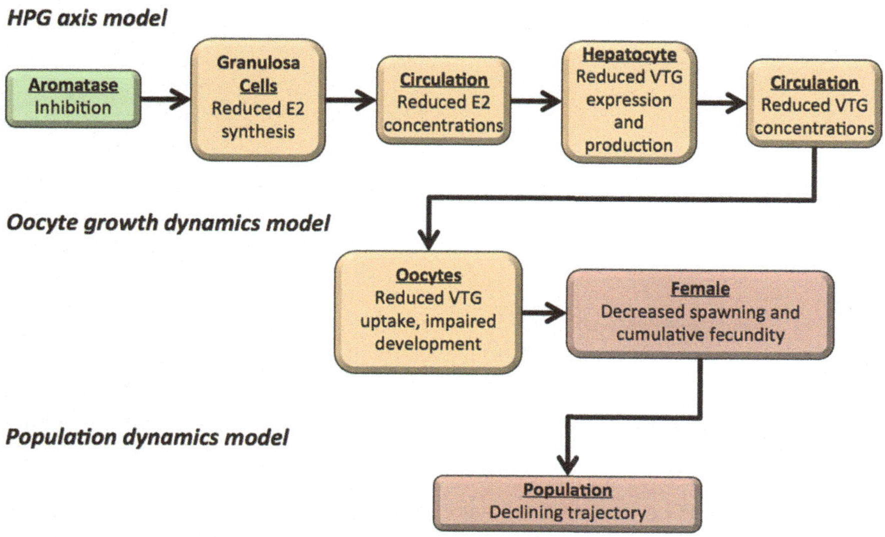

Fig. 13.4 Quantitative AOP for aromatase inhibition formed by linking previously developed computational models that represent different scales

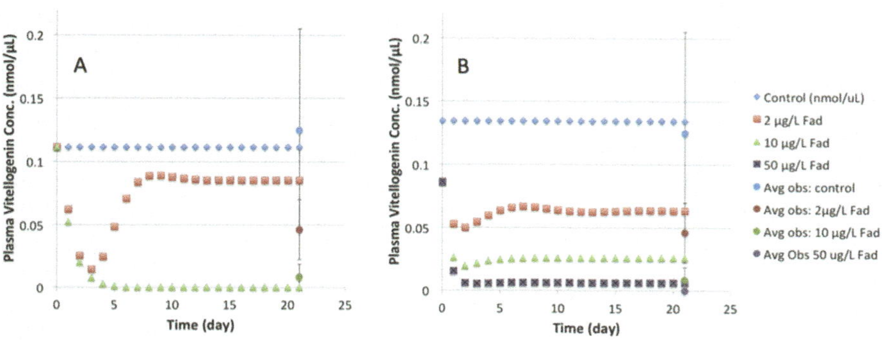

Fig. 13.5 (**a**, **b**) Hypothalamus-pituitary-gonadal axis model predictions of plasma vitellogenin concentrations vs. time (Reproduced from Watanabe et al. (2014). (**a**) Mayo et al. (2012); (**b**) Cheng et al. (2016))

collaborative effort of model developers and experimentalists, the three models were linked to construct a qAOP for aromatase inhibition in female fathead minnows as shown in Fig. 13.4.

Fadrozole hydrochloride (Fad, CASRN 102676-31-3) is a highly specific, reversible inhibitor of the aromatase enzyme that was used in Japan to treat breast cancer (Sainsbury 2004). In a 21-day reproduction study, Ankley et al. (2002) exposed fathead minnows to 2, 10, or 50 μg Fad/L in a group spawning design study. For each treatment concentration, two HPG axis models where run to predict plasma VTG concentrations as a function of time (Fig. 13.5a, b).

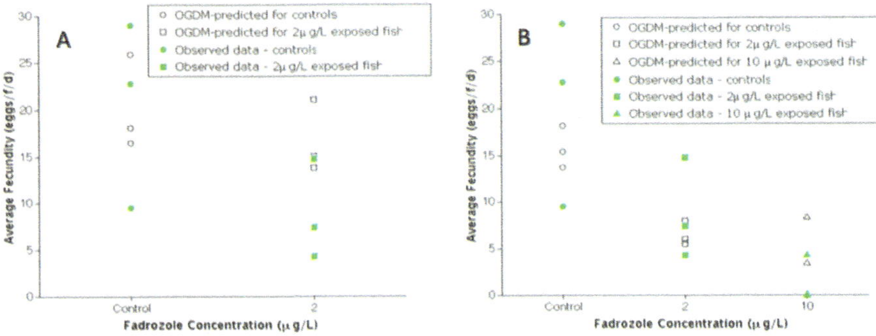

Fig. 13.6 (**a, b**) Predicted average fecundity calculated from oocyte growth dynamics model simulation results (Reproduced from Watanabe et al. (2014). Plasma vitellogenin concentrations as a function of time predicted by two HPG axis models were used as input: (**a**) the Mayo et al. (2012) model; (**b**) Cheng et al. (2016) model)

The plasma VTG concentrations were used as input into a modified version of the oocyte growth dynamics model (Watanabe et al. 2016) that accepts time varying plasma VTG concentrations. Three groups of four fathead minnows (12 fish total) were simulated to reproduce the experimental design, then the results were used as input into the population dynamics model. Figure 13.6 shows average fecundity calculated from the oocyte growth dynamics model predictions. Figure 13.6a does not show results for the 10 or 50 μg Fad/L treatments because the Mayo et al. (2012) model predicted plasma VTG concentrations essentially equal to zero (see Fig. 13.5a). The predicted average fecundity values are plotted with the experimentally observed average fecundity from Ankley et al. (2002), and shows good agreement for control (unexposed) fish. The 2 μg Fad/L treatment predictions from the Mayo et al. model yielded average fecundity values higher than results from the Cheng et al. model, and on average higher than the experimentally observed values. This results from the higher plasma VTG concentration profile predicted by the Mayo et al. model.

Mean cumulative fecundity (average cumulative fecundity for three groups vs time) over 21 days was used to inform recruitment in the population dynamics model. Figure 13.7 show changes in population size vs. time (over 20 years). A dramatic difference in population trajectory is predicted for the 2 μg Fad/L treatment due to the different HPG axis models. Yet, both HPG axis models did a reasonably good job of predicting plasma VTG measured at the end of the 21-day reproduction study, and average fecundity values calculated from the oocyte growth dynamics model matched experimentally observed values well with a slightly better "fit" obtained using the Cheng et al. model plasma VTG concentrations.

While it may appear contradictory to have two different HPG axis models that predict plasma VTG concentrations there is value in comparing predictions from different models. A computational model cannot represent all the mechanistic processes within biological systems – in part because of a lack of knowledge, but also due to computational limitations – and different approaches to modeling a system

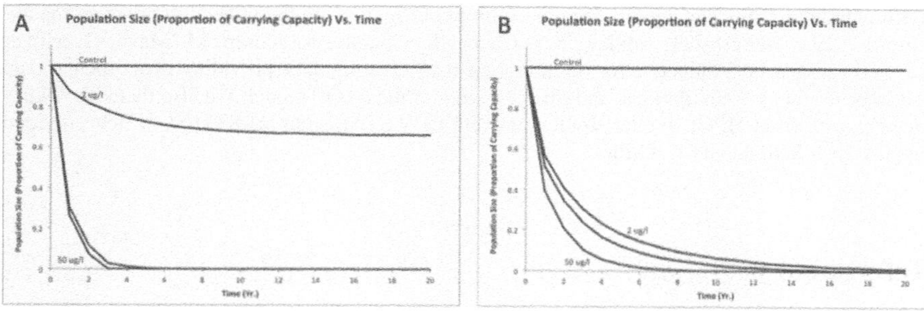

Fig. 13.7 (**a**, **b**) Population effects predicted from changes in fecundity (Reproduced from Watanabe et al. (2014). Fecundity predictions from oocyte growth dynamics model using plasma vitellogenin predicted by: (**a**) Mayo et al. (2012) model; and (**b**) Cheng et al. (2016) model)

are valid as long as they fit the available data. When these models are used to make predictions that were not constrained by measured data, differences in model formulation can produce vastly different model predictions. These case studies illustrate how existing models can be used to create qAOP models, and some of the challenges that arise.

13.4 Summary

This chapter describes methods to create qAOP models for a variety of adverse outcomes and species. Two approaches to qAOP model development were highlighted in the Case Studies section: *ab initio* model development and linking existing models that span different KEs in an AOP. Both approaches are valid and using one or the other depends upon the research question or application, e.g., risk assessment, and available resources. Research applications of qAOP models include testing hypotheses about biological processes and how an adverse outcome changes, facilitating experimental design, and predicting outcomes for different levels of molecular initiation. qAOP models can also support ecological risk assessments by predicting adverse outcomes from diverse sources of data including those made from in vitro assay measurements. To our knowledge, none of the models described here have been used in a formal ecological risk assessment. However, the AOP concept is still relatively new and the need for qAOP models is slowly being recognized. A similar process occurred for mammalian biologically based dose-response models and human risk assessments. There was a delay of many years between the description of models such as physiologically based pharmacokinetic models and their use in risk assessments (Andersen et al. 1987; Friess 1987; Bois et al. 1989; Clewell et al. 1995). In future years, we expect many different types of qAOP models to be developed and become essential tools for ecological risk assessments.

Acknowledgements The authors are grateful for contributions to the fathead minnow aromatase inhibition case study by G. Ankley, W.Y. Cheng, R. Conolly, K. Jensen, M. Mayo, D. Miller, E.J. Perkins, and D. Villeneuve for running their respective models, providing experimental data that were used in the simulations, and for the design of the qAOP model. We also thank K. Gillies for help with trout HPGL model simulations and EPA STAR grant #R835167, which provided support for I. Schultz and K. Gillies.

References

Ananthasubramaniam B, McCauley E, Gust KA, Kennedy AJ, Muller EB, Perkins EJ, Nisbet RM (2015) Relating suborganismal processes to ecotoxicological and population level endpoints using a bioenergetic model. Ecol Appl 25(6):1691–1710

Andersen ME, Clewell HJ III, Gargas ML, Smith FA, Reitz RH (1987) Physiologically based pharmacokinetics and the risk assessment process for methylene chloride. Toxicol Appl Pharmacol 87:185–205

Andersen ME, Clewell IHJ, Frederick CB (1995) Applying simulation modeling to problems in toxicology and risk assessment – a short perspective. Toxicol Appl Pharmacol 133(2):181–187. doi:10.1006/taap.1995.1140

Ankley GT, Kahl MD, Jensen KM, Hornung MW, Korte JJ, Makynen EA, Leino RL (2002) Evaluation of the aromatase inhibitor fadrozole in a short-term reproduction assay with the fathead minnow (*Pimephales promelas*). Toxicol Sci 67(1):121–130

Ankley GT, Bennett RS, Erickson RJ, Hoff DJ, Hornung MW, Johnson RD, Mount DR, Nichols JW, Russom CL, Schmieder PK, Serrrano JA, Tietge JE, Villeneuve DL (2010) Adverse outcome pathways: a conceptual framework to support ecotoxicology research and risk assessment. Environ Toxicol Chem 29(3):730–741

Bois FY, Zeise L, Tozer TN (1989) The use of pharmacokinetic models in the determination of risks for regulatory purposes. In: Bonin JJ, Stevenson DE (eds) Proceedings of the 1987 annual meeting of the society for risk analysis. advances in risk analysis, vol 7. Plenum, New York, pp 573–583

Campbell B, Dickey JT, Swanson P (2003) Endocrine changes during onset of puberty in male spring Chinook salmon, Oncorhynchus tshawytscha. Biol Reprod 69(6):2109–2117. doi:10.1095/biolreprod.103.020560

Cheng W-Y, Zhang Q, Schroeder A, Villeneuve DL, Ankley GT, Conolly R (2016) Computational modeling of plasma vitellogenin alterations in response to aromatase inhibition in fathead minnows. Toxicol Sci. doi:10.1093/toxsci/kfw142

Cilfone NA, Kirschner DE, Linderman JJ (2015) Strategies for efficient numerical implementation of hybrid multi-scale agent-based models to describe biological systems. Cell Mol Bioeng 8(1):119–136. doi:10.1007/s12195-014-0363-6

Clewell HJ, Gentry PR, Gearhart JM, Allen BC, Andersen ME (1995) Considering pharmacokinetic and mechanistic information in cancer risk assessments for environmental contaminants: examples with vinyl chloride and trichloroethylene. Chemosphere 31(1):2561–2578

Crim LW, Wilson CE, So YP, Idler DR, Johnston CE (1992) Feeding, reconditioning, and rematuration responses of captive atlantic salmon (Salmo salar) kelt. Can J Fish Aquat Sci 49(9):1835–1842

Friess SL (1987) Risk assessment: historical perspectives. In: National Research Council (ed) Pharmacokinetics in risk assessment, drinking water and health, vol 8. The National Academies Press, Washington, DC, pp 3–7

Gillies K, Krone S, Nagler JJ, Schultz IR (2016) A computational model of the rainbow trout hypothalamus-pituitary-ovary-liver axis. PLoS Comput Biol 12(4). doi:10.1371/journal.pcbi.1004874

Groh KJ, Carvalho RN, Chipman JK, Denslow ND, Halder M, Murphy CA, Roelofs D, Rolaki A, Schirmer K, Watanabe KH (2015) Development and application of the adverse outcome pathway framework for understanding and predicting chronic toxicity: I. Challenges and research needs in ecotoxicology. Chemosphere 120:764–777. doi:10.1016/j.chemosphere.2014.09.068

Jusko WJ, Ko HC (1994) Physiologic indirect response models characterize diverse types of pharmacodynamic effects. Clin Pharmacol Ther 56(4):406–419

Kim J, Hayton WL, Schultz IR (2006) Modeling the brain-pituitary-gonad axis in salmon. Mar Environ Res 62(Supplement 1):S426–S432

Krauss M, Schaller S, Borchers S, Findeisen R, Lippert J, Kuepfer L (2012) Integrating cellular metabolism into a multiscale whole-body model. PLoS Comput Biol 8(10). doi:10.1371/journal.pcbi.1002750

Le Novere N (2015) Quantitative and logic modelling of molecular and gene networks. Nat Rev Genet 16(3):146–158. doi:10.1038/nrg3885

Li Z, Villeneuve DL, Jensen KM, Ankley GT, Watanabe KH (2011) A computational model for asynchronous oocyte growth dynamics in a batch-spawning fish. Can J Fish Aquat Sci 68(9):1528–1538. doi:10.1139/f2011-066

Luckenbach JA, Kusakabe M, Swanson P, Young G (2008) Unilateral ovariectomy increases egg size and reduces follicular atresia in the semelparous coho salmon, Oncorhynchus kisutch. J Exp Zool A Ecol Genet Physiol 309(8):468–476. doi:10.1002/jez.476

Mayo M, Habib T, Garcia-Reyero N, Villeneuve DL, Ankley GT, Perkins EJ (2012) Androgen receptor mediated compensation of estradiol in response to aromatase inhibition: a mathematical model. In: Presented at the SETAC North America 33rd annual meeting, Long Beach, CA, Nov 11–15. In SETAC North America 33rd annual meeting abstract book, p 362

Miller DH, Ankley GT (2004) Modeling impacts on populations: fathead minnow (*Pimephales promelas*) exposure to the endocrine disruptor 17b-trenbolone as a case study. Ecotoxicol Environ Saf 59(1):1–9

Miller DH, Tietge JE, McMaster ME, Munkittrick KR, Xia X, Griesmer DA, Ankley GT (2015) Linking mechanistic toxicology to population models in forecasting recovery from chemical stress: a case study from Jackfish Bay, Ontario, Canada. Environ Toxicol Chem 34(7):1623–1633. doi:10.1002/etc.2972

Nichols JW, McKim JM, Lien GJ, Hoffman AD, Bertelsen SL (1991) Physiologically based toxicokinetic modeling of three waterborne chloroethanes in rainbow trout (Oncorhynchus mykiss). Toxicol Appl Pharmacol 110(3):374–389

Pring SR, Owen M, King JR, Sinclair KD, Webb R, Flint APF, Garnsworthy PC (2012) A mathematical model of the bovine oestrous cycle: simulating outcomes of dietary and pharmacological interventions. J Theor Biol 313:115–126

Riedl J, Schreiber R, Otto M, Heilmeier H, Altenburger R, Schmitt-Jansen M (2015) Metabolic effect level index links multivariate metabolic fingerprints to ecotoxicological effect assessment. Environ Sci Technol 49(13):8096–8104. doi:10.1021/acs.est.5b01386

Roblitz S, Stotzel C, Deuflhard P, Jones HM, Azulay DO, van der Graaf PH, Martin SW (2013) A mathematical model of the human menstrual cycle for the administration of GnRH analogues. J Theor Biol 321:8–27

Sainsbury R (2004) Aromatase inhibition in the treatment of advanced breast cancer: is there a relationship between potency and clinical efficacy? Br J Cancer 90(9):1733–1739. doi:10.1038/sj.bjc.6601731

Schultz IR, Orner G, Merdink JL, Skillman A (2001) Dose-response relationships and pharmacokinetics of vitellogenin in rainbow trout after intravascular administration of 17 alpha-ethynylestradiol. Aquat Toxicol 51(3):305–318

Seamons TR, Quinn TP (2010) Sex-specific patterns of lifetime reproductive success in single and repeat breeding steelhead trout (Oncorhynchus mykiss). Behav Ecol Sociobiol 64(4):505–513. doi:10.1007/s00265-009-0866-7

Sturla SJ, Boobis AR, FitzGerald RE, Hoeng J, Kavlock RJ, Schirmer K, Whelan M, Wilks MF, Peitsch MC (2014) Systems toxicology: from basic research to risk assessment. Chem Res Toxicol 27(3):314–329. doi:10.1021/tx400410s

Sundling K, Craciun G, Schultz I, Hook S, Nagler J, Cavileer T, Verducci J, Liu Y, Kim J, Hayton W (2014) Modeling the endocrine control of vitellogenin production in female rainbow trout. Math Biosci Eng 11(3):621–639. doi:10.3934/mbe.2014.11.621

Thorgaard GH, Bailey GS, Williams D, Buhler DR, Kaattari SL, Ristow SS, Hansen JD, Winton JR, Bartholomew JL, Nagler JJ, Walsh PJ, Vijayan MM, Devlin RH, Hardy RW, Overturf KE, Young WP, Robison BD, Rexroad C, Palti Y (2002) Status and opportunities for genomics research with rainbow trout. Comp Biochem Physiol B Biochem Mol Biol 133(4):609–646

Tyler CR, Sumpter JP, Witthames PR (1990) The dynamics of oocyte growth during vitellogenesis in the rainbow trout (Oncorhynchus mykiss). Biol Reprod 43(2):202–209. doi:10.1095/biolreprod43.2.202

Watanabe KH, Andersen ME, Basu N, Carvan MJ III, Crofton KM, King KA, Suñol C, Tiffany-Castiglioni E, Schultz IR (2011) Defining and modeling known adverse outcome pathways: Domoic acid and neuronal signaling as a case study. Environ Toxicol Chem 30(1):9–21

Watanabe KH, Mayo M, Miller DH, Cheng WY, Conolly RB, Villeneuve DL, Perkins EJ (2014) Modeling a molecular initiating event to population effects. In: Presented at the SETAC North America 35th annual meeting, Vancouver, BC, Canada, Nov 9–13. In SETAC North America 35th annual meeting abstract book, p 111

Watanabe KH, Mayo M, Jensen KM, Villeneuve DL, Ankley GT, Perkins EJ (2016) Predicting fecundity of fathead minnows (*Pimephales promelas*) exposed to endocrine-disrupting chemicals using a Matlab®-based model of oocyte growth dynamics. PLoS One. doi:10.1371/journal.pone.0146594

Wu SM Tsai, JW Tzeng WN, Chen WY, Shih WY (2015) Analyzing the effectiveness of using branchial NKA activity as a biomarker for assessing waterborne copper toxicity in tilapia (Oreochromis mossambicus): a damage-based modeling approach. Aquat Toxicol 163:51–59

Yamaoka K, Terumichi N, Uno T (1978) Application of akaike's information criterion (aic) in the evaluation of linear pharmacokinetic equations. J Pharm Biopharm 6:165–175

Yang A (2013) On the common conceptual and computational frameworks for multiscale modeling. Ind Eng Chem Res 52(33):11451–11462. doi:10.1021/ie303123s

Chapter 14
Linking Adverse Outcome Pathways to Dynamic Energy Budgets: A Conceptual Model

Cheryl A. Murphy, Roger M. Nisbet, Philipp Antczak, Natàlia Garcia-Reyero, Andre Gergs, Konstadia Lika, Teresa Mathews, Erik B. Muller, Diane Nacci, Angela Peace, Christopher H. Remien, Irvin R. Schultz, and Karen H. Watanabe

Abstract Ecological risk assessment quantifies the likelihood of undesirable impacts of stressors, primarily at high levels of biological organization. Data used to inform ecological risk assessments come primarily from tests on individual organisms or from suborganismal studies, indicating a disconnect between primary data and protection goals. We know how to relate individual responses to population dynamics using individual-based models, and there are emerging ideas on how to make connections to ecosystem services. However, there is no established methodology to connect effects seen at higher levels of biological organization with suborganismal dynamics, despite progress made in identifying Adverse Outcome Pathways (AOPs) that link molecular initiating events to ecologically relevant key events. This chapter is a product of a working group at the National Center for Mathematical and Biological Synthesis (NIMBioS) that assessed the feasibility of using dynamic energy budget (DEB) models of individual organisms as a "pivot"

C.A. Murphy (✉)
Department of Fisheries and Wildlife, Michigan State University, East Lansing, MI, USA
e-mail: camurphy@msu.edu

R.M. Nisbet
Department of Ecology, Evolution and Marine Biology, University of California, Santa Barbara, CA, USA

P. Antczak
Institute for Integrative Biology, University of Liverpool, Liverpool, UK

N. Garcia-Reyero
Environmental Laboratory, US Army Engineer Research & Development Center, Vicksburg, MS, USA

A. Gergs
gaiac – Research Institute for Ecosystem Analysis and Assessment, Aachen, Germany

N. Garcia-Reyero, C.A. Murphy (eds.), *A Systems Biology Approach to Advancing Adverse Outcome Pathways for Risk Assessment*,
DOI 10.1007/978-3-319-66084-4_14

connecting suborganismal processes to higher level ecological processes. AOP models quantify explicit molecular, cellular or organ-level processes, but do not offer a route to linking sub-organismal damage to adverse effects on individual growth, reproduction, and survival, which can be propagated to the population level through individual-based models. DEB models describe these processes, but use abstract variables with undetermined connections to suborganismal biology. We propose linking DEB and quantitative AOP models by interpreting AOP key events as measures of damage-inducing processes in a DEB model. Here, we present a conceptual model for linking AOPs to DEB models and review existing modeling tools available for both AOP and DEB.

14.1 Introduction

The adverse outcome pathway (AOP) framework conceptualizes the transfer of information from molecular to organismal levels of organization as the first step in scaling up to inform human and ecological risk assessment (Ankley et al. 2010). The AOP framework is an effective tool for arranging information at the sub-organismal levels of organization, and may aid in interpreting data from high-throughput screening methods for the purpose of risk assessment. With these methods, the potential of thousands of chemicals to interact with molecular and

K. Lika
Department of Biology, University of Crete, Voutes University Campus, Heraklion, Greece

T. Mathews
Environmental Sciences Division, Oak Ridge National Laboratory, Oak Ridge, TN, USA

E.B. Muller
Marine Science Institute, University of California, Santa Barbara, CA, USA

D. Nacci
US Environmental Protection Agency, Office of Research and Development, Narragansett, RI, USA

A. Peace
Department of Mathematics and Statistics, Texas Tech University, Lubbock, TX, USA

C.H. Remien
Department of Mathematics, University of Idaho, Moscow, ID, USA

I.R. Schultz
Pacific NW National Laboratory – Marine Sciences Lab, Sequim, WA, USA

K.H. Watanabe
School of Mathematical and Natural Sciences, Arizona State University, Glendale, AZ, USA

cellular processes can be determined rapidly and cost efficiently (see Chap. 2), but how those interactions translate into impact on organismal performance and ecological processes remains obscure (Margiotta-Casaluci et al. 2016). For human health risk assessment, which focuses on the protection of the individual, AOPs that link lower level events to organismal level responses may be sufficient. Thus, adverse outcomes may include cellular endpoints such as skin sensitization, or cell proliferation (see Chap. 11). Ecological risk assessments are generally concerned with the protection of populations, food webs, or ecosystems, so there is a critical need to develop AOPs that inform these higher levels of biological organization. The approach thus far has focused on adverse outcomes on individuals that relate to the general processes of growth, survival and reproduction, which are processes that are included in population-level assessments (Kramer et al. 2011).

In an AOP framework, existing knowledge is organized around the causal linkages between observable biological changes or key events (KEs) that are integral to the progression from a molecular initiating event (MIE) to an adverse outcome (AO) and are considered relevant to regulatory decision making (Ankley et al. 2010). These predictive causal linkages between KEs are called key event relationships (KERs). The AOPs describe biological relationships between KEs, and how stressors or contaminants perturb those relationships. As such, AOPs are not chemically specific. But, to understand how a chemical could impact an AOP, understanding of chemical-specific properties including potency and pharmacokinetic factors (i.e., absorption, distribution, metabolism and excretion (ADME)) is critical because these properties ultimately define the magnitude and duration of perturbation at the MIE (Villeneuve et al. 2014).

More specifically for our purposes, a quantitative AOP (qAOP) mathematically describes the processes of an AOP from an MIE to an adverse outcome (Chap. 13). A qAOP can be a series of quantitative response-response relationships that describe transitions between key events (Conolly et al. 2017); these response-response relationships are ideally mechanistic, but correlative information can also be useful.

Obtaining the mechanistic detailed information necessary to develop qAOPs on a single species is challenging (e.g. Chap. 13; Margiotta-Casaluci et al. 2016), and so it is infeasible to repeat this exercise for thousands of species. Also, existing AOPs generally converge on a single biological endpoint as if independent (e.g., growth *or* reproduction), and thus ignore the trade-offs implied by resource limitation, e.g., through the competition for energy among physiological processes. Further, linking a molecular chain of events to effects on growth and reproduction, requires the use of a generalizable theory at the organismal level of organization, one that can explain the effects of toxicants on an organism's acquisition of resources from the environment and the consequences for energy demanding traits such as growth and reproduction (Jager et al. 2016).

Dynamic Energy Budget (DEB) theory (Kooijman 2010; Nisbet et al. 2000) offers such a mechanistic framework, and is the best-tested and most comprehensive theory for the energy budget of organisms (Jager et al. 2016). DEB models have been developed for many species for over three decades, have impressive calibration and validation, and have been used to determine toxic effects on organisms and populations (DEBTox: http://www.debtox.info/). The DEB approach employs unifying metabolic theory that can theoretically be applied to any species using a small number of parameters; this approach can potentially be used to determine effects of chemicals on many more species (e.g. Jager et al. 2006, 2011; Muller et al. 2010; Martin et al. 2013). Moreover, the approach offers the possibility to extrapolate population and higher level dynamics from individual level energy budget by the means of individual based modelling (Martin et al. 2013; Gergs et al. 2014, 2016) which would provide a connection to the AOP beyond the organism level if we could relate the AOP to some feature of a DEB model.

Dynamic energy budget models describe the process of assimilation of energy or mass and internal use for physiological functions. Although the term "Dynamic Energy Budget (DEB)" theory is sometimes used to refer to any dynamic representation of energy budgets, it is most commonly used to refer to the comprehensive theory of metabolic organization due to Kooijman (1986, 2010; Sousa et al. 2008). The standard DEB model (Kooijman et al. 2008), uses a small number of differential equations and parameters to describe individual life history processes that are based on energy fluxes: organisms assimilate resources from the environment and subsequently allocate energy, through a reserve compartment, to maintenance, growth (increase in structure) and the reproduction system, i.e., they mature and after reaching puberty stop maturation and start reproduction. Both structure and reserve contribute to biomass, whereby only structure requires maintenance and only reserve fuels metabolic processes. These state variables link to observable traits such as body size or time to first reproduction, but, because of their rather abstract nature, none of them can be measured directly.

Kooijman's DEB theory captures the metabolic dynamics of an individual organism through its entire life-cycle, be it ectothermic or endothermic, autotrophic or heterotrophic, and is explicitly tied to food/substrate availability and temperature. Basic to DEB theory is the coherence between levels of biological and ecological organization, using the life cycle of an individual as primary focus, from which sub- and supra-organismic levels are considered. Thus, DEB theory bears promise to serve as a pivotal framework for building process-based models that link molecular, cellular, and tissue level responses to apical endpoints, such as survival, growth, and reproduction, subsequently to those at higher levels of ecological organization.

The qAOP and DEB models have complementary strengths and weaknesses (Rohr et al. 2016). The qAOPs offer tight connections with known biochemistry at the cost of parameter richness, and have limited ability to quantitatively predict whole organism level responses to stressors with consideration of energetic tradeoff between growth, development, and reproduction. DEB models are parameter sparse (Kearney et al. 2015), and describe performance of complete organisms, but their variables are defined abstractly and are only implicitly related to measured physiological and biochemical endpoints. Identifying linkages between qAOP and DEB models could transform risk assessment, but the challenge lies in the different computational and mathematical approaches used by qAOPs and DEB. Computational AOP approaches use pathway analyses, network systems and statistics to link the molecular responses to ODEs (Watanabe et al. 2017). The

Table 14.1 Proposed variables and equations used to characterize response to toxicity within an organism

Variables
$\mathbf{K} = \{K_1, K_2, \ldots \ldots \ldots\}$= set of sub-organismal key events from AOP
$\mathbf{R} = \{R_1, R_2, \ldots \ldots \ldots\}$= set of damage-related variables – may overlap with **K**.
$\mathbf{Q} = \{Q_1, Q_2, \ldots \ldots \ldots\}$= set of internal toxicant-related concentrations
$\mathbf{B} = \{B_1, B_2, \ldots \ldots \ldots\}$= set of DEB model variables
$\mathbf{E} = \{E_1, E_2, \ldots \ldots \ldots\}$ = set of environment variables
Dynamics
$\frac{d\mathbf{B}}{dt} = \text{functions of } \mathbf{B}, \mathbf{R}, \mathbf{E}, \mathbf{Q}$ $\frac{d\mathbf{R}}{dt} = \text{functions of } \mathbf{Q}, \mathbf{R}, \mathbf{K} \left(\text{occasionally } \mathbf{B}\right)$ $\frac{d\mathbf{Q}}{dt} = \text{functions of } \mathbf{B}, \mathbf{Q} - \text{TK model}$

systems of ODEs in some AOP-based models have state variables that represent specific enzymes or metabolites; compounds associated with measured response to a stressor (e.g. Table 14.1 in Murphy et al. 2009). DEB models use a small number of more abstract state variables such as "reserve" or "structure" that represent loosely defined combinations or "metabolic clusters" of compounds (e.g. Table 14.2 of Nisbet et al. 2000), but make immediate predictions of quantities that are critical for population dynamics, such as reproduction rates. Linking the two approaches will involve the mathematical problem of model order reduction, as well as biological interpretation of the variables in a reduced qAOP model in the context of those used in DEB theory.

14.2 DEB and Ecotoxicology

The appealing simplicity and generality of DEB theory comes with a price; model quantities and processes have a relatively high level of abstraction. This makes the application of DEB theory challenging because the state variables and parameters may not be directly observable. Auxiliary assumptions (that may be organism specific) link these variables to quantities that can be measured directly such as length, wet or dry weight, respiration, time to/length at first brood, egg output, and so on (Lika et al. 2011). There is a large body of literature on methods for estimating DEB model parameters, including routine multivariate, nonlinear regression (or analogous likelihood) methods (Kooijman et al. 2008), a computer-intensive state-space method (Fujiwara et al. 2005), a Bayesian approach (Johnson et al. 2013) and an innovative, heuristic "pseudo-Bayesian" approach (Lika et al. 2011).

Table 14.2 A "toy" model that illustrates the proposed classification of model variables

Variables	
K_1 = level of upregulation of gene	K_1
$R_1 = K_2$ = intracellular ROS concentration	Z
$R_2 = K_3$ = intracellular antioxidant concentration	A
R_3 = intracellular concentration of damaged proteins	P
Q_1 = intracellular concentration of contaminant	Q
B_1 = total structural biomass of bacteria	M_V
B_2 = reserve density in bacteria	e
E_1 = substrate concentration in environment	S
E_2 = contaminant concentration in the environment	C (assumed constant here)

Dynamics of R-variables (damage related)

Dynamics of oxidative stress:

$$\frac{dZ}{dt} = \underbrace{\Phi_M(M_V)}_{\text{ROS production from normal metabolism}} + \underbrace{\Phi_S(Q)}_{\text{ROS production caused by contaminant}} - \underbrace{\Phi_A(A,Z)}_{\text{antioxidant action}} - \underbrace{\Phi_Z(Z)}_{\text{neutralization of ROS}}$$

$$\frac{dA}{dt} = \text{antioxidant production and loss}$$

$$\frac{dP}{dt} = \underbrace{\gamma Z}_{\text{ROS induced damage}} - \text{dilution due to growth}$$

Dynamics of B-variables (bioenergetics)

DEB dynamics:

$$\frac{dM_V}{dt} = \underbrace{\frac{(ve - mg)}{e + g}}_{\text{production of structure from reserve}} - \underbrace{h(P)M_V}_{\text{mortality}}$$

$$\frac{de}{dt} = \underbrace{f - e}_{\text{reserve homeostasis}}$$

Dynamics of Q-variables (TK)

$$\frac{dQ}{dt} = k_i C - k_e Q \quad \text{one compartment kinetics}$$

Dynamics of E-variable (environment)

$$\frac{dS}{dt} = \underbrace{-\alpha f M_V}_{\text{uptake of substrate by cells}}$$

The variable names that would match the proposed classification in Table 14.1 are on the left, more familiar names used in the dynamic equations are on the right.

Currently, the most widely used approach to the estimation of DEB parameters (called "add-my-pet" http://www.bio.vu.nl/thb/deb/deblab/add_my_pet/index.html) is based on the implicit assumption that a single parameter set can explain the life history of a cohort of organisms that is followed in empirical experiments over time. Inter-individual differences within a cohort of a species are usually treated as measurement error. These differences might, however, be caused by differences among the individuals in terms of their physiology and in initial conditions; recognizing this would require more sophisticated statistical approaches incorporating process error. Intra-species variations might not only have consequences for the statistical power of the parameter estimation and model validity (Jager 2013), but are the basis of differential sensitivity to chemical exposure (Gergs et al. 2015) and are drivers for selection in an evolutionary time scale.

Research using DEB models to interpreting toxicity data started with a simplified DEB model (Kooijman and Metz 1984); continuing research over the following decade lead to a suite of models (Kooijman and Bedaux 1996) that are still being developed. Collectively these models are referred to as DEBtox. Within DEBtox models several options for an effect are considered and are referred to as "physiological modes of action" (Álvarez et al. 2006). The concept of physiological modes of action summarizes how a stressor might interfere with processes along the cascade of energy assimilation and allocation as represented in the DEB model. Within this framework, a chemical might act on the assimilation of food, increase the costs for somatic and maturity maintenance rate, increased costs for structure or the costs for reproduction or pose a hazard to embryo (Jager and Zimmer 2012). A toxicokinetic model (see next subsection) may link the exposure concentration and the effect, and a stress function is applied to the different DEB parameters, thus, parameter values change proportional to the (internal) toxicant concentration.

DEBtox and related models have been used to analyze toxicity data such as derived from growth and reproduction tests under constant exposure conditions (Jager et al. 2006; Jager and Selck 2011; Goussen et al. 2015), or time-varying exposure (Pieters et al. 2006) and effects resulting from chemical mixtures (Jager et al. 2010). These models have usually been applied to organismal growth and reproduction data to derive the suitable physiological mode of action. However, in many cases, the analyzed data could be well described by any of the physiological modes of action and additional data on respiration and feeding are needed for identification (Muller et al. 2010; Jager et al. 2016). Furthermore, sublethal chemical effects might not be adequately described by a single mode of action, rather it is conceivable that organisms respond in multiple ways to the same toxicant.

Recently, a simplified version of standard DEB (isomorphic animal) was proposed for use in ecotoxicology (Jager and Zimmer 2012). Compared to the standard formulation, this version has a reduced number of input parameters. This simplification, however, has several limitations: (1) reserve is assumed to be in a steady state with the food level, which is only realistic at conditions of constant food, (2) reproduction is initiated at a constant size irrespective of food availability and chemical exposure and (3) the energetic costs for producing one egg is constant, thus, maternal effects are not considered. More flexibility with regard to variable environmental conditions is possible by the use of the somewhat more complex standard DEB model (Kooijman et al. 2008) which explicitly accounts for maturation and uses maturation thresholds, e.g., for puberty, thus, modelled individuals are allowed to reproduce at smaller size when food is not at *ad libitum*. However, the concept of physiological modes of action is also applicable to the standard DEB model and the applicability of these models in an individual based population modelling framework, for the purpose of effect extrapolation of higher biological levels, have been demonstrated for chemical stress (Martin et al. 2014) and stress due to food limitation and crowding (Gergs et al. 2014).

14.3 Toxicokinetic-Toxicodynamic Models

Our descriptions of the qAOP and DEB models do not yet consider specific toxicants since AOPs and DEB models are chemical agnostic. To connect with toxicity, we assume that response to contaminants is related to internal (i.e., within organism) concentrations of the contaminant. Thus, modeling toxicity requires coupling the model representation of physiological processes (qAOP or DEB) to toxicokinetic (TK) and toxicodynamic (TD) sub-models. TK models describe the dynamics of bioaccumulation, elimination, and chemical transformations of chemical contaminants within an organism. Toxicodynamic (TD), sometimes called “toxic effect”, models describe processes leading from toxicant interation with a biological target to effects.

The literature contains many approaches to modeling *survival* that use TK-TD approaches but do not consider detailed physiological processes that cause mortality. These have, to a great extent, been reconciled within the General Unified Threshold model for Survival (GUTS, Jager et al. 2011). Current GUTS implementations consider two TD alternatives: death due to variation in individual tolerance and stochastic death. Within stochastic death models, the hazard rate, i.e., the event rate at time t, conditional on survival until time t, for an individual thereby increases linearly with the dose metric beyond a threshold, and an organism has an increased probability of mortality. In contrast, in individual tolerance models, the threshold follows a frequency distribution within a population and death is instantaneous for an organism when dose metric exceeds the individual survival threshold.

In GUTS, there is a dose metric, assumed proportional to the hazard (i.e., per capita mortality) rate, that may include processes such as bioaccumulation, distribution within the organism, biotransformation and elimination, damage accrual and recovery, and physiological compensation processes. These dose metrics are also applicable in the context of *sublethal* effects (e.g., Muller et al. 2015). The most simple dose metric is the scaled internal concentration when information on body residues is lacking. A TD stage of damage can be used to link internal concentrations (full body residues or concentration at target site, depending on TK model complexity) and effects. The complexity of the damage model can considerably differ among studies, depending on the processes considered in the approach. One approach is to use a one parameter scaled damage model (Jager et al. 2011), while other approaches may include damage accrual and recovery (Ashauer et al. 2007) although in the latter approach the rate constant for damage accrual is interpreted as a combined parameter for damage and effect. In one example. Jager and Kooijman (2005) consider receptor kinetics in the damage process for analyzing of survival of organisms exposed to organophosphorus pesticides. Here, they assume that functional receptors are knocked out by the chemical, and functional receptors are turned into non-functional ones. Veltman et al. (2014) extend this approach to predict sodium loss and acute mortality in several aquatic species. Enzyme (acetylcholinesterase) inhibition was also considered in the time-dependent accrual of damage on the molecular level to explain differential sensitivity at the organism level (Kretschmann et al. 2011, 2012). We will consider the TK-TD definition of damage to aid our conceptual model that links AOP to DEB.

14.4 Linking AOP to DEB

We propose a conceptual model to initiate exploration into mechanistically linking AOP to DEB (Fig. 14.1).

Essentially an AOP represents a pathway that is integral to a DEB, and we propose that translations from one to other are possible because of mutual constraints (Fig. 14.1). Within the AOP, KEs and adverse outcomes that occur at the molecular, cellular, organ and whole organism level can influence DEB parameters, by imposing *damage*, a term common to both the TKTD and DEB literature. This potentially

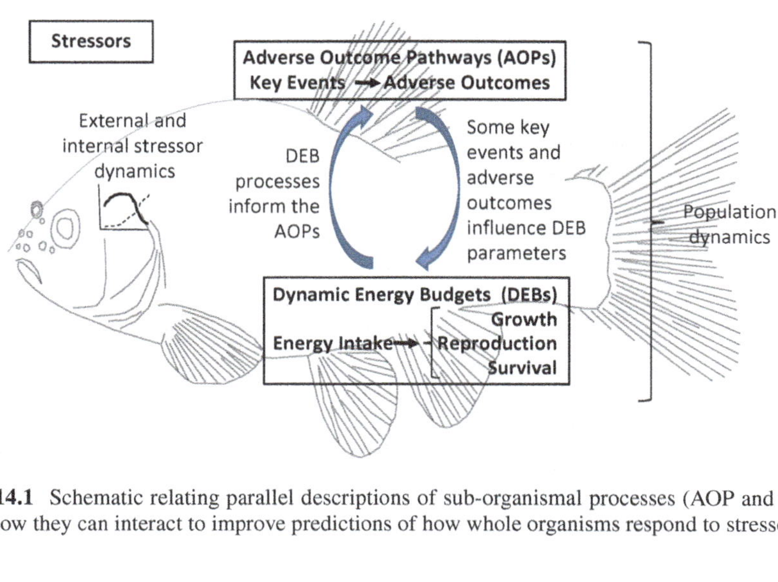

Fig. 14.1 Schematic relating parallel descriptions of sub-organismal processes (AOP and DEB) and how they can interact to improve predictions of how whole organisms respond to stressors

integrates the AOP into a more holistic DEB model that uses a small number of variables to describe the integrated effect of all metabolic processes, allows for tradeoffs of energy between growth, reproduction and survival, which are population relevant outputs. Feedbacks from the DEB to AOP (whole organism processes to molecular response) allow for mass balance constraints and ground AOPs in realistic scenarios that allow for energetic tradeoffs within an organism.

14.4.1 Mathematical Formalism

A model of organism-level dynamics should not simply allow everything to be connected to everything else. Figure 14.2 suggests a mathematical structure that recognizes 5 types of variables (summarized in Table 14.1) and their causal connections.

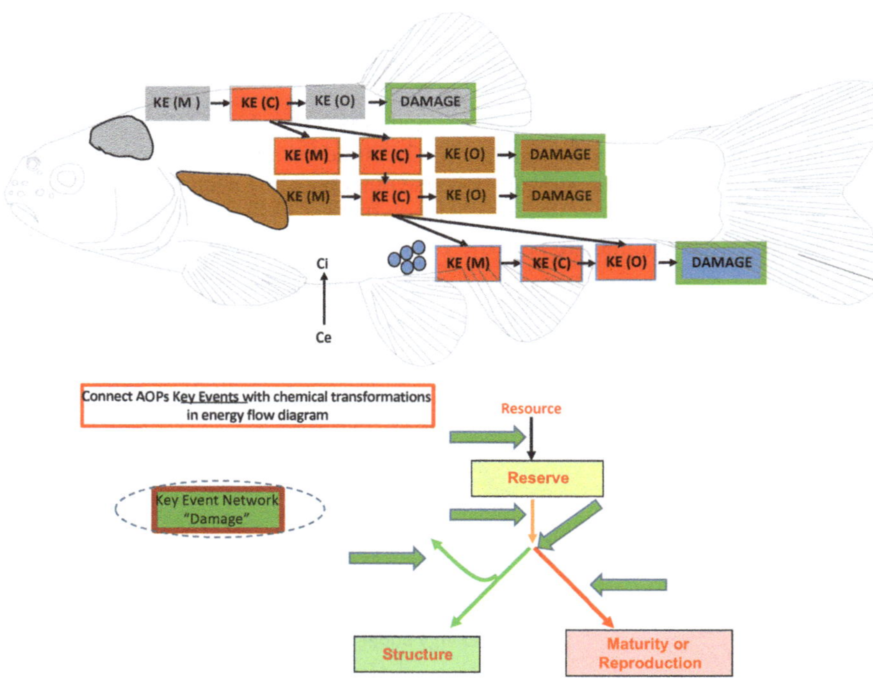

Fig. 14.2 A conceptual framework to link AOPs to DEB for a particular stressor/contaminant scenario will first require an inventory of the key events (KEs) affected in different organs and at different levels of biological organization, – such as molecular, cellular, organ level responses, noted as KE(M), KE(C), KE(O) this key event network then would translate to a damage term that would have to be related to key rates or allocation "decisions" (indicated by *green arrows* in lower panel) in dynamic energy budget models. Ce refers to concentration of toxicant external to the organism, and Ci is the concentration inside the organism

KEs are measureable endpoints in an AOP. These key events either represent, or are themselves, measures of *damage* that is manifest at the level of cells, organs or of the entire organism. Damage is caused, directly or indirectly, by *internal toxicant concentrations*– the accumulation of damage being described by some TK/TD representation. Damage impacts the processes in a *bioenergetic* model. In this chapter we have in mind Kooijman's (2010) DEB model with abstract variables (structure, reserve etc.), but the conceptual links are appropriate for other bioenergetics models that have variables corresponding to quantities such as biomasses, or measures of chemical concentration that are directly observable.

The connections described in Fig. 14.2 restrict the form of dynamic equations we can use (see Table 14.1 for a list of variables and functional dependencies). In applications that we can currently envision, the model of the complete organism will use dynamic (differential or difference) equations to describe "performance" – growth, development, reproduction, risk of mortality etc. These are described by the DEB model which is tightly coupled to the body burden dynamics (growth impacts internal toxicant concentration and vice versa), and also to the damage dynamics. Changes in internal concentrations are described by a TK submodel. The damage variables will commonly be decoupled from the higher level processes, but connection is possible, e.g. oxidative stress involves production or reactive oxygen species (ROS) from both "routine" metabolism and from toxicity.

To illustrate this process we describe a "toy" model suggested by previous work (Hanegraaf and Muller 2001; Klanjscek et al. 2012, 2016). We consider a batch culture of bacteria as a multicellular super-organism exposed to a soluble toxicant. The toxicant generates intracellular ROS that in turn stimulates production of an anti-oxidant enzyme for which there is a genetic marker. Up-regulation of this gene constitutes a key event. We assume that all rates are directly proportional to structural biomass (V1-morphy in the terminology used in DEB papers); then the above structure describes both a single organism and a population.

The toy model describes one way to link AOP to DEBs, but there are numerous ways to accomplish these linkages, and the numerous linkages should be explored in full to determine if there are generalizable processes. For different species and different stressors, there are different types of key events in the AOP that are measureable or observable. For example, in *Daphnia*, much of the effects of contaminants is measured at either the molecular level, or the whole organism level, and so to link AOP to DEB in Daphnia, the KE(M) would have to be directly translated to DEB rates. In other organisms, for example fish, key events have been measured at molecular, cellular and organ level responses, and some integration has been done through physiological models (Murphy et al. 2005; Watanabe et al. 2009; Gillies et al. 2016). The key event network would then integrate the responses of KE(M), KE(C) and KE(O) and translate that to DEB rates. There are numerous ways to integrate data at the various levels of biological organization before trying to link to DEB parameters, and we describe a few ideas below.

14.4.2 Gene Expression Analysis

There are a multitude of statistical methods available to analyze and interpret OMICs (eg. transcriptomics, proteomics, lipidomics, metabolomics) based data. Most commonly the initial analysis aims at identifying features such as genes, metabolites or proteins, which are significantly changed as a result of a given perturbation. These gene lists can then be used as inputs to functional enrichment approaches to identify biological processes which are altered, which is then used for interpretation of the response of the organism to the given perturbation. The biological processes could be represented by DEB rates and fluxes. To increase statistical power and reduce the overall complexity of OMICs data, data summarization techniques such as component analyses (Principal (PCA) and Independent (ICA)), or clustering of profiles can be applied and then a standard analysis pipeline resumed. In relation to AOPs, time and dose components play a large role. Analysis methodologies for such data are still limited, but can be approached via linear modelling techniques to identify changing features over the course of the exposures. More applicable in the context of AOPs is a time-delay correlation analysis which can provide a directed network indicating which features are influencing other features over time. With the inclusion of additional measurements, such as phenotypic characterizations, the neighborhood of these can be explored and interpreted in a biological context and related to rates and parameters relevant for DEB models (Villaverde et al. 2014; Zoppoli et al. 2010).

14.4.3 Data Integration

In recent years, with the advent of OMICs technologies and its ever reducing cost, it has become feasible to integrate multiple levels of biological organization and analyze those in a more comprehensive manner. While the computational approaches for true integration of multi-level data are still scarce, a simpler integration across these levels has been approached in a number of publications (Rohart et al. 2017; Antczak et al. 2006; Van Aggelen et al. 2010; Williams et al. 2011; Joyce and Palsson, 2006). AOPs in particular would benefit greatly from multi-level analysis approaches as an AOP inherently represents the combination of many biological levels. Metaboanalyst 3.0 (http://www.metaboanalyst.ca/faces/home.xhtml) is an excellent example of the current state of multi-level integration, allowing for metabolite and gene expression level integration to understand the possible affected pathways within a given organism. With the increase of available metabolic models for more and more species, a more integrated methodology could be developed for understanding effects across multiple levels.

14.5 Other Modeling Approaches that Complement the AOP-DEB Modeling Framework

There are a number of different quantitative tools available to augment qAOPs from MIE to physiological processes and we describe them here along with their limitations if they were to be the only tool used for chemical risk assessment. These tools can define key events or key event relationships which could eventually be linked to DEB to improve their utility. Some of these tools include toxicant dynamics within, even though an AOP itself is chemical agnostic.

14.5.1 Metabolic Network Models

Several methods have been developed to probe the relationship between genotype and phenotype using the large volumes of high-throughput molecular data and are broadly classified as metabolic network models (Henry et al. 2010; Lewis et al. 2012). Some of these methods are more quantitative and mechanistic than others, such as constraint-based models, stochastic and deterministic kinetic models. Metabolic network models break down metabolic pathways into respective reactions and enzymes. As a starting point, the more popular techniques are constraint-based models such as flux-balance analysis (FBA), which require very little kinetic information, and can be done on a large scale. FBA can calculate steady-state metabolic fluxes for very large models with over 2000 reactions and do this by describing a system of linear equations related to concentration changes in a metabolic network using matrix algebra (Kauffman et al. 2003; Raman and Chandra 2009; Orth et al. 2010). Some assumptions for FBA include steady state or exponential growth (for simple linear equations) and that the system is optimized for maximum growth (Kauffman et al. 2003). Constraints, related to boundary constraints (for example nutrient uptake/and excretion), or internal constraints that limit the rates or direction of reactions within an organism, can be added to the flux rates of the reactions within the network (Price et al. 2004). A major drawback of this approach is the lack of inherent dynamic or regulatory predictions and no explicit representation of metabolic concentrations (Bordbar et al. 2014). More predictive models include stochastic and deterministic kinetic models because they require detailed understanding of kinetic parameters, but these models are often challenging to parameterize. Furthermore, kinetic models that capture biological stochasticity and biophysics, are challenging to model across various timescales (Bordbar et al. 2014). Ideally, progression of models would move from constraint based → deterministic → stochastic kinetic models as information and hypotheses progress, and much research has been focused on improving these models for future use. Recently published reviews (Ghaffari et al. 2015; Yizhak et al. 2015) survey studies of cancer metabolism by reconstructed metabolic model approaches and discusses challenges such approaches face.

Predictive metabolic network models are restrictive and challenging to apply to ecological risk assessment because of the vast quantities of data needed to parameterize all the kinetic processes in these models. Very few of these models include post-translational processes, allow for feedback, are truly mechanistic and make broad assumptions that limit the system to assumptions of steady state and optimal growth. However, once connections between qAOPs and DEB are established, it is possible that the fluxes and rates from the DEB model could provide additional constraints that could realistically ground the FBA model within an entire organism.

14.5.2 Mechanistic Physiological Models

The U.S. National Research Council's report on toxicity testing in the twenty first century (NRC 2007) advocated for the development of various types of biologically based dose response (BBDR) models that more closely link toxicant induced cellular or sub-cellular perturbations with whole organism effects. This is in recognition that biological systems are organized along several interconnected scales and that no single scale can be fully considered in isolation. In contrast to empirical, statistical based techniques, BBDR or physiologically-based models link molecular initiating events to apical endpoints by incorporating mathematical descriptions of processes at different levels of biological organization. An important feature of any physiological model is the ability to describe the system in its natural or undisturbed state prior to injury. Physiological and BBDR types of mathematical models can be linked with diverse types of biological data to allow more complex descriptions of a cell, tissue or whole organism processes. Typically, these models are formulated as ordinary differential equations with measureable biological quantities or key events as state variables. Many physiologically-based models aim to predict the temporal dynamics of injury following exposure to a stressor. Analysis of these models can uncover critical features, such as thresholds in exposure beyond which an organism can no longer recover. Interactions between scales such as hormonal signaling and feedback can also be incorporated. A major challenge in developing physiologically-based models includes choosing state variables and functional relationships among those variables appropriately and wisely, especially given existing knowledge about physiological function and details of the molecular initiating event may be incomplete.

Recently, several physiological models were developed to characterize the reproductive effects of exposure to endocrine disrupters in fish, including Atlantic croaker (Murphy et al. 2005, 2009) salmon and trout (Kim et al. 2006; Sundling et al. 2014; Gillies et al. 2016) and fathead minnows (Watanabe et al. 2009; Shoemaker et al. 2010; Li et al. 2011; Connolly et al. 2017). These modeling efforts reflect the importance of predicting stressor effects on fecundity and success of spawning. Opportunities for future model development not yet explored include additional endocrine signaling, neurological function and linking changes in neuronal activity and damage with cognitive effects and behavioral changes. There are also examples of advanced multiscale models for mammalian systems that could be adapted to lower vertebrates, including models developed for cardiac function, which incorporate diverse processes such as cellular electrical activation, dynamic tissue mechanics associated with contraction, and fluid mechanic properties of blood passing through the heart (ten Tusscher et al. 2004; Crowcombe et al. 2016).

BBDR models can be quite useful for DEBs. The dynamics of particular processes, for example hormone production, can potentially be linked to specific DEB parameters. We have explored linking dynamics of estradiol and its role in reproduction, to the kappa variable in DEB and have been able to successfully mechanistically connect the two processes (Muller et al. in prep). These connections can be particularly valuable because standard DEB has the control mechanisms implied and cannot directly deal with process disturbances due to the toxicant action.

14.5.3 Physiologically Based Toxicokinetic Models (PBTK)

Physiologically based toxicokinetic (PBTK) models include compartments for organs and tissues that have an influence upon the absorption, distribution, metabolism and elimination (ADME) of a toxicant, and they are used to predict target tissue chemical concentrations and effects, respectively. Numerous PBTK models have been developed for human health risk assessment beginning with a model for styrene in rats and humans (Ramsey and Andersen 1984), and for waterborne chemicals in rainbow trout (Nichols et al. 1991). Since the emphasis of this chapter is upon integrating AOPs with DEB models, the example that follows focuses upon a PBTK model developed for female fathead minnows; an example for rainbow trout is described in Chap. 16.

A next-generation PBTK model for female fathead minnows (*Pimphales promelas*) was developed to predict the ADME of estrogenic and androgenic chemicals (Li et al. 2011). Where a traditional PBTK model focuses entirely upon the disposition of a xenobiotic chemical(s) in the body, Li et al. integrated xenobiotic chemical disposition (i.e., 17α-ethinylestradiol and 17β-trenbolone) with their interaction and effect upon endogenous steroid hormone levels (e.g., 17β-estradiol) through estrogen and androgen receptor binding in certain tissue compartments. The model contains six tissue compartments (Fig. 14.1, Chap. 13): gill, brain, gonad, liver, blood, and "other" (a composite of the remaining tissues). Mathematically, a system of ordinary differential equations describes the dynamics of endogenous compounds (e.g., 17β-estradiol, testosterone, estrogen receptor, and vitellogenin), and xenobiotic compounds (17α-ethinylestradiol and 17β-trenbolone). This model robustly predicted plasma concentrations of 17β-estradiol, testosterone and vitellogenin in unexposed female fathead minnows and females exposed to 17α-ethinylestradiol and 17β-trenbolone. In a qAOP context, a PBTK model such as this can be used to span multiple key events and make predictions of changes in plasma vitellogenin that can then be used as input into an oocyte growth dynamics model to predict changes in spawning and fecundity. A review of different approaches to quantitative AOP model development is described in Chap. 13.

PBTK and BBDR models require large amounts of data about the physiological, biochemical, and physicochemical processes that occur in biological systems. The fact that these data are often not available from only one source raises concerns about the accuracy and validity as well as the comparability of the different types of data used to develop each model. Crump et al. (2010) identified several concerns on the use of BBDR models (and by extension PBTK) to predict low-dose toxicity. According to them, these models do not eliminate the need for empirical modeling of the relationship between dose and effect, but only move it from the whole organism to a lower level of biological organization while introducing significant sources of uncertainty. They concluded that BBDR models are unlikely to be fruitful in reducing uncertainty in quantitative estimates of risk from low-level exposures. Additionally, they believe that the use of *in vitro* data in these models will introduce new issues regarding extrapolation of data from *in vitro* systems. As some processes are well characterized and others are not, information gaps may exist. These information gaps may cause the model to fail to optimally predict the outcomes of a specific behavior or stressor. It is therefore crucial to emphasize that the quality of the simulations depends on the model, the data and their purpose, and the uncertainty of the data used to build the model should be properly reflected. Simulation results should be supported by experimental data and should not replace reliable data as primary degree of evidence. Furthermore, it should be noticed that poor quality modeling practices could lead to biased predictions or overestimation of the predictive power of the model. In order to minimize bias, it would be necessary to perform extensive and continuous evaluation of the model.

PBTK models, similar to physiologically-based models, could be linked to DEBtox models, but in this case the toxicant dynamics would be included. Although we have not explored this, we can envision that these linkages would greatly improve the DEBtox predictions because detailed mechanistic information from existing toxicity tests could then be included into DEBtox and would greatly improve predictions of whole organism response and could also inform population or community predictions.

14.5.4 Quantitative Structure Activity Relationships (QSAR)

Structure Activity Relationship, SARs, and Quantitative SARs (QSARs), by definition, link specific chemical structural features to biological responses or outcomes. The most basic form of SAR models have been developed to predict hazard identification based on physico-chemical properties and measured whole animal toxicity. For example, the Verhaar modeling scheme (Verhaar et al. 1992, now implemented in ToxTree http://Toxtree.sourceforge.net, Ellison et al. 2015) predicts the toxicity of pollutants classified by broad mechanisms of action (MOA), such as narcosis, from simple chemical properties such as hydrophobicity. The utility of this approach has been expanded as models are being built using much more extensive data (ECOTOX http://www.epa.gov//ecotox/) and sophisticated computational chemistry approaches that resolve much finer MOA assignments (e.g., Barron et al. 2015). However, the impossibility of assessing risks on a chemical-by-chemical basis and the mandate to reduce whole animal testing necessitates the need to develop predictive toxicological approaches that rely more heavily on information derived from *in silico* and *in vitro* methodologies. Largely driven by human health concerns, programs like the US's ToxCast (http://www.epa.gov/ncct/toxcast) and Tox21 (http://www.epa.gov/ncct/Tox21), and Japan's Toxicogenomics Project (http://toxico.nibio.go.jp/english/index.html) produce data, databases and tools for computational toxicology. These analytical and predictive tools can include QSARs that relate, for example, chemical features to fine-scale outcomes such as the likelihood of off-target effects to pharmacological agents (QSAR Toolbox from OECD, e.g., Sullivan et al. 2014). Through the availability of detailed chemical information and novel analyses, QSAR is providing increasingly sophisticated compound descriptions which have the potential to link structural features to mechanisms (review, Garcia-Serna et al. 2015). Despite challenges associated with linking QSAR with genomics data, interactions between chemical features and molecular networks have been modeled successfully to predict toxic mechanisms and outcomes for certain systems (e.g., Antczak et al. 2010). And because few mechanisms of action are well known, this approach seems especially valuable as a discovery tool: for example, providing strong evidence for putative adverse outcome pathways (AOPs) associated with narcosis by lipophilic compounds (Antczak et al. 2015).

The AOP concept provides a linear representation of the linkages between MIEs, KEs and adverse outcomes which are causally linked and are inherently independent of specific chemical stressor. QSAR/SARs on the other hand are inherently chemical specific, linking chemical structural features to response or activity of a given biological process. By design QSARs aim to provide a predictive framework for read-across of the chemical space to link to the biological process. The main limitation is the narrow set of features with highly cross-correlated variables which provide a challenge to the development and interpretation of these types of models. This can also result in false correlations and model overfitting which reduces the predictive power and their applicability in the quest to understanding biological variation. While QSARs have been designed to be mechanistically driven, recent applications have become more correlative to explain certain biological behavior. The best-known application of QSAR is its involvement in the prediction of toxicity from Log Kow values, which has been shown to be highly predictive for a large amount of chemicals across multiple species (Cronin and Schultz 2001; Patlewicz et al. 2003; Claeys et al. 2013). Inclusion of additional variables, such as gene expression, can improve this fit but also provide a basis for biological interpretation (Antczak 2015).

Due to the different application of the use of chemicals within QSAR and the AOP concepts, the inclusion of QSAR in qAOPs (for example PBTK) might be ill placed but could provide a new dimension for read-across and association of chemicals for DEBtox.

14.6 Conclusion

In this chapter we described a conceptual model for linking AOP to DEB. We used the damage concept from TKTD models to translate the key event effects from AOP to DEB parameters and rates, and are embracing some concepts inherent in DEBtox. The case study we presented is a first attempt and relates oxidative stress to DEB parameters in a single cell organism. Our working group is currently working on two case studies based on *Daphnia* and rainbow trout that merge information collected for AOPs with DEB. For rainbow trout, we focus on endocrine disruption for which there are quantitative AOPs that integrate molecular, cellular and organ level responses to predict effects on reproduction. Connecting with a DEB representation required modifying the "standard" DEB model to include feedbacks that characterize the integrated effects of hormonal control mechanisms (Muller et al. in prep). With *Daphnia*, there is little organ level data, so we seek correlative connections with transcriptomic data. *Daphnia* were exposed to a gradient of food rations and contaminant concentrations versus time with measurements of gene expression and contaminant body burdens along with routine measurements of size, survival and reproduction. Gene expression data (interpreted as key events), summarized using statistical analyses appear to exhibit some molecular responses that may correlate with parameters controlling relevant DEB fluxes. We hope, that with continued exploration, we could select a final set of key events, or a key event network that could inform specified DEB parameters and rates. Eventually we envision a system where AOPs link to DEB rates, and the DEB is then used to constrain AOPs within the construct of a whole organism where energetic tradeoffs between physiological processes are considered. Such a system would improve the predictive power of suborganismal key events, and place such KEs into a framework that would allow for extrapolation to population (via IBMS) and up to population, community, and ecosystem effects. Also, by accounting for energetic tradeoffs within an organism, we can use the multitude of existing DEB models for other species or employ evolutionary life history theory to modify the AOP-DEB system to make more realistic predictions from KEs for species where the extensive information needed to build qAOPs is lacking.

References

Ankley GT, Bennett RS, Erickson RJ et al (2010) Adverse outcome pathways: a conceptual framework to support ecotoxicology research and risk assessment. Environ Toxicol Chem 29:730–741

Antczak P, Ortega F, Chipman JK, Falciani F (2010) Mapping drug physio-chemical features to pathway reveals molecular networks linked to toxicity outcome. PLoS One 5(8):e12385

Álvarez OA, Jager T, Redondo EM, Kammenga JE (2006) Physiological modes of action of toxic chemicals in the nematode Acrobeloides nanus. Environmental Toxicology and Chemistry 25(12):3230-7

Antczak P, White TA, Giri A, Michaelangeli F, Viant MR, Cronin MTD, Vulpe C, Falciani F (2015) Systems biology approach reveals a calcium-dependent mechanism for basal toxicity in *Daphnia magna*. Environ Sci Technol 49:11132–11140

Ashauer R, Boxall ABA, Brown CD (2007) New ecotoxicological model to simulate survival of aquatic invertebrates after exposure to fluctuating and sequential pulses of pesticides. Environ Sci Technol 41:1480–1486

Álvarez OA, Jager T, Redondo EM, Kammenga JE (2006) Physiological modes of action of toxic chemicals in the nematode Acrobeloides nanus. Environ Toxicol and Chem 25(12):3230–7

Barron MG, Lilavois CR, Martin TM (2015) MOATox: a comprehensive mode of action and acute aquatic toxicity database for predictive model development. Aquat Toxicol 161:102–107

Bordbar A, Monk JM, King ZA, Palsson BO (2014) Constraint-based models predict metabolic and associated cellular functions. Nat Rev Genet 15:107–120

Claeys L, Iaccino F, Janssen CR, Van Sprang P, Verdonck F (2013) Development and validation of a quantitative structure-activity relationship for chronic narcosis to fish. Environ Toxicol Chem 32:2217–2225

Connolly RB, Ankley GT et al (2017) Quantitative adverse outcome pathways and their application to predictive toxicology. Environ Sci Technol 51(8):4661–4672

Cronin MTD, Schultz TW (2001) Development of quantitative structure-activity relationships for the toxicity of aromatic compounds to *Tetrahymena pyriformis*: comparative assessment of the methodologies. Chem Res Toxicol 14:1284–1295

Crowcombe J, Dhillon SS, Hurst RM, Egginton S, Muller F, Sik A, Tarte E (2016) 3D finite element electrical model of larval zebrafish ECG signals. PLoS One 11:e0165655

Crump KS, Chen C, Chiu WA, Louis TA, Portier CJ, Subramaniam RP, White PD (2010) What role for biologically based dose-response models in estimating low-dose risk? Environ Health Perspect 118:585–588

Ellison CM, Madden JC, Cronic MTD, Enoch SJ (2015) Investigation of the Verhaar scheme for predicting the acute aquatic toxicity. Chemosphere 139:146–154

Fujiwara M, Kendall BE, Nisbet RM, Bennett WA (2005) Analysis of size trajectory data using an energetic –based growth model. Ecology 86:1441–1451

Garcia-Serna R, Vidal D, Remez N, Mestres J (2015) Large-scale predictive drug safety: from structural alerts to biological mechanisms. Chem Res Toxicol 28:1875–1887

Gergs A, Preuss TG, Palmqvist A (2014) Double trouble at high density: cross-level test of resource-related adaptive plasticity and crowding-related fitness. PLoS One 9:e91503

Gergs A, Kulkarni D, Preuss TG (2015) Body size-dependent toxicokinetics and toxicodynamics could explain intra- and interspecies variability in sensitivity. Environ Pollut 206:449–455

Gergs A, Gabsi F, Zenker A, Preuss TG (2016) Demographic toxicokinetic-toxicodynamic modeling of lethal effect. Environ Science Technol 50:6017–6024

Ghaffari P, Mardinoglu A, Nielsen J (2015) Cancer metabolism: a modeling perspective. Front Physiol 6:382

Gillies K, Krone SM, Nagler JJ, Schultz IR (2016) A computational model of the rainbow trout hypothalamus-pituitary-ovary-liver Axis. PLoS Comput Biol 12:e1004874

Goussen B, Beaudouin R, Dutilleul M, Buisset-Goussen A, Bonzom J-M, Péry ARR (2015) Energy-based modelling to assess effects of chemicals on *Caenorhabditis elegans*: a case study on uranium. Chemosphere 120:507–514

Hanegraaf PPF, Muller EB (2001) The dynamics of the macromolecular composition of biomass. J Theor Biol 212:237–251

Henry CS, DeJongh M, Best AA, Frybarger PM, Linsay B, Stevens RL (2010) High-throughput generation, optimization and analysis of genome-scale metabolic models. Nat Biotechnol 28:977–982

Jager T (2013) All individuals are not created equal; accounting for interindividual variation in fitting life-history responses to toxicants. Environ Sci Technol 47:1664–1669

Jager TJ, Kooijman SALM (2005) Modeling receptor kinetics in the analysis of survival data for organophosphorous pesticides. Environ Sci Technol 39:8307–8314

Jager T, Selck H (2011) Interpreting toxicity data in a DEB framework: a case study for nonylphenol in the marine polychaete Capitella Teleta. J Sea Res 66:456–462

Jager T, Zimmer EI (2012) Simplified dynamic energy budget model for analysing ecotoxicity data. Ecol Model 225:74–81

Jager T, Heugens EHW, Kooijman SALM (2006) Making sense of ecotoxicological test results: towards application of process-based models. Ecotoxicology 15:305–314
Jager T, Vandenbrouck T, Baas J, De Coen WM, Kooijman SA (2010) A biology-based approach for mixture toxicity of multiple endpoints over the life cycle. Ecotoxicology 19:351–361
Jager T, Albert C, Preuss TG, Ashauer R (2011) General unified threshold model of survival – a toxicokinetic-toxicodynamic framework for ecotoxicology. Environ Sci Technol 45:2529–2540
Jager T, Ravagnan E, Dupont S (2016) Near-future ocean acidification impacts maintenance costs in sea-urchin larvae: identification of stress factors and tipping points using a DEB modelling approach. J Exp Mar Bio Ecol 474:11–17
Johnson LR, Pecquerie L, Nisbet RM (2013) Bayesian inference for bioenergetic models. Ecology 94:882–894
Joyce AR, Palsson BO (2006) The model organism as a system: integrating 'omics' data sets. Nat Rev Mol Cell Biol 7:198–210
Kauffman KJ, Prakash P, Edwards J (2003) Advances in flux balance analysis. Curr Opin Biotechnol 14:491–496
Kearney MR, Domingos T, Nisbet R (2015) Dynamic energy budget theory: an efficient and general theory for ecology. Bioscience 65:341–341
Kim J, Hayton WL, Schultz IR (2006) Modeling the brain–pituitary–gonad axis in salmon. Mar Environ Res 62:S426–S432
Klanjscek T, Nisbet RM, Priester JH, Holden PA (2012) Modeling physiological processes that relate toxicant exposure and bacterial population dynamics. PLoS One 7:e26955
Klanjscek T, Muller EB, Nisbet RM (2016) Feedbacks and tipping points in organismal response to oxidative stress. J Theor Biol 404:361–374
Kooijman SALM (2010) Dynamic energy budget theory for metabolic organization. Cambridge University Press, Cambridge
Kooijman SALM, Bedaux JJM (1996) Analysis of toxicity tests on Daphnia survival and reproduction. Water Res 30:1711–1723
Kooijman SALM, Metz JAJ (1984) On the dynamics of chemically stressed populations: the deduction of population consequences from effects on individuals. Ecotoxical Environ Safe 8:254–274
Kooijman SALM (1986) Energy budgets can explain body size relations. J Theor Biol 121:269–282
Kooijman SALM, Sousa T, Pecquerie L, Van der Meer J, Jager T (2008) From food-dependent statistics to metabolic parameters, a practical guide to the use of dynamic energy budget theory. Biol Rev 83:533–552
Kramer VJ, Etterson MA, Hecker M, Murphy CA, Roesijadi G, Spade DJ, Spromberg JA, Wang M, Ankley GT (2011) Adverse outcome pathways and ecological risk assessment: bridging to population-level effects. Environ Toxicol Chem 30:64–76
Kretschmann A, Ashauer R, Hitzfeld K, Spaak P, Hollender J, Escher BI (2011) Mechanistic toxicodynamic model for receptor-mediated toxicity of diazoxon, the active metabolite of diazinon, in *Daphnia magna*. Environ Sci Technol 45:4980–4987
Kretschmann A, Ashauer R, Hollender J, Escher BI (2012) Toxicokinetic and toxicodynamic model for diazinon toxicity-mechanistic explanation of differences in the sensitivity of *Daphnia magna* and *Gammarus pulex*. Environ Toxicol Chem 31:2014–2022
Lewis NE, Nagarajan H, Palsson BO (2012) Constraining the metabolic genotype–phenotype relationship using a phylogeny of in silico methods. Nat Rev Micro 10:291–305
Li Z, Kroll KJ, Jensen KM, Villeneuve DL, Ankley GT, Brian JV, Sepúlveda MS et al (2011) A computational model of the hypothalamic-pituitary-gonadal axis in female fathead minnows (*Pimephales promelas*) exposed to 17α-ethynylestradiol and 17β-trenbolone. BMC Syst Biol 5:63
Lika K, Kearney MR, Freitas V, van der Veer HW, van der Meer J, Wijsman JWM, Pecquerie L, Kooijman SALM (2011) The "covariation method" for estimating the parameters of the standard dynamic energy budget model I: philosophy and approach. J Sea Res 66:270–277

Margiotta-Casaluci L, Owen SF, Huerta B, Rodríguez-Mozaz S, Kugathas S, Barceló D, Rand-Weaver M, Sumpter JP (2016) Internal exposure dynamics drive the adverse outcome pathways of synthetic glucocorticoids in fish. Sci Rep 6: 21978

Martin BT, Jager T, Preuss TG, Nisbet R, Grimm V (2013) Predicting population dynamics from the properties of individuals: a cross-level test of dynamic energy budget theory. Am Nat 181:506–519

Martin B, Jager T, Nisbet RM, Preuss TG, Grimm V (2014) Limitations of extrapolating toxic effects on reproduction to the population level. Ecol Appl 24:1972–1983

Muller EB, Nisbet RM, Berkley HA (2010) Sublethal toxicant effects with dynamic energy budget theory: model formulation. Ecotoxicology 19:38–47

Muller EB, Lin S, Nisbet RM (2015) Quantitative adverse outcome pathway analysis of hatching in zebrafish with CuO nanoparticles. Environ Sci Technol 49:11817–11824

Murphy CA, Rose KA, Thomas P (2005) Modeling vitellogenesis in female fish exposed to environmental stressors: predicting the effects of endocrine disturbance due to exposure to a PCB mixture and cadmium. Reprod Toxicol 19:395–409

Murphy CA, Rose KA, Rahman MS, Thomas P (2009) Testing and applying a fish vitellogenesis model to evaluate laboratory and field biomarkers of endocrine disruption in Atlantic croaker exposed to hypoxia. Environ Toxicol Chem 28:1288–1303

Nichols JW, McKim JM, Lien GJ, Hoffman AD, Bertelsen SL (1991) Physiologically based toxicokinetic modeling of three waterborne chloroethanes in rainbow trout (*Oncorhynchus mykiss*). Toxicol Appl Pharmacol 110:374–389

Nisbet RM, Muller EB, Lika K, Kooijman SALM (2000) From molecules to ecosystems through dynamic energy budgets. J Anim Ecol 69:913–926

NRC (2007) Toxicity testing in the 21st century: a vision and a strategy. The National Academies Press, Washington, DC

Orth JD, Thiele I, Palsson BØ (2010) What is flux balance analysis? Nat Biotechnol 28:245–248

Patlewicz G, Rodford R, Walker JD (2003) Quantitative structure-activity relationships for predicting mutagenicity and carcinogenicity. Environ Toxicol Chem 22:1885–1893

Pieters BJ, Jager T, Kraak MHS, Admiraal W (2006) Modeling responses of *Daphnia magna* to pesticide pulse exposure under varying food conditions: intrinsic versus apparent sensitivity. Ecotoxicology 15:601–608

Price ND, Reed JL, Palsson BØ (2004) Genome-scale models of microbial cells: evaluating the consequences of constraints. Nat Rev Microbiol 2:886–897

Raman K, Chandra N (2009) Flux balance analysis of biological systems: applications and challenges. Brief Bioinform 10:435–449

Ramsey JC, Andersen ME (1984) A physiologically based description of the inhalation pharmacokinetics of styrene in rats and humans. Toxicol Appl Pharmacol 73:159–175

Rohart F, Gautier B, Singh A, Le Cao KA (2017) mixOmics: an R package for'omics feature selection and multiple data integration. BioRxiv 108597

Rohr JR, Salice CJ, Nisbet RM (2016) The pros and cons of ecological risk assessment based on data from different levels of biological organization. Crit Rev Toxicol 46:756–784

Shoemaker JE, Gayen K, Garcia-Reyero N, Perkins EJ, Villeneuve DL, Liu L, Doyle FJ (2010) Fathead minnow steroidogenesis: in silico analyses reveals tradeoffs between nominal target efficacy and robustness to cross-talk. BMC Syst Biol 4:89

Sousa T, Domingos T, Kooijman SALM (2008) From empirical patterns to theory: a formal metabolic theory of life. Philos Trans R Soc Lond B: Biol Sci 363:2453–2464

Sullivan K, Manuppello JR, Willett CE (2014) Building on a solid foundation: SAR and QSAR as a fundamental strategy to reduce animal testing. SAR QSAR Environ Res 25:357–365

Sundling K, Craciun G, Schultz I, Hook S, Nagler J, Cavileer T et al (2014) Modeling the endocrine control of vitellogenin production in female rainbow trout. Math Biosci Eng 11:621

ten Tusscher KH, Noble D, Noble PJ, Panfilov AV (2004) A model for human ventricular tissue. Am J Physiol Heart Circ Physiol 286:H1573–H1589

Van Aggelen G, Ankley GT, Baldwin WS, Bearden DW, Benson WH, Chipman JK et al (2010) Integrating omic technologies into aquatic ecological risk assessment and environmental monitoring: hurdles, achievements, and future outlook. Environ Health Perspect 118:1–5

Veltman K, Hendriks AJ, Huijbregts MA, Wannaz C, Jolliet O (2014) Toxicokinetic toxicodynamic (TKTD) modeling of ag toxicity in freshwater organisms: whole-body sodium loss predicts acute mortality across aquatic species. Environ Sci Technol 48:14481–14489

Verhaar HJM, van Leeuwen CJ, Hermens JLM (1992) Classifying environmental pollutants. 1. Structure activity relationships for prediction of aquatic toxicity. Chemosphere 25:471–491

Villaverde AF, Ross J, Morán F, Banga JR (2014) MIDER: network inference with mutual information distance and entropy reduction. PLoS One 9:e96732

Villeneuve DL, Crump D, Garcia-Reyero N, Hecker M, Hutchinson TH, LaLone CA, Landesmann B, Lettieri T, Munn S, Nepelska M, Ottinger MA (2014) Adverse outcome pathway (AOP) development I: strategies and principles. Toxicol Sci 142:312–320

Watanabe KH, Li Z, Kroll KJ, Villeneuve DL, Garcia-Reyero N, Orlando EF et al (2009) A computational model of the hypothalamic-pituitary-gonadal axis in male fathead minnows exposed to 17α-ethinylestradiol and 17β-estradiol. Toxicol Sci p.kfp069 109(2):180–192

Watanabe KH, Aladjov H, Bell SM, Burgoon L, Cheng W-Y, Conolly R, Edwards SW et al (2017) Big data integration and inference. In: Neagu D, Richarz A (eds) Big data in predictive toxicology. The Royal Society of Chemistry, Cambridge

Williams TD, Turan N, Diab AM, Wu H, Mackenzie C, Bartie KL et al (2011) Towards a system level understanding of non-model organisms sampled from the environment: a network biology approach. PLoS Comput Biol 7:e1002126

Yizhak K, Chaneton B, Gottlieb E, Ruppin E (2015) Modeling cancer metabolism on a genome scale. Mol Syst Biol 11:817

Zoppoli P., Morganella S., Ceccarelli M. (2010) An Information Theoretic Approach to Reverse Engineering of Regulatory Gene Networks from Time–Course Data. In: Masulli F., Peterson L.E., Tagliaferri R. (eds) Computational Intelligence Methods for Bioinformatics and Biostatistics. CIBB 2009. Lecture Notes in Computer Science, vol 6160. Springer, Berlin, Heidelberg

Chapter 15
Weight of Evidence Frameworks in Evaluation of Adverse Outcome Pathways

Taylor Rycroft, Olivia Massey, Christy M. Foran, and Igor Linkov

Abstract Weight of Evidence (WoE) is an assessment mechanism used to systematically consider a collection of scientific data that addresses a specific hypothesis. WoE frameworks are used to help form a reasonable conclusion based on all available information, and are commonly utilized in risk assessment. They have recently been applied to toxicological assessments that seek to understand Adverse Outcome Pathways (AOPs), or the cascade of physiological events that link toxicant exposure to a downstream adverse health outcome. In case studies, WoE methods have proven useful in assessing AOPs and estimating pathway-based risk. However, WoE approaches vary considerably and have received criticism for their lack of transparency, reproducibility, and quantitative rigor. The subjective nature of qualitative WoE constructs has led to a push for a quantitative methodology that is consistent, objective, and robust. The purpose of this chapter is to provide a brief background on WoE methodology and its historical use in AOP discovery, as well as highlight progress in the development of a standardized quantitative WoE framework. An example of a newly proposed standardized quantitative WoE framework for AOPs is discussed, and gaps and suggested improvements are examined in order to identify next steps towards making quantitative WoE methods for AOPs more objective, transparent, and reproducible.

T. Rycroft • O. Massey
Contractor to the Environmental Laboratory, U.S. Army Engineer Research and Development Center, Concord, MA, USA

C.M. Foran (✉) • I. Linkov
Environmental Laboratory, U.S. Army Engineer Research and Development Center, Concord, MA, USA
e-mail: christy.m.foran@usace.army.mil

N. Garcia-Reyero, C.A. Murphy (eds.), *A Systems Biology Approach to Advancing Adverse Outcome Pathways for Risk Assessment*,
DOI 10.1007/978-3-319-66084-4_15

15.1 Introduction

Weight of Evidence (WoE) is an assessment mechanism used across an array of disciplines to systematically consider a collection of scientific data that addresses a specific hypothesis (Linkov et al. 2011, 2015; Gough 2007). Commonly used in risk assessment as a method of interpreting epidemiological and toxicological data, WoE evaluations seek to summarize multiple lines of evidence (LoE) and form a reasonable conclusion based on all available information (Weed 2005). Typically, each scientific data source (e.g. a scientific paper) serves as an independent LoE that either supports or opposes the hypothesis, and the relevance and quality of each LoE is scored using a set of criteria. LoE scores are then integrated into an aggregate WoE that provides the justification for a final conclusion about the hypothesis.

WoE frameworks have been utilized in toxicological assessments that seek to understand the cascade of physiological events that link toxicant exposure to a downstream adverse health outcome, a sequence known as an Adverse Outcome Pathway (AOP) (Ankley et al. 2010). Once each element in an AOP's chain of events has been proposed, WoE techniques are incorporated to help evaluate the degree of certainty for each individual linkage and the overall pathway. This structured analysis ultimately aids an assessor in deciding whether sufficient evidence exists to confidently describe *how* a substance causes harm to humans or the environment. By systematically scoring and comparing proposed linkages and pathways, AOP development using WoE can help predict potential hazards and inform regulatory measures for substances with limited toxicology data.

Establishing a toxicity profile for a substance using traditional in vivo toxicology studies is time- and resource-intensive (Krewski et al. 2010). As the library of synthetic chemicals continues to expand, AOPs have emerged as a novel alternative for mechanistically assessing a wide array of substances with limited toxicity data (Villeneuve et al. 2014a, b). Simply, an AOP is a map that connects Key Events (KEs), or changes in biological state that are measurable and essential to the progression of a defined biological disturbance (Vinken et al. 2013). As shown in Fig. 15.1, the sequence begins with a molecular initiating event (MIE), or the initial point of interaction between the stressor and the biological receptor within the organism, and advances through a string of higher order biological events – the KEs – culminating in the adverse outcome (AO). KEs are connected to one another via linkages defined as Key Event Relationships (KERs) (Becker et al. 2015).

While useful in assessing AOPs and estimating pathway-based risk, WoE methods vary considerably and have received criticism for their lack of transparency, reproducibility, and quantitative rigor. The very definition of WoE is inconsistent and ambiguous in the literature. Researchers commonly use the term in a metaphorical sense – where "weight of evidence" is taken to mean that a collection of studies supports a hypothesis – but provide no further explanation of the specific methodology used to weigh each line of evidence (Weed 2005). Current WoE practice mostly underutilizes quantitative methods, and risk assessors rely heavily on qualitative WoE methods to make a decision, combining empirical evidence with

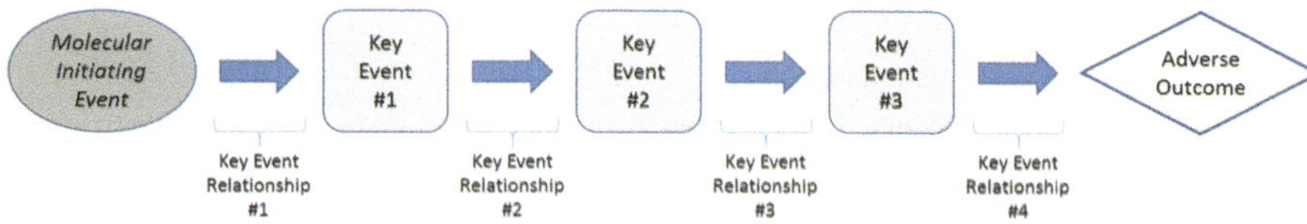

Fig. 15.1 Adverse outcome pathway (AOP) (Adapted from OECD 2014)

expert judgment (Linkov et al. 2011, 2015). The subjective nature of qualitative WoE constructs has led to a push for a quantitative methodology that is consistent, objective, and robust.

The purpose of this chapter is to provide a brief background on WoE methodology and its historical use in toxicological assessments and AOP discovery. Additionally, this section aims to highlight progress in the development of a consistent quantitative WoE framework, as well as underscore gaps and propose potential improvements that can make quantitative WoE methods more objective, transparent, and reproducible.

15.2 WoE Approaches

Contemporary WoE methodologies have advanced with the evolution of statistical science. In the 1960s, it was proposed that WoE processes should follow an inherently Bayesian statistical approach in which "prior" beliefs for or against a particular hypothesis are updated after evaluation of information or evidence in order to achieve a "posterior" belief (Good 1960). Regulatory frameworks of WoE for AOPs, however, have not historically utilized Bayesian techniques. Reviews conducted in 2005 and 2009 (Weed 2005; Linkov et al. 2009) found WoE methods to be primarily qualitative in nature with a general lack of agreement in methodology. In 2014, the National Research Council (NRC) reviewed EPA's Integrated Risk Information System (IRIS) assessments of formaldehyde and methanol and found a primary "challenge" faced during review was determining "how the phrase weight of evidence is used by EPA and others" (National Research Council 2014). The NRC declared the subjective nature of WoE in its historical application to be "far too vague" and "of little scientific use." In response, a number of WoE models tailored to AOPs were set in rigid frameworks and presented (Collier et al. 2016). These models were more quantitative and deemed less subjective than either traditional WoE methods or the methods recommended by NRC to be used in place of WoE.

Historically, a number of WoE applications have been applied to hazard assessments. Linkov et al. performed a literature review of 114 papers in which WoE methods were employed and described the range of approaches and categorized them by their quantitative rigor (shown in Table 15.1) (Linkov et al. 2009).

Listing evidence is the simplest WoE application and the only method that does not attempt to integrate the individual LoEs. Instead, assessors present LoEs for or

Table 15.1 Categories of existing WoE methods

WoE method	Explanation	Rigor
Listing evidence	Single LoEs presented without integration or weighting	Least quantitative
Best professional judgment	Lines of evidence qualitatively integrated without numerical evaluation	↓
Causal criteria	Cause and effect relationships evaluated using criteria-based methods	
Logic construct	Qualitative logic methods provide standardized evaluation of single LoEs	
Scoring	Simple weighting or ranking techniques integrate LoEs quantitatively	
Indexing	Empirical models integrate LoEs into a single value	
Quantification	Mathematical models used to weigh the body of evidence	Most quantitative

Adapted from Linkov et al. (2009)

against a certain hypothesis but do not draw conclusions or discuss the lines in detail. A major drawback of this approach is that by omitting a conclusion with a logical justification, readers may reach different determinations depending on their unique interpretation of the evidence (Linkov et al. 2011).

Best Professional Judgment utilizes authoritative, professional opinions to integrate the LoEs and form a conclusion. To justify a decision, the subject matter expert may offer background from previous experiences that are similar to the case at hand. There is inherent subjectivity and risk of bias with this approach, but estimations can be made more rigorous if additional expert opinions are solicited and convergence towards a common decision emerges.

Causal criteria and **logic constructs** both employ best professional judgement to combine LoEs but are considered more rigorous and transparent because they specify a consistent analytical structure that is reproducible for future studies. The method of **causal criteria** involves using a preset structure to evaluate cause and effect relationships. The Bradford Hill causal criteria (Hill 1965) generally provide this structure, but assessors have also formulated their own case-specific causal criteria. Generally, an assessor provides an overview of the criteria used for evaluating causality and then proceeds to describe evidence that the criteria have been fulfilled, thus "proving" the cause and effect relationship.

Assessors also use a standard set of criteria or an agreed protocol when applying **logic constructs.** With this method, the set of criteria against which evidence is measured is more constrained than in a causal criteria assessment because the logic construct follows a firm set of guidelines or best practices (often prescribed by an agency or professional organization) with little flexibility. For example, if exposure to a toxicant was the first KE in an AOP and the toxicant was not present at the site of concern, then a logic method would determine no risk of adverse outcome but a causal criteria assessment would still consider episodic exposures or other factors that could lead to toxicant exposure. Causal criteria and logic constructs are more mechanistic and systematic than listing evidence and best professional judgement, but they are still inherently subjective and qualitative.

The **scoring** and **indexing** methods both assign scores to individual LoEs and are of similar quantitative strength. Multiple **scoring** methods exist, but most assign weights based on best professional judgement for characteristics of the evidence such as consistency, specificity, or strength of association. The assigned weight represents the relative importance of the characteristic or criterion, and the score is an assessment of the LoE's performance on that criterion. Once a score is determined, individual LoEs can be compared to one another or summed into categories that help the assessor reach a conclusion. **Indexing** differs from scoring in that it aims to roll-up the scores from the LoEs into a single overarching confidence score for the AOP. This confidence score is then compared to a predetermined threshold value that informs the conclusion. Despite their inherent quantitative nature, both scoring and indexing are of limited value because they fail to quantify judgments in a systematic way, such as with formal probability techniques or decision analysis strategies, and therefore lack reproducibility and transparency.

Quantitative WoE frameworks are the most rigorous of the range of methods and make use of established mathematical models to evaluate the body of evidence and project the likelihood and uncertainty of a conclusion. Probability distributions, correlations, and other statistical methods are employed under strict guidelines, making these frameworks transparent and reproducible in nature. While large amounts of data are necessary for quantitative WoE frameworks to project useful estimations, these techniques are essential when analyzing complex systems with interrelated LoEs. Examples of such methods are multi-criteria decision analysis (MCDA) and Bayesian modeling. MCDA is a set of methods designed to ensure that the synthesis of multiple sources of information is documented and directed toward a stated goal (Linkov et al. 2006). The MCDA approach uses the evaluation and integration of subjective priorities to compare alternatives; in the context of AOP it has been proposed as a way to characterize the relative confidence in KEs within the same AOP (Collier et al. 2016) or in alternative AOPs between a MIE and AO (Becker et al. 2017). Bayesian modeling focuses on updating probability distributions based on prior information. While Bayesian models are the most rigorous WoE method, practical applications are often limited by data availability (National Research Council (NRC); Committee to Review the IRIS Process; Board on Environmental Studies and Toxicology; Division on Earth and Life Studies 2014).

Not all WoE evaluations for AOPs require intensive quantitative analysis. When determining the quantitative rigor of the WoE method, Linkov, et al. suggests the general rule of selecting "the most quantitative method that satisfies the problem scale and complexity." This allows one to achieve the most robust possible decision and gain access to the benefits of employing quantitative techniques while ensuring a valid conclusion (Linkov et al. 2011).

15.3 Trend Towards a Standardized Quantitative WoE Framework for Comparing AOPs

In response to the critiques that traditional, more-qualitative, WoE methods lack rigor, transparency and reproducibility, the regulatory science community is moving towards a standard framework for quantitative, comparative WoE assessments that can be used consistently by AOP developers and risk assessors (OECD 2013; Becker et al. 2017). Such a framework is intended to enable assessors to better convey the sufficiency of the scientific evidence that supports a hypothesized AOP. This sufficiency, in turn, can be compared to that of other hypothesized AOPs and can provide justification for selecting one AOP alternative over another in order to answer a specific research question. Depending on the assessor's objective, the hypothesized AOP that is best supported by the scientific evidence may be used to inform efforts to limit the occurrence of the AO, such as regulation of the chemical that induced the MIE or drug development to disrupt the pathway, or simply to direct further toxicology research toward a particular KE.

In a step towards improving the consistency of AOP assessment, the World Health Organization (WHO), through its International Program on Chemical Safety (IPCS), has developed the *Mode of Action Framework* and *Human Relevance Framework* (Boobis et al. 2006), and the Organization for Economic Cooperation and Development (OECD) has designed the *Users' Handbook Supplement to the Guidance Document for Developing and Assessing AOPs* (OECD 2014). The OECD guidance promotes increased consistency for WoE determinations by prompting an assessor to answer defining questions about the underlying LoEs and to measure the empirical support as "strong," "moderate," or "weak" based on general definitions.

One example of a newly proposed standardized quantitative WoE framework that augments the structures set forth by WHO/IPCS and OECD is the concept of *quantitative confidence scoring*. This method employs tailored Bradford-Hill (BH) considerations (Becker et al. 2015) to compare the degree of confidence in hypothesized AOPs. Briefly, the recommended procedure for quantitative confidence scoring is as follows (adapted from Becker et al. 2017):

1. From a supported MIE to AO relationship, hypothesize multiple AOP alternatives and their associated KEs and KERs.
2. Qualitatively evaluate the scientific evidence relevant to each AOP alternative.

3. Quantitatively score the strength of evidence for each KE in the AOP path using tailored BH considerations for causality.
4. Derive the weighted score for each KE by multiplying the quantitative rating of the evidence for the KE by the weight of each BH criterion.
5. Derive the overall "confidence score" for the AOP by summing the scores of the KEs and dividing by the total possible score.
6. Compare the resultant confidence scores of the AOP alternatives.

The tailored BH considerations (Meek et al. 2014a, b) referenced in step three are a truncated subset of the initial list of nine BH criteria for causation (Hill 1965). Deemed the most relevant criteria for use in AOP development, the following five considerations were selected for the quantitative confidence scoring method:

- *Biological Plausibility*, which answers the question "Are the KEs accepted by the scientific community to be consistent with biological understanding?"
- *Essentiality*, which answers the question "If the upstream event is blocked, will the downstream event still occur?"
- *Dose response and temporal concordance*, which answers the question "does the KE occur at a dose lower than the AO, and does the KE occur in the proposed chronology?"
- *Consistency*, which answers the question "do the observed events occur in other test systems?"
- *Analogy*, which answers the question "do the observed events occur for a broader set of substances?"

Each of the five criteria have associated defining questions that are posed for the collective body of evidence and a qualitative score – "strong," "moderate," or "weak" – is assigned to the evidence. This score is translated to a quantitative rating from −3 to 3, where a larger absolute value represents stronger evidence and a negative rating indicates counter-evidence for the relationship.

In step four of the quantitative confidence scoring process, the BH considerations are given a numerical weight based on their perceived importance. The weight of each criteria may vary depending on the assessor and the context of the assessment, but an initial proposal in the regulatory science community is to assign 40% weight of importance to both *essentiality* and *dose response and temporal concordance* and 10% weight of importance to both *consistency* and *analogy*. While *biological plausibility* has been recognized as a critical BH criterion, its weight of importance has not been proposed and requires additional consideration (Becker et al. 2017). Scores for a KE are then derived by multiplying the weight of the BH consideration by the evidence ranking to get a score for that specific criteria (Fig. 15.2). The overall score for the KE is derived by summing the scores from each of the five BH criteria. Summing the scores for the individual KEs and dividing by the maximum possible score establishes the "confidence score" for the hypothesized AOP, which can be used to compare the strength of the AOP to other hypothesized AOPs.

Although still in its proof of concept stage, quantitative confidence scoring represents a practical and reliable approach for standardizing quantitative WoE analysis

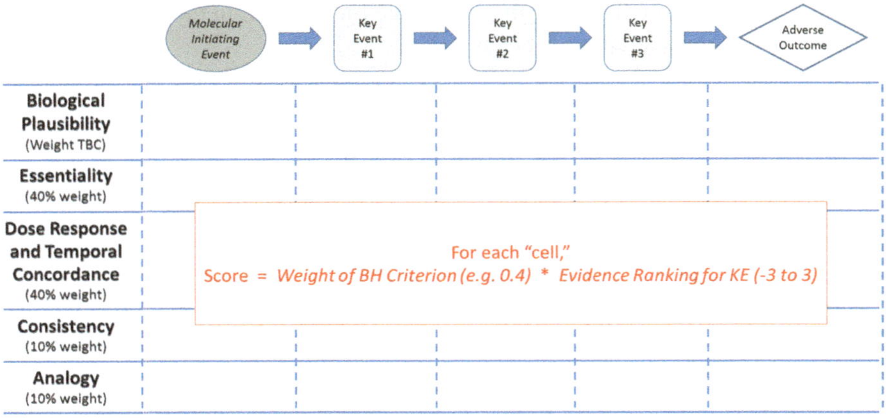

Fig. 15.2 Example WoE scoring using Tailored BH considerations (Adapted from Becker et al. 2017)

in AOP development. Compared to qualitative methods, a framework like quantitative confidence scoring allows for a structured assessment that results in a numerical representation of the degree of scientific confidence in an AOP. The final confidence score allows an assessor to determine the best-supported AOP relative to other hypothesized AOPs for the same MIE-AO relationship, highlight uncertainty, and clearly communicate findings to a non-technical audience. An additional benefit of a quantitative WoE framework is that it allows for sensitivity analyses. By changing criteria weighting schemes, an assessor can determine the sensitivity of the WoE for each KE which can be used to highlight areas where additional research would be most impactful to developing a greater degree of confidence in the AOP (Becker et al. 2015).

15.4 Potential Improvements in WoE Approaches

As with any maturing methodology, certain considerations should be addressed to improve the robustness and increase the adoption of a standardized quantitative WoE framework. There are three integral elements of a standard WoE framework that can be refined and highlighted by assessors to ensure relevant and consistent outcomes: the grading of LoEs (including the incorporation of new evidence), the context of WoE analysis, and the sufficiency of the AOP. Acknowledging and addressing these factors will, respectively, ensure the evaluation framework is transparent and repeatable and allow revision as new studies are completed, provide material evidence to justify a decision, and provide empirical support for the AOP.

15.4.1 Grading of Evidence

Current quantitative WoE frameworks require a *qualitative* evaluation of the collected scientific evidence relevant to the AOP. This collective evaluation of evidence informs the assignment of scores for individual KEs, such as "strong" (S), "moderate" (M), or "weak" (W). Utilization of expert judgement to complete this task is the most subjective component of the WoE analysis because two expert assessors may reach different conclusions as to which descriptor is appropriate (S, M, or W) when presented with the same set of evidence. The OECD identified this limitation and attempted to limit variability in evidence scoring by providing refined guidance for deriving evidence descriptors (OECD 2014). Even if a consistently measurable rubric for scoring the scientific evidence is applied, the relevant, collected scientific evidence that supports or refutes the AOP is constantly changing.

Each scientific study should serve as a LoE, and each LoE should be evaluated *individually* when deriving the score for the KE, rather than collectively. The U.S. Environmental Protection Agency (EPA) adopted this approach in its final version of *Guidelines for Carcinogen Risk Assessment*, advocating for an assessment of all of the individual LoEs followed by a single integrative step to achieve the overall WoE for a cancer determination (US EPA Risk Assessment Forum 2005). This improvement from the original EPA *Guidelines* allows each LoE to be measured objectively before the importance of that measurement is considered.

To evaluate individual LoEs, "rules of evidence," in the form of a consistent rubric, should be defined that state how quality, technical merit, and applicability should be graded for each LoE. This way, assessors are guided toward the same "strong," "moderate," or "weak" conclusion about a piece of evidence. The challenge with guiding assessors to the same objective conclusion about an individual LoE rather than a subjective conclusion about the overall collection of evidence, however, is that a standard scoring rubric must be developed and accepted. Currently, no standard LoE scoring rubric exists, although many AOP assessors use similar criteria to measure the reliability and quality of LoEs. Juberg et al. (2013) demonstrated how individual LoEs in an AOP evaluation can be evaluated for reliability and quality using the Klimisch criteria (Klimisch et al. 1997) and the Toxicological Data Reliability Assessment Tool (Schneider et al. 2009). By measuring each LoE against these guidelines, the authors were able to establish a more objective and thorough WoE narrative for the AOP, increasing the likelihood that an assessor seeking to reproduce their results would reach the same conclusions (Juberg et al. 2013). Despite their proven utility in the assessment of in vivo and in vitro toxicology studies, neither the Klimisch criteria nor the ToxRTool serve as a universal standard in LoE evaluation for AOPs.

To establish a standard LoE scoring rubric for AOPs, all AOP assessors must agree on the attributes that define a quality study (e.g. extent of peer review, use of GLP or an accepted testing guideline, clarity and reproducibility, etc.). These attributes will form the backbone of the consistent rubric. Assuming agreement can be reached, the rubric must also contain an element that evaluates how well the study applies to the KE being scored (within the context of the specific AOP being assessed) and the specific criteria (e.g. BH criteria) that it supports or refutes. All assessors – regulatory scientists, chemical screeners, etc. – should be able to be easily trained in using this standard rubric, and user testing must show that the same "strong", "moderate," or "weak" conclusion is consistently achieved for the individual LoE. Integration of components from each of the aforementioned assessment criteria - the OECD guidance for deriving WoE descriptors, the Klimisch criteria, the ToxRTool - and several evaluative criteria utilized by government agencies such as the General Assessment Factors established by the EPA (US EPA Science Policy Council 2003; Collier et al. 2016) may serve as the logical foundation for constructing an AOP-specific LoE scoring rubric.

Once scored using the standard rubric, LoEs should be systematically combined into an overall evaluation for the collective evidence (the WoE), rather than qualitatively aggregated, so that assessors reach the same objective conclusions about the WoE for a given KE. A variety of methods have been used for this systematic aggregation process, and which method is appropriate depends on the way(s) the output will be used. A review by Burton et al. (2002) analyzed eight WoE approaches for their advantages and disadvantages and identified *Tabular Decision Matrices* as being more robust, sensitive, and transparent than qualitative combination (Burton et al. 2002). Using a tabular decision matrix, individual LoEs and their respective scores can be arranged as line items in a table and a logic model can be applied that rapidly computes the WoE. Logic for combining LoE scores into an aggregate WoE may take the form of a simple summation or an average, where the output value falls into a spectrum of "strong," "moderate," or "weak" evidence categories. Alternatively, a threshold number of studies may be preferred, such that as long as two "strong" studies are present in the set of LoEs, the overall WoE can be considered "strong." The number of studies must also be factored into this assessment, so that, for example, four studies with an average score of "medium" would carry a stronger WoE than a single "medium" study. Logic constructs for aggregating LoEs should be tested using AOP case studies and a preferred method should be recommended for adoption as the standard that will facilitate the consistent grading of evidence for AOPs.

An additional advantage to scoring LoEs separately rather than as a collective is that assessors can easily and transparently add newly published scientific information and recalculate an existing AOP score. The above-described systematic method of combining LoEs allows an assessor to insert a new LoE into the WoE calculation without having to subjectively reevaluate the collection of LoEs. Infrastructure that supports the incorporation of new evidence has already been constructed in the form of an AOP Wiki (AOP-Wiki 2014), a knowledgebase built by the EPA and sanctioned by the OECD. AOPs are submitted to the AOP Wiki where they undergo OECD approval. Leveraging this AOP repository, a wiki data manager could monitor suggested changes to an *approved* AOP and bring stakeholders – those that developed the AOP and those that identified the new relevant LoE – together to gain consensus for integration of the new data. The modified AOP could then be re-published to the wiki for collaborative discussion, peer-review, and OECD re-approval.

15.4.2 Context of WoE Analyses

The LoE scoring rubric and aggregation method suggested above is designed to steer all assessors to a consistent conclusion of strength of the WoE. The implications of the WoE conclusion may be different across assessors, however, and the application of different criteria weight schemes may lead to variation in AOP scores despite the same underlying evidence. For example, for the same WoE score (S, M, W), a regulatory scientist may not be convinced that a criterion is fulfilled while a pharmaceutical researcher may feel they have enough evidence to continue exploring the KE. The same context-dependence applies when evaluating the complete AOP. A weight scheme assigned to the criteria (such as the 40%, 40%, 10%, 10% weighting scheme assigned to the BH criteria in the quantitative confidence scoring example) may make sense for regulatory scientists but may not accurately portray the priorities of a pharmaceutical researcher trying to infer toxicity from one cellular receptor to another.

Such context-driven flexibility in criteria weighting is beneficial at this stage of the AOP analysis because it empowers a variety of stakeholders to engage in AOP evaluation by providing the freedom to adjust the weights of criteria according to their unique priorities. The advantage to enabling broader stakeholder participation in AOP assessment is that it may lead to increased KE discovery by bringing together a greater knowledge of chemicals and their effects, a chief objective of the OECD AOP Knowledge Base project (OECD 2016). The disadvantage to a non-standardized weight scheme is that it restricts comparison of AOP scores to assessments that use the identical weight scheme. The same kind of flexibility has been necessary in applying WoE in the context of the Sediment Quality Triad (SQT; Chapman et al. 2002). As adoption of a standardized WoE framework increases, discipline-specific (e.g. pharmaceutical science, regulatory science, etc.) weighting schemes may emerge, allowing for broader AOP score interpretation. Until then, AOP assessors should always highlight the weight scheme used and should only compare AOP scores derived from the same weight scheme (Collier et al. 2016).

15.4.3 Sufficiency of the AOP

The AOP concept addresses questions pertaining to whether and how a particular MIE can cause an AO (Ankley et al. 2010). In utilizing a standardized WoE framework to analyze AOPs, an assessor seeks a deeper understanding of the AOP, asking *what is known about the MIE-AO relationship, and is there sufficient scientific evidence to justify proceeding with my research objective?* The weight scheme applied and the degree of evidence deemed "sufficient" depends on the unique objectives of the assessor and the research question posed, and greater confidence is required for applications with greater potential impact (Meek et al. 2011). For a pharmaceutical researcher, the research objective may be to develop a therapy that disrupts a link in the AOP chain. An AOP evaluation using the standardized WoE framework can point the researcher to the priority AOP alternative to allocate R&D resources towards. For a regulatory scientist at a governing body, the objective may be to impose protective restrictions in an effort to prevent exposure to the chemical that prompted the MIE. An AOP evaluation using the standardized WoE framework can help a regulator substantiate claims of hazard and causality and identify areas where more research would bolster their claim. In both applications, a standardized quantitative WoE framework for AOP assessment is best suited for research questions that can be answered with comparative analysis (Becker et al. 2017), where multiple hypothesized AOP alternatives are evaluated using a standard method and their resulting scores are compared. If the research question calls for an absolute characterization of the degree of empirical evidence, the standardized quantitative WoE framework falls short because the overall score for an AOP cannot be interpreted without context.

While AOPs provide a useful simplification of intricate and complex biochemical processes, they are not intended to represent the complete set of interactions that take place when a system responds to a chemical stimulus. A standardized quantitative WoE framework can help identify the AOP that is best supported by current scientific knowledge, but it cannot confirm the magnitude of the influence of that pathway. There may be several branches that connect the same MIE to the same AO (Villeneuve et al. 2014a), with one branch serving as the most influential pathway. Inhibition of that operative pathway, however, may not prevent the AO from occurring via secondary pathways. The reality that there may be multiple important pathways must be emphasized when assessors present results of AOP comparisons.

Finally, there must be acknowledgment of the tenet *a chain is only as strong as its weakest link*. Like other biological sequences of events, demonstrating that the majority of individual links in the AOP chain are strong does not imply that the entire chain is strong (Rosenblum et al. 1996). A single KE with weak empirical support indicates that the AOP requires either more research (Collier et al. 2016) or an adjustment to the pathway, such as a substitution of the weak KE or insertion of a pathway around it. It is important for assessors to communicate the restrictions and limitations of their evaluations so audiences do not misinterpret conclusions and implications of AOP discovery.

15.5 Concluding Remarks

In summary, a standardized quantitative WoE framework can provide a consistent method for evaluating scientific evidence relevant to AOPs. Although proposed frameworks like quantitative confidence scoring are still novel and relatively untested, their application holds significant promise. AOPs developed as proofs of concept have shown that a standardized quantitative WoE framework improves upon traditional qualitative methods by enhancing the rigor, transparency, and reproducibility of WoE determinations. Additionally, quantitative methods excel at identifying areas where further research would have the greatest impact on an AOP's score. A standardized quantitative WoE framework can help justify management decisions across fields such as toxicology research, drug development, and regulatory science, and can be refined and improved as it becomes a more commonly employed element of the analytical toolbox used by risk assessors and AOP developers.

References

Ankley GT, Bennett RS, Erickson RJ, Hoff DJ, Hornung MW, Johnson RD, Mount DR, Nichols JW, Russom CL, Schmieder PK (2010) Adverse outcome pathways: a conceptual framework to support ecotoxicology research and risk assessment. Environ Toxicol Chem 29(3):730–741

AOP-Wiki (2014) Users' handbook supplement to the guidance document for developing and assessing AOPs. From https://aopkb.org/common/AOP_Handbook.pdf

Becker RA, Ankley GT, Edwards SW, Kennedy SW, Linkov I, Meek B, Sachana M, Segner H, Van Der Burg B, Villeneuve DL (2015) Increasing scientific confidence in adverse outcome pathways: application of tailored Bradford-Hill considerations for evaluating weight of evidence. Regul Toxicol Pharmacol 72(3):514–537

Becker RA, Dellarco V, Seed J, Kronenberg JM, Meek B, Foreman J, Palermo C, Kirman C, Linkov I, Schoeny R, Dourson M (2017) Quantitative weight of evidence to assess confidence in potential modes of action. Regul Toxicol Pharmacol 86:205–220

Boobis AR, Cohen SM, Dellarco V, McGregor D, Meek ME, Vickers C, Willcocks D, Farland W (2006) IPCS framework for analyzing the relevance of a cancer mode of action for humans. Crit Rev Toxicol 36(10):781–792

Burton GA, Chapman PM, Smith EP (2002) Weight-of-evidence approaches for assessing ecosystem impairment. Hum Ecol Risk Assess 8(7):1657–1673

Chapman PM, McDonald BG, Lawrence GS (2002) Weight-of-evidence issues and frameworks for sediment quality (and other) assessments. Hum Ecol Risk Assess 8(7):1489–1515

Collier ZA, Gust KA, Gonzalez-Morales B, Gong P, Wilbanks MS, Linkov I, Perkins EJ (2016) A weight of evidence assessment approach for adverse outcome pathways. Regul Toxicol Pharmacol 75:46–57

Good IJ (1960) Weight of evidence, corroboration, explanatory power, information and the utility of experiments. J R Stat Soc Series B (Methodol) 22(2):319–331

Good, I. J. (1991). Weight of evidence and the Bayesian likelihood ratio. The use of statistics in forensic science, 85–106.

Gough D (2007) Weight of evidence: a framework for the appraisal of the quality and relevance of evidence. Res Pap Educ 22(2):213–228

Hill AB (1965) The environment and disease: association or causation? Proc R Soc Med 58(5):295

Juberg DR, Gehen SC, Coady KK, LeBaron MJ, Kramer VJ, Lu H, Marty MS (2013) Chlorpyrifos: weight of evidence evaluation of potential interaction with the estrogen, androgen, or thyroid pathways. Regul Toxicol Pharmacol 66(3):249–263

Klimisch H-J, Andreae M, Tillmann U (1997) A systematic approach for evaluating the quality of experimental toxicological and ecotoxicological data. Regul Toxicol Pharmacol 25(1):1–5

Krewski D, Acosta D Jr, Andersen M, Anderson H, Bailar JC III, Boekelheide K, Brent R, Charnley G, Cheung VG, Green S Jr (2010) Toxicity testing in the 21st century: a vision and a strategy. J Toxicol Environ H, Part B 13(2–4):51–138

Linkov I, Satterstrom F, Kiker G, Batchelor C, Bridges T, Ferguson E (2006) From comparative risk assessment to multi-criteria decision analysis and adaptive management: recent developments and applications. Environ Int 32(8):1072–1093

Linkov I, Loney D, Cormier S, Satterstrom FK, Bridges T (2009) Weight-of-evidence evaluation in environmental assessment: review of qualitative and quantitative approaches. Sci Total Environ 407(19):5199–5205

Linkov I, Welle P, Loney D, Tkachuk A, Canis L, Kim J, Bridges T (2011) Use of multicriteria decision analysis to support weight of evidence evaluation. Risk Anal 31(8):1211–1225

Linkov I, Massey O, Keisler J, Rusyn I, Hartung T (2015) From "weight of evidence" to quantitative data integration using multicriteria decision analysis and Bayesian methods. ALTEX 32(1):3

Meek ME, Boobis AR, Crofton KM, Heinemeyer G, Raaij MV, Vickers C (2011) Risk assessment of combined exposure to multiple chemicals: a WHO/IPCS framework. Regul Toxicol Pharmacol 60(Suppl 1):S1–S14

Meek M, Boobis A, Cote I, Dellarco V, Fotakis G, Munn S, Seed J, Vickers C (2014a) New developments in the evolution and application of the WHO/IPCS framework on mode of action/species concordance analysis. J Appl Toxicol 34(1):1–18

Meek M, Palermo CM, Bachman AN, North CM, Jeffrey Lewis R (2014b) Mode of action human relevance (species concordance) framework: evolution of the Bradford Hill considerations and comparative analysis of weight of evidence. J Appl Toxicol 34(6):595–606

National Research Council (NRC); Committee to Review the IRIS Process; Board on Environmental Studies and Toxicology; Division on Earth and Life Studies (2014) Review of EPA's Integrated Risk Information System (IRIS) process. National Academies Press (US), Washington, DC

OECD (2013). Guidance document on developing and assessing adverse outcome pathways. Series on testing and assessment, No. 184, Vol. ENV/JM/MONO(2013)6. Organisation for Economic Cooperation and Development, Environment Directorate, Paris

OECD (2014) Users' handbook supplement to the guidance document for developing and assessing AOPs [ENV/JM/MONO(2013)6]. From: https://aopkb.org/common/AOP_Handbook.pdf

OECD (2016) Adverse outcome pathways, molecular screening and toxicogenomics. www.oecd.org/chemicalsafety/testing/adverse-outcome-pathways-molecular-screening-and-toxicogenomics.htm. Accessed 4 Oct 2016

Rosenblum JS, Gilula NB, Lerner RA (1996) On signal sequence polymorphisms and diseases of distribution. Proc Natl Acad Sci 93(9):4471–4473

Schneider K, Schwarz M, Burkholder I, Kopp-Schneider A, Edler L, Kinsner-Ovaskainen A, Hartung T, Hoffmann S (2009) "ToxRTool", a new tool to assess the reliability of toxicological data. Toxicol Lett 189(2):138–144

US EPA Risk Assessment Forum (2005) Guidelines for Carcinogen risk assessment. EPA/630/P--03/001F. https://www.epa.gov/sites/production/files/2013-09/documents/cancer_guidelines_final_3-25-05.pdf

US EPA Science Policy Council (2003) A summary of general assessment factors for evaluating the quality of scientific and technical information. Science Policy Council, U.S. Environmental Protection Agency, Washington, DC

Villeneuve DL, Crump D, Garcia-Reyero N, Hecker M, Hutchinson TH, LaLone CA, Landesmann B, Lettieri T, Munn S, Nepelska M, Ottinger MA (2014a) Adverse outcome pathway (AOP) development I: strategies and principles. Toxicol Sci 142(2):312–320

Villeneuve D, Volz DC, Embry MR, Ankley GT, Belanger SE, Léonard M, Schirmer K, Tanguay R, Truong L, Wehmas L (2014b) Investigating alternatives to the fish early-life stage test: a strategy for discovering and annotating adverse outcome pathways for early fish development. Environ Toxicol Chem 33(1):158–169

Vinken M, Landesmann B, Goumenou M, Vinken S, Shah I, Jaeschke H, Willett C, Whelan M, Rogiers V (2013) Development of an adverse outcome pathway from drug-mediated bile salt export pump inhibition to cholestatic liver injury. Toxicol Sci 136(1):97–106

Weed DL (2005) Weight of evidence: a review of concept and methods. Risk Anal 25(6):1545–1557

Chapter 16
Using a Vitellogenesis Model to Link in vitro Neurochemical Effects of Pulp and Paper Mill Effluents to Adverse Reproductive Outcomes in Fish

Brandon M. Armstrong, Cheryl A. Murphy, and Niladri Basu

Abstract Many environmental contaminants may cause adverse reproductive effects through disruption of the neuroendocrine system. Pulp and paper mill effluent can contain up to 250 different contaminants and has been linked to reproductive impairment in fish. We used data from an in vitro study that characterized the potential neurochemical effects of pulp and paper mill effluent, to model vitellogenin production after exposure to different effluent fractions. We hypothesized that disrupted in vitro γ-aminobuytric acid and dopamine signaling could be modeled to simulate reproductive impairment in fish, specifically reduced vitellogenin production. These neurotransmitters are involved in the release of gonadotropin releasing hormone from the hypothalamus, which stimulates downstream processes related to vitellogenin production in the liver of female fish. In our approach, we integrated the in vitro results into a fish vitellogenesis model to predict adverse reproductive outcomes at the individual level. Our model results indicate that exposure to toxicants within the pulp and paper mill effluent, which interferes with neurotransmission, may cause harmful reproductive effects by impairing vitellogenin production. While the model has yet to be validated, our proof of principle approach highlights the use of computational models as a means to integrate results from in vitro studies that assess complex mixtures to potentially adverse effects on fish reproduction.

B.M. Armstrong (✉) • C.A. Murphy
Department of Fisheries and Wildlife, Michigan State University, East Lansing, MI 48824, USA
e-mail: armstl16@msu.edu

N. Basu
Faculty of Agricultural and Environmental Sciences, McGill University, Montreal H9X 3V9, QC, Canada

Department of Environmental Health Sciences, University of Michigan School of Public Health, Ann Arbor, MI, USA
e-mail: niladri.basu@mcgill.ca

N. Garcia-Reyero, C.A. Murphy (eds.), *A Systems Biology Approach to Advancing Adverse Outcome Pathways for Risk Assessment*,
DOI 10.1007/978-3-319-66084-4_16

16.1 Introduction

16.1.1 Environmental Mixtures & Their Effects

Aquatic organisms are exposed to complex mixtures of stressors in their environment derived from both natural and anthropogenic sources. Predatory cues, food limitations, temperature fluctuations, hypoxia and salinity are just a few of the stressors that exist within aquatic environments (Holmstrup et al. 2010; Hooper et al. 2013). Anthropogenic sources can also include the release of chemical compounds through household and industrial wastes and agricultural practices (Kolpin et al. 2002; Adams 2005). For example, in a recent survey of 139 streams across the United States, over 80 organic wastewater contaminants, including personal care products, agricultural, industrial and household compounds and pharmaceuticals were detected and at least one of these compounds was detected in 80% of the streams sampled (Kolpin et al. 2002). Understanding how these stressors interact with one another in a complex mixture to produce biological effects remains a difficult yet pressing task for ecotoxicology.

Complex mixtures may cause long-term adverse effects on communities inhabiting aquatic environments. A commonly measured adverse effect is mortality; however sublethal effects may occur following exposures to doses lower than those that induce mortality. Often, contaminants impair behavior, gene expression and physiological function (Laws 2000), which may interfere with bioenergetics, endocrine and neuroendocrine functions and immunity of organisms. Chemical interactions within complex mixtures may result in adverse effects even below their individual no-observable-adverse-effect concentration (NOAEC; Monosson 2005). For example, Armstrong et al. (2015) determined that a mixture of unionized ammonia (NH_3; 0.03 mg/L) and the pharmaceutical 17α-ethinylestradiol (EE2; 0.25 ng/L) at their respective NOAEC resulted in increased fathead minnow mortality during a 21 day exposure.

16.1.2 Future Direction of Environmental Toxicology

Traditionally, toxicological studies have assessed the effects of a single contaminant on a single organismal species. Many of these studies were conducted offering optimal environmental conditions for a chosen organism that could survive in a laboratory setting. Additionally, these studies focused on laboratory model species, which aren't necessarily predictive of wild species. Lastly, many of them did not assess the interactions that may occur between stressors and that these interactions may culminate in cumulative, synergistic, antagonistic or additive effects. Therefore, effect concentrations derived from single contaminant exposures may fall short in protecting aquatic organisms inhabiting environments where complex mixtures of environmental stressors exist.

Assessing the biological effects of chemical mixtures is a challenge as there are currently over 85,000 registered chemicals on the U.S. Environmental Protection Agency's Toxic Substances Control Act (TSCA) chemical substance inventory (U.S. EPA 2015). Fully assessing the biological effects of each chemical on every species inhabiting aquatic environment is not attainable. It is also impractical to determine the effects of every possible combination of chemicals found in a complex mixture using traditional approaches. Chemical instrumentation analyses can determine the types and quantities of each chemical associated with a chemical mixture; however this technique offers little or no information on biological effects stemming from an exposure to that mixture. Biological methods can be used to derive effect concentrations for various endpoints however these methods cannot determine which chemical in the mixture (or their interactions) are producing the observed biological effect.

Exposure to contaminants that evoke sublethal effects within individuals can indicate exposure through identifiable profiles, known as biomarkers, which can be measured at the molecular, biochemical or cellular level (U.S. EPA 2012). Biomarkers are quantifiable biological responses used to indicate the biological state of an organism or cell (Carvan et al. 2008). These biomarkers are generally easier to measure than conducting whole-organism and/or population-level studies, and so, a goal of toxicology today is to extrapolate from mechanistically relevant molecular and subcellular biomarkers to whole individual and population/community level effects.

Biomarkers are not without their own set of limitations. Biomarkers only provide a measurement at a single time point and due to the complexity of organismal systems one must consider the temporal dependence of a biomarker from the onset of exposure (Forbes et al. 2006). Measuring multiple biomarkers at any given time may only lead to ambiguity as individual biomarkers likely have different dose-response curves (Depledge 1994; Forbes et al. 2006). Forbes et al. (2006) suggested that suites of biomarkers must be linked to a mechanistic model tied to some measurement of fitness in order to be useful. When coupled with a mechanistic model anchored to an exposure or toxicologically relevant phenotype, biomarkers measured in an individual can be useful predictors of population risk (Carvan et al. 2008).

For ecotoxicological purposes, sublethal effects must be translated to population level impacts. There are significant limitations to obtaining population-level data to evaluate the thousands of anthropogenic chemicals, many have been recognized by the U.S. National Research Council (NRC 2007) who suggested that new, predictive approaches be developed to examine toxicant effects. These recommendations range from molecular level changes in individuals to impacts on entire populations (NRC 2007). The NRC suggested that researchers and regulators move away from the reliance on in vivo studies to in vitro model systems. One approach that can help meet these goals is the adverse outcome pathway (AOP), a conceptual framework linking molecular initiating events caused by contaminant exposures to adverse outcomes at higher levels of organization considered relevant to risk assessment (Ankley et al. 2010).

An AOP framework approach can be used to assess the risks of environmental contaminant mixtures to fish and wildlife by scaling molecular data to the individual and also to the population level (Watanabe et al. 2011). This framework could help population modeling and ecological risk assessment efforts by linking ecotoxicology data with mechanistic predictors of effects to reproduction and population survival (Kramer et al. 2011). Currently, AOPs are being developed to relate chronic toxicity to impaired fish growth (Groh et al. 2015) and reproduction (Ankley et al. 2010). As further AOPs are established, modeling approaches will be needed to link several AOPs into an integrated key event network (Chap. 14). This linkage process is necessary as it is unlikely that exposure to a chemical will result in only a single AOP being activated, especially considering the wide range of molecular targets within complex organisms.

In this chapter we investigated a pulp and paper mill effluent (PPME) case study to show, as proof of principle, a way to integrate in vitro neurochemical data into a fish vitellogenesis model to predict adverse reproductive effects following exposure to complex mixtures. The vitellogenesis model was adapted from Murphy et al. (2005) to include the release of neurotransmitters specific to fish reproduction and their binding to associated receptors, which influence the downstream production of sex steroids. We included processes for γ-aminobutyric acid (GABA) and dopamine (DA) as they are the two primary neurotransmitters involved in controlling gonadotropin production in fish. Neurotransmitters binding to their respective receptors and the activity of their associated degrading enzymes can be measured using cell-free in vitro assays. We leveraged data from an in vitro study (Basu et al. 2009) which exposed common goldfish whole brain extracts to fractionated PPME and characterized subsequent impairments to GABA and DA receptor binding and enzyme activity. Our objective was to incorporate the in vitro neurochemical data into the mathematical model and simulate the hypothalamic-pituitary-gonadal-liver (HPGL) axis of a generic female fish by predicting cumulative vitellogenin production over one spawning season. Model simulations for each effluent fraction were then compared against a control to determine if an impairment of vitellogenin production would result from PPME exposure.

16.1.3 Neurotransmitters & Fish Reproduction

Vertebrate reproduction is controlled by many hormones and compounds within the HPGL axis, Fig. 16.1 (Van Der Kraak et al. 1998; Trudeau et al. 2000).

Environmental cues, such as temperature and photoperiod, stimulate the release of neurotransmitters. Neurotransmitters facilitate the body's communication system, including reproduction, by acting as chemical messengers and activating specific receptors in post synaptic cells, Fig. 16.2 (Lauder 1993).

Neural cells transmit information from one cell to another at the synapse, the narrow space between the axon of initiating cell and recipient cell (McGeer et al. 2013). The initiating cell generates an action potential that travels through the axon to the

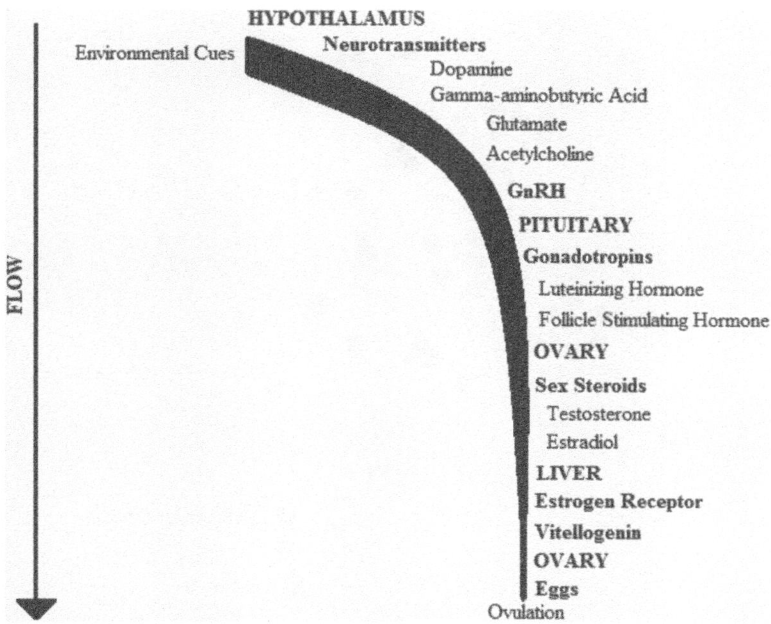

Fig. 16.1 The hormonal cascade of the fish hypothalamic-pituitary-gonadal-liver axis and subsequent reproduction

terminal triggering the release of neurotransmitters from the axon to the synapse. The neurotransmitters then bind to post-synaptic receptors of the recipient cell and depolarize the cell membrane. The chemical signal is converted back to an electrical signal and either destroyed or cleared. There are three major classes of neurotransmitters: biogenic amines, amino acids and neuropeptides (Kurreck and Stein 2015).

There are a number of mechanisms in which toxicants can cause neurotoxicity. Once released, neurotransmitters must then bind to receptors to initiate the signaling process. Toxicants can impede the cell signaling by binding or activating these receptors and/or inhibiting neurotransmitter release. Inactivation of free (unbound) neurotransmitters within the synaptic cleft occurs through enzymatic degradation. Toxicants may disrupt enzyme activity within the hypothalamus (Basu et al. 2009). Disruption of neurotransmitter receptor binding or enzyme activity may cause downstream effects including altered sex hormone dynamics, which may impair reproduction (Fig. 16.1).

Dopamine is a biogenic amine neurotransmitter that acts directly at the pituitary cell level to inhibit the release of gonadotropin through interaction with the gonadotrope D2 receptor (Zohar et al. 2010). In some species such as the goldfish, *Carassius auratus*, DA has been shown to inhibit both gonadotropin release directly as well as gonadotropin releasing hormone (GnRH) mediated gonadotropin secretion (Popesku et al. 2008). This is thought to be an evolutionary mechanism to prevent spawning during periods of poor environmental conditions (Zohar et al. 2010).

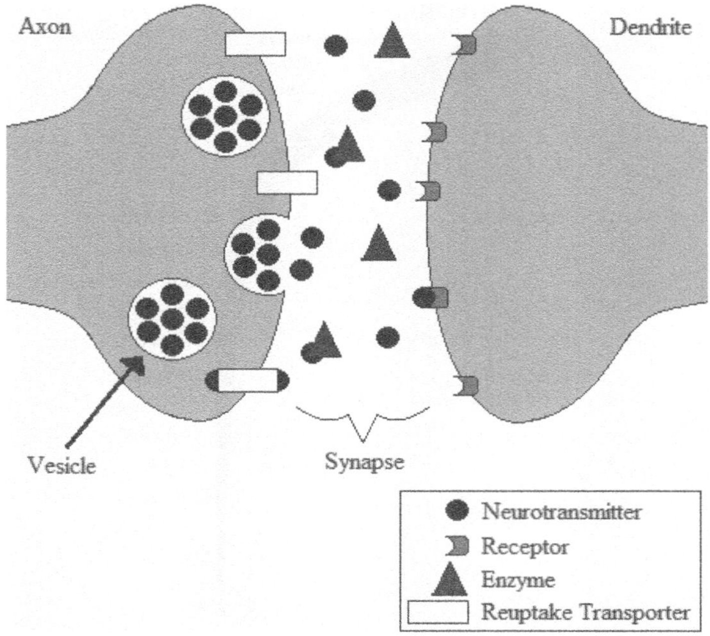

Fig. 16.2 Simplified schematic of a chemical synapse. Neurotransmitters are released from the presynaptic neuron terminal into the synaptic cleft. The neurotransmitters then are able to bind to open receptor sites on the postsynaptic terminal to elicit a response. Unbound neurotransmitters are subjected to enzymatic degradation and transporter reuptake

Once released, free DA is subject to reuptake into the presynaptic terminal by DA transporters or degradation by the enzyme monoamine oxidase (MAO) (Bortolato et al. 2008). The role of DA is dependent on the fish species. For example, DA inhibition of GnRH was found to be very high in cyprinids and was much less pronounced in salmonids and nearly absent in Atlantic croaker, *Micropogonias undulates* (Van Der Kraak 2009) and percids (Zakes and Demska-Zakes 2005; Dabrowski et al. 1994; Żarski et al. 2015) which suggests dopamine's role in fish reproduction is a species-dependent process (Levavi-Sivan et al. 2010).

The γ-aminobutyric acid is an amino acid neurotransmitter found in the brain of vertebrates (Zohar et al. 2010). Once released it is reabsorbed by the presynaptic terminal, degraded by the enzyme GABA-transaminase (GABA-T) or bound to post synaptic terminal receptors (Treiman 2001). In mammals, GABA acts as an inhibitory neurotransmitter within the brain and, through interactions with GnRH neurons, GABA inhibits GnRH release (Smith and Jennes 2001). However, in fish, GABA has a stimulatory role in reproduction by promoting GnRH secretion from the hypothalamus through exciting GnRH neurons and inhibiting DA release (Popesku et al. 2008; Watanabe et al. 2014). For example, rainbow trout injected with a single dose of GABA exhibited increased LH release from the pituitary (Mananos et al. 1999; Levavi-Sivan et al. 2010) likely through actions on the hypothalamic GnRH neurons which directly innervate the pituitary (Peter et al. 1990). There are two types of GABA receptors localized throughout the hypothalamus, $GABA_A$ and $GABA_B$ (Maffucci and Gore 2009). While the specific functions of these two receptor types aren't completely understood, mammalian research has shown that activation of the $GABA_A$ receptor blocks the proestrus LH surge and release whereas in fish, $GABA_A$ receptor activation excites GnRH neurons (Watanabe et al. 2014).

16.1.4 Pulp and Paper Mill Effluent

The conversion of wood fibers into paper products generates a large amount of pollution (Ali and Sreekrishnan 2001) with up to 100 million kg of pollutants released into the environment each year (Dey et al. 2013). Additionally, the industry is one of the largest in terms of water consumption, as the formation of paper products requires an expansive amount of freshwater (Thompson et al. 2001). Pulp and paper mill effluents can contain over 250 different chemicals at various stages of the treatment process (Ali and Sreekrishnan 2001) including chlorinated compounds, fatty acids, tannins, organic polymers and sulfuric compounds (Zayas et al. 2011).

Several studies have shown that exposure to PPME can impair fish reproduction. For example, female largemouth bass exposed to PPME ($\geq$ 20% effluent) for 56 days exhibited reduced 17β-estradiol and vitellogenin production (Sepulveda et al. 2003). Additionally, fathead minnow egg production was significantly reduced following a 5-day exposure to 100% PPME (Waye et al. 2014b). The chemicals found in PPME can also alter in vitro neurochemical signaling in fish (Basu et al. 2009), which may be a mechanism for disrupted reproduction in fish exposed to the effluent (Kovacs et al. 2013). In this case study, we will focus on the neurochemical effects of PPME reported by Basu et al. (2009).

Basu et al. (2009) assessed the potential neurochemical effects of PPME in goldfish brain tissue by measuring changes in GABA and DA neurotransmitter receptor binding and associated enzyme activities (Basu et al. 2009). The PPME was collected following primary treatment (clarifier) and secondary treatment (conventional activated sludge) from a facility in Eastern Canada and fractionated into different chemical components using classic solvent polarity (hexane, ethyl acetate and water) and polyphenolic extraction (Polyvinylpolypyrrolidone, PVPP) methods. While the fractionation process used did not identify the exact active compounds within the effluent, it did separate out chemicals into different classes based on unique properties of the individual compounds. Using an in vitro technique which measured changes in receptor binding and enzyme activity, Basu et al. (2009) reported that chemicals in both the primary and secondary PPME extracts can disrupt the function of neurotransmission in fish brains in vitro (Table 16.1). The data collected by Basu et al. (2009), by itself, is hard to interpret on higher levels of biological organization, such as an individual's reproductive potential, as some receptors were increased while their respective enzymes were also increased.

Table 16.1 Summary of in vitro mean neurochemical results expressed as a % of control samples following PPME exposure (Basu et al. 2009)

Neurochemical		Primary effluent					Secondary effluent				
		Solvent series			PVPP		Solvent series			PVPP	
		Ethyl Acetate	*Water*	*Hexane*	*Water*	*Ethanol*	*Ethyl Acetate*	*Water*	*Hexane*	*Water*	*Ethanol*
Receptors	D2	126.0*	148.3*	65.2	74	124.4	61.4	121.7*	99.9	144.9*	71.7
	$GABA_A$	104.8	92.1	288.9*	71.4	50.8	73	34.9*	73	33.3*	42.9
Enzymes	MAO	72.3*	103.7	53.3*	98.8	52.1*	96.7	85.6*	99.6	63.5*	100.1
	GABA-T	120.9*	66.6*	168.6*	158.1*	150.5*	108.3	123.7	112.7	67.1*	80.3

Samples were incubated with either primary or secondary stage treated pulp and paper mill effluent. In vitro assays were conducted on male and female pooled whole brains collected from 100 common goldfish. N = 3 assay runs per neurochemical

D2 dopamine receptor, $GABA_A$ γ-aminobutyric acid A receptor; MAO monoamine oxidase, GABA-T γ-aminobutyric acid transaminase, PVPP polyvinylpolypyrrolidone

* indicates significant differences ($p < 0.05$) from controls

For example, the hexane fraction of the primary effluent increased GABA binding to the $GABA_A$ receptor by 289%, which may indicate increased gonadotropin production. However, GABA-transaminase activity was also increased 169%, which may inhibit gonadotropin production. It is difficult to determine if reproduction would be impaired based on this in vitro information alone.

Linking neurochemical changes with vitellogenin production may provide a scheme to better interpret the aforementioned work. Vitellogenin is a precursor protein for egg yolk in oviparous organisms (Arukwe and GoskØyr 2003) and is directly related to fecundity and egg quality in an individual fish (Miller et al. 2007). Thus, vitellogenin is an important biomarker of exposure to endocrine disrupting contaminants. The neurochemical effects of PPME reported by Basu et al. (2009) may lead to downstream effects along the HPG axis and disrupt reproduction, specifically vitellogenin production. The results from Basu et al. (2009) suggest that dopamine in the brain may be increased due to the general trend of decreased MAO activity. Female rainbow trout exposed to 0.01 mg/L hydrogen cyanide for 12 days exhibited increased brain dopamine levels which correlated with reduced plasma vitellogenin levels and smaller oocytes (Ruby et al. 1986; Szabo et al. 1991). The GABA results reported by Basu et al. (2009) are hard to determine if there would be an increase or decreased effect on GnRH production as there was a general increase in GABA-T activity but decrease in $GABA_A$ receptor binding. In another study, injection of the GABA receptor agonist, muscimol (0.1 μg/g) increased serum LH in female goldfish after 30 min (Trudeau et al. 1993). Similarly, a single injection of the GABA-T inhibitor, γ-vinyl-GABA (300 μg/g), in female goldfish increased serum LH at days 1, 7 and 14 which returned to baseline levels by day 21 (Trudeau et al. 1993). We assume that both of these results would lead to increased vitellogenin production.

Based on these prior studies, we hypothesized that exposure to PPME and its subsequent GABA and DA signaling disruption (as determined using cell-free in vitro methods) within the hypothalamus will lead to reduced liver vitellogenin production in female fish. Therefore, the goal of this study was to incorporate a neurochemical compartment to an existing fish vitellogenesis computational model (Murphy et al. 2009) to link neurochemical changes to impaired fish vitellogenin production using an AOP framework.

16.2 Computational Model of the Fish HPGL Axis

The model consisted of a series of differential equations which determined the rate of change of 11 state variables within five compartments; Hypothalamus, Pituitary, Ovary, Liver and Blood. Hypothalamic state variables included GABA and DA (nmol/mg) neurotransmitters, $GABA_A$ and D2 receptors (nM) and GnRH (ng/ml). LH release into circulation (ng/ml) was the sole state variable in the pituitary compartment. The gonad compartment consisted of T and E2 sex steroid concentration (ng/ml). The liver compartment contained the state variables estrogen receptor (nM) and the blood compartment contained the state variables steroid binding proteins (nM) and vitellogenin (mg/ml).

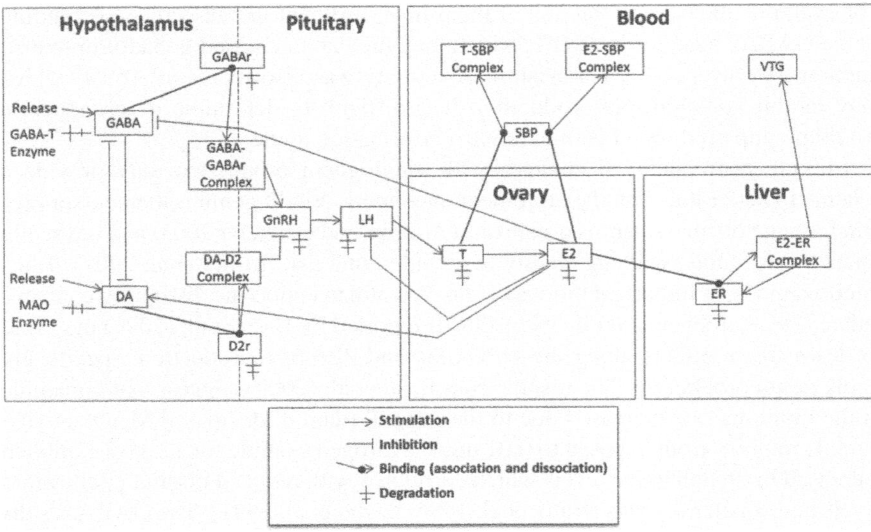

Fig. 16.3 Conceptual compartmentalized computational model of the fish HPGL axis

The model schematic showing relationships between state variables is presented in Fig. 16.3. A description of all variables, parameters, initial values and associated units used in the model are listed in Table 16.2. The simulations ran for 6 months to correspond to the period of vitellogenesis in a generic female fish (Murphy et al. 2005). Fourth-order Runge-Kutta integration with a time step of 0.0001 h was used to solve the model. The model was developed in FORTRAN 90 with the Lahey Fujitsu compiler (version 7.3) to perform simulation experiments.

The model was driven by the release of the neurotransmitters GABA (*Syn[GABA]*; Eq. 16.1, nmol/mg/h) and DA (*Syn[DA]* Eq. 16.2, nmol/mg/h). Many neurochemicals are released in a pulsatile manner (Terasawa 1994) including GABA (Maffucci and Gore 2009) and DA (van den Pol 2010). Additionally, the circadian rhythm of many hormones is believed to be caused by a diurnal circle of hypothalamic neurotransmitters (Macho et al. 1986) and GABA and DA have been shown have a circadian rhythm in other brain regions (Castaneda et al. 2004). Therefore we characterized the neurotransmitter release using a slightly modified diurnal cycling function (Murphy et al. 2005) which included neurotransmitter inhibition parameters. Neurotransmitters were released into the system for the first 2 months of the simulation to correspond to the period of gonadal recrudescence (Murphy et al. 2005). The release of GABA and DA were subjected to either a DA (inh_{DA}) or GABA (inh_{GABA}) inhibition rate, respectively (Popesku et al. 2008). During vitellogenesis, E2 has been shown to stimulate the production of DA in salmonids (Zohar et al. 2010) and therefore our model included an E2 derived DA stimulation rate ($Stim_{E2DA}$). Additionally, it was reported that GABA transmission to GnRH neurons is reduced by E2 (Watanabe et al. 2014), therefore we included an E2 inhibition (Inh_{E2GABA}) rate in the GABA release function.

Table 16.2 Baseline values and definitions for all parameters used in the generic female fish vitellogenesis model

C	Compound	Parameter	Definition	Value	Units	Species	Reference
H	GABA	GABA	Basal GABA concentration	24	nmol/mg	*Rattus norvegicus* (male)	Smolders et al. (1997)
		V_{GABA}	Maximum uptake rate	0.402	nmol/mg/h	*Rana temporaria* (unsexed)	Voaden et al. (1974)
		K_{GABA}	Half-saturation of uptake	25,000	nmol/mg/h	*Rana temporaria* (unsexed)	Voaden et al. (1974)
		Inh_{GABA}	Inhibition of DA synthesis	0.58	NA	NA	Assumed
	$GABA_A$	$GABA_A$	Basal receptor concentration	0.125	nM	NA	Assumed
		ka_{GABA}	Association rate to receptor	17.2	1/10^6 M/h	*Rattus norvegicus* (male)	Chu et al. (1990)
		kd_{GABA}	Dissociation rate from receptor	5.796	1/h	*Rattus norvegicus* (male)	Chu et al. (1990)
		$kelim_{GABA\text{-}A}$	Elimination rate of receptor	80	1/h	NA	Assumed
		$kind_{GABA\text{-}A}$	Induction rate of receptor	0.024	1/h	NA	Assumed
	GABA-T	V_{GABAT}	Maximum degradation rate	4.05	nmol/mg/h	*Carassius auratus* (unsexed)	Lam (1972)
		K_{GABAT}	Half saturation of degradation	1670	nmol/mg/h	*Mus domesticus* (unsexed)	Bardakdjian et al. (1979)
	DA	DA	Basal DA concentration	2.2	nmol/mg	*Rattus norvegicus* (male)	Smolders et al. (1997)
		V_{DA}	Maximum uptake rate	3.96	nmol/mg/h	*Carassius auratus* (unsexed)	Sarthy and Lam (1979)
		K_{DA}	Half-saturation of uptake	261	nmol/mg/h	*Carassius auratus* (unsexed)	Sarthy and Lam (1979)
		Inh_{DA}	Inhibition of GABA synthesis	0.8	NA	NA	Assumed
	D2	D2	Basal receptor concentration	0.125	nM	NA	Assumed
		ka_{DA}	Association rate to receptor	12.84	1/10^6 M/h	*Coturnix japonica* (male)	Kubikova et al. (2009)
		kd_{DA}	Dissociation rate from receptor	3.42	1/h	*Coturnix japonica* (male)	Kubikova et al. (2009)
		$kelim_{D2}$	Elimination rate of receptor	10	1/h	NA	Assumed
		$kind_{D2}$	Induction rate receptor	0.048	1/h	NA	Assumed
	MAO	V_{MAO}	Maximum degradation rate	39.9	nmol/mg/h	*Carassius auratus* (female)	Hall et al. (1982)
		K_{MAO}	Half saturation of degradation	60,630	nmol/mg/h	*Carassius auratus* (female)	Hall et al. (1982)
	GnRH	$kdeg_{GH}$	Degradation rate	1.032	1/h	NA	Assumed
P	LH	$kdeg_{LH}$	Degradation rate	2.032	1/h	NA	Assumed

(continued)

Table 16.2 (continued)

C	Compound	Parameter	Definition	Value	Units	Species	Reference
G	T	T	Basal T concentration	0.0822	ng/ml	NA	Assumed, 1.2% of total
		V_T	Maximum rate production	0.12	ng/ml/h	NA	Assumed
		K_T	Half saturation of production	0.3	ng/ml	*Micropogonias undulatus* (model)	Murphy et al. (2009)
		H_T	Hill coefficient	1.8	Unitless	*Micropogonias undulatus* (model)	Murphy et al. (2009)
		$kdeg_T$	Degradation rate	1.386	1/h	*Micropogonias undulatus* (model)	Murphy et al. (2009)
	E2	E2	Basal E2 concentration	0.00816	ng/ml	NA	Assumed, 1.2% of total
		V_E	Maximum rate production	1.5	ng/ml/h	NA	Assumed
		K_E	Half saturation of production	0.052	ng/ml	*Micropogonias undulatus* (model)	Murphy et al. (2009)
		H_E	Hill coefficient	4	Unitless	NA	Assumed
		$kdeg_E$	Degradation rate	1.386	1/h	*Micropogonias undulatus* (model)	Murphy et al. (2009)
		$Stim_{E2DA}$	Stimulation rate of DA	0.1	1/h	NA	Assumed
		Inh_{E2GABA}	Inhibition rate of GABA	0.1	1/h	NA	Assumed
B	SBP	SBP	Basal SBP concentration	400	nM	*Micropogonias undulatus* (model)	Murphy et al. (2009)
		ka_T	Association rate of T to SBP	5.6687	1/10^9 M/h	*Micropogonias undulatus* (model)	Murphy et al. (2009)
		kd_T	Dissociation rate of T to SBP	27.72	1/h	*Micropogonias undulatus* (model)	Murphy et al. (2009)
		ka_E	Association rate of E2 to SBP	17.743	1/h	*Micropogonias undulatus* (model)	Murphy et al. (2009)
		kd_E	Dissociation rate of E2 to SBP	5.6687	1/10^9 M/h	*Micropogonias undulatus* (model)	Murphy et al. (2009)
L	ER	E2	Basal ER concentration	0.125	nM	*Micropogonias undulatus* (model)	Murphy et al. (2009)
		Kdegu	Degradation rate of free ER	0.00058	1/h	*Micropogonias undulatus* (model)	Murphy et al. (2009)
		Kdega	Degradation rate of activated ER	0.012	1/h	*Micropogonias undulatus* (model)	Murphy et al. (2009)
		k1	Association rate of E2 to ER	7.43	1/10^8 M/h	*Micropogonias undulatus* (model)	Murphy et al. (2009)
		k_1	Dissociation rate of E2 to ER	0.81	1/h	*Micropogonias undulatus* (model)	Murphy et al. (2009)
	VTG	k3	VTG production rate	3.465	1/h	*Micropogonias undulatus* (model)	Murphy et al. (2009)

Compartments of the model included the hypothalamus (H), pituitary (P), gonad (G), liver (L), and blood (B)

$$Syn[GABA] = \frac{500\left[1 - 1.0^{*} \frac{\cos 2\pi(t-6.0)}{24.0}\right]}{1.0 + [DA]^{*} inh_{DA} + E2^{*} inh_{E2GABA}} \tag{16.1}$$

$$Syn[DA] = \frac{500\left[1 - 1.0^{*} \frac{\cos 2\pi(t-6.0)^{*}}{24.0} [E2]^{*} Stim_{E2DA}\right]}{1.0 + [GABA]^{*} inh_{GABA}} \tag{16.2}$$

Once released, the concentrations of free (unbound) GABA and DA neurotransmitters (*[GABA]*, Eq. 16.3; *[DA]*, Eq. 16.4, nmol/mg) within the hypothalamus compartment were bound to a receptor, degraded by an enzyme or cleared from the synapse via a reuptake transporter. GABA was bound to $GABA_A$ receptors or degraded by the GABA-T enzyme whereas DA was bound to D2 receptors or degraded by MAO. The concentration of free neurotransmitters was calculated based on the amount being released, adding the concentration disassociating from the receptor and subtracting the concentration bound to a receptor, degraded by an enzyme or undergoing reuptake by a transporter.

$$\frac{dGABA}{dt} = Syn[GABA] - ka_{GABA}[GABA][GABA-A] + kd_{GABA}[GABA\_GABA-A] - Up_{GABA} - Deg_{GABA} \tag{16.3}$$

$$\frac{dDA}{dt} = Syn[DA] - ka_{DA}[DA][D2] + kd_{DA}[DA\_D2] - Up_{DA} - Deg_{DA} \tag{16.4}$$

Enzymatic degradation (*Deg*; Eqs. 16.5 and 16.6; nmol/mg/h) and synaptic reuptake (*Up*; Eqs. 16.7 and 16.8, nmol/mg/h) of both GABA and DA followed Michaelis-Menten kinetics as has been reported in other studies (Wheeler and Hollingsworth 1979, Venton et al. 2003) using the parameters *V* (maximum rate) and *k* (half saturation rate). Enzyme degradation multipliers ($Mult_{GABA\text{-}T}$ and $Mult_{MAO}$) were incorporated to simulate PPME exposure, further detailed in Sect. 16.2.6. We assumed that in vitro enzyme activity was directly related to degradation rates of the neurotransmitters.

$$\deg_{GABA} = \frac{V_{GABAT}^{*}[GABA]^{*}}{k_{GABAT} + [GABA]} Mult_{GABA-T} \tag{16.5}$$

$$\deg_{DA} = \frac{V_{MAO}^{*}[DA]^{*}}{k_{MAO} + [DA]} Mult_{MAO} \tag{16.6}$$

$$Up_{GABA} = \frac{V_{GABA}{}^{*}[GABA]}{k_{GABA} + [GABA]} \tag{16.7}$$

$$Up_{DA} = \frac{V_{DA}{}^{*}[DA]}{k_{DA} + [DA]} \tag{16.8}$$

Additionally, we assumed that ligand-receptor binding followed a generalized mathematical formula (Murphy et al. 2005). For example, the concentration of ligand bound receptors (*[GABA_GABA-A],* Eq. 16.9; *[DA_D2]*, Eq. 16.10, nmol/mg/h) was calculated as the association rate constant (*ka;* Table 16.2) multiplied by the number of open receptors ([GABA-A] or [D2]) and the concentration of unbound ligand ([GABA] or [DA]). The concentration dissociating from the receptor was calculated by multiplying the dissociation rate constant (*kd;* Table 16.2) by the concentration of ligand bound receptors.

$$\begin{aligned}\frac{dGABA\_GABA-A}{dt} &= ka_{GABA}[GABA][GABA-A] \\ &\quad - kd_{GABA}[GABA\_GABA-A]\end{aligned} \tag{16.9}$$

$$\frac{dDA\_D2}{dt} = ka_{DA}[DA][D2] - kd_{DA}[DA\_D2] \tag{16.10}$$

The concentration of open neurotransmitter receptors *[GABA-A*; Eq. 16.11*]* and *[DA-D2*; Eq. 16.12*]* were calculated based on a basal receptor induction rate (*kind*, 1/h), the amount of ligand associating or disassociating to the receptor and a basal elimination rate of the receptor (*kelim*, 1/h). Neurotransmitter receptor binding multipliers ($Mult_{GABA\text{-}A}$ and $Mult_{D2R}$) were incorporated to simulate PPME exposure, further detailed in Sect. 16.2.6.

$$\frac{dGABA-A}{dt} = \begin{pmatrix} kind_{GABA-A}[GABA-A] \\ -ka_{GABA}[GABA][GABA-A] \\ +kd_{GABA}[GABA_{GABA}-A] - kelim_{GABA-A} \end{pmatrix}^{*} Mult_{GABA-A} \tag{16.11}$$

$$\frac{dD2}{dt} = \left(kind_{D2}[D2] - ka_{DA}[DA][D2] + kd_{DA}[DA_{D2}] - kelim_{D2}{}^{*}\, Mult_{D2R}\right) \tag{16.12}$$

The release of GnRH was a ratio of *GABAStim* to *DAInhib* (Eq. 16.13, ng/ml) where *GABAStim* (Eq. 16.14) and *DAInhib* (Eq. 16.14) were second order Michaelis-Menten kinetic equation, similar to what has been used in LH simulations (Blum et al. 2000). The amount of free GnRH in the system (*[GnRH]*, Eq. 16.16, ng/ml) was calculated by subtracting the amount being converted into luteinizing hormone (LH) from the amount of GnRH being released.

$$Syn[GnRH] = GabaStim / DAInhib \tag{16.13}$$

$$GABAStim = \frac{5^{*}[GABA\_GABA - BZ]2}{2.0 + [GABA\_GABA - BZ]2} \tag{16.14}$$

$$DAInhib = \frac{5^{*}[DA\_D2]2}{2.0 + [DA\_D2]2} \tag{16.15}$$

$$\frac{dGnRH}{dt} = Syn[GnRH] - Syn[LH] \tag{16.16}$$

16.2.1 Pituitary Compartment

The GnRH stimulated LH release rate (*GnRHStim*, Eq. 16.17, ng/ml) was calculated as a second-order Michaelis-Menten kinetic equation (Blum et al. 2000). Circulating E2 can diffuse through pituitary tissue and inhibit the release of LH (Van Der Kraak 2009). Therefore the actual amount of LH being released into the system (*SynLH*, Eq. 16.18, ng/ml) included an E2 inhibition rate (Murphy et al. 2005). The concentration of free LH (*[LH]*, Eq. 16.19, ng/ml) was calculated by subtracting the amount of LH being converted into testosterone (T) from the concentration of LH being synthesized.

$$GnRHStim = \frac{5^{*}[GnRH]^{2}}{2.0 + [GnRH]^{2}} \tag{16.17}$$

$$Syn[LH] = \frac{GnRHStim}{1 + \frac{[E2]}{10}} \tag{16.18}$$

$$\frac{dLH}{dt} = Syn[LH] - Syn[T] \tag{16.19}$$

16.2.2 Gonad Compartment

Once LH was released into the blood it traveled to the ovary to promote T synthesis (*Syn[T]*, Eq. 16.20, ng/ml) within the gonad which was calculated as a Hill function (Murphy et al. 2005). The concentration of T in the system (*[T]*, Eq. 16.21, ng/ml) was calculated by subtracting the enzymatic degradation rate (*kdegT*, ng/ml/h), the concentration being bound by steroid binding proteins (SBP) and the amount of T being converted into E2 (*Syn[E2]*, ng/ml) from the amount of T being synthesized and dissociated from SBP (Murphy et al. 2005).

$$Syn[T] = \frac{V_T{}^*[LH]^{H_T}}{k_T{}^{H_T} + [LH]^{H_T}} \tag{16.20}$$

$$\frac{dT}{dt} = Syn[T] - kdeg_T[T] + kd_T[SBP\_T] - Syn[E2] - ka_T[T][SBP] \tag{16.21}$$

The synthesis of E2 from the aromatization of T (*Syn[E2]*, Eq. 16.22, ng/ml) was defined by a Hill function (Murphy et al. 2005). The concentration of free E2 in the system (*[E2]*, Eq. 16.23, ng/ml) was dependent upon *Syn[E2]* and the concentration being degraded by enzymes, associated to and dissociated from the liver estrogen receptors (ER) and the concentration associating to and disassociating from blood SBP (Murphy et al. 2005). It is important to note that we only characterized the initial surge of E2 production, which occurs during the early vitellogenic period. We did not model the ovulatory surge that occurs prior to spawning.

$$Syn[E2] = \frac{V_{E2}{}^*[T]^{H_{E2}}}{k_{E2}{}^{H_{E2}} + [T]^{H_{E2}}} \tag{16.22}$$

$$\begin{aligned}\frac{dE2}{dt} = {} & syn[E2] - kdeg_{E2}[E2] + k\_1[ER\_E2] \\ & + kd_{E2}[SBP_{E2}] - k1[E2][ER] - ka_{E2}[E2][SBP]\end{aligned} \tag{16.23}$$

16.2.3 Blood Compartment

SBP are located in the plasma of teleost fish, which bind to free sex steroid hormones (T and E2) to prevent their metabolic degradation (Murphy et al. 2005). We assumed free and bound steroid hormones are at equilibrium in the blood, however, only the free hormone is physiologically active (Hammond 2016). In the model, the concentration of unbound SBP in the blood (*[SBP]*; Eq. 16.24, nM) was calculated by subtracting the concentration of E2 and T being associated to unbound SBP from the concentration of E2 and T dissociating from bound SBP (Murphy et al. 2005). The concentration of bound SBP to T and E2 (*[SBP-T]* & *[SBP-E2]*, Eqs. 16.25 and 16.26, nM, respectively) was calculated by subtracting the concentration of ligands being dissociated from the amount of ligands associating to unbound SBP (Murphy et al. 2005).

$$\frac{dSBP}{dt} = kd_T[SBP_T] + kd_E[SBP_{E2}] - ka_T[T][SBP] - ka_E[E2][SBP] \quad (16.24)$$

$$\frac{dSBP\_T}{dt} = ka_T[SBP][T] - kd_T[SBP\_T] \quad (16.25)$$

$$\frac{dSBP\_E2}{dt} = ka_{E2}[SBP][E2] - kd_{E2}[SBP\_E2] \quad (16.26)$$

16.2.4 Liver Compartment

Once the free E2 reached the liver it bound to an open estrogen receptor (ER), forming the ER-E2 complex (Murphy et al. 2005). Unbound ER concentration (*[ER]*, Eq. 16.27, nM) was calculated by subtracting the amount of E2 being associated to unbound ER from the concentration of E2 dissociating from bound ER. k1 and k_1 are the association and dissociation rate constants for E2 to ER (Murphy et al. 2005). Additionally, we assumed a background degradation rate (*kdegu*) of unbound ER and an induction rate (*k2*) where activated ER lead to further production of unbound ER (Murphy et al. 2005). Fish have three distinct ERs, ERα, ERβ and ERγ (Sabo-Attwood et al. 2004). Each receptor type has its own distinct physiological action and its concentration depends on the species and tissue (Leaños-Castañeda and Van Der Kraak 2007). In our model, we assume all of the vitellogenin production comes from the binding of E2 to the ERβ subtype (Leaños-Castañeda and Van Der Kraak 2007; Nelson and Habibi 2013).

$$\frac{dER}{dt} = 1.15^{*} k2[ER_{E2}] - kdegu[ER] + k\_1[ER\_E2] - k1[ER][E2] \quad (16.27)$$

$$\frac{dER\_E2}{dt} = k1[E2][ER] - kdega[ER\_E2] - k1[ER\_E2] - k2[ER\_E2] \quad (16.28)$$

Once ER was bound by E2, the model assumed a forward rate constant (*k3*) of vitellogenin production (*[VTG]*, Eq. 16.29, ng/ml) (Murphy et al. 2005). Vitellogenin is a good indicator of egg production because as the oocyte is growing, it is continually being filled with vitellogenin derived yolk proteins. Therefore we assumed a direct relationship between cumulative vitellogenin predicted by the model and fecundity (Miller et al. 2007).

$$\frac{dVTG}{dt} = k3[ER\_E2] \quad (16.29)$$

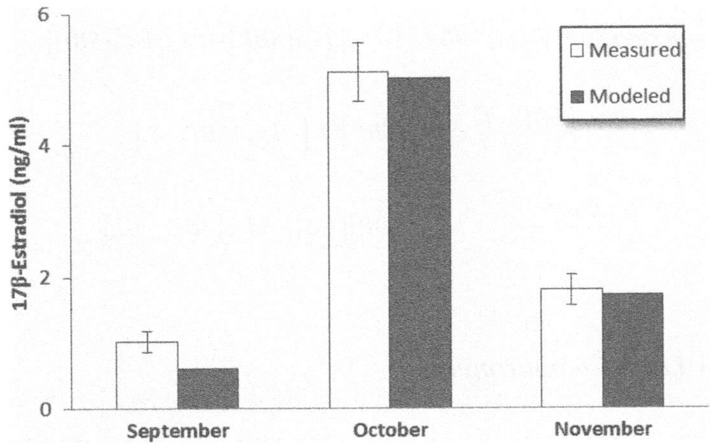

Fig. 16.4 Sexually mature female yellow perch (*Perca flavescens*) were raised in an indoor laboratory environment. Plasma 17β-estradiol concentrations during early vitellogenesis were obtained over a 3 month period using a radioimmunoassay technique and used to calibrate computational HPGL model. Error bars ±1 standard error, N = 12

16.2.5 Model Calibration

The generic female vitellogenesis model was calibrated for yellow perch (*Perca flavescens*) using sex steroid data collected over their early vitellogenic period, which runs from September through December and peaks in October (Fig. 16.4; Debofsky et al. 2015). Concentrations of in vivo plasma E2 and T was measured in mature female yellow perch using a radioimmunoassay technique (Jensen et al. 2011). Parameter values specific to neurotransmitter inhibition and receptor production, kelimG, kelimD, kindG, kindD, inhGABA, inhDA (Armstrong 2016), were incrementally adjusted during repeated simulations until the modeled sex steroid concentrations roughly mimicked the laboratory measurements.

16.2.6 PPME Simulation Conditions

Basu et al. (2009) assessed the potential neurochemical effects of both primary and secondary effluent from an Eastern Canadian pulp and paper mill. Both effluents were fractionated into four extracts by two methods of extraction; classic polarity and polyphenolic extraction. Whole brain tissues from male and female common goldfish were incubated with the effluent extracts for 30 min. Radioligand binding to both GABA and DA receptors was measured in vitro for each effluent extract using male and female common goldfish whole brain tissue. MAO and GABA-T proteins concentrations were determined in whole brain tissue. GABA-T enzyme activity was determined by incubating 5 μg of protein homogenates in 100 mM potassium pyrophosphate buffer containing 5 mM α-ketoglutarate, 4 mM NAD, 3.5 mM 2-mercaptoethanol, 10 μM pyridoxal-5′-phosphate, pH 8.6 for 15 min at 37 °C. 10 mM GABA was added and the absorbance was observed every 10 s for 2 min and maximal velocity of the enzyme reaction calculated. A final effluent concentration of 0.5 mg/ml was added and the enzyme activity measured. Similarly, MAO activity was measured by mixing 5 μg of protein with 100 μM 10-acetyl-3,7-dihydroxyphenoxazine, 200 mU horseradish peroxidase and 100 mM tyramine. Samples were incubated for 30 min and the production of resorufin was monitored between 30 and 50 min.

The neurotransmitter receptor binding and enzyme activity data obtained from Basu et al. (2009) were incorporated into the model as multipliers, expressed as a percent change from the control (Table 16.1). Multipliers for enzyme activity were incorporated in the enzyme degradation Eqs. 16.5 and 16.6 for GABA-T and MAO, respectively. Additionally, neurochemical receptor binding multipliers for $GABA_A$ and D2 were added to the state variable Eqs. 16.11 and 16.12, respectively. Five separate simulations were run for both the primary and secondary PPME, one for each fraction.

16.3 Modeled Effects of PPME on the Fish HPGL Axis

16.3.1 Neurochemicals

The model predicted that free GABA would reach its highest concentration on day 14 following the onset of gonadal recrudescence at 2120 nmol/mg in the control brain tissue (Fig. 16.5). The PVPP water and PVPP ethanol fractions from the primary effluent increased the maximum concentration of free GABA by 29.1% and 80.3%, respectively, compared to the control (Fig. 16.5a). The increase was due to the decreased number of receptor sites available for GABA binding, 71.4 and 50.8% of the control, respectively. The hexane fraction decreased the maximum concentration of free GABA by 68.1%. This decrease in free GABA resulted due to the increased number of receptor sites (288.9% of the control) and increased GABA-T enzyme activity (168.6% of the control). The remaining primary effluent fractions, ethyl acetate and water, slightly reduced the maximum concentration of free GABA by 10.5% and 0.1%, respectively. The secondary PPME resulted in increased GABA concentrations during all modeled fraction simulations (Fig. 16.5b). The highest concentration of free GABA was increased by 23.9%, 41.9%, 216.0%, 266.7% and 278.3% nmol/mg in the ethyl acetate, hexane, PVPP ethanol, water only and PVPP water fractions, respectively, compared to the control. These increases were due to

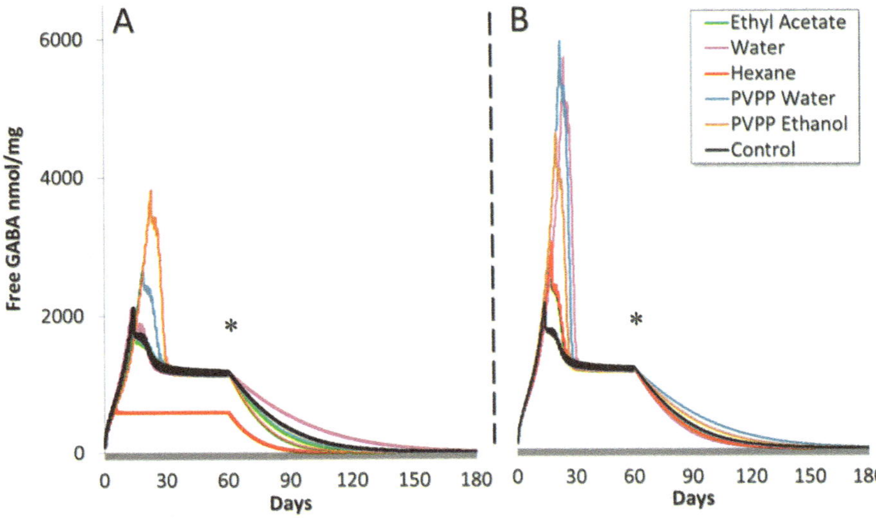

Fig. 16.5 Effects of primary (**a**) and secondary (**b**) pulp and paper mill effluent on modeled free γ-aminobuytric acid (GABA) in a generic female fish during vitellogenesis. * indicates when the neurotransmitter release function was turned off

the fact that all fractions reduced in vitro GABA receptor binding compared to the control (Table 16.1).

Modeled free DA reached a maximum concentration by day 3 at 127 nmol/mg in the control brain tissue. This maximum concentration was followed by a quick decline to 39.8 nmol/mg at day 20 in the control due to the increased free GABA concentration and subsequent DA inhibition. As E2 increased in circulation, which inhibited GABA release, the free DA then increased to 58.4 nmol/mg in the control brain tissue by day 60. The model predicted that exposure to the primary effluent would result in a slight increase in the maximum concentration of free DA by 103.2%, 104.6%, 106.8%, 107.2% and 107% nmol/mg in the ethyl acetate, PVPP water, PVPP ethanol, and hexane fractions, respectively, compared to the control (Fig. 16.6a). The water fraction resulted in a 2.4% decrease in maximum concentration of DA. The low concentration of free DA at day 30 due to GABA inhibition was reduced by 12.1%, 23.1%, and 47.5% nmol/mg in the water, PVPP water and PVPP ethanol fractions, respectively and increased by 17.8% and 38.2% in the hexane and ethyl acetate fractions, respectively, compared to the control. Interestingly, even after the neurotransmitter release was shut off at day 60, dopamine was still present in the hexane fraction for an additional 60 days. Overall, the highest concentration of the free DA increased by 0.02%, 1.1% and 3.1% in the ethyl acetate, hexane and water only simulations and decreased by 0.4% and 1.8% in the PVPP water and PVPP ethanol simulations, respectively. However, because all of the secondary PPME effluent fractions increased the amount of free GABA, all fraction simulations predicted a lower concentration at day 30 in free DA (Fig. 16.6b). The subse-

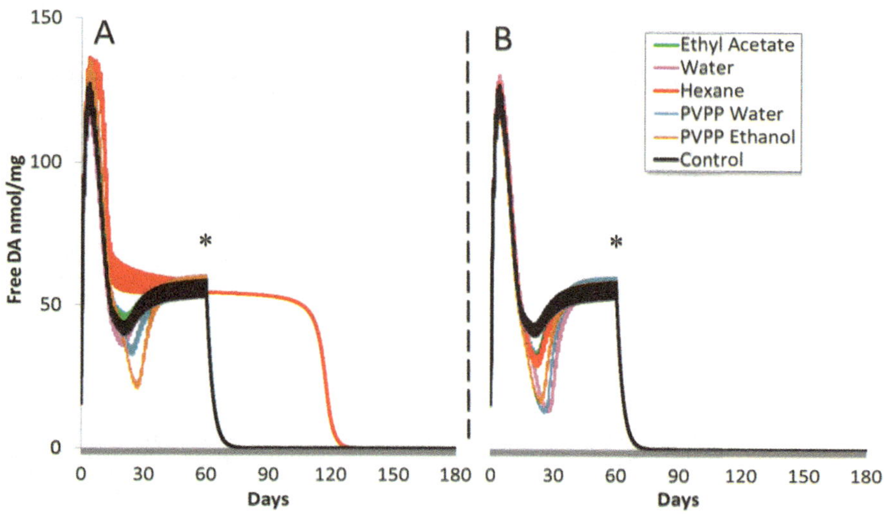

Fig. 16.6 Effects of primary (**a**) and secondary (**b**) pulp and paper mill effluent on modeled free dopamine (DA) in a generic female fish during vitellogenesis. * indicates when the neurotransmitter release function was turned off

quent free DA low concentration was reduced by 22.3%, 26.6%, 52.0%, 63.3% and 66.38% in the ethyl acetate, hexane, PVPP ethanol, PVPP water and water only simulation, respectively, compared to the control.

16.3.2 *Sex Steroid and Vitellogenin Production*

Free plasma E2 production reached the highest concentration of 5.0 ng/ml in the control simulation. The maximum concentration was slightly increased by 2.3%, 4.5% and 5.7% in the water only, PVPP water and PVPP ethanol fractions of the primary PPME, respectively. The ethyl acetate and hexane fractions reduced the maximum concentration of free E2 compared to the control by 10.0 and 81.2%, respectively, in the primary PPME. All simulated fractions from the secondary PPME slightly increased free E2 compared to the control simulation. The maximum concentration of E2 increased 3.9%, 4.5%, 5.1%, 5.6%, and 6.0% ng/ml in the ethyl acetate, hexane, PVPP ethanol, PVPP water and water only fractions of the secondary PPME, respectively.

Modeled cumulative vitellogenin was 459.4 mg/ml in the control after the simulation (Fig. 16.7). Only the water only fraction from the primary PPME increased modeled cumulative vitellogenin, which was 111.5% of the control. The remaining fractions from the primary PPME reduced modeled cumulative vitellogenin, by 7.8%, 11.9%, 16.9% and 90.1% in the PVPP water, PPVP ethanol, ethyl acetate,

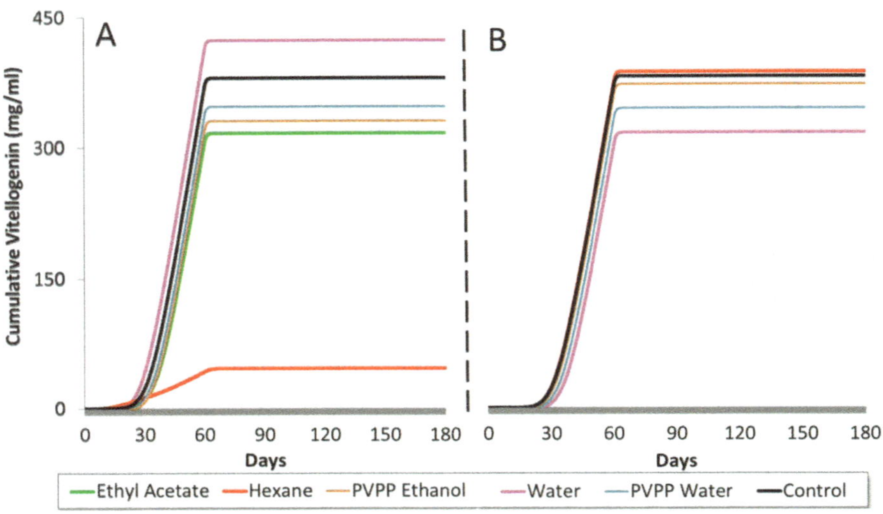

Fig. 16.7 Effects of primary (**a**) and secondary (**b**) pulp and paper mill effluent on modeled cumulative vitellogenin production in a generic female fish during vitellogenesis

and hexane fraction simulations, respectively (Fig. 16.7a). There was a lesser effect on cumulative vitellogenin from the secondary PPME simulations. In comparison to the control simulation, the hexane and ethyl acetate fractions slightly increased cumulative vitellogenin by 1.2% while the PVPP ethanol, PVPP water and water only fractions reduced cumulative vitellogenin by 1.7%, 8.8% and 16.9%, respectively (Fig. 16.7b).

16.3.3 Relating Effects on Vitellogenin to Egg Production

If we assume that vitellogenin production is directly related to egg production as described in other studies (Miller et al. 2007; Murphy et al. 2009), we can convert the cumulative vitellogenin results into an estimated fecundity effect (Fig. 16.8). Using the control simulation as our baseline, the model predicts a 111.5% increase in fecundity in the water only fraction from the primary PPME. The PVPP water, PVPP ethanol, ethyl acetate and hexane fractions each would reduce fecundity to 92.2%, 88.1%, 83.1% and 9.9%, respectively, of the control. The secondary PPME ethyl acetate and hexane fractions would slightly increase fecundity 101.9%, while the PVPP ethanol, PVPP water and water only fractions would reduce fecundity to 98.3%, 91.2% and 84.1%, respectively, compared to the control. Additionally, assuming that the effects caused by individual fractions are additive to one another, we would expect that fecundity would be reduced by 92.6 and 6.9% in the primary and secondary PPME, respectively, compared to the control.

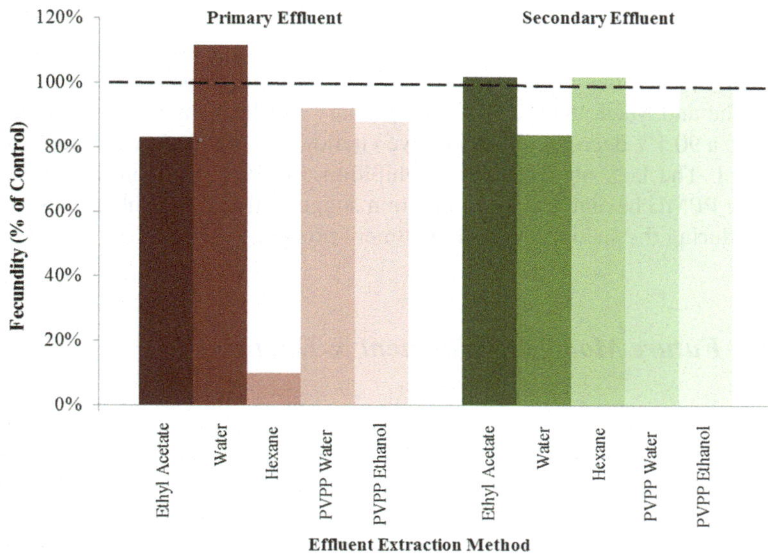

Fig. 16.8 Extrapolating the effects of primary and secondary pulp and paper mill effluent on vitellogenin production to egg production in a generic female fish during vitellogenesis. Dashed line represents an unexposed female fish's egg production

16.3.4 Comparison of Primary and Secondary PPME

Using a computational vitellogenesis model, we linked changes in neurochemical in vitro data for a complex mixture within PPME to adverse effects of vitellogenin production in a generic adult female fish to demonstrate how this could be done. While we don't have chemistry data to know the exact contaminants within the PPME – which may be over 250 different compounds (Ali and Sreekrishnan 2001), the in vitro work conducted by Basu et al. (2009) determined that at least 4 different molecular initiating events, changes to GABA and DA receptor binding and MAO and GABA-T enzyme activity, can occur following exposure. The model suggests that disrupted GABA and DA receptor binding and associated enzyme activity due to exposure to both primary and secondary treated PPME would reduce vitellogenin production, with the primary PPME having a more pronounced (13x) effect. Secondary treatment of the PPME reduced the modeled reproductive toxicity of the effluent, vitellogenin production increased 10x and comparable to the level of the control (102%).

Pulp and paper mill effluent can contain phytochemicals such as terpene, a product derived from conifer resin (Waye et al. 2014a; Basu et al. 2009). Terpene is a non-polar lipophilic, organic chemical, which can be extracted with hexane and other similar solvents (Basu et al. 2009). Terpenoids and other phytochemicals are hormonally active compounds present in bleached kraft pulp mill effluent (Belknap et al. 2006), which has been shown to reduce testosterone production in mummichog (Dube and MacLatchy 2000). The primary PPME hexane extract simulation resulted in a 90.1% decrease in cumulative vitellogenin production in comparison to the control. The lack of vitellogenin inhibition (102% of the control) during the secondary PPME hexane extract simulation suggests that these substances may be removed during the secondary stage treatment process.

16.3.5 Future Model Development & Data Gaps

This model is still in the early stages of development and its intent is not to predict future population changes nor recommend any regulatory changes. Research needs to be conducted to test the validity of the assumptions made in this model, most importantly the Michaelis-Menten kinetic values for the neurotransmitters, many of which were derived from the mammalian and avian literature. Additionally, the model was developed for a generic single batch synchronous spawning generic female fish and calibrated using sex steroid data collected from a percid laboratory study. The in vitro work conducted by Basu et al. (2009) used common goldfish as a model species, which is an asynchronous spawner capable of spawning many times over the course of their breeding season.

The work by Basu et al. (2009) used pooled data from both male and female whole brains for their in vitro assays. Future work assessing the potential of PPME to affect GABA and DA dynamics and subsequent vitellogenin production should focus on only female brains, specifically the hypothalamus region, and preferably from individuals. To test the validity of the model, vitellogenin and egg production should be measured in vivo from female fish exposed to PPME in order to provide a population relevant 2nd anchor to the integrated AOP vitellogenesis model (Fig. 16.9). These data would also allow us to test the assumption that reductions in vitellogenin directly relate to a reduction in egg production for fish exposed to PPME. Modeling exercises such as this can greatly advance the environmental toxicology field by synthesizing available information and directing future research towards addressing data gaps.

Future research should also look at testing in vitro effects of whole effluent on GABA and DA neurochemical processes. The Basu et al. (2009) paper fractionated the effluent into different chemical classes; however, it did not assess the effects of the whole effluent nor its predicted concentration in receiving waters. Without these data, we do not have a way to determine if these chemicals within the PPME are acting synergistically, antagonistically or additively to one another. Additional analysis of the specific chemical compounds found within the effluent fractions would also be beneficial in determining the active compounds of the effluent.

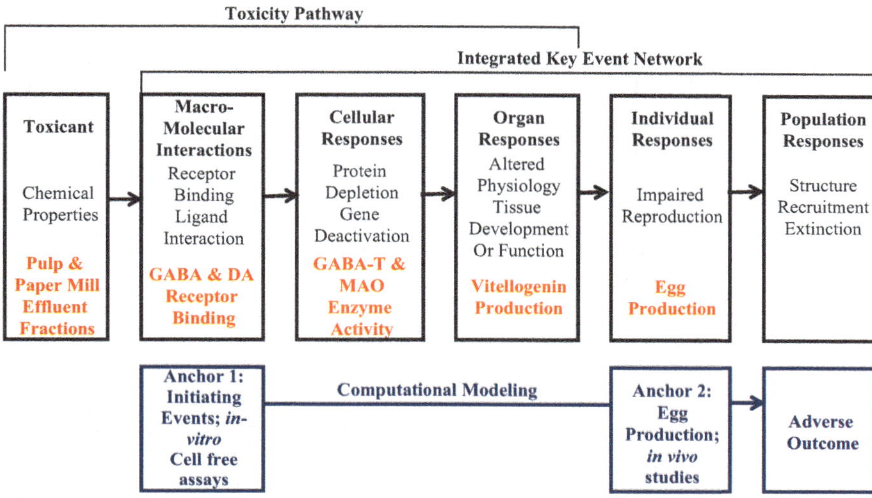

Fig. 16.9 Future integrated AOP and how computational modeling can be used to link in vitro data to estimate in vivo effects

16.3.6 Proof of Principle and Future Intent

While not yet validated, these simulations demonstrate proof of principle for using computational modeling to link molecular effects of complex mixtures to adverse effects on fish populations. Once validated, this model can be used as a screening tool using in vitro data to help direct in vivo exposures towards potentially more harmful contaminants. The traditional fish short-term reproductive assay for assessing estrogenic chemical effects takes 35 days to conduct and is only capable of testing either just a few concentrations of a single contaminant or a single concentration of a mixture of just a couple of contaminants (Ankley et al. 2001; Armstrong et al. 2015). With over 85,000 registered chemicals on the U.S. Environmental Protection Agency's TSCA Chemical Substance Inventory (U.S. EPA 2015), it would be impractical to conduct in vivo methods on every chemical and even more impractical to test complex chemical mixtures. The effects of thousands of chemicals at varying concentrations and their mixtures can be collected in that same 35 day time span using in vitro methods. High-throughput screening can currently screen 100,000 compounds per day with the potential to test up to one million samples per day (Szymański et al. 2012). The data from these assays can be extremely variable and hard to relate to an in vivo response (Basu et al. 2009; Knudsen et al. 2011), however when coupled with a computational model using an AOP framework, quantifiable linkages can be made relating in vitro data to adverse in vivo effects.

Taking an Integrated AOP approach, we hypothetically linked together several molecular initiating events that may result in the same adverse outcome – impair-

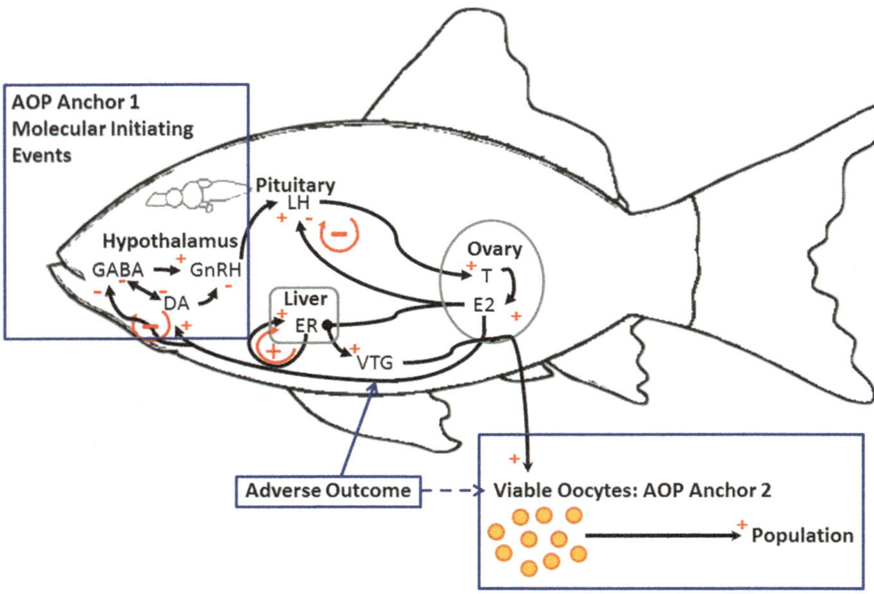

Fig. 16.10 Conceptual vitellogenesis model of an integrated adverse outcome pathway (AOP) framework to link multiple initiating events to adverse effects on egg production. Black arrows indicate either a inhibitory (-) or stimulatory (+) relationship between two state variables. Red arrows indicate positive (+) or negative (-) feedback loops. GABA: γ-aminobutyric acid; DA: dopamine; GnRH: gonadotropin releasing hormone; LH: luteinizing hormone; T: testosterone; E2: 17β-estradiol; ER: estrogen receptor; and VTG: vitellogenin

ment of vitellogenin production. Using a computational model to create quantitative linkages within the AOP framework, we extrapolated from in vitro neurochemical data to an adverse population outcome stemming from PPME exposure (Fig. 16.10).

It is likely that several AOPs will be activated due to exposure to complex mixtures. In order for computational models and AOPs to be used for risk assessment purposes they must undergo strict developmental guidelines, which include testing for reliability and robustness (OECD 2013). Weight of evidence analyses then need to be conducted for each key element of the AOP(s) (Becker et al. 2015). These analyses can include a simple qualitative method by assigning a level of confidence (very strong, strong moderate, weak, very weak) to the amount of data available linking key events within an AOP to an apical endpoint (OECD 2013). Additionally, a quantitative weight of evidence approach can be applied where an expert panel uses specific criteria to apply weights and scores to individual lines of evidence for each key event within an AOP. This criterion would include mechanistic relationships between key events, evidence of a downstream key event being impaired if an upstream key event is blocked, and consistency across a wide range of taxa and stressors for which the key event(s) occur. A mathematical or statistical model would then evaluate the weights and scores to determine an overall conclusion for support of the new AOP (Becker et al. 2015). The AOPs linked in this model are strictly hypothetical as none have been directly linked to vitellogenin production, nor has a weight of evidence approach been applied. Further research is needed to determine if GABA receptor antagonism, D2 receptor agonism, increased GABA-T activity and/or decreased MAO activity results in reduced vitellogenin production. As more single AOPs are tested and become available, future computational models could be updated to create a stronger integrated AOP framework.

With this research, we have a better understanding of how contaminant mixtures can affect fish populations, which is invaluable for developing guidelines for acceptable contaminant loads for healthy ecosystems. Once validated, we can use this approach in combination with the U.S. Environmental Agency's ToxCast™ program to help prioritize chemicals for further toxicological analyses. The ToxCast™ program consists of data collected from high-throughput screening assays and uses computational toxicology methods to predict toxicity of specific compounds (Knudsen et al. 2011).

References

Adams S (2005) Assessing cause and effect of multiple stressors on marine systems. Mar Pollut Bull 51:649–657

Ali M, Sreekrishnan TR (2001) Aquatic toxicity from pulp and paper mill effluents: a review. Adv Environ Res 5(2):175–196

Ankley GT, Jensen KM, Kahl MD et al (2001) Description and evaluation of a short-term reproductive test with the fathead minnow (*Pimephales promelas*). Environ Toxicol Chem 20(6):1276–1290

Ankley GT, Bennett RS, Erickson RJ et al (2010) Adverse outcome pathways: a conceptual framework to support ecotoxicology research and risk assessment. Environ Toxicol Chem 29(3):730–741

Armstrong BM (2016) Using computational models to scale sublethal effects of stressors to adverse population outcomes in fish. Dissertation. Michigan State University

Armstrong BM, Lazorchak JM, Murphy CA et al (2015) Determining the effects of a mixture of an endocrine disrupting compound, 17α-ethinylestradiol, and ammonia on fathead minnow (*Pimephales promelas*) reproduction. Chemosphere 120:108–114

Arukwe A, GoskØyr A (2003) Eggshell and egg yolk proteins in fish: hepatic proteins for the next generation: oogenetic, population, and evolutionary implications of endocrine disruption. Comp Hepatol 2:1–21

Bardakdjian J, Tardy M, Pimoule C et al (1979) GABA metabolism in cultured glial cells. Neurochem Res 4(4):518–527

Basu N, Ta CA, Waye J et al (2009) Pulp and paper mill effluents contain neuroactive substances that potentially disrupt neuroendocrine control of fish reproduction. Environ Sci Technol 43:1635–1641

Becker RA, Ankley GT, Edwards SW et al (2015) Increasing scientific confidence in adverse outcome pathways: application of tailored bradford-hill considerations for evaluating weight of evidence. Regul Toxicol Pharmacol 72:514–537

Belknap AM, Solomon KR, MacLatchy DL (2006) Identification of compounds associated with testosterone depressions in fish exposed to bleached kraft pulp and paper mill chemical recovery condensates. Environ Toxicol Chem 25(9):2322–2333

Blum JJ, Reed MC, Janovick JA, Conn PM (2000) A mathematical model quantifying GnRH-induced LH secretion from gonadotropes. Am J Physiol Endocrinol Metab 278:263–272

Bortolato M, Chen K, Shih JC (2008) Monoamine oxidase inactivation: from pathophysiology to therapeutics. Adv Drug Deliv Rev 60:1527–1533

Carvan MJ III, Incardona JP, Rise ML (2008) Meeting the challenges of aquatic vertebrate ecotoxicology. Bioscience 58(11):1015–1025

Castaneda TR, Marquez de Prado B, Prieto D et al (2004) Circadian rhythms of dopamine, glutamate, and GABA in the striatum and nucleus accumbens of the awake rate: modulation by light. J Pineal Res 36(3):177–185

Chu DCM, Albin RL, Young AB, Penney JB (1990) Distribution and kinetics of GABAB binding sites in rat central nervous system: a quantitative autoradiographic study. Neuroscience 34(2):341–357

Dabrowski K, Ciereszko A, Ramseyer I et al (1994) Effects of hormonal treatment on induced spermiation and ovulation in the yellow perch (*Perca flavescens*). Aquaculture 120:171–180

Debofsky AR, Klingler RH, Mora FX et al (2015) Female reproductive impacts of dietary methylmercury in yellow perch (*Perca flavescens*) and zebrafish (*Danio rerio*)

Depledge MH (1994) The rational basis for the use of biomarkers as ecotoxicological tools. In: Fossi MC, Leonzio C (eds) Nondestructive biomarkers in vertebrates. Lewis, Boca Raton, pp 271–295

Dey S, Choudhurry MD, Das S (2013) A review on toxicity of paper mill effluent on fish. Bull Environ, Pharmacol Life Sci 2(3):17–23

Dube MG, MacLatchy DL (2000) Endocrine responses of *Fundulus heteroclitus* to effluent from a bleached-kraft pulp mill before and after installation of reverse osmosis treatment of a waste stream. Environ Toxicol Chem 19(11):2788–2796

Forbes VE, Palmqvist A, Bach L (2006) The use and misuse of biomarkers in ecotoxicology. Environ Toxicol Chem 25(1):272–280

Groh KJ, Carvalho RN, Chipman JK et al (2015) Development and application of the adverse outcome pathway framework for understanding and predicting chronic toxicity: II. Focus Growth Impair fish Chemosphere 120:778–792

Hall TR, Urena G, Figueroa HR (1982) In vivo and in vitro effects of temperature on monoamine oxidase activity in brain and other tissues of the goldfish, *Carassius auratus L.* Comp Biochem Physiol 73C:175–180

Hammond GL (2016) Plasma steroid-binding proteins; primary gatekeepers of steroid hormone action. J Endocrinol 230:R13–R25

Holmstrup M, BindesbØl A-M, Oostingh GJ et al (2010) Interactions between effects of environmental chemicals and natural stressors: a review. Sci Total Environ 408:3746–3762

Hooper MJ, Ankley GT, Cristol DA et al (2013) Interactions between chemical and climatic stressors: a role for mechanistic toxicology in assessing climate change risks. Environ Toxicol Chem 32(1):32–48

Jensen KM, Korte JJ, Kahl MD et al (2011) Aspects of basic reproductive biology and endocrinology in the fathead minnow (*Pimephales promelas*). Comp Biochem Physiol Part C 128:127–141

Knudsen TB, Houck KA, Sipes NS et al (2011) Activity profiles of 309 ToxCast™ chemicals evaluated across 292 biochemical targets. Toxicology 282:1–15

Kolpin D, Furlong E, Meyer M (2002) Pharmaceuticals, hormones, and other organic wastewater contaminants in U.S. streams, 1999-2000: a national reconnaissance. Environ Sci Technol 36:1202–1211

Kovacs TG, Martel PH, O'Connor BI et al (2013) A survey of Canadian mechanical pulp and paper mill effluents: insights concerning the potential to affect fish reproduction. J Environ Sci Health A 48(10):1178–1189

Kramer VJ, Etterson MA, Hecker M et al (2011) Adverse outcome pathways and ecological risk assessment bridging to population-level effects. Environ Toxicol Chem 30(1):64–76

Kubikova L, Vyboh P, Kost'al L (2009) Kinetics and pharmacology of the D1 and D2 like dopamine receptors in Japanese quail brain. Cell Mol Neurobiol 29:961–970

Lam DMK (1972) The biosynthesis and content of gamma-aminobutyric acid in the goldfish retina. J Cell Bio 54:225–231. https://doi.org/10.1083/jcb.54.2.225

Kurreck J, Stein CA (2015) Molecular medicine: an introduction. Wiley-Blackwell, Weinheim

Lauder JM (1993) Neurotransmitters as growth regulatory signals: role of receptors and second messengers. Trends Neurosci 16(6):233–240

Laws EA (2000) Aquatic pollution: an introductory text, 3rd edn. Wiley-Interscience, New York

Leaños-Castañeda O, Van Der Kraak G (2007) Characterization of estrogen receptor subtypes, ERα and ERβ, mediating vitellogenesis production in the liver of rainbow trout. Toxicol Appl Pharmacol 224:116–125

Levavi-Sivan B, Bogerd J, Mananos EL et al (2010) Perspectives on fish gonadotropins and their receptors. Gen Comp Endocrinol 165:412–437

Macho L, Kvetnansky R, Culman J et al (1986) Neurotransmitter levels in the hypothalamus during postnatal development in rates. Exp Clin Endocrinol 88(2):142–150

Maffucci JA, Gore AC (2009) Hypothalamic neural systems controlling the female reproductive life cycle: gonadotropin-releasing hormone, glutamate and GABA. Int Rev Cell Mol Biol 274:69–127

Mananos EL, Anglade I, Chyb J et al (1999) Involvement of gamma-aminobutyric acid in the control of GTH-1 and GTH-2 secretion in male and female rainbow trout. Neuroendocrinology 69(4):269–280

McGeer PL, Eccles JC, McGeer EG (2013) Molecular neurobiology of the mammalian brain. Plenum Press, New York

Miller DH, Jensen KM, Villeneve DL et al (2007) Linkage of biochemical responses to population-level effects: a case study with vitellogenin in the fathead minnow. Environ Toxicol Chem 26:521–527

Monosson E (2005) Chemical mixtures: considering the evolution of toxicology and chemical assessment. Environ Health Perspect 113(4):383–390

Murphy CA, Rose KA, Thomas P (2005) Modeling vitellogenesis in female fish exposed to environmental stressors; predicting the effects of endocrine disturbance due to exposure to a PCB mixture and cadmium. Reprod Toxicol 19:395–409

Murphy CA, Rose KA, Rahman MS et al (2009) Testing and applying a fish vitellogenesis model to evaluate laboratory and field biomarkers of endocrine disruption in Atlantic croaker exposed to hypoxia. Environ Toxicol Chem 28:1288–1303

National Research Council (2007) Toxicity testing in the 21st century: a vision and a strategy. National Academies Press, Washington, DC

Nelson ER, Habibi HR (2013) Estrogen receptor function and regulation in fish and other vertebrates. Gen Comp Endocrinol 192:15–24

Organisation for Economic Co-operation and Development (2013) Guidance document on developing and assessing adverse outcome pathways: series on testing and assessment No. 184. http://www.oecd.org/officialdocuments/publicdisplaydocumentpdf/?cote=env/jm/mono(2013)6&doclanguage=en

Peter RE, Yu KL, Marchant TA et al (1990) Direct neural regulation of the teleost adenohypophysis. J Exp Zool A Ecol Genet Physiol 256(S4):84–89

Popesku JT, Martyniuk CJ, Mennigen J et al (2008) The goldfish (*Carassius auratus*) as a model for neuroendocrine signaling. Mol Cell Endocrinol 293:43–56

Ruby SM, Idler DR, So YP (1986) The effect of sublethal cyanide exposure on plasma vitellogenin levels in rainbow trout (*Salmo gairdneri*) during early vitellogenesis. Arch Environ Contam Toxicol 15:603–607

Sabo-Attwood T, Kroll KJ, Denslow ND (2004) Differential expression of largemouth bass (*Micropterus salmoides*) estrogen receptor isotypes alpha, beta and gamma by estradiol. Mol Cell Endocrinol 218:107–118

Sarthy PV, Lam DMK (1979) The uptake and release of [3H] dopamine in the goldfish retina. J Neurochem 32:1269–1277

Sepulveda MS, Quinn BP, Denslow ND et al (2003) Effects of pulp and paper mill effluents on reproductive success of largemouth bass. Environ Toxicol Chem 22(1):205–213

Smith MJ, Jennes L (2001) Neural signals that regulate GnRH neurons directly during the oestrous cycle. Reproduction 122:1–10

Smolders I, Bogaert L, Ebinger G et al (1997) Muscarinic modulation of striatal dopamine, glutamate, and GABA release, as measured with in vivo microdialysis. J Neurochem 68:1942–1948

Szabo A, Ruby SM, Rogan F et al (1991) Changes in brain dopamine levels, oocyte growth and spermatogenesis in rainbow trout, *Oncorhynchus mykiss*, following sublethal cyanide exposure. Arch Environ Contam Toxicol 21:152–157

Szymański P, Markowicz M, Mikiciuk-Olasik E (2012) Adaptation of high-throughput screening in drug discovery – toxicological screening tests. Int J Mol Sci 13(1):427–452

Terasawa E (1994) Steroid modulation of pulsatile LHRH release in the Rheses monkey. Horm Behav 28:408–416

Thompson G, Swain J, Kay M et al (2001) The treatment of pulp and paper mill effluent: a review. Bioresour Technol 77:275–286

Treiman DM (2001) GABAergic mechanisms in epilepsy. Epilepsia 42(3):8–12. https://doi.org/10.1046/j.1528-1157.2001.042suppl.3008.x

Trudeau VL, Sloley BD, Peter RE (1993) GABA stimulation of gonadotropin-II release in goldfish: involvement of GABAA receptors, dopamine, and sex steroids. Am J Physiol 265(2):R348–R355

Trudeau VL, Spanswick D, Fraser EJ et al (2000) The role of amino acid neurotransmitters in the regulation of pituitary gonadotropin release in fish. Biochem Cell Biol 78:241–259

United States Environmental Protection Agency (2012) Research in action: exposure biomarkers in aquatic organisms. http://www.epa.gov/nerleerd/research/exposurebiomarkers.html

United States Environmental Protection Agency (2015) The toxic substances control act (TSCA) chemical substance inventory. https://www.epa.gov/tsca-inventory

van den Pol AN (2010) Excitatory neuromodulator reduces dopamine release, enhancing prolactin secretion. Neuron 65(2):147–149. https://doi.org/10.1016/j.neuron.2010.01.013

Van Der Kraak G (2009) The GnRH system and the neuroendocrine regulation of reproduction. Fish Neuroendocrinol 28:115–149

Van Der Kraak G, Zacharewski T, Janz DM et al (1998) Comparative endocrinology and mechanisms of endocrine modulation in fish and wildlife. In: Principles and processes for evaluating endocrine disruption in wildlife. SETAC Press, Pensacola, pp 97–119

Venton BJ, Zhang H, Garris PA et al (2003) Real-time decoding of dopamine concentration changes in the caudate-putamen during tonic and phasic firing. J Neurochem 87:1284–1295

Voaden MJ, Marshall J, Murani N (1974) The uptake of [3H]-amino butyric acid and [3H] glycine by the isolated retina of the frog. Brain Res 67:115–132

Watanabe K, Anderson M, Basu N et al (2011) Defining and modeling known adverse outcome pathways: Domoic acid and neuronal signaling as a case study. Environ Toxicol Chem 30(1):9–21

Watanabe M, Fukuda A, Nabekura J (2014) The role of GABA in the regulation of GnRH neurons. Front Neurosci 8:1–9

Waye A, Annal M, Tang A et al (2014a) Canadian boreal pulp and paper feedstocks contain neuroactive substances that interact in vitro with GABA and dopaminergic systems in the brain. Sci Total Environ 468-139:315–325

Waye A, Lado WE, Martel PH et al (2014b) Ovulation but not milt production is inhibited in fathead minnows (*Pimephales promelas*) exposed to a reproductively inhibitory pulp mill effluent. Reprod Biol Endocrinol 12:1–43

Wheeler DD, Hollingsworth RG (1979) A model of GABA transport by cortical synaptosomes from the long-Evans rat. J Neurosci Res 4:265–289

Zakes Z, Demska-Zakes K (2005) Artificial spawning of pikeperch (*Sander lucioperca (L.)*) stimulated with human chorionic gonadotropin (HCG) and mammalian GNRH analogue with a dopamine inhibitor. Arch Pol Fish 13(1):63–75

Żarski D, Horváth A, Held JA et al (2015) Artificial reproduction of percid fishes. In: Kestemont P, Dabrowski K, Summerfelt CR (eds) Biology and culture of percid fishes: principles and practices. Springer Netherlands, Dordrecht, pp 123–161

Zayas T, Picazo M, Salgado L (2011) Removal of organic matter from paper mill effluent by electrochemical oxidation. J Water Resour Prot 3:32–40

Zohar Y, Munoz-Cueto JA, Elizur A et al (2010) Neuroendocrinology of reproduction in teleost fish. Gen Comp Endocrinol 165:438–455

Chapter 17
Use of Adverse Outcome Pathways in Human Risk Assessment and Toxicology

Catherine Willett, Suzanne Fitzpatrick, Bette Meek, and Carl Westmoreland

Abstract Mechanistic information has been used for many years to inform chemical hazard and risk assessments. NRC reports and several agency strategic plans in recent years promote the large-scale use of mechanistic information, organized in the form of pathways at different levels of biological organization as a basis to underpin a dramatic change in the way chemical assessment is performed. As a result, there now exist international collaborations to develop the data and knowledge bases, guidance and principles for development and use of "Adverse Outcome Pathways" (AOPs). Many of the principles for developing and using pathways are based on experience with Mode of Action frameworks for human health risk assessment. Expert groups within the Organization for Economic Cooperation and Development (OECD) are publishing guidance and partnering with the US EPA and European Commissions Joint Research Centre (JRC) to develop a public knowledge base for building AOPs on a large scale. Although this direction is fairly new, there are many pathways already in development. In addition, pathway-based approaches are increasingly being applied to a variety of assessments of hazard in a number of sectors. This chapter describes the genesis of the AOP concept, the development of the necessary tools based on international collaborations, and provides some examples of the use of AOPs in human health risk assessment.

C. Willett (✉)
The Humane Society of the United States, Washington, DC 20037, USA
e-mail: kwillett@humanesociety.org

S. Fitzpatrick
Center for Food Safety and Applied Nutrition, Food and Drug Administration, College Park, MD 20740–3835, USA

B. Meek
McLaughlin Centre for Population Health Risk Assessment, University of Ottawa, Ottawa, ON, Canada, K1N

C. Westmoreland
Safety & Environmental Assurance Centre, Unilever, Colworth Science Park, Sharnbrook, Bedfordshire, UK, MK44 1LQ

N. Garcia-Reyero, C.A. Murphy (eds.), *A Systems Biology Approach to Advancing Adverse Outcome Pathways for Risk Assessment*,
DOI 10.1007/978-3-319-66084-4_17

17.1 Introduction: Use of Mechanistic Information in Determining Hazard for Human Health

Consideration of mechanistic information in hazard assessment makes sense from the point of view that understanding the biological process(es) affected by chemical exposure will increase confidence in decisions and predictions made about the use of that chemical. It would also increase confidence in grouping similar chemicals, and in predicting potential outcomes of cumulative exposure of similar chemicals. From this reckoning came the realization that the better our understanding of biology generally, and the greater our knowledge about chemical-biological interactions, the better informed and less uncertain hazard and risk assessments will become overall. Hence, the desire for a systematic approach to collect, evaluate and organize this information has increased dramatically over the past few years, and the use of this information is becoming more wide-spread. This chapter describes some of these efforts.

The incorporation of mechanistic information in risk assessment has a long history, including dose-response modeling efforts (for example, see Clewell et al. 1995) and mode-of-action frameworks, such as those developed by the International Life Sciences Institute (Meek et al. 2003; Seed et al. 2005) and evolved and extended more recently by the International Program on Chemical Safety (IPCS) to determine human relevance of mode- of-action(s) of pesticides and industrial chemicals (Boobis et al. 2006, 2008), and the creation of mode-of-action pathways in drug development (e.g., Iorio et al. 2010) and their application (e.g., Schadt and Lum 2006). Indeed, these efforts were novel in considering the systematic application of molecular and chemical mechanistic information to the interpretation of empirical data on complex endpoints and responded to increasing early recognition of the importance of such information to increase predictivity and decrease uncertainty in risk assessment for regulatory purposes (See, for example, US EPA 2005a).

17.1.1 Early Applications of Mechanistic Information in Hazard and Risk Characterization

Mechanistic information has been described as the entirety of "critical biological factors that regulate particular biological processes and their interrelationships… occurring at all levels of organization: population, organism, organ, cell and molecular" (Becking 1995). Mechanistic information has been used for decades in all steps of risk assessment (hazard identification, dose-response modeling, exposure modeling, and overall risk characterization) to decrease uncertainty and increase predictivity in comparison to traditionally adopted application of default assumptions to empirical data . One of the first areas of toxicology in which mechanistic understanding was applied was carcinogenicity, due in part to the mechanistic information available (in particular, that on mutagenicity) and the importance of the endpoint. An example of an early application of mechanistic information is the adoption of differential defaults for dose-response modeling for compounds acting via a mutagenic mode of action using linear models) versus those for which early mutation isn't the key event for the critical effect where models assume a threshold). This results in the development of "safe" doses for all endpoints except cancer, where for compounds which are mutagenic, the incidence of disease in the population is estimated. Mechanistic information has also been incorporated into more sophisticated pharmacokinetic modeling to estimate the "effective dose" of the active form of a substance in the target tissue, for example in the liver after exposure of a chemical by inhalation, by including information about physiological and metabolic processes that occur in the relevant tissues (Conolly and Andersen 1993; Clewell et al. 1995). Mechanistic information has also been used to support interspecies extrapolation and coverage of sensitive populations by including specific physiological differences between the test species and the target species (e.g. species differences in the efficiency of uptake, clearance or metabolism of the chemical by relevant tissues, differences in tissue and blood volume), thus decreasing uncertainly about the extrapolation and estimation of safe exposure (reviewed in Becking 1995 and Haber et al. 2001; Meek et al. 2002a, b; IPCS 2005).

17.1.2 Mode-of-Action (MoA) Frameworks

The formal consideration or mechanistic information in the hazard assessment of chemicals is commonly referenced as mode of action analysis. Early work by the International Life Sciences Institute and more recently IPCS in the development of cancer and non-cancer MoA species concordance frameworks outline a systematic process of describing and documenting mechanistic support for chemical specific MoA in animals and comparing those with likely MoA in humans to determine human relevance (Meek et al. 2003; Seed et al. 2005; Boobis et al. 2006, 2008). They include reference to several founding principles of pathway-based approaches:

- MoA is defined as a series of key events along a biological pathway from the initial chemical interaction through to the toxicological outcome, with key events being defined as measurable necessary precursor events to the adverse outcome (for a list of terms and definitions, see Table 17.1);
- that a MoA does not need to reflect complete mechanistic understanding to be useful and that its use depends on level of completeness (e.g., incomplete MoA can inform testing strategies but is likely not sufficient to support hazard classification);
- focus on the most likely mode of action which meaningfully integrates information on potential early key events but considers the plausibility of alternative modes;

Table 17.1 Definitions of some common and important terms relating to adverse outcome pathways

Adverse outcome Pathway (AOP)	OECD (2013a)	An AOP is a sequence of events from the exposure of an individual or population to a chemical substance through a final adverse (toxic) effect at the individual level (for human health) or population level (for ecotoxicological endpoints). The key events in an AOP should be definable and make sense from a physiological and biochemical perspective. AOPs incorporate the toxicity pathway and mode of action for an adverse effect. AOPs may be related to other mechanisms and pathways as well as to detoxification routes
Integrated approach to testing and assessment (IATA)	OECD (2015c)	A structured approach that strategically integrates and weights all relevant data to inform regulatory decisions regarding potential hazard and/or risk and/or the need for further targeted testing and therefore optimizing and potentially reducing the number of tests that need to be conducted
Intermediate event	OECD (2013a)	Biological events that lie "between the molecular initiating event and the apical outcome" from which the key events are identified
Key event (KE)	Meek et al. (2014a)	"An empirically observable step or its marker, which is a necessary element of the mode of action critical to the outcome (i.e., necessary, but not necessarily sufficient in its own right); key events are measurable and reproducible"
Key event relationship (KER)	OECD (2014)	A scientifically-based relationship that connects one key event to another, defines a directed relationship between the two (i.e., identifies one as upstream and the other as downstream), and facilitates inference or extrapolation of the state of the downstream key event from the known, measured, or predicted state of the upstream key event
Mode of action (MoA)	Meek et al. (2014a)	"A biologically plausible series of key events leading to an effect"
Initiating event (MIE)	OECD (2011)	The initial point of chemical-biological interaction within the organism that starts the pathway

- definition of a "key event" as a step in the pathway that is critical to development of the toxicological outcome and is measurable; a requirement to systematically establish causation between key events;
- the importance of quantitation in the application of MoA to risk assessment; and the need to establish relevance to human biology,
- taking into account both chemical specific toxico-kinetics and -dynamics.

Establishing evidence for the MoA hypothesis is based on considerations modified from those of Bradford-Hill and introduced originally for establishing the causality of observed associations in epidemiological studies (Bradford-Hill 1965). The IPCS frameworks recommend determining human relevance by answering four key questions: (1) is there sufficient weight-of-evidence (WoE) for the MoA in animals? (2) can human relevance be excluded on the basis of qualitative differences in key events? (3) can human relevance be excluded on the basis of quantitative differences in key events? and (4) do the quantitative differences affect the default uncertainty factors applied in risk assessment?

The MoA framework has been updated to accommodate insights from the expanding application of pathway-based approaches to risk assessment in general (Meek et al. 2014a and b). The updated framework references more explicit description of the application of information on mode of action in a more predictive context (i.e., predicting later key events and adverse outcomes from earlier key events) and the contribution of chemical-specific information such as metabolism and chemical agnostic information at different levels of biological complexity. In this framework, MoA and Adverse Outcome Pathways (see below) are considered conceptually similar, with a distinction that MoA does not necessarily imply adversity, though an adverse outcome has often been included as a key event in MoA descriptions; it can also refer, for example, to a description of key events documenting the basis for the therapeutic efficacy of drugs (Table 17.2). Two applications of the updated framework are presented; for observed (in vivo) effects and for hypothesized events, with several case studies being presented for each, as concrete examples of regulatory application taking into consideration of the weight of integrated evidence using modified Bradford- Hill considerations, to transparently document the level of confidence.

MoA/human relevance framework has been incorporated into international guidance (EFSA 2006; EC 2003; IPCS 2006; JMPR 2006; OECD 2002; UNECE 2007) and is routinely used in toxicological assessments by the US EPA (organic arsenic: U.S. EPA 2005b, 2007; chloroform: US EPA 1999; atrazine: US EPA 2000, and Dellarco and Baetcke 2005), the United Kingdom (COC 2004), Health Canada (see, for example, Liteplo and Meek 2003), Australia (see, for example, NICNAS 2006), and others (US EPA 2007a and b; Meek et al. 2008; Carmichael et al. 2011). Some limitations to the use of this approach have been that traditional animal tests do not provide data to support MoA development or assessment, therefore, such assessment requires a fair amount of additional specialized animal experimentation as well as in vitro analyses. In addition, weight of evidence has been inconsistently documented in part due to limited experience evaluating adequate mechanistic information. Other barriers have included a lack of harmonized terminology and assessment methods as well as human information for comparison (note, these issues currently pertain to all pathway based approaches) (Carmichael et al. 2011). Creation and use of a common knowledge base may address these issues (see below). This experience prompted development of the original prototype of the AOP wiki (previously described as the MOA wiki).

17.1.3 Adverse Outcome Pathways

In its seminal 2007 report Toxicity Testing for the twenty-first Century: a Vision and a Strategy, the National Academies of Science panel describes a "new toxicity-testing system that evaluates biologically significant perturbations in key toxicity

Table 17.2 Modified Bradford-Hill considerations as guidance for evaluating relative confidence in AOP KERs and overall

Biological Plausibility Defining question: is there a mechanistic (i.e., structural or functional) relationship between the $KE_{upstream}$ and the $KE_{dowstream}$ consistent with established biological knowledge?[a, b, c]		
High (strong) confidence: extensive understanding of the KER based on extensive previous documentation and broad acceptance (e.g., mutation leading to tumors), i.e., an established mechanistic basis	Moderate confidence: the KER is plausible based on analogy to accepted biological relationships, but scientific understanding is not completely established	Low (weak) confidence: there is empirical support for an association between KEs (see empirical evidence below), but the structural or functional relationship between them is not understood
Essentiality[d] Defining question: are downstream KEs and/or the adverse outcome prevented if an upstream K E is blocked?		
High (strong) confidence: direct evidence from specifically designed experimental studies illustrating essentiality for at least one of the important key events (e.g., stop/reversibility/recovery studies, antagonism, knockout models, etc.)	Moderate confidence: indirect evidence that sufficient modification of an expected modulating factor attenuates or augments a KE (e.g., augmentation of proliferative response in the $KE_{upstream}$ leading to an increase in $KE_{downstream}$ or in the AO)	Low (weak) confidence: no or contradictory experimental evidence of the essentiality of any of the K E
Empirical Evidence[e, f] Defining questions: does the empirical evidence support that a change in $KE_{upstream}$ leads to an appropriate change in $KE_{downstream}$? Does $KE_{upstream}$ occur at lower doses and earlier time points than $KE_{downstream}$ and is the incidence of $KE_{upstream}$ greater than that for the $KeyEvent_{downstream}$? Are there inconsistencies in empirical support across taxa, species and stressors that don't align with an expected pattern for the hypothesized AOP? (Note: in many cases, evidence that contributes to quantitative understanding of a KER also provides empirical support for the relationship, and such relevant information should be considered as part of the overall weight-of-evidence evaluation of the concordance of empirical observations and consistency of the KER)		
High (strong) confidence: multiple studies showing dependent change in both events following exposure to a wide range of specific stressors. Extensive evidence for temporal, dose-response and incidence concordance and no or few critical data gaps or conflicting data	Moderate confidence: demonstrated dependent change in both events following exposure to a small number of specific stressors and some evidence inconsistent with an expected pattern that may be explained by factors such as experimental design, technical considerations, differences among laboratories, etc	Low (weak) confidence: limited or no studies reporting dependent change in both events following exposure to a specific stressor (i.e., endpoints never measured in the same study or not at all); and/or significant inconsistencies in empirical support across taxa and species that don't align with expected pattern for hypothesized AOP

(continued)

Table 17.2 (continued)

From OECD (2016)

[a] The guidance for "high", "moderate" and "low" draws on limited current experience. Additional delineation of the nature of relevant evidence in these broadly defined categories requires more experience with larger numbers of documented AOPs

[b] "Direct evidence" implies specifically designed experiments to consider the relevant element. "Indirect evidence" normally relates to empirical support and is largely duplicative of Element 3 [empirical evidence]

[c] To the extent possible, each of the relevant Bradford Hill considerations is addressed for each of the KERs (biological plausibility and empirical support) and KEs (essentiality) and separate rationales provided

[d] While the essentiality of each of the KEs is addressed separately, delineation of the degree of confidence is based on consideration of evidence for all of the KEs within the AOP and therefore, only one rationale is required

[e] This is normally considered on the basis of tabular presentation of available data on temporal and dose-response aspects, in a template that documents the extent of support. See, for example, Meek and Klaunig (2010)

[f] Note that this relates to concordance of dose response, temporal and incidence relationships for KERs rather than the KEs; the defining question is not whether or not there is a dose response relationship for the KE but rather there is concordance with that for earlier and later KEs. This is normally demonstrated in studies with different types of stressors

pathways by using new methods in computational biology and a comprehensive array of in vitro tests based on human biology" (NRC 2007). In this context, a "toxicity pathway" is a normal biological pathway that becomes perturbed beyond the point of homeostatic correction leading to toxicity. A description of toxicity begins with chemical characterization, progresses through elucidation of the chemical interaction with the biological system (the pathway), involves targeted testing to query effects at critical steps of the pathway, and dose-response extrapolation to estimate human exposures required to elicit the effect. Additional population-based modeling is required to predict ecological effects.

The related concept of "adverse outcome pathways", or AOPs, arose from the field of ecotoxicology as a way of improving the efficiency of chemical assessment by effectively integrating information on various levels of biological organization for risk assessment for an increasing number of chemicals and endpoints (Ankley et al. 2010). An AOP describes the events that occur following exposure to chemicals or other stressors, beginning with the molecular interaction of the chemical with a biomolecule (e.g., a protein, receptor, etc.) – the molecular initiating event (MIE) – followed by a description of the sequential cellular and tissue perturbations (intermediate events, IE) that lead to the eventual toxicological effect – or adverse outcome (AO) – which is at the individual level for most human health endpoints or at the population level for environmental endpoints. The AOP framework allows for the integration of all types of information at these different levels of biological organization, from molecular to population level, to provide a rational, biologically based argument (or series of hypotheses) to predict the outcome of an initiating event. In this description, the AOP builds on the MoA concepts and includes the "toxicity pathways" as described in the 2007 NRC report. Notable distinctions between MoA and AOP is that MoA is more chemical or group specific; MoA includes critical metabolic key events and MoA species concordance analysis takes into account both qualitative and quantitative toxicokinetic and toxicodynamic variations within and between species. AOPs are, then, building blocks to which chemical or group specific space and metabolism and toxicokinetics need to be added to conduct MoA analysis (Table 17.1).

The Organization for Economic Coordination and Development (OECD) held a Workshop on Using Mechanistic Information in Forming Chemical Categories in 2010, one of the first workshops to gather scientific expertise to guide further AOP development (OECD 2011). The 2010 workshop focused on the use of AOPs in building chemical categories for read-across, how AOPs might be used in Integrated Approaches to Testing and Assessment (IATA) and in identifying Key Events (KEs) (KEs are intermediate events that are (1) essential for the progression to the adverse outcome and (2) measurable; OECD 2008).

Since the 2010 workshop, OECD has published guidance that includes a template for development (OECD 2013a) and updated guidance in the form of a Handbook (OECD 2016). The Guidance/Handbook includes a description of the elements and uses of AOPs, a glossary of terms, and a template for developing an AOP. The goals of this guidance are to provide consistency in structure and facilitate harmonized use of AOPs.

According to the OECD Guidance/Handbook, an AOP consists of three main types of elements: one MIE leading to one adverse outcome with any number of intermediate KEs (Fig. 17.1). Although a MIE can be associated with a number of

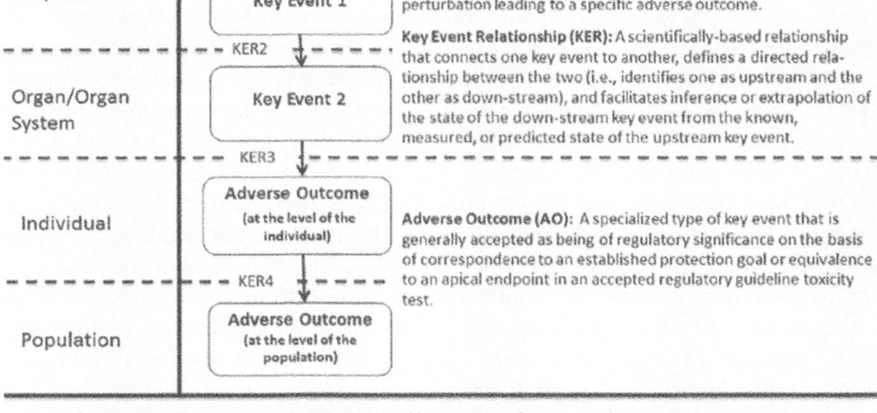

Fig. 17.1 General outline and description of AOP elements (From Becker et al. 2015)

different adverse effects, and similarly an adverse effect can result from a number of different MIEs, OECD has defined an AOP as being a linear pathway from one MIE to one adverse outcome, to streamline the development and use of AOPs,. A full description of an MIE should include cellular/tissue location – as immediately elicited intermediate events may be similar in two different AOPs, but differ in cell-type or tissue location (for example, metabolic transformation of a chemical to an electrophilic species may occur in both skin sensitization and liver fibrosis – only in keratinocytes for the former and hepatocytes for the latter). In order for an intermediate event to be identified as a "key event (KE)," it must be able to be evaluated experimentally and causally linked to the adverse outcome. A key event may be shared between two or more AOPs. The OECD Handbook describes the process for evaluating WoE for each step in development of an AOP.

The discrepancy between the artifice of the OECD's practical definition of an AOP, and the reality of the biology it is intended to model, has caused some consternation; however, there is full realization that this is an over-simplification; biological complexity is more accurately described by an interdependent network of pathways (Villeneuve et al. 2014a). The OECD nomenclature and guidance is intended as an initial practical step to initiate the process of pathway development and use.

The OECD is collaborating with the European Commission's Joint Research Centre, the US Army Corps of Engineers and the US Environmental Protection Agency to develop the infrastructure necessary for the advancement of AOP approaches. An important element of the AOP infrastructure is the Adverse Outcome Pathway Knowledge Base (AOP KB).[1] The AOP KB is an information technology system to capture, manage and share AOP information and currently consists of four modules: (1) the AOP-WIKI, a text-based tool allowing the management of AOP-related knowledge [AOPs, KEs, and relationships between them (key event relationships or KERs)] in a Wikipedia-like environment, (2) Effectopedia, a graphical tool for implementing quantitative models depicting the qualitative and quantitative relationship between two events in an AOP, (3) AOP explorer, a computational tool that will automatically generate graphical representations of AOPs and AOP networks and (4) an Intermediate Effects Database to manage information about chemical-specific MIE and intermediate effects (information here can be linked to the AOP Wiki; AOPs themselves are not chemical-specific).

The AOP Wiki, which became publically available in 2014, leads AOP developers through the steps to capture the scientific information needed to document an AOP, The AOP-Wiki follows and implements OECD guidance on how to describe AOPs. The Wiki also provides a collaborative space for groups to develop AOPs independent of geography or organizational boundaries. The OECD Handbook includes detailed instructions for developing and documenting AOPs and uploading the relevant information into the AOP Wiki (OECD 2016). The conventions and best practices of describing the various AOP elements are further explained by Villeneuve et al. (2014a and b).

[1] http://aopkb.org/

There are currently dozens of projects in the OECD AOP work plan (53 as of September, 2017), and several recent workshops and publications have advanced the theory and addressed practical issues of developing and evaluating AOPs and have begun to explore the potential use of AOPs in regulatory decision-making (see for example SEURAT-1 2014; Willett et al. 2014a, b; Garcia-Reyero 2015; Groh et al. 2015a and b; Villeneuve et al. 2014a; Becker et al. 2015; Perkins et al. 2015; Tollefsen et al. 2014). Regulatory application of AOPs is specific to the regulatory need being addressed and by necessity requires engagement and consideration from the regulatory community (further discussion below). Some current examples, as well as theoretical considerations, of the use of AOPs to improve the confidence in chemical hazard and risk decision-making, focusing on human health endpoints, are explored in Sect. 17.2.

17.2 How AOPs Can be Applied Within Various Regulatory Decision Contexts

The information necessary for chemical assessment decision-making depends on the context of the decision, with regard to both the level (regulatory vs non-regulatory, e.g. prioritization, read-across, classification and labelling, risk assessment or management) and the regulatory/legislative framework (e.g. industrial chemicals, pesticides, consumer products, or pharmaceuticals. Similarly, the potential of an AOP to support various types of decisions is related to its completeness, the confidence in the underlying information (and extent of its documentation), and the strength of information supporting the relationships between the AOP elements (between the MIE and Key Events and the eventual AO, termed Key Event Relationships, or KERs). It is important to recognize that AOPs at any level of development can be useful in supporting potential application since they provide a framework within which to organize and relate biological information, thereby bringing as much information to bear as possible for potential use. However, it is critically important to consider the overall confidence in the various supporting elements in determining the extent to which a given AOP can be relied upon (or AOP network) to within a particular decision context.

17.2.1 Confidence in an AOP is Related to Its use in a Given Decision Context

There are many different types of decisions to which AOPs can contribute, including (1) supporting chemical category formation and "read-across,"(allowing information from one chemical to be used for another, related chemical) (2) screening and priority setting for further testing, (3) hazard identification (4) classification and

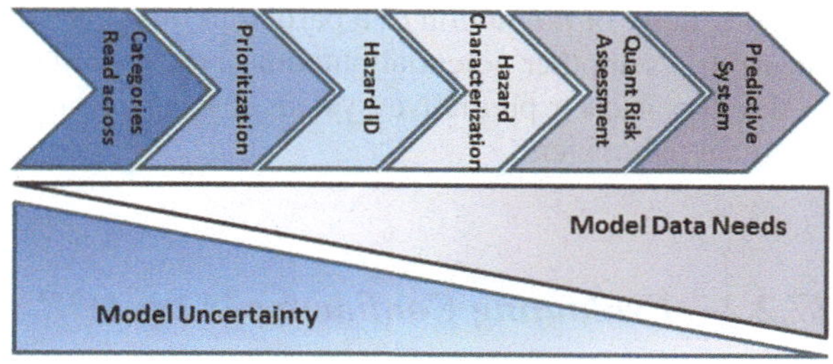

Fig. 17.2 Relationship between the certainty of the AOP and the extent to which it can be relied upon within different decision contexts

labeling, (5) designing integrated testing strategies (ITS) or integrated approaches to testing and assessment (IATA), and (6) risk assessment. Necessarily, chemical specific information on exposure, metabolism and toxicokinetics (i.e. mode of action analysis) and quantitation relevant to dose-response analysis is also taken into account to varying extents in these different applications.

Different types of decisions require different levels of confidence, depending on the consequences of the decision. For example, in screening a large number of chemicals to identify those that require more attention, it isn't necessary to be absolutely certain of the hazard that chemical pay present – only that it is likely to present more of a hazard than the other chemicals. Whereas, if you are making a risk assessment about a particular chemical, you need to be more certain about the levels of safe exposure, and the potential consequences of exposure to that chemical. Therefore, the extent of resource allocation can be tailored to the needs of the particular decision to be made.

As the intended application progresses from the decisions listed in 1–5 above, a corresponding increase in the level of evidence and certainty is normally required to ensure adequate confidence in decision- making (put another way, the consequences of a wrong decision are greater) (Fig. 17.2). For example, to inform structure activity relationship (SAR) modeling or to prioritize chemicals for further testing and/or assessment, there should be good evidence of biological plausibility linking the screening assay to the adverse outcome; similarly hazard identification requires that the KEs being measured are linked to the AO with fairly high confidence. To decrease uncertainty for higher level decision contexts some quantitative understanding of the relationships between the events being measured and with the adverse outcome, is likely desirable depending on specific decision context. For example, to increase confidence in a risk decision for a specific chemical, it is important to have some information about the level of certainty that a KE, measured at a certain activity, will lead to the particular AO. Similarly, there are corollary hierarchies of increasingly informed chemical specific exposure, toxicokinetic (including metabolic) and dose-response data (i.e. potency) that are necessarily combined with information on AOPs to inform application. As more information is gathered about the relationships between intersecting pathways (and potential feedback loops), the prediction of the level of response becomes less uncertain as does the specificity of prediction of a particular outcome (the likelihood of one outcome occurring vs another potential outcome); the eventual goal of the AOP approach is to develop such a predictive system to characterize likely response for specific chemical parameters.

17.2.2 Evaluating Confidence in an AOP

Confidence in an AOP involves two general aspects: the quality, nature (e.g. qualitative or quantitative) and amount of supporting information used to inform various elements of the pathway (e.g. KEs and KERs), and the availability, quality and appropriateness of assays and prediction models used to query the pathway and predict the AO. The OECD guidance, template and handbook largely address the former aspect, including a template for collecting information and references and for evaluating the WoE of the information supporting each element, and for capturing quantitative information about KERs. The OECD Handbook also includes a table for evaluating the overall confidence in the AOP, based on modified Bradford-Hill considerations for evaluating MoA, mentioned in Sect. 17.1.2 (Meek et al. 2014a, b), and gives guidance on evaluating the WoE of each considerations for each element (KEs and KERs) and for the AOP overall [and evaluating each as high (strong), moderate, or low (weak)]. Examples of early evaluations of AOPs using these considerations can be found in Becker et al. (2015). These authors also propose a preliminary multi-criteria decision analysis (MCDA) model for quantifying the WoE analysis of AOPs.

Briefly, the Bradford-Hill considerations have been modified to apply to evaluation of AOPs (which represent the toxicodynamic components of a mode of action) and have been rank ordered as: biological plausibility of the KER, essentiality of the KE (within the context of the AOP) and empirical support for the KER (including dose-response, temporal relationship of responses, and consistency of supporting information) (Table 17.2 OECD 2016; Becker et al. 2015). The OECD handbook provides a template for capturing this information and evaluation for each KE and KER; such documentation is intended to provide transparency critical to communicating and increasing consistency in documentation of and resulting confidence in WoE evaluations of the information supporting AOPs, as a basis for increasing the confidence in the application of AOPs in decision making.

The second aspect of confidence in the use of AOPs – evaluating the quality and appropriateness of AOPs for a purpose-specific application has received less attention; however, a "scientific confidence framework (CFI)" for such an evaluation has been proposed (Cox et al. 2014) and applied to the use of AOPs (Patlewicz et al. 2015). The CFI is based on OECD Quantitative Structure-Activity Relationship (QSAR) validation principles and Institute of Medicine biomarkers guidance and consists of three basic elements: (1) analytic validation of query assays, (2) qualification of prediction models and (3) WoE evaluation of the use of the prediction model in a specific decision context. Patlewicz et al. (2015) outline how these evaluations could be done for a number of examples; prioritization of chemicals for endocrine evaluation, read across for skin sensitization, and development of an integrated testing approach for skin sensitization. Evaluation of AOPs, their constituent elements and potential applications is still at an early stage, and is likely to evolve with case studies and experience.

17.3 Application of AOPs within IATA; Theory and Examples

OECD defines Integrated approaches to testing and assessment (IATA) as "a structured approach that strategically integrates and weights all relevant data to inform regulatory decisions regarding potential hazard and/or risk and/or the need for further targeted testing and therefore optimizing and potentially reducing the number of tests that need to be conducted" (Fig. 17.3; OECD 2015c). IATA are envisioned as an iterative hypothesis generating and testing process that defines how to assess or test strategically based on regulatory needs, which is generally framed in problem formulation for different assessment and management objectives (Meek et al. 2014a). The use of IATA in hazard and risk assessment has been explored by OECD (2008) and the US EPA (2011), and the use of AOPs to support IATA was the topic of a recent OECD workshop (OECD 2015c). IATA and Integrated Testing Strategies (ITS) are similar; however ITS have tended to relate principally to hazard (particularly in Europe), while IATA may involve considerations of exposure, since this is also a critical driver of testing strategies in some decision contexts.

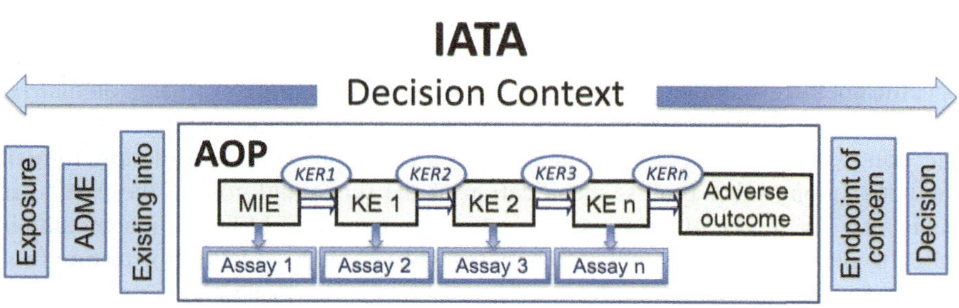

Fig. 17.3 The Role of an AOP within an IATA. An IATA begins with problem formulation, which includes the decision context. This will dictate what kind of information needs to be gathered and assessed (for example, if the decision is hazard-base, there is no need for exposure information). AOP information can support an IATA in several ways; by informing interpretation of existing information, by supporting chemical grouping and read-across; by informing testing strategies to obtain additional needed information; and to support weight-of-evidence evaluations performed in the process of decision-making

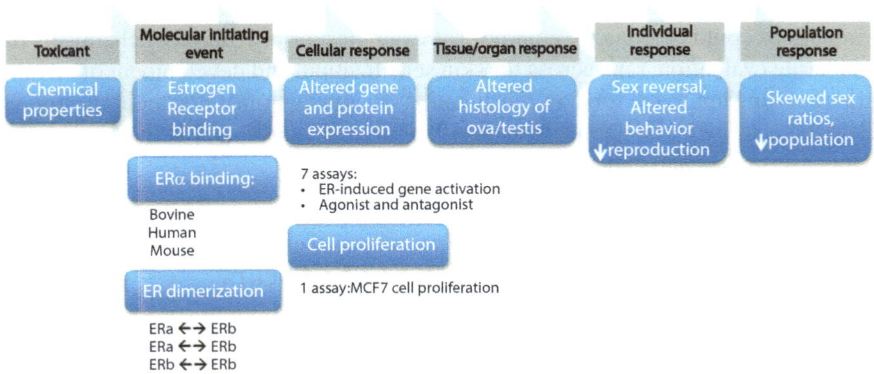

Fig. 17.4 Estrogen receptor-mediated reproductive impairment AOP with ToxCast assays. KEs along the ER pathway include gene activation, cell proliferation (in estrogen-responsive tissues) and resulting organ weight changes, as well as other modifications of gonadal tissues that can lead to reproductive impairment, and, for ecotoxicological application, to changes in populations. EPA's ToxCast program includes 17 assays that directly measure estrogen activity via a number of outputs (ER binding, ER dimerization, ER-dependent gene activation, cell proliferation) as well as via a number of read-out technologies (protein displacement, fluorescence, luminescence, cell number). Using combined output from these assays, EPA has developed an interpretation algorithm that predicts ER activity with high confidence (Browne et al. 2015; Judson et al. 2015; see these references for a detailed description of the assays and prediction model)

While AOPs can be used as a basis upon which to formulate testing strategies, AOPs themselves do not consider the context of the decision addressed by a testing strategy, e.g. the regulatory (or non-regulatory) context, exposure, level of decision (e.g. individual or population), etc.; rather, they provide the mechanistic building blocks on dynamic key events to which chemical or group specific toxicokinetic information can be added to address testing and/or assessment needs relevant to specific chemicals and/or groups. IATA envisages construction of a decision matrix, combined with iterative data generation as necessary, to address testing needs within the context of a particular application, based on problem formulation (which includes among other things, specification of the scope and goal of the particular decision) (Fig. 17.4). AOPs are an important component of IATA, which are anticipated to provide a consistently documented knowledge foundation for testing biological hypotheses, thereby informing testing strategies. Principles for developing an AOP-supported IATA have been suggested (Tollefsen et al. 2014): (a) define the endpoint of regulatory concern being assessed; (b) define the purpose/application for which the IATA is proposed; (c) describe the rationale, including mechanistic basis (e.g. AOP), according to which the IATA is constructed; (d) describe the individual information sources constituting the IATA; (e) characterize the predictive performance and applicability domain of the IATA, or IATA subcomponent(s) that can be expressed as a prediction model(s).

17.3.1 Examples of Application of an AOP Supported IATA for Chemical Categorization and Read-Across

17.3.1.1 Skin Sensitization

The AOP for sensitization has been well-described (OECD 2012a), with several KEs identified and assays for many KEs developed and validated, diagrammed in Fig. 17.4 (e.g. OECD 2015a, b). Since there are well-understood KE's along the pathway, and the links to the AO are strongly supported, this AOP has potential to be suitable for several applications, including informing chemical categories, hazard identification and hazard characterization (see below). The MIE has been thoroughly described and there is sufficient supporting data for the MIE to allow the creation of several predictive models that assist in chemical grouping and read-across. Well-established assays exist for most of the KEs (listed in Fig. 17.4) that allow the design of strongly predictive integrated testing strategies. In fact, the amount of data that has been collected from KE assays has allowed the creation of profilers for each KE that have been incorporated into the OECD QSAR toolbox (Patlewicz et al. 2014, 2015; http://www.oecd.org/chemicalsafety/risk-assessment/theoecdqsartoolbox.htm). (The OECD QSAR toolbox is a collection of predictive models, chemical profilers, and supporting databases to assist with chemical characterization). A number of profilers are available to identify structural alerts or to derive quantitative mechanistic models for predicting sensitization potential and/or potency (Patlewicz et al. 2014, 2015).

To support chemical categories, profilers would be used to identify structural alerts for electrophilic properties, indicating potential sensitizing activity. Based on the wealth of information supporting the sensitization MIE and the strong linkages to the AO, results from a mechanistic test addressing the MIE (e.g. the DPRA, or other models that incorporate kinetics, e.g. glutathione depletion; Schultz et al. 2005) should be sufficient to support read-across for this endpoint; this potential is currently being explored by the European Commission's Joint Research Centre.[2] To effectively use OECD QSAR toolbox profilers and prediction models, care must be taken to ensure that the chemical in question is within the domain of the models being used.

17.3.1.2 REACH

Arguably the world's largest chemicals testing program, the European Regulation for Registration, Evaluation, Authorisation and Restriction of Chemicals (REACH; EC 2006), provides ample opportunity, in theory at any rate, for use of AOP-supported IATA to streamline testing. The European Chemicals Agency (ECHA)

[2] https://eurl-ecvam.jrc.ec.europa.eu/validation-regulatory-acceptance/topical-toxicity/skin-sensitisation

has provided extensive guidance on the use of QSAR for grouping of chemicals and the use of ITS for specific endpoints (ECHA 2008, 2015). The potential use of IATA (including QSAR for read across or ITS for specific endpoints) has thus far been limited, due to some part to the nascent state of AOP-supported IATA as well as lack of familiarity with these novel approaches by both registrants and regulators. Use is also hampered by regulatory and legislative requirements for traditional test data.

For example, sensitization potential (not potency) is required in Annex VII (for all chemicals produced or imported in amounts of one tonne per year or more), and is therefore one of the mostly broadly required endpoints. REACH Annex VII was updated in 2016 to indicate a preference for in vitro methods to satify skin sensitization requirments (EC 2016a), largely based on the OECD AOP (OECD 2012a, c) and proposed EURL-ECVAM/OECD AOP-based IATA (EC 2017). The timely acceptance of ITS (and the associated IATA) for skin sensitization is important to allow use for the 2018 registration deadline for chemicals produced or imported in volumes of 1–100 tonnes per year.

Other endpoints that could be prioritized for application of AOP-based IATA in the near term based on the extent of their development are acute fish toxicity (Ankley et al. 2010), which is required in Annex VIII (chemicals at 10 tonnes per year or greater), and carcinogenicity (Benigni 2014), required in Annex X (chemicals at 1000 tonnes or greater; mutagenicity is required in Annex VIII, and a positive finding in vitro or in vivo can trigger a rodent cancer bioassay).

17.3.1.3 US EPA Pesticide Registration

Registration of pesticides requires extensive toxicological information, including a broad array of acute and chronic endpoints (e.g. in the US, pesticide registration requirements can be found in the Federal Code of Regulations 40 Part 158: http://www.gpo.gov/fdsys/granule/CFR-2012-title40-vol25/CFR-2012-title40-vol25-part158). Initial assessment of pesticide actives and formulations involves generating information on six acute endpoints (often called the "six pack") that is used for classification and handling instructions: acute oral, dermal and inhalation toxicity; dermal and eye irritation; and sensitization potential. Non-animal methods and ITS are available for some of these endpoints, including dermal and eye irritation and sensitization (described above). The US EPA provides guidance on waiving or bridging acute endpoints (US EPA 2012a); waiving of an endpoint is generally acceptable when physical or chemical characteristics make testing moot (e.g. oral acute toxicity can be waived for gases or highly volatile substances), bridging is generally based on use similarity or composition (for example, if a registered substance for which there is toxicological information is used in a different product, but at a lesser concentration than a registered product, information for that substance can be bridged). The US EPA is working with the National Institutes of Health's National Toxicology Program's NTP Interagency Center for the Evaluation of Alternative Methods (NICEATM) on predicting acute dermal toxicity from acute oral data (https://ntp.niehs.nih.gov/pubhealth/evalatm/test-method-evaluations/

acute-systemic-tox/index.html). As described above, identifying chemical categories can also be used to read-across toxicological information from a tested substance to a similar, non-tested substance.

To date, these approaches are limited for traditional toxicity endpoints considered in repeat-dose studies; however, addressing these endpoints through the use of AOP-supported ITS is part of EPA's long-term strategic plan (http://www.epa.gov/pesticides/science/testing-assessment.html).

17.3.2 *Examples of Use of an AOP-Supported IATA for Chemical Prioritization and Initial Hazard Identification*

17.3.2.1 Endocrine Disruptors in the EU

In the EU, the identification of endocrine active substances is of increasing priority. For REACH purposes, proven endocrine disrupting chemicals (EDCs) are considered equivalent to substances of very high concern (SVHC) and therefore subject to authorization. The World Health Organization (WHO) defines "An endocrine disruptor is an exogenous substance or mixture that alters function(s) of the endocrine system and consequently causes adverse health effects in an intact organism, or its progeny, or (sub) populations." (IPCS 2002); the European Commission has adopted this definition. Based on this definition, identification of an EDC requires proving both the mechanism (alters function of the endocrine system) and an adverse outcome (adverse health effects in an intact organism, progeny or subpopulations). The European Commission recently issued criteria for identifying EDCs in the context of EU pesticide legislation (EU 2016b). These criteria closely follow the WHO definition and require establishing an endocrine mode of action linked to an adverse oucome in an intact animal. The WHO definition does not specify which endocrine systems are included, but a 2012 WHO report suggests the intention is to broaden the definition beyond estrogen, androgen and thyroid pathways, perhaps to include metabolism, fat storage, bone development and the immune system (UNEP/WHO 2012). To adequately cover the breath of possible chemical effects in such a wide array of potential adverse outcomes, it is clear that reliance on AOP-based IATA will be critical for development of a practicable assessment program.

17.3.2.2 The US EDSP

As a result of a legislative mandate, the US EPA has invested resources for nearly twenty years to design a program to test for endocrine activity (the Endocrine Disruptor Screening Program (EDSP); http://www.epa.gov/endo). EPA estimates that approximately 10,000 chemicals (pesticides and drinking water contaminants) fall under the remit of the EDSP as possible endocrine active substances, and therefore are looking for ways to identify priority chemicals for testing (US EPA 2012b). In addition, the current EDSP screening battery (Tier 1 of the EDSP), which consists of 11 assays, is expensive (approximately 1 million USD), time consuming (taking several years to perform the assays and assess results), and uses more than 500 animals per chemical (Industry comments to EPA, http://www.regulations.gov/#!documentDetail;D = EPA-HQ-OPP-2012-0818-0027; Willett et al. 2011). Positive results in the screen will likely lead to more extensive testing in single or multi-generation tests in one or more species (Tier 2 of the EDSP), testing that may cost millions more and use thousands of animals. EPA has outlined a strategic plan for moving away from such resource intensive screening and testing that incorporates the spirit of the recommendations made by the National Academies of Science (NRC 2007; US EPA 2009a). As part of this process, the EPA is developing a high-throughput screening approach for endocrine activity, as a subset of its ToxCast™ program (http://www.epa.gov/ncct/toxcast), starting with estrogen, as proof of principle for the approach, before developing similar approaches for androgen and thyroid activity (US EPA 2014; Browne et al. 2015).

An AOP for estrogen receptor mediated reproductive impairment has been proposed based on a large amount of supporting data from fish (this AOP is limited to nuclear receptor form of estrogen receptor alpha (and perhaps beta) (Fig. 17.4; OECD 2009; OECD 2011); and rodents (OECD 2007). This AOP is well supported and has formed the basis of a decision tree to screen large libraries of mostly inactive chemicals (US EPA 2009c).

The immediate decision context for EPA is prioritization of chemicals under remit of the EDSP, and, if possible, initial hazard identification (possibly replacing Tier 1 screening). The ToxCast battery includes 17 ER-related assays (Fig. 17.4), 16 of which address either the MIE or the first KE (ER binding, ER dimerization, or transcriptional activation), and one cell proliferation assay that has been shown to be estrogen-responsive (Wilson et al. 2004). EPA has published a number of evaluations of estrogen activity based on ToxCast assay results (Reif et al. 2010; Rotroff et al. 2013; US EPA 2014) and has recently proposed an associated computational model, the performance of which has been compared with results from accepted in vitro and in vivo assays (Browne et al. 2015). Based on these results, EPA is proposing to use this computational model to prioritize chemicals for the EDSP and for preliminary identification of estrogenic activity, thereby replacing three of the Tier 1 assays, the ER binding, ER transcriptional activation and uterotrophic assays. Since the AOP is well supported with evidence linking the MIE of ER binding through the KEs of transcriptional activation and gonadal cellular changes through to the adverse outcomes of altered sex ratio, altered sexual behavior and population effects (in fish), the AOP supports the application of these assays to hazard identification. The OECD has developed a conceptual framework and extensive guidance to suggest possible additional considerations, including additional testing, to further characterize hazard and risk (OECD 2012b).

In addition, the US EPA's ToxCast program is developing methods for prioritizing chemicals based on a broad battery of assays (in addition to those related to estrogen activity) combined with pathway information and consideration of potential for exposure (Wambaugh et al. 2014). Such approaches could potentially be applied to other aspects of EPA's regulatory programs or for example, in identifying "greener" (in this case, less toxic) pesticide substances (Sanderson 2011; Richard 2014).

17.3.2.3 Prioritizing Drug Candidates

Mode of action information, combined with high-throughput screening, has been used for decades by pharmaceutical companies to identify and prioritize promising chemicals for further development from libraries of thousands of potential drug candidates. The practical considerations of screening potential drug candidates have generally been more streamlined than those for other classes of chemicals: drug candidates are usually chosen or designed with a particular biological mechanism or activity in mind; drugs are intended to be highly bioactive (in contrast with other classes of chemicals whose biological activities are incidental and associated exposure generally several orders of magnitude lower); chemical libraries of drug candidates are generally highly related chemicals, and therefore applicability of the chemical domain of individual tests is of less concern; and finally, there is a single organism of concern. In spite of improvements in the efficiency of preclinical screening, successful identification of efficacious therapeutics with few off-target toxicities remains rare. Because the high-throughput, mechanistic assays are used for prioritization and early activity identification, the assays focus primarily on MIE and early KEs. The prediction of drug candidates with improved efficacy and safety would be facilitated by further elaboration of AOPs and disease pathways, along with the design of IATA based on them (Langley 2011; Langley et al. 2015).

Pharmaceutical companies are uniquely suited to take advantage of this approach in that they have access to human data as well as a broad spectrum of animal data for many different chemistries. Thus far, 6 pharmaceutical companies have donated a total of 135 failed drugs (with associated data) to the Tox21 program for screening (http://www.epa.gov/comptox/dsstox/sdf_toxcst.html). This is in addition to some 2800 NCGC inventory of marketed, withdrawn and investigational drugs contained in the NCGC chemical inventory (Huang et al. 2011; http://www.epa.gov/comptox/dsstox/sdf_tox21s.html#Description). Data from Tox21 screening of these drugs should provide additional mechanistic information that can be used to inform AOPs.

17.3.2.4 Skin Sensitization

As mentioned in Sect. 17.3.1.1, the AOP for sensitization has been well-described, with several validated assays mapped to KEs (Fig. 17.5; OECD 2012a). Several integrated testing strategies (ITS) have been evaluated for identifying skin sensitizing potential (Maxwell et al. 2011; Bauch et al. 2012; Natsch et al. 2013; Jaworska et al. 2013; van der Veen et al. 2014) and potency (McKim et al. 2012; Nukada et al. 2013; Tsujita-Inoue et al. 2014). A composite ITS has been proposed and shown to

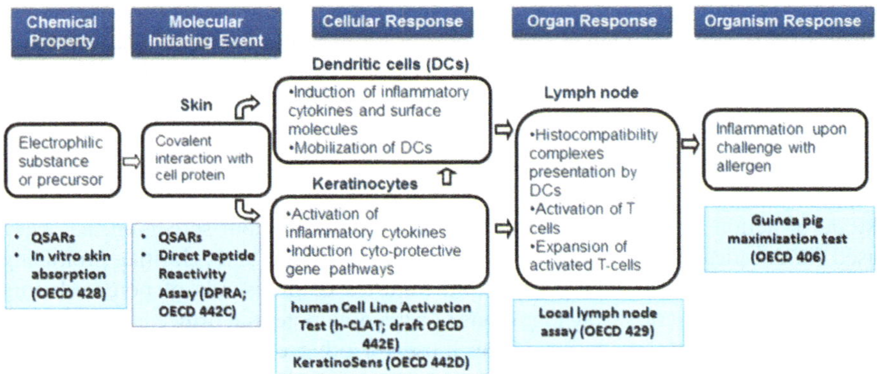

Fig. 17.5 Skin sensitization AOP with validated query assays. The MIE for the vast majority of chemicals is protein reactivity. This MIE is strongly linked to sensitization, and can be predictive of this AO on its own for many chemicals. Chemical-protein modification in the skin causes gene- and cytokine-induction in both keratinocytes and dendritic cells (immune cells located in the skin) (cellular responses). Activation of dendritic cells also causes them to process the modified protein and migrate to the local lymph nodes, where they present this antigen to immature T cells (organ response). The immature T cells then become activated to recognize this antigen and migrate into the body to cause an allergy the next time the skin is exposed (organism response)

perform favorably compared with results from animal tests (Urbisch et al. 2015). Although these in vitro methods and integrated strategies were being developed simultaneously with the AOP for skin sensitization, the AOP nonetheless provides biological rationale and support for the use of these ITS for assessing sensitizing potential, and in some cases, potency. The European Union Reference Laboratory for alternatives to animal testing (EURL-ECVAM) is currently working with an expert group at OECD to develop an AOP-based IATA that provides a framework for applying a collection of ITS in different regulatory contexts (Urbisch et al. 2015; EC 2017).

17.3.2.5 Personal Care Product Safety Assessment

The need for AOP-based IATA for assessing human health endpoints is particularly great in the area of assuring the safety of ingredients in cosmetics and personal care products due to the European Cosmetics Regulation,[3] which prohibits the testing and sale of cosmetics containing ingredients that have been tested on animals. A full ban on marketing cosmetics containing ingredients tested on animals came into

[3] European Commission Directive 76/768/EEC covering regulation of cosmetics as amended by Directive 2003/15/EC introducing a test ban from 2004 and sales ban from 2009, later postponed until 2013. http://ec.europa.eu/growth/sectors/cosmetics/index_en.htm

force in March 2013; ingredients tested on animals after this date cannot be used in cosmetics marketed within the EU. Similar legislation has been passed in other countries (e.g. India,[4] Israel,[5] Norway[6]and New Zealand[7]).

Due to the prohibition on testing on animals, the European Commission (EC) and cosmetics industry have invested in the development of alternative approaches to assessing the safety of cosmetics ingredients, work that has included development of in silico prediction of toxicity, in vitro methods, mathematical modelling approaches, integrated strategies and AOPs.[8] Several alternative methods, based not so much on AOPs or mechanism, but on direct replacement methods that mimic the animal tests, are available for skin and eye irritation and have widespread regulatory acceptance (OECD 2013b, c, 2015d, e, f, g, h, i; US EPA 2015). As described above, the IATA for skin sensitization were in part, a result of these efforts (e.g. Maxwell et al. 2011; Urbisch et al. 2015). Remaining challenges include evolving these qualitative tools, which were developed for hazard identification, to an approach that will allow quantitation of sensitizer potency as a basis for consideration of "safe" levels of ingredients within cosmetic products. Despite understanding of the key events that drive skin sensitization, the quantitative characterization of hazard data from non-animal tests to establish safe levels of human exposure for sensitizing chemicals remains a key gap (MacKay et al. 2013).

In addition, several joint EC Framework Program projects have focused on repeat-dose systemic toxicity for which there are currently no non-animal approaches, most recently the SEURAT-1 project (Safety Evaluation Ultimately Replacing Animal Testing) which is developing AOPs for mitochondrial and liver toxicity, among others (SEURAT-1 2014). Development and application of AOP-based IATAs aspects of systemic toxicity necessitate an evolution in thinking from toxicity testing for specific endpoints to more mechanistically driven approaches, as illustrated by the SEURAT-1 strategy. Once again, a key challenge in this area is to develop methodologies that will allow robust assessment of risk based on both novel exposure data (understanding the kinetics of systemic exposure to ingredients) and evolving understanding of more predictive quantitation of hazard.

---

[4] Testing ban in force from May 2014: Drugs & Cosmetics Act, Rule 148-C; sales ban in force from October 2014: Drugs & Cosmetics Act, Rule 135-B (import ban).

[5] Testing ban in force since Jan 2013: Amendment to the Animal Experimentation Law; Sales ban in force since Jan 2015: Draft Pharmacists' Regulations (Cosmetics) 2012–5773.

[6] Act relating to cosmetic products and body care products, etc. 2005: http://app.uio.no/ub/ujur/oversatte-lover/data/lov-20051221-126-eng.pdf, see also http://www.mattilsynet.no/language/english/cosmetics/import_of_cosmetics_to_norway/importing_cosmetics_to_norway.8321

[7] New Zealand: Animal Welfare Legislation Recognizes Animals as Sentient, Bans Cosmetic Testing: http://www.loc.gov/lawweb/servlet/lloc_news?disp3_l205404408_text

[8] E.g. Framework Programme 6 projects ACuteTox, Sens-it-iv, Re-Pro-Tect; Framework Programme 7 projects SEURAT-1).

17.4 Summary and Future Directions

There is widespread interest in developing systems biology approaches for improving chemical assessment. In recent years, there has been progress in developing tools to implement pathway-based mechanistic information, including internationally-coordinated development of an AOP knowledgebase, extensive guidance and publications on developing and evaluating AOPs and AOP-based IATA.

To evolve more mechanistically based hazard and risk assessment further, critical areas that should be prioritized include development of more predictive consideration of exposure based on key parameters, as well as incorporation of metabolism and chemical-specific toxicokinetic and toxicodynamic information. This will contribute additionally to increased understanding of the conditions under which adaptation is likely to occur vs those which result in adversity.

The information used to develop pathway-based approaches may be quite different from the information ultimately generated by assays that appropriately query the pathway with specific chemicals. Understanding the metabolism and toxicokinetics of test chemicals within query assays is equally important, as is an understanding of the actual in vitro exposure associated with alterations in pathways (rather than reliance on nominal concentrations). When choosing or developing assays or combinations of assays and mathematical models for use within IATA, consideration should be given to how well the assays represent key events, as a basis for characterization of AOPs for use in hazard and risk assessment and how well the mathematical models represent the underlying biology.

There may be hundreds or even thousands of biological "pathways" that potentially are affected by chemicals; however, these pathways intersect and overlap to form a relational web of biological processes. Up to this point, development of AOPs has been voluntary and pathways chosen largely based on specific expertise and interest. To develop a useful systems knowledgebase, it would be helpful to prioritize AOP development, based on for example, regulatory or human or environmental health priorities as well as those pathways most frequently implicated from understanding the chemistry of MIEs (e.g. Allen et al. 2014). This necessarily requires early input from the regulatory community, which is, in part, being addressed in collaborative international programs involving both researchers and regulators.

Since pathway development is continually enhanced by additional information, it will be necessary to discern when a pathway is sufficiently mature for a particular application. Since the purpose of incorporating pathway understanding into chemical evaluation is to decrease uncertainty in decision-making, the (potentially reduced) uncertainty of alternative approaches will need to be transparently described and compared with that of traditional less predictive approaches.

Success of a systems biology approach to toxicology improves with additional information and additional expertise. The approach is also not limited to scientists working in the field of toxicology but would benefit by synergism with other disciplines such as medicine and disease research. It is important to engage stakeholders as broadly as possible, including additional regulatory sectors (e.g. pharmaceutical) as well as regions (e.g. other non-OECD countries like Brazil, China and Russia). Equally important to building high confidence pathway networks and reliable testing strategies to fully implement this strategy is to engage stakeholders that will use these approaches (regulators, risk assessors in industry) and those affected by the decisions (the public, groups concerned with human and environmental heath). Regulators need to be engaged at many levels, during pathway and evaluation strategy development, finding effective ways to use these new approaches within the context of existing regulations or by modifying regulations as necessary, and by relating this information to the regulated community. Outreach to interested public stakeholders is critical to address concerns early on and to transparently communicate intentions, methods, and results.

All of this work will require high level coordination and leadership from all countries participating in this process. It will also be important to identify sources of dedicated funding, through government, industry, and public-private partnerships. If enough resources are applied in a coordinated fashion, we can look forward to a not-so-distant future where predictive modeling, informed by well-documented AOPs, can support IATA that will provide more certain decisions regarding human and environmental health, more quickly, more effectively, and using fewer animals that our current system allows.

References

Allen TEH, Goodman JM et al (2014) Defining molecular initiating events in the adverse outcome pathway framework for risk assessment. Chem Res Toxicol 27(12):2100–2112

Ankley GT, Bennett RS et al (2010) Adverse outcome pathways: a conceptual framework to support ecotoxicology research and risk assessment. Environ Toxicol Chem 29:730–741

Bauch C, Kolle SN et al (2012) Putting the parts together: combining in vitro methods to test for skin sensitizing potentials. Regul Toxicol Pharmacol 63:489–504

Becker RA, Ankley GT et al (2015) Increasing scientific confidence in adverse outcome pathways: application of tailored Bradford-Hill considerations for evaluating weight of evidence. Reg Toxicol Pharmacol 72:514–537

Becking GC (1995) Use of mechanistic information in risk assessment for toxic chemicals. Toxicol Lett 77:15–24

Benigni R (2014) Predicting the carcinogenicity of chemicals with alternative approaches: recent advances. Expert Opin Drug Metab Toxicol 10(9):1199–1208

Boobis AR, Cohen SM et al (2006) IPCS framework for analyzing the relevance of a cancer mode of action for humans. Crit Rev Toxicol 36:781–792

Boobis AR, Doe JE et al (2008) IPCS framework for analyzing the relevance of a noncancer mode of action for humans. Crit Rev Toxicol 38:87–96

Bradford-Hill A (1965) The environment and disease: association or causation? Proc R Soc Med 58:295–300

Browne P, Judson RS et al (2015) Screening chemicals for estrogen receptor bioactivity using a computational model. Environ Sci Technol Environ Sci Technol 49:8804–8814

Carmichael N, Bausen M et al (2011) Using mode of action information to improve regulatory decision making: an ECETOC/ILSI/HESI workshop overview. Crit Rev Toxicol 41:175–186

Clewell HJ, Gentry PR et al (1995) Considering pharmacokinetic and mechanistic information in cancer risk assessments for environmental contaminants: examples with vinyl chloride and trichloroethylene. Chemosphere 31:2561–2578

Committee on Carcinogenicity (COC) (2004) Guidance on a strategy for the risk assessment of chemical carcinogens. Department of Health, London

Conolly RB, Andersen ME (1993) An approach to mechanism barred cancer risk assessment for formaldehyde. Environ Health Perspect 101(Suppl. 6):169–176

Cox LA, Douglas D et al (2014) Applying a scientific confidence framework to a HTS-derived prediction model for endocrine endpoints: lessons learned from a case study. Regul Toxicol Pharmacol 69:443–450

Dellarco VL, Baetcke K (2005) A risk assessment perspective: application of mode of action and human relevance frameworks to the analysis of rodent tumor data. Toxicol Sci 86:1–3

EC (2006) EC Regulation No 1907/2006 of the European parliament and of the council of 18 December 2006 concerning the Registration, Evaluation, Authorisation and Restriction of Chemicals (REACH), establishing a European Chemicals Agency, amending Directive 1999/45/EC and repealing Council Regulation (EEC) No 793/93 and Commission Regulation (EC) No 1488/94 as well as Council Directive 76/769/EEC and Commission Directives 91/155/ EEC, 93/67/EEC, 93/105/EC and 2000/21/EC. Off J Eur Union L396/1 of 30.12.2006

EC (2014) Defining criteria for identifying endocrine disruptors in the context of the implementation of the plant protection product regulation and biocidal products regulation. DG Environment and DG SANCO. Available via http://ec.europa.eu/dgs/health_food-safety/dgs_consultations/food/consultation_20150116_endocrine-disruptors_en.htm. Accessed 28 Nov 2015

ECHA (2015) Guidance on information requirements and chemical safety assessment Chapter R.7a and R.7b: endpoint specific guidance. Available via http://echa.europa.eu/guidance-documents/guidance-on-information-requirements-and-chemical-safety-assessment. Assessed 28 Nov 2015 (note that both of these documents are undergoing revision in 2015)

EC (2016a) Commission Regulation (EU) 2016/1688 of 20 September 2016 amending Annex VII to Regulation (EC) No 1907/2006 of the European Parliament and of the Council on the Registration, Evaluation, Authorisation and Restriction of Chemicals (REACH) as regards skin sensitization. Available via: http://eur-lex.europa.eu/legalcontent/EN/TXT/?uri=uriserv%3AOJ.L_.2016.255.01.0014.01.ENG. Accessed 5 Sept 2017

ED (2016b) Communication from the Commission to the European Palriment and the Council on endocrine disruptors and the draft Commission acts setting out scientific criteria for their determination in the context of the EU legislation on plant protection products and biocidal products. COM(2016) 350 European Commission, Brussels. Available via: https://ec.europa.eu/health/sites/health/files/endocrine_disruptors/docs/com_2016_350_en.pdf. Accessed 5 Sept 2016

EC (2017) EURL ECVAM Recommendation on the use of nonanimal approaches for skin sensitisation assessment (March 2017 Draft). Available from: https://eurl-ecvam.jrc.ec.europa.eu/eurl-ecvam-recommendations/publiccomments/draft-recommendation-on-the-use-of-non-animal-approaches-for-skin-sensitisation-testing. Accessed 6 Sept 2017

European Chemicals Agency (ECHA) (2008) Guidance on information requirements and chemical safety assessment. Chapter R.6: QSARs and grouping of chemicals. Available via http://echa.europa.eu/documents/10162/13632/information_requirements_r6_en.pdf. Accessed 28 Nov 2015

European Commission (EC) (2003) Technical guidance document on risk assessment. European Commission Joint Research Centre, Ispra

European Food Safety Agency (EFSA) (2006) Opinion of the scientific panel on plant health, plant protection products and their residues on the scientific principles in the assessment and guidance provided in the field of human toxicology between 2003 and 2006. EFSA J 346:1–13

Garcia-Reyero N (2015) Are adverse outcome pathways here to stay? Environ Sci Technol 49:3–9

Groh K, Carvalho RN et al (2015a) Development and application of the adverse outcome pathway framework for understanding and predicting chronic toxicity: I. Challenges and research needs in ecotoxicology. Chemosphere 120:764–777

Groh K, Carvalho RN et al (2015b) Development and application of the adverse outcome pathway framework for understanding and predicting chronic toxicity: II. A focus on growth impairment in fish. Chemosphere 120:778–792

Haber LT, Maier A et al (2001) Applications of mechanistic data in risk assessment: the past, present, and future. Toxicol Sci 61:32–39

Huang R, Southall N et al (2011) The NCGC pharmaceutical collection: a comprehensive resource of clinically approved drugs enabling repurposing and chemical genomics. Sci Transl Med 3:80ps16

International Programme on Chemical Safety (IPCS) (2002) In: Damstra T, Barlow S, Bergman A, Kavlock RJ, van der Kraak G (eds) Global assessment of the state-of-the-science of endocrine disruptors. World Health Organization, Geneva, p 2002

International Programme on Chemical Safety (IPCS) (2005) Chemical-specific adjustment factors for interspecies differences and human variability: Guidance document for use of data in dose/concentration-response assessment. Harmonization Project document no. 2. World Health Organization, Geneva

IPCS (2006) Concise international chemical assessment document 68: tetrachloroethylene. World Health Organization, International Programme on Chemical Safety. World Health Organization, Geneva, Geneva

Iorio F, Bosotti R et al (2010) Discovery of drug mode of action and drug repositioning from transcriptional responses. Proc Natl Acad Sci U S A 107:14621–142626

Jaworska J, Dancik Y et al (2013) Bayesian integrated testing strategy to assess skin sensitization potency: from theory to practice. J Appl Toxicol 33:1353–1364

Joint Meeting on Pesticide Residues (JMPR). 2006. Report of the joint meeting of the FAO panel of experts on pesticide residues in food and the environment and WHO the core assessment group. FAO plant production and protection paper 187 (Thiacloprid). Available via ftp://ftp.fao.org/docrep/fao/010/a0888e/a0888e00.pdf. Accessed 28 Nov 2015

Judson RS, Magpantay FM et al (2015) Integrated model of chemical perturbations of a biological pathway using 18 in vitro high-throughput screening assays for the estrogen receptor. Toxicol Sci 148(1):137–154

Knapen D, Vergauwen L et al (2015) The potential of AOP networks for reproductive and developmental toxicity assay development. Reprod Toxicol 56:52–55

Langley GR (2011) Considering a new paradigm for Alzheimer's disease research. Drug Discov Today 19:1114–1124

Langley G, Christopher P et al (2015) Lessons from toxicology: developing a 21st-century paradigm for medical research. Environ Health Perspect 123:A268–A272

Liteplo RG, Meek ME (2003) Inhaled formaldehyde: exposure estimation, hazard characterization and exposure-response analysis. J Toxicol Environ Health B 6:85–114

MacKay C, Davies M et al (2013) From pathways to people: applying the adverse outcome pathway (AOP) for skin sensitization to risk assessment. ALTEX 30:473–486

Maxwell G, Aeby P et al (2011) Skin sensitisation: the Colipa strategy for developing and evaluating non-animal test methods for risk assessment. ALTEX 28:50–55

McKim JM Jr, Keller DJ 3rd et al (2012) An in vitro method for detecting chemical sensitization using human reconstructed skin models and its applicability to cosmetic, pharmaceutical, and medical device safety testing. Cutan Ocul Toxicol 31:292–305

Meek ME, Renwick A et al (2002a) Guidelines for application of chemical-specific adjustment factors in dose/concentration-response assessment. Toxicology 27:115–120

Meek B, Renwick A et al (2002b) Guidance for derivation of chemical-specific adjustment factors (CSAF). Dev implement Hum Ecol Risk Assess 8:769

Meek M, Bucher J et al (2003) A framework for human relevance analysis of information on carcinogenic modes of action. Crit Rev Toxicol 33:591–653

Meek ME, Berry C et al (2008) Mode of action frameworks: a critical analysis. J Toxicol Environ Health B 11:681–685
Meek ME, Klaunig JE (2010) Proposed mode of action of benzene-induced leukemia: Interpreting available data and identifying critical data gaps for risk assessment. Chem Biol Interact 184(1-2):279–285
Meek ME, Boobis A et al (2014a) New developments in the evolution and application of the WHO/IPCS framework on mode of action/species concordance analysis. J Appl Toxicol 34:1–18
Meek ME, Palermo CM et al (2014b) Mode of action human relevance (species concordance) framework: evolution of the Bradford Hill considerations and comparative analysis of weight of evidence. J Appl Toxicol 34:595–606
National Industrial Chemical Notification and Assessment Scheme (NICNAS) (2006) Priority existing chemical assessment report 21: Benzene. National Industrial Chemical Notification and Assessment Scheme, Commonwealth of Australia. Available via http://www.nicnas.gov.au/chemical-information/pec-assessments. Accessed 28 Nov 2015
National Research Council (2007) Toxicity testing in the 21st century: a vision and a strategy. National Academy of Sciences, Washington, DC
Natsch A, Ryan CA et al (2013) A dataset on 145 chemicals tested in alternative assays for skin sensitization undergoing prevalidation. J Appl Toxicol 33:1337–1352
Nukada Y, Miyazawa M et al (2013) Data integration of non-animal tests for the development of a test battery to predict the skin sensitizing potential and potency of chemicals. Toxicol In Vitro 27:609–618
OECD (2007) Test guideline No. 440: uterotrophic bioassay in rodents A short-term screening test for oestrogenic properties. OECD, Paris, France. Available via http://www.oecd-ilibrary.org/environment/test-no-440-uterotrophic-bioassay-in-rodents_9789264067417-en. Accessed 28 Nov 2015
OECD (2008) Report of the second survey on available Omics tools. OECD environment, health and safety publications series on testing and assessment no. 100. OECD, Paris, France. Available via http://www.oecd.org/chemicalsafety/testing/seriesontestingandassessmentpublicationsbynumber.htm. Accessed 28 Nov 2015
OECD (2009) Report of the expert consultation to evaluate an estrogen receptor binding affinity model for hazard identification. OECD series on testing and assessment no. 111 OECD, Paris, France. Available via http://www.oecd.org/chemicalsafety/testing/seriesontestingandassessmentpublicationsbynumber.htm. Accessed 28 Nov 2015
OECD (2011) Report of the workshop on using mechanistic information on forming chemical categories. OECD series on testing and assessment no. 138. OECD, Paris, France. Available via http://www.oecd.org/chemicalsafety/testing/seriesontestingandassessmentpublicationsbynumber.htm. Accessed 28 Nov 2015
OECD (2012a) The adverse outcome pathway for skin sensitisation initiated by covalent binding to proteins, part 1: scientific evidence. Series on testing and assessment, no.168. OECD, Paris, France. Available via http://www.oecd.org/chemicalsafety/testing/seriesontestingandassessmentpublicationsbynumber.htm. Accessed 28 Nov 2015
OECD (2012b) Guidance document on standardized test guidelines for evaluating chemicals for endocrine disruption. Series on testing and assessment, no. 150. OECD, Paris, France. Available via http://www.oecd.org/chemicalsafety/testing/seriesontestingandassessmentpublicationsbynumber.htm. Accessed 28 Nov 2015
OECD (2012c) The adverse outcome pathway for skin sensitisation by covalent binding to proteins. Part 2: use of the AOP to develop chemical categories and integrated testing and assessment approaches. Series on testing and assessment, no.168. OECD, Paris, France. Available via http://www.oecd.org/chemicalsafety/testing/seriesontestingandassessmentpublicationsbynumber.htm. Accessed 28 Nov 2015
OECD (2013a) Guidance document on developing and assessing adverse outcome pathways. Series on testing and assessment, no. 184. OECD, Paris, France. Available via http://search.

oecd.org/officialdocuments/displaydocumentpdf/?cote=env/jm/mono%282013%296&doclanguage=en. Accessed 28 Nov 2015

OECD (2013b) Test guideline no. 438: isolated chicken eye test method for identifying (i) Chemicals inducing serious eye damage and (ii) Chemicals not requiring classification for eye irritation or serious eye damage. OECD Paris, France. Available via 10.1787/9789264203860-en. Accessed 28 Nov 2015

OECD (2013c) Test guideline no. 437: bovine corneal opacity and permeability test method for identifying (i) Chemicals inducing serious eye damage and (ii) Chemicals not requiring classification for eye irritation or serious eye damage. OECD Paris, France. Available via 10.1787/9789264203846-en. Accessed 28 Nov 2015

OECD (2015a) Test guideline no. 442C: In Chemico skin sensitisation: Direct Peptide Reactivity Assay (DPRA). OECD Paris, France. Available via 10.1787/9789264229709-en. Accessed 28 Nov 2015

OECD (2015b) Test guideline no. 442D: in vitro skin sensitisation : ARE-Nrf2 Luciferase test method. OECD Paris, France. Available via 10.1787/9789264229822-en. Accessed 28 Nov 2015

OECD (2015c) Report of the workshop on a framework for the development and use of integrated approaches to testing and assessment. Series on testing and assessment no. 215. OECD Paris, France. Available via http://www.oecd.org/env/ehs/testing/reportoftheworkshoponaframeworkforthedevelopmentanduseofintegratedapproachestotestingandassessment.htm. Accessed 28 Nov 2015

OECD (2015d) Test no. 439: in vitro skin irritation: reconstructed human epidermis test method. OECD, Paris, France. Available via 10.1787/9789264242845-en. Accessed 28 Nov 2015

OECD (2015e) Test guideline no. 435: in vitro membrane barrier test method for skin corrosion, OECD. OECD, Paris, France. Available via 10.1787/9789264242791-en. Accessed 28 Nov 2015

OECD (2015f) Test guideline no. 431: in vitro skin corrosion: reconstructed human epidermis (RHE) test method. OECD, Paris, France. Available via 10.1787/9789264242753-en. Accessed 28 Nov 2015

OECD (2015g) Test guideline no. 430: in vitro skin corrosion: transcutaneous electrical resistance test method (TER). OECD, Paris, France. Available via 10.1787/9789264242739-en. Accessed 28 Nov 2015

OECD (2015h) Test guideline no. 492: reconstructed human Cornea-like Epithelium (RhCE) test method for identifying chemicals not requiring classification and labelling for eye irritation or serious eye damage. OECD, Paris, France. Available via 10.1787/9789264242548-en. Accessed 28 Nov 2015

OECD (2015i) Test guideline no. 491: short time exposure in vitro test method for identifying (i) Chemicals inducing serious eye damage and (ii) Chemicals not requiring classification for eye irritation or serious eye damage. OECD, Paris, France. Available via 10.1787/9789264242432-en. Accessed 28 Nov 2015

OECD (2016) User's handbook supplement to the guidance document for developing and assessing AOPs. OECD Paris, France. Available via https://aopkb.org/common/AOP_Handbook.pdf. Accessed 28 Nov 2015

Organization for Economic Coordination and Development (OECD) (2002) Guidance notes for analysis and evaluation of chronic toxicity and carcinogenicity studies. OECD series on testing and assessment no. 35 OECD, Paris, France. Available via http://www.oecd.org/chemicalsafety/testing/seriesontestingandassessmentpublicationsbynumber.htm. Accessed 28 Nov 2015

Patlewicz G, Kuseva C et al (2014) Towards AOP application – implementation of an integrated approach to testing and assessment (IATA) into a pipeline tool for skin sensitization. Reg Toxicol Pharmacol 69:529–545

Patlewicz G, Simon TW et al (2015) Proposing a scientific confidence framework to help support the application of adverse outcome pathways for regulatory purposes. Regul Toxicol Pharmacol 71:463–477

Perkins EJ, Antczak P et al (2015) Adverse outcome pathways for regulatory applications: examination of four case studies with different degrees of completeness and scientific confidence. Toxicol Sci 148:14–25
Reif DM, Martin MT et al (2010) Endocrine profile and prioritization of environmental chemicals using ToxCast data. Environ Health Perspect 118:1714–1720
Richard A (2014) Mining the ToxCast chemical-data landscape for green chemical design: an SAR-Chem-Informed approach (GCE18). Presented at: the 18th annual green chemistry & engineering conference (June 17–19, 2014). Available via http://presentations.acs.org/common/media-player.aspx/GCE2014/GCESC/GCE004/GCE18. Accessed 28 Nov 2015
Rotroff DM, Dix DJ et al (2013) Using in vitro high throughput screening assays to identify potential endocrine disrupting chemicals. Environ Health Perspect 121:7–14
Sanderson K (2011) It's not easy being green. Nature 469:18–20
Seed J, Carney EW et al (2005) Overview: Using mode of action and life stage information to evaluate the human relevance of animal toxicity data. Crit Rev Toxicol 35(8-9):664–672
Schultz TW, Yarbrough JW et al (2005) Structure–activity relationships for reactivity of carbonyl compounds with glutathione. SAR QSAR Environ Res 16:313–322
Schadt EE, Lum PY (2006) Reverse engineering gene networks to identify key drivers of complex disease phenotypes. J Lipid Res 47:2601–2613
SEURAT-1 (Safety Evaluation Ultimately Replacing Animal Testing) (2014) Alternative testing strategies: towards the replacement of in vivo repeated dose systemic toxicity testing, annual report volume 4. Available via http://www.seurat-1.eu/pages/library/seurat-1-annual-reports.php. Accessed 28 Nov 2015
Tollefsen KE, Scholz S et al (2014) Applying Adverse Outcome Pathways (AOPs) to support integrated approaches to testing and assessment (IATA). Reg Toxicol Pharmacol 70:629–640
Tsujita-Inoue K, Hirota M et al (2014) Skin sensitization risk assessment model using artificial neural network analysis of data from multiple in vitro assays. Toxicol In Vitro 28:626–639
United Nations Economic Commission for Europe (UNECE) (2007) Amendments to the globally harmonized system of classification and labelling of chemicals (GHS). Available via http://www.unece.org/trans/danger/publi/ghs/ghs_rev02/02files_e.html. Accessed 28 Nov 2015
United Nations Environment Programme (UNEP)/World Health Organization (WHO) (2012) State of the science of endocrine disrupting chemicals. Available via http://www.who.int/ipcs/publications/new_issues/endocrine_disruptors/en/. Accessed 28 Nov 2015
Urbisch D, Mehling A et al (2015) Assessing skin sensitization hazard in mice and men using non-animal test methods. Regul Toxicol Pharmacol 71:337–351
US Environmental Protection Agency (EPA) (1999) Minutes of the chloroform risk assessment review subcommittee meeting, October 27–28, 1999, U.S. Environmental protection agency, science advisory board. U.S. Environmental Protection Agency, Washington, DC. Available via http://yosemite.epa.gov/sab/sabproduct.nsf/ec0009.pdf. Accessed 28 Nov 2015
US EPA (2000) Scientific advisory panel report no. 2000–05: atrazine: hazard and dose-response assessment and characterization. Office of pesticide programs, U.S. Environmental Protection Agency, Washington, DC. Available via www.thecre.com/pdf/finalatrazine.pdf. Accessed 28 Nov 2015
US EPA (2005a) Guidelines for carcinogen risk assessment office of pesticide programs, U.S. Environmental Protection Agency, Washington, DC. Available via http://www3.epa.gov/airtoxics/cancer_guidelines_final_3-25-05.pdf. Accessed 28 Nov 2015
US EPA (2005b) Science issue paper: mode of carcinogenic action for cacodylic acid (dimethylarsinic acid, DMAV) and recommendations for dose response extrapolation. July 26. Health Effects Division, Office of Pesticide Programs, U.S. Environmental Protection Agency, Washington, DC. Available via http://users.physics.harvard.edu/~wilson/arsenic/EPA%20DMA%20mode.pdf. Accessed 28 Nov 2015
US EPA (2007a) Scientific advisory panel report: advisory on EPA's assessments of carcinogenic effects of organic and inorganic arsenic: report of the US EPA science advisory board.

U.S. Environmental Protection Agency, Washington, DC. Available via http://yosemite.epa.gov/sab/sabproduct.nsf/EPA-SAB-07-008.pdf. Accessed 28 Nov 2015

US EPA (2007b) Framework for determining a mutagenic mode of action for carcinogenicity: using EPA's 2005 cancer guidelines and supplemental guidance for assessing susceptibility from early-life exposure to carcinogens. Available via http://www3.epa.gov/airtoxics/childrens_supplement_final.pdf. Accessed 28 Nov 2015

US EPA (2009a) The U.S. Environmental protection agency's strategic plan for evaluating the toxicity of chemicals. U.S. Environmental Protection Agency, Washington, DC. Available via http://nepis.epa.gov/Exe/ZyPDF.cgi/P1003351.PDF?Dockey=P1003351.PDF. Accessed 28 Nov 2015

US EPA (2009b) Integrated bioactivity and exposure ranking: a computational approach for the prioritization and screening of chemicals in the endocrine disruptor screening program. December 2014. Docket ID No. EPA–HQ–OPP–2014–0614–0003. Office of Pesticide Programs, U.S. Environmental Protection Agency, Washington, DC. Available via http://www.regulations.gov. Accessed 28 Nov 2015

US EPA (2009c) The use of structure activity relationships of estrogen binding affinity to support prioritization of pesticide inert ingredients and antimicrobial pesticides for screening and testing, Office of Pesticide Programs, U.S. Environmental Protection Agency, Washington, DC. Available via http://archive.epa.gov/scipoly/sap/meetings/web/pdf/082509minutes.pdf. Accessed 28 Nov 2015

US EPA (2011) Integrated approaches to testing and assessment strategy: use of new computational and molecular tools us environmental protection agency. FIFRA Scientific Advisory Panel Consultation May 24–26, 2011. Office of Pesticide Programs, U.S. Environmental Protection Agency, Washington, DC. Available via http://yosemite.epa.gov/sab/sabproduct.nsf/373C1DB0E0591296852579F2005BECB3/$File/OPP+SAP+document-May2011.pdf. Accessed 28 Nov 2015

US EPA (2012a) Guidance for waiving or bridging of mammalian acute toxicity tests for pesticides and pesticide products (acute oral, acute dermal, acute inhalation, primary eye, primary dermal, and dermal sensitization). Office of Pesticide Programs, U.S. Environmental Protection Agency, Washington, DC. Available via http://www2.epa.gov/sites/production/files/documents/acute-data-waiver-guidance.pdf. Accessed 28 Nov 2015

US EPA (2012b) Universe of chemicals and general validation principles. Office of Pesticide Programs, U.S. Environmental Protection Agency, Washington, DC. Available via http://www2.epa.gov/sites/production/files/2015-07/documents/edsp_chemical_universe_and_general_validations_white_paper_11_12.pdf. Accessed 28 Nov 2015

US EPA (2014) Integrated bioactivity and exposure ranking: a computational approach for the prioritization and screening of chemicals in the endocrine disruptor screening program. Docket ID Number EPA-HQ-OPP-2014-0614-0003Office of Pesticide Programs, U.S. Environmental Protection Agency, Washington, DC. Available via http://www.regulations.gov. Accessed 28 Nov 2015

US EPA (2015) Use of an alternate testing framework for classification of eye irritation potential of EPA pesticide products. Office of Pesticide Programs, U.S. Environmental Protection Agency, Washington, DC. Available via http://www2.epa.gov/sites/production/files/2015-05/documents/eye_policy2015update.pdf. Accessed 28 Nov 2015

Van der Veen JW, Rorije E et al (2014) Evaluating the performance of integrated approaches for hazard identification of skin sensitizing chemicals. Regul Toxicol Pharmacol 69:371–379

Villeneuve DL, Crump D et al (2014a) Adverse Outcome Pathway (AOP) development I: strategies and principles. Toxicol Sci 142:312–320

Villeneuve DL, Crump D et al (2014b) Adverse Outcome Pathway (AOP) development II: best practices. Toxicol Sci 142:321–330

Wambaugh J, Wang A et al (2014) High throughput heuristics for prioritizing human exposure to environmental chemicals. Environ Sci Technol 48:12760–12767

Willett C, Bishop P et al (2011) Application of an integrated testing strategy to the US EPA endocrine disruptor screening program. Toxicol Sci 123:15–25

Willett C, Rae JC et al (2014a) Pathway-based toxicity: history, current approaches and liver fibrosis and steatosis as prototypes. ALTEX 31:407–421

Willett C, Rae JC et al (2014b) Building shared experience to advance practical application of pathway – based toxicology: liver toxicity mode-of-action. ALTEX 31:500–519

Wilson VS, Bobseine K et al (2004) Development and characterization of a cell line that stably expresses an estrogen-responsive luciferase reporter for the detection of estrogen receptor agonist and antagonists. Toxicol Sci 81(1):69–77

Chapter 18
Use and Acceptance of AOPs for Regulatory Applications

Clemens Wittwehr

Abstract Adverse Outcome Pathways (AOPs) play an increasingly important role in risk and hazard assessment, giving scientists across many disciplines and organisations the opportunity to jointly collect and discuss their knowledge about the biological and toxicological processes leading to an Adverse Outcome. While this knowledge becomes more distributed and accepted by the scientific mainstream, its acceptance in a regulatory context is still not well developed. A promising avenue for increasing visibility and applicability of AOPs in the regulatory context is the role they play in the development of Integrated Approaches to Testing and Assessment (IATA), where the mechanistic knowledge AOPs provide will underpin testing strategies accepted by regulators and relying less and less on animal testing. The development of a standardised exchange format for AOPs (AOP-XML) and of an OECD-endorsed regulatory reporting format for results of non-classical test methods (OHT 201) are further enabling factors for any upcoming adoption of AOPs in the regulatory world.

18.1 Regulatory Acceptance and Regulatory Use

Modern society embraces risk avoidance; people tend to be better informed than ever about intrinsic hazards of the products they are using every day or the risk that comes from exposure with environmental chemicals. This has led and still leads to a wide range of regulations that are supposed to guarantee safety of chemicals and other related products. The majority of these regulations rely on animal tests, and about a fifth to a quarter of all animal experiments is done for regulatory purposes. Many initiatives underline the need to replace, reduce, or refine ("3R" principle) laboratory animal use, but the regulatory acceptance and use of existing and emerging 3R models is not yet wide spread.

C. Wittwehr (✉)
European Commission Joint Research Centre, Ispra, Italy
e-mail: clemens.wittwehr@ec.europa.eu

N. Garcia-Reyero, C.A. Murphy (eds.), *A Systems Biology Approach to Advancing Adverse Outcome Pathways for Risk Assessment*,
DOI 10.1007/978-3-319-66084-4_18

EURL ECVAM (European Union Reference Laboratory – European Centre for the Validation of Alternative methods) defines Regulatory Acceptance of a test method as

> its formal acceptance by regulatory authorities indicating that the test method may be used to provide information to meet a specific regulatory requirement. This includes, but is not limited to, a formal adoption of a test method by EU and/or OECD as an EU test method and included in the EU Test Methods Regulation and/or as an OECD Test Guideline, respectively. EURL ECVAM (2016)

Schiffelers et al. (2012) define Regulatory Acceptance as referring

> [...] to the written or unwritten adoption of testing strategies by regulatory authorities. Regulatory acceptance in this context is defined as the formal adoption of a (validated) test method by a regulatory agency/authority. Depending on the product sector, regulatory acceptance can be accomplished at a national, European, and/or a global level.

The same authors define Regulatory Use as referring

> [...] to the actual uptake of a method by a regulatory authority or a manufacturer for quality and/or safety testing purposes. This step is often also referred to as implementation. In the field of policy science, however, implementation would cover the whole process from the initial intention to work towards alternatives to the actual use.

What is common to all these definitions is that Regulatory science should be kept in pace with technological developments, and early involvement of regulators in international initiatives is crucial to achieve progress in this rapidly evolving field, especially as well working feedback mechanisms between regulator and the regulated industry are crucial. Regulated organisations (e.g. chemical companies) often have a knowledge advantage, because they are the ones promoting technical progress in their field and are intimately familiar with emerging issues. Regulators do not have this advantage of using new methods (3R methods, in this case) in the research phase or in the production process. It is therefore a fact that regulatory authorities have problems properly assessing an alternative method.

Risk aversion therefore leads to a certain degree of inertia in the field of 3R models acceptance, a typical phenomenon in all technology transitions. The paradigm shift from observational toxicology towards predictive approaches – an underlying goal of the application of alternative methods – therefore requires intensive dissemination and communication efforts.

A novel approach for this communication and dissemination effort between the stakeholders (regulators, industry, and research) is the joint effort to create and apply AOPs, which can in that context be conceived as a general framework that allows the placement of available information on a particular biological pathway into an organized, usable format. Information in an AOP format can then be used for assessing chemical risks in a several ways, including:

- Informing Integrated Approaches to Testing and Assessment (IATA, see below)
- Hazard identification and characterisation
- Read-across and chemical classification
- Priority setting/screening/ranking for further testing

- Quantitative considerations

The AOP works then as a wireframe onto which existing knowledge is pinned in a globally agreed format giving an overview of the extent to which the Molecular Initiating Events (MIE), the Key Events (KE), the Key Event Relationships (KER) and the Adverse Outcomes (AO) are understood.

18.2 IATA: The Missing Link Between AOPs and Regulatory Acceptance

The Organisation for Economic Co-operation and Development (OECD) is actively working towards the adoption of methods to replace animal tests where possible, or to refine existing tests to reduce the number of animals used and minimise suffering.

A number of OECD Test Guidelines are already based on non-animal tests, including but not limited to skin corrosion/irritation, phototoxicity and skin absorption, eye damage/irritation, genotoxicity and endocrine disruption.

To be comfortable when using alternative methods for complex endpoints, like carcinogenicity or developmental/reproductive toxicity, science has to provide evidence for how toxicity is brought about at the molecular level resulting subsequently in an effect at organ or organism level. Once the mechanism underlying the toxicity is understood, the adverse outcome can be predicted based on sound scientific evidence.

An AOP is an objective and systematic mechanistic based framework that provides the biological context to facilitate the interpretation of results from alternative testing and non-testing approaches in predicting an adverse effect and facilitates their application in regulatory decision-making: AOPs help to organise and analyse relevant mechanistic data on a given substance or group of substances.

Structured approaches are used for (a) chemical(s) or group(s) of chemicals, and they strategically integrate and weight all available data and guide the targeted generation of new data where required (hypothesis driven) to inform regulatory decisions regarding the hazard identification (potential), hazard characterisation (potency) and/or safety assessment (potential/potency and exposure).

While AOPs are focused on toxicodynamics phenomena from the moment of the MIE, they are chemical-independent and do not give information about the chemical-related steps that precede the MIE: exposure, bioavailability, external vs internal dose, ADME (absorption, distribution, metabolism, excretion) considerations etc. are not subject of an AOP.

In order for an AOP to fully support the application of alternative methods in the regulatory context, it must therefore be embedded in a larger framework taking into account all aspects of toxicity, i.e. not only toxicodynamics but also toxicokinetics.

This larger framework is the OECD-led initiative for Integrated Approaches to Testing and Assessment, or IATA, as shown in Fig. 18.1 taken from OECD (2015):

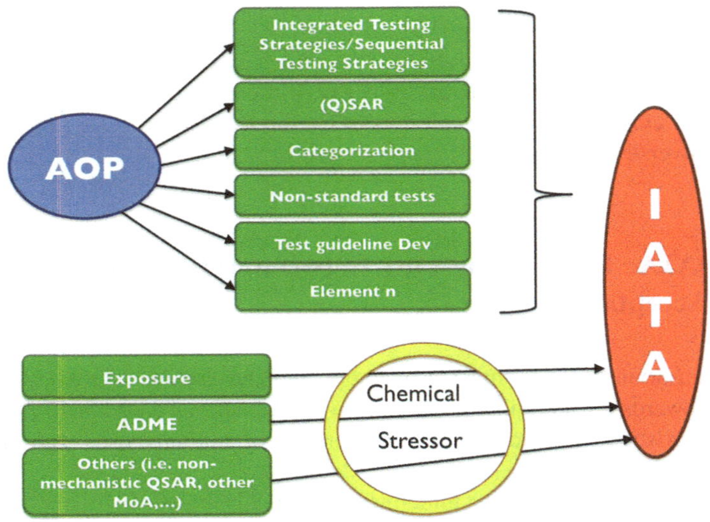

Fig. 18.1 IATA and AOPs

Current thinking is that the degree to which an IATA needs to be populated by a full complement of information will be dependent on the ultimate purpose it is being used for. Thus flexibility is foreseen in the construction of IATA depending on the regulatory need and tailored to the substance(s) under consideration. However, there is also a need to provide regulators with some degree of consistency and understanding of the assumptions on which the IATA is based, with the AOP concept (at least for the toxicodynamics part) being an ideal and transparent organising framework.

Tollefsen et al. (2014), when focusing on how IATA and AOP can work together, suggest

> a framework that is driven by the problem formulation, which involves a consideration of the risk management scope, the data requirements and the level of acceptable uncertainty associated with the decision being made.

Depending on the regulatory application the necessary level of AOP confidence is determined. Tollefsen et al. (2014) continues:

> If the outcome derived from the framework is of insufficient confidence, then additional data might need to be generated through new testing and assessment. The new information derived will then be passed back into the framework for re-evaluation. Indeed a decision outcome could result in more thorough regulatory follow up or implementation of measures to reduce use and/or exposure. Any new information generated will also be used to augment the corresponding AOP.

To summarize: In the past, the Adverse Outcome, or (eco-)toxicological endpoint, was the focus of all testing strategies (mostly animal-based), as shown in Fig. 18.2.

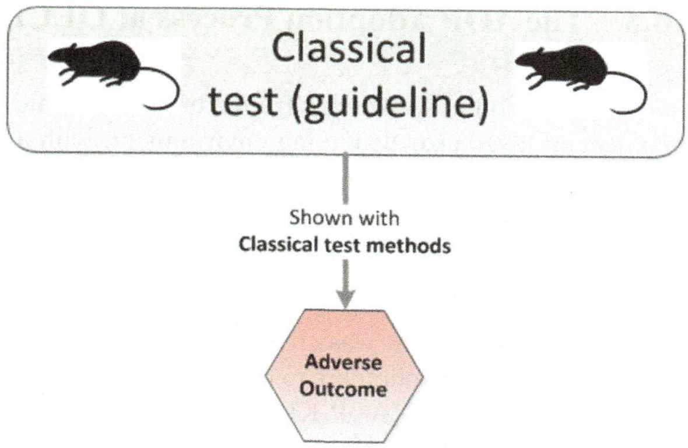

Fig. 18.2 Classical test guidelines focussing on adverse outcome (endpoint)

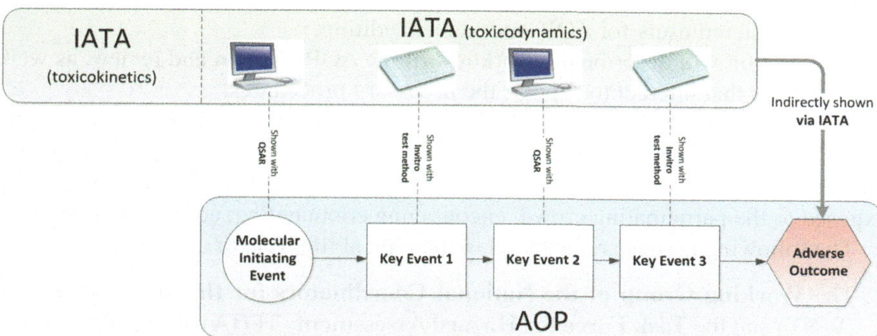

Fig. 18.3 IATA underpinned by AOP

The IATA-based approach will, as shown in Fig. 18.3, be a game changer: With regulatory-relevant Adverse Outcomes obviously still at the centre of risk assessment, they will be treated indirectly via methods (mostly non-classical) that are brought together in an IATA:

The main contribution AOPs bring to IATA is the mechanistic knowledge that is needed to underpin any regulatory decision about the acceptance of any non-classic test method. The better the AOP used in an IATA describes the reasoning behind the way toxicity is triggered in a special case, the more likely is the regulatory acceptance of the methods suggested to produce the evidence necessary to proof the presence or absence of a certain Key Event and subsequently the Adverse Outcome.

It is therefore not 100% correct to talk about "regulatory acceptance" of AOPs as such, but about the increased probability of regulatory acceptance and use of 3R methods illustrated in an IATA underpinned by a reliable, trusted AOP:

Confidence in an AOP is therefore crucial for its applicability in the regulatory context; and confidence can only be generated by a broad consensus, which requires a formal lifecycle procedure for each AOP, ideally governed by an international body.

18.3 The AOP Adoption Process at OECD

In order to elicit the knowledge from a broad scientific basis, AOPs are best created in a decentralised crowdsourcing environment, with a strong steering element at a central level. Crowdsourcing is

> the practice of obtaining needed services, ideas, or content by soliciting contributions from a large group of people and especially from the online community rather than from traditional employees or suppliers. (Merriam-Webster 2016)

Such an environment was created by the OECD in 2012: The OECD AOP program encourages stakeholders to contribute and peer-review their AOP knowledge. The relevant ICT tools (AOP-KB, AOP-KB Wiki etc.) are developed in a collaborative fashion between major scientific and regulatory bodies (see Chaps. 1 and 12) and at time of print the AOP-KB Wiki is open to the public for AOP browsing and open to registered users for AOP entering and editing.

This section will describe the workflow of the AOP creation and review, as well as the bodies that interact to support the necessary procedures.

AOPs are ideally created in the AOP-KB Wiki (https://aopwiki.org) from scratch, i.e. no paper or other electronic formats (Word etc.) are necessary; the principle of crowdsourcing facilitates an approach in which even first draft ideas should be exposed to the participating crowd, encouraging commenting, edits, and additions.

The following parties are involved in the typical lifecycle of an AOP:

- The **Working Group of the National Coordinators for the Test Guidelines** (WNT) and the **Task Force for Hazard Assessment** (TFHA) as the OECD bodies that steer the AOP and IATA processes.
- The **OECD Extended Advisory Group on Molecular Screening and Toxicogenomics** (EAGMST), a group of experts from various areas of toxicology, designated by governmental or non-governmental affiliations (academia, agencies, industry, animal welfare groups, scientific societies, etc.). The EAGMST meets twice per year (one face to face meeting and one teleconference) to keep pace with new developments.
- The EAGMST is the governing body of the OECD AOP program and decides about the addition of an AOP to its work program.
- The **Society for the Advancement of AOPs** (SAAOP, http://www.saaop.org/), a group of professionals in the AOP and ICT arena, who are responsible for the hosting, development and maintenance of the AOP-KB Wiki, following guidelines issued by the EAGMST. SAAOP provides summary reports on the status and use of the AOP-KB Wiki for the semi-annual meetings of the EAGMST. These bi-annual reports also provide a summary of all SAAOP responses to any EAGMST requests for information on, or requests for changes to, the AOP-KB Wiki. Figure 18.4 shows the logo of SAAOP.
- The "**crowd**", i.e. the scientific community interested in browsing or contributing AOP knowledge.

Fig. 18.4 SAAOP Logo

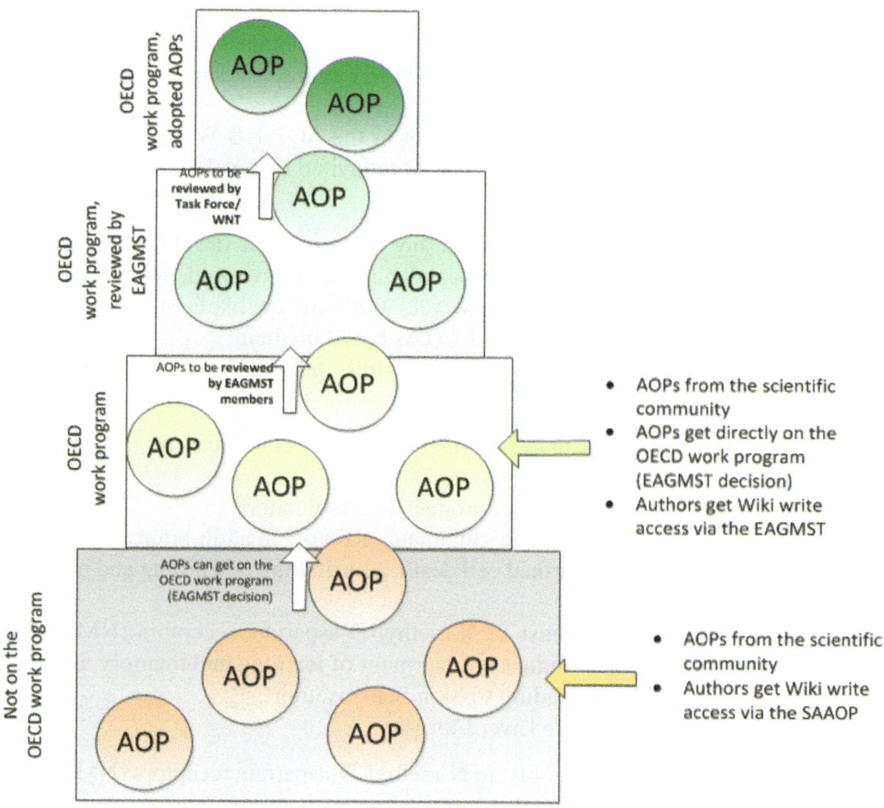

Fig. 18.5 AOP lifecycle

Figure 18.5 shows the typical AOP lifecycle (from bottom to top):
The following rules and principles apply:

- AOPs can get into the AOP-KB Wiki via two channels:
  - Members of the crowd (scientific community) request write access to the AOP-KB Wiki from the SAAOP and start entering their AOP knowledge.

- Members of the crowd (scientific community) request the addition of their AOP to the OECD work program (see http://www.oecd.org/chemicalsafety/testing/adverse-outcome-pathways-molecular-screening-and-toxicogenomics.htm for guidance); write access to the AOP-KB Wiki is automatic.

- AOPs in the lowest tier (not on the OECD work program) can, but do not have to go through the EAGMST procedure to get onto the OECD work program.
- Once an AOP is considered mature enough by its authors, it can undergo an internal review by members of the EAGMST.
- Once the AOP has successfully passed the EAGMST review, it is handed over for external review to experts nominated by the WNT and the TFHA.
- Once an AOP is endorsed by the WNT/TFHA it is considered adopted by the OECD.
- Adopted AOPs are conserved as snapshot in the AOP-KB Wiki for later reference, but the AOP authors and the crowd are invited to develop the AOP further as science evolves.

While AOPs can be used in IATAs at any stage of their development, it is of course recommended to use them especially in their reviewed, ideally in their adopted stage. The further up in their lifecycle AOPs are, the likelier it is that regulatory authorities will be inclined to trust IATAs based on them.

At time of print, the following AOPs are fully adopted and endorsed by the OECD:

- Alkylation of DNA in male pre-meiotic germ cells leading to heritable mutations
- Aromatase inhibition leading to reproductive dysfunction
- Binding of agonists to ionotropic glutamate receptors in adult brain causes excitotoxicity that mediates neuronal cell death, contributing to learning and memory impairment
- Chronic binding of antagonist to N-methyl-D-aspartate receptors (NMDARs) during brain development induces impairment of learning and memory abilities
- Covalent Protein binding leading to Skin Sensitisation
- Protein Alkylation leading to Liver Fibrosis
  - Chronic binding of antagonist to N-methyl-D-aspartate receptors (NMDARs) during brain development leads to neurodegeneration with impairment in learning and memory in aging
  - Androgen receptor agonism leading to reproductive dysfunction

18.4 Method References in the MIE, KE and AO Wiki Entries

In addition to promoting AOP usage via IATAs, an important element in the regulatory acceptance of 3R methods via AOPs is mentioning them explicitly in the AOP-KB Wiki entries for MIEs, KEs and AOs.

Overview table: How it is measured or detected [edit]

Method(s)	Reference	URL	Regulatory Acceptance	Validated	Non Validated
h-CLAT	draft TG under discussion at OECD	[1]		X	
	DB-ALM	[2]			
	EURL ECVAM Recommendation	[3]			
	Ashiga et al., 2015	[4]			
Genomic Allergen Rapid Detection test (GARD)	Johansson et al., 2013	[5]			X
VitroSens	Hooyberghs et al., 2008	[6]			X

Fig. 18.6 Reference to 3R methods in AOP MIE article

Figure 18.6 shows a well written entry for the MIE "Dendritic Cells, Activation" (https://aopkb.org/aopwiki/index.php/Event:398) in the AOP "Covalent Protein binding leading to Skin Sensitisation" (https://aopkb.org/aopwiki/index.php/Aop:40):

The verbal and tabular description of all test methods available to show that a chemical is implicated in an AOP-relevant event is the link between the AOP thinking and its possible application in a regulatory context.

18.5 AOP-XML

While AOPs are normally created in the AOP-KB Wiki, making it possible to create them with any third party tool would further promote the concept. In order to facilitate this, the OECD has started a project to create an AOP-XML standard. XML stands for Extensible Markup Language (XML), which is

> a simple, very flexible text format [...] Originally designed to meet the challenges of large-scale electronic publishing, XML is also playing an increasingly important role in the exchange of a wide variety of data on the Web and elsewhere. (W3C 2015)

Any third party whose application can produce an export file compatible with this AOP-XML Schema would then be able to share its AOPs with the official website. At time of print, the Schema is scheduled to be officially adopted by OECD in late 2017 or early 2018. An import mechanism that will allow importing AOP-XML-formatted files into the AOP-KB Wiki has already been developed and will be published in parallel to the AOP-XML format itself.

With the AOP-XML Schema in place, information exchange and submission of AOPs as part of regulatory dossiers will become easier as regulators would receive AOP information in a format they are familiar with and cumbersome standardising effort would be no longer necessary. The AOP-XML Schema will become part of the OECD series of global standardized formats that aim at making world-wide data exchange between stakeholders a straightforward task. AOP-XML is actually inspired by the OECD Harmonised Templates (OHT, see http://www.oecd.org/ehs/

templates/), which are standard reporting formats for 100+ phys/chem, human health and ecological hazard properties, already now the prescribed way to submit chemical dossiers to regulatory authorities, esp. to the European Chemicals Agency (ECHA) under REACH (Registration, Evaluation, Authorisation and Restriction of Chemicals), which also promotes alternative methods for the hazard assessment of substances in order to reduce the number of tests on animals. Being able to attach AOP information in the AOP-XML format to a chemical dossier will increase the visibility and usefulness of the AOP concept.

18.6 OECD Harmonised Template 201: Intermediate Effects

Consideration and ultimately acceptance of AOP-derived results in regulatory contexts can be further promoted by using a standardised data format to report non-classical test method results. AOP knowledge, if properly captured and presented, leads to a better understanding of toxicity mechanisms, and ultimately the AOP knowledge – derived from testing several chemicals – may be extrapolated to predict the toxicity of all chemicals that trigger the same MIE. Until recently, the absence of a template to report Intermediate Effects (like MIEs and KEs) was a blocking factor.

The OECD had already designed and published 114 OECD Harmonised Templates (OHTs) to report test results concerning:

- physical/chemical properties (e.g. boiling point, density, flammability, …),
- human toxicity (e.g. carcinogenicity, acute toxicity, …)
- environmental toxicity (e.g. aquatic toxicity, terrestrial toxicity, …) and
- other properties describing degradation, accumulation etc.

These templates are geared towards results derived from classical (mostly OECD guideline) studies, focusing on apical endpoints, i.e. Adverse Outcomes.

However, reporting MIEs or KEs with such a classical OHT would tie them inseparably to the one Adverse Outcome the one template covers, which is undesirable, as the (in-vitro, in-silico mechanistic) information is then not easily accessible for building AOPs leading to other Adverse Outcomes: A Key Event can be relevant not only for one AOP, but several. Reporting the Intermediate Effect in an "AO-neutral" template makes the data available for all kinds of AOPs.

A new, AO-neutral OHT was therefore needed that would allow reporting observations from mechanistic (in-vitro and in-silico) tests, without immediately locking into one of several AOs the Intermediate Effect could lead to.

Knowing not only about results of animal tests (classical OHTs), but being able to cross-reference these test results with the intermediate effect observations (new OHT) has the potential to lead the way towards a less animal-centred hazard assessment.

OECD therefore started an initiative to come up with a stable, stakeholder-endorsed OHT for reporting on "intermediate effects" – being observed via in vitro assays and possibly other non-animal test methods (computational predictions etc.). The template was titled “OHT 201 – Intermediate effects”.

OHT 201 was endorsed by the OECD Joint Meeting in 2015 and was finally published in August 2016, see http://www.oecd.org/ehs/templates/harmonised-templates-intermediate-effects.htm for more details.

The basic principle of OHT 201 is that:

- one or several objective observation(s) (= results from non-classical test methods)
- lead(s) to one subjective conclusion (= Intermediate Effect present, yes or no).

A properly filled in OHT 201 template therefore conveys a clear statement:

- Based on Observations O1, O2, …,
- a certain Chemical
- triggers/does not trigger
- a certain Intermediate Effect
- on a certain Biological Level
- at a certain Effect Concentration.

With OHT 201 being implemented in IUCLID (the International Uniform ChemicaL Database, see http://iuclid.eu/) the ICT system used by industry to fulfil reporting obligations under more and more legislative programmes (e.g. REACH), the notion of Intermediate Effects (and implicitly AOPs and predictive toxicology) has started to get attention in the regulatory world. This is a first step towards the acceptance of results from alternative tests for regulatory purposes, with the ultimate goal of replacing in-vivo-centred Adverse Outcome observations with alternative-methods-centred IATA/AOP considerations as the basis for risk assessment.

18.7 Conclusion

AOPs provide mechanistic knowledge on how, when and to what extent biological processes are disturbed so that an Adverse Outcome results. While this mechanistic knowledge is of paramount importance to better understand toxicology, it can only be “accepted” in the regulatory context in combination with testing strategies based on exactly that knowledge. The regulatory acceptance of AOPs will therefore be an indirect one: AOP knowledge (ideally endorsed by the OECD in a dedicated review process) will inform Integrated Approaches to Testing and Assessment (IATA) leading to a mix of (less) classical and (more) non-classical, (more) in-vitro/in-silico and (less) in-vivo test methods fulfilling the needs of any specific hazard or risk assessment case. AOPs and non-classical test results will be stored in OECD-published data exchange and reporting formats, thereby further facilitating their uptake by regulators.

References

EURL ECVAM (2016) https://eurl-ecvam.jrc.ec.europa.eu/validation-regulatory-acceptance. Accessed 9 Aug 2016

OECD (2015) Report of the workshop on a framework for the development and use of integrated approaches to testing and assessment, ENV/JM/MONO(2015)22, Series on testing and assessment no 215

Marie-Jeanne W. A. Schiffelers, Bas J. Blaauboer, Coenraad F. M. Hendriksen, Wieger E. Bakker (2012) Regulatory acceptance and use of 3R models: a multilevel perspective. ALTEX 29(3), 287-300. doi:10.14573

Merriam-Webster (2016) http://www.merriam-webster.com/dictionary/crowdsourcing. Accessed 9 Aug 2016

Tollefsen KE, Scholz S, Cronin MT, Edwards SW, de Knecht J, Crofton K, Garcia-Reyero N, Hartung T, Worth A, Patlewicz G (2014) Applying adverse outcome pathways (AOPs) to support integrated approaches to testing and assessment (IATA). Regul Toxicol Pharmacol 70(3):629–640

W3C (2015) https://www.w3.org/XML/. Accessed 9 Aug 2016

Chapter 19
The Future of Adverse Outcome Pathways: Analyzing their Social Context

Kevin C. Elliott, Cheryl A. Murphy, and Natàlia Garcia-Reyero

Abstract This chapter places the development of adverse outcome pathways (AOPs) in their social context. It begins by highlighting the intense social and political polarization that currently exists around environmental regulations. Given this context, any gaps, assumptions, or uncertainties associated with AOPs are likely to receive intense scrutiny whenever they have regulatory implications that could generate adverse consequences for particular stakeholder groups. Therefore, the chapter argues that in the near future, AOPs are likely to be much more fruitful when they are employed in "win-win" contexts, such as in the design of safer chemicals or the assessment of alternative products and methods. Moreover, AOPs are likely to be more useful and more widely accepted if their development process is characterized by two principles: engagement and transparency. Following these principles has the potential to alleviate some of the conflict that has characterized recent chemical regulatory policy.

19.1 Introduction

The adverse outcome pathway (AOP) framework organizes available biological information along levels of biological organization and has the potential to revolutionize the chemical risk-assessment process by helping to elucidate the causal relationships between molecular level processes (Molecular Initiating Events; MIEs); cellular, organ level, organ system and organism-level processes (generically termed

---

K.C. Elliott (✉)
Lyman Briggs College, Department of Fisheries and Wildlife, and Department of Philosophy, Michigan State University, East Lansing, MI, USA
e-mail: kce@msu.edu

C.A. Murphy
Department of Fisheries and Wildlife, Michigan State University, East Lansing, MI, USA

N. Garcia-Reyero
Environmental Laboratory, US Army Engineer Research & Development Center, Vicksburg, MS, USA

N. Garcia-Reyero, C.A. Murphy (eds.), *A Systems Biology Approach to Advancing Adverse Outcome Pathways for Risk Assessment*,
DOI 10.1007/978-3-319-66084-4_19

Key Events; KEs); and Adverse Outcomes (AOs) that are informative for risk assessment (Garcia-Reyero 2015). The AOP framework was born as a response to the release of a National Research Council report in 2007 entitled *Toxicity Testing in the 21st Century* (NRC 2007). This report envisioned a revolution of toxicity testing, moving from whole-organism testing to a set of in vitro, *in silico*, and pathway-based tools to predict toxicity. This vision was rapidly embraced by the ecotoxicology community and led to the conception of the AOP concept (Ankley et al. 2010).

Currently, risk assessments are based on slow, expensive toxicity tests of individual chemicals (typically administered at high doses) on specific species of experimental animals. In contrast, our social goals would be better served if we could quickly assess the direct and indirect ways in which mixtures of chemicals at realistic doses affect human health and the ecosystem services on which we depend (Forbes and Calow 2012). Furthermore, there is ever-increasing social and ethical pressure to minimize the use of animals for toxicity testing due to concerns about animal welfare ("Animal Rights and Wrongs" 2011). Therefore, AOPs could potentially give regulators the causal understanding to predict the effects of chemicals at multiple levels of biological organization without performing slow, expensive, socially and ethically controversial, and sometimes unreliable animal tests.

Despite this potential, AOPs face a number of challenges. Perhaps most importantly, it is not obvious how the analysis of causal pathways at the molecular level can reliably predict emergent toxic properties at much higher biological levels that arise from extremely complex interactions and feedback processes (Forbes and Calow 2012). This has the potential to limit the usefulness of AOPs in the context of risk assessment, because stakeholders are likely to exploit these uncertainties to cast doubt on results that conflict with their interests (Sarewitz 2004). Of course, our knowledge is constantly evolving and improving, and partial AOPs can still generate valuable information for regulatory purposes (Perkins et al. 2015). Nevertheless, failure to recognize the polarized social context surrounding chemical regulations and its implications for the development of AOPs has the potential to generate conflict and prevent them from reaching their full potential.

This chapter explores the social context surrounding AOPs in order to determine how they can be developed in a manner that is most likely to yield fruitful results. Section 19.2 highlights the polarization and conflict that surrounds contemporary chemical regulations. Given this context, Sect. 19.3 argues that in the near future AOPs are likely to be most useful in "win-win" settings where stakeholders do not have incentives to emphasize their uncertainties and limitations. Section 19.4 proposes that the quality and social utility of AOPs is most likely to be maximized if the development process is characterized by efforts at engagement and transparency. In sum, the chapter emphasizes that AOPs are being developed in a very sensitive social and political context, and therefore they are much more likely to play a fruitful role in future regulatory policy if those developing AOPs remain cognizant of that context.

19.2 The Setting

To appreciate how AOPs are likely to be received by different stakeholders, it is important to consider the current social context surrounding chemical regulations in the United States and around the world. One of the central features of this social context is *conflict*. Since at least the time of Rachel Carson, industry and citizen groups have clashed over the assessment of risks associated with industrial chemicals. A number of books have recently appeared with titles like *Doubt Is Their Product* (Michaels 2008), *Merchants of Doubt* (Oreskes and Conway 2010), *Global Spin* (Beder 2000), and *Deceit and Denial* (Markowitz and Rosner 2002). These books express the concern that the chemical industry has engaged in many of the same strategies as the tobacco industry, seeking to manufacture public uncertainty about the harmfulness of its products.

In *Merchants of Doubt*, for example, historians Naomi Oreskes and Erik Conway argued that many of the same scientists, institutions, and strategies appeared over and over in a number of different regulatory conflicts over the past 60 years. When the tobacco industry began to face evidence that their products were harmful during the 1950s, they enlisted the influential public relations (PR) firm Hill and Knowlton to help them craft a plan for responding. They developed a number of strategies for deflecting public attention from the harmfulness of tobacco: (1) funding "decoy" research designed to avoid obtaining evidence that tobacco was harmful; (2) attacking scientists who publicized evidence about the health effects of tobacco; (3) withholding research findings that put tobacco in an unfavorable light; and (4) casting doubt on studies that appeared to provide evidence that tobacco was harmful (Elliott 2016). The tobacco companies enlisted very influential scientists, including Fred Seitz, a former president of the National Academy of Sciences and Rockefeller University, to administer grant programs on their behalf and to launch organizations like The Advancement of Sound Science Coalition (TASSC) (Oreskes and Conway 2010). Many of these same scientists and organizations, including TASSC and Hill and Knowlton, were later active in casting doubt on the scientific evidence for numerous other environmental and public-health hazards, including acid rain, the ozone hole, industrial chemicals, fast food, and climate change (Elliott 2016; Oreskes and Conway 2010).

These activities have contributed to public skepticism about the chemical industry. In his book *Doubt Is Their Product,* David Michaels (2008) argues that the industry has employed a variety of the same questionable strategies as the tobacco industry in an effort to defend harmful chemicals like lead, asbestos, chromium, benzene, beryllium, and dioxin. Recently, bisphenol-A (BPA) has become a hot-button subject of controversy, because independently funded studies appear to identify a number of harmful effects that have not been found in industry-funded studies (Myers 2009; Vandenburg and Prins 2016). Those associated with the industry-funded research have responded that their studies were performed according to the standardized test guidelines approved by the Organization for Economic Cooperation and Development (OECD) and other regulatory bodies (Tyl 2009). But critics of the industry-funded studies argue that they had design flaws and that the standardized protocols do not incorporate cutting-edge techniques for identifying effects from endocrine disrupting chemicals (Myers et al. 2009; Vandenburg and Prins 2016).

These cases illustrate the considerable distrust that has arisen between different stakeholders over risk assessments of industrial chemicals. Industry contends that environmental organizations and citizen groups are quick to call for regulating and banning chemicals based on preliminary data and without paying adequate attention to the costs of regulations. Critics of industry-funded studies argue that those studies are often designed in ways that minimize the potential for finding harmful effects (Myers et al. 2009). They also contend that industry pays product-defense companies to perform questionable re-analyses of independently-funded studies in order to "manufacture uncertainty" about evidence that their products are harmful (Michaels 2008). As a result, the regulatory process has been slowed to a virtual standstill by litigation and conflict (Cranor 2011).

One of the motivations behind the development of AOPs is to speed up the regulatory process and alleviate this gridlock. By facilitating the use of in vitro and *in silico* techniques, the use of AOPs has the potential to generate results much more quickly and cheaply than with animal or epidemiological studies. Moreover, by uncovering underlying mechanisms, they could help regulators to address non-monotonic dose responses and chemical mixtures. Nevertheless, this potential could easily be lost if AOPs become subject to the same controversy that has plagued the existing chemical regulatory process.

Literature in the field of science and technology studies (STS) has shown that when there are high stakes surrounding scientific information, any uncertainties or assumptions or methodological concerns become subject to intense scrutiny and debate (Funtowicz and Ravetz 1992; Hackett et al. 2007; Sarewitz 2004). At the early stages of their development, AOPs will be subject to a host of uncertainties and assumptions, including questions about the quality of the data underlying them, the causal factors linking key events (KEs) in AOPs, the relationships among multiple AOPs in networks, the potential for variations in effects across species and members of species, the effects of feedback loops within and between levels of biological organization, the linear construct of AOPs, and other factors. While existing whole-animal tests are also fraught with uncertainties (Greek and Menache 2013; Hartung 2013), companies are still likely to fear that the use of new in vitro and *in silico* methods could come back to "bite" them if they are not as predictive as expected. In contrast, many animal-welfare organizations strongly support the use of the new methods, but this is unlikely to ease the concerns of many scientists who already feel threatened by these organizations and question their motives ("Animal Rights and Wrongs" 2011).

Because of these many uncertainties and accompanying values, the potential for AOPs to help alleviate gridlock could easily be lost if they are employed in ways that aggravate existing conflicts over chemical regulation and risk assessment. For example, if an AOP or a network of AOPs were to serve as the primary basis for arguing that particular chemicals should be subject to increased or decreased regulation, stakeholders opposed to the decisions would be likely to highlight the

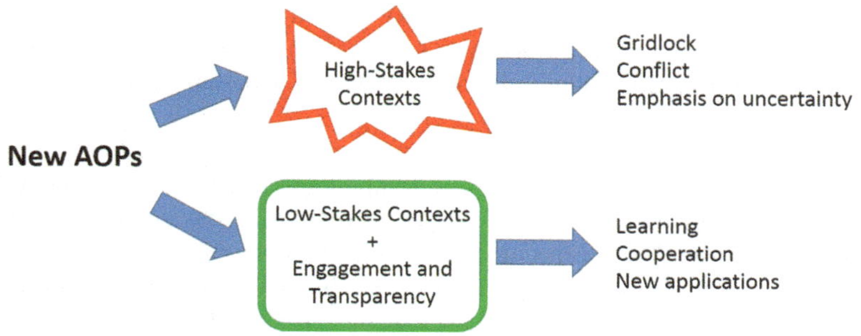

Fig. 19.1 An illustration of how new AOPs are likely to be received differently, depending on whether they are applied to high-stakes contexts (e.g., contentious risk assessments) or low-stakes contexts (e.g., chemical prioritization or alternatives assessment)

limitations of AOPs in excruciating detail (Sarewitz 2004). While it could be fruitful in some contexts to clarify the strengths and weaknesses of these techniques, it is unlikely that the criticisms that they would receive in a polarized regulatory environment would actually serve a fruitful purpose. Instead, the criticisms would simply contribute to ongoing gridlock and obfuscation. Therefore, it would be wise to develop AOPs in a manner that minimizes the potential for them to become subject to intense social controversy (see Fig. 19.1). The next section considers contexts in which they could be developed and implemented without generating significant conflict, and the following section proposes principles for developing AOPs in a manner that is likely to maximize their usefulness and acceptance.

19.3 The Importance of Context

Given the highly conflicted social setting surrounding the development of AOPs and the uncertainties associated with them, they are initially likely to be most useful in contexts where stakeholders do not have significant incentives to create gridlock by questioning their results. In other words, the best contexts for employing AOPs in the near future are likely to be those in which they generate "win-win" scenarios or at least those in which no stakeholders have a great deal to lose. For example, there are increasing calls for industry to find ways to engineer "safety by design" into new chemicals (Colvin 2003; McDonough and Braungart 2002). Many contemporary conflicts over chemical regulations occur because industry is threatened with having to pull very profitable products from the market. But if better techniques were available for predicting which chemical classes or structures were least likely to display toxic properties, companies could choose to focus on developing chemicals that were less likely to be toxic. This would be beneficial to all stakeholders, so AOPs would receive widespread support if they could assist in this process of designing safer products. As AOPs are developed, they will vary in terms of their scientific completeness and the uncertainty associated with all their components (e.g., the characteristics of specific KEs), but even relatively incomplete AOPs can still provide valuable information about correlations between MIEs and AOs (Perkins et al. 2015).

Closely related to the goal of designing safer products is the practice of "alternatives assessment," in which chemicals are compared to find the safest ones that can be used to perform a given task. Performing alternatives assessment is another way to generate "win-win" scenarios in which industry finds new chemicals that are comparable in terms of cost and convenience but that have less toxic properties than previous ones. For example, the state of Massachusetts passed a Toxic Use Reduction Act in 1989 that required companies to perform alternatives assessment. It created a list of chemicals that raised health concerns, but rather than banning them it required companies using large quantities of the chemicals to report their annual usage and to investigate potential alternatives. In the years following the legislation, companies found that in many cases they could actually save money by using alternative chemicals that were less toxic (Tickner 1999, 2003). The use of AOPs could potentially help to provide information about alternatives that would be difficult or much slower to obtain otherwise. For example, when an MIE or KE has already been linked to an AO, it can facilitate the development of high-throughput or in vitro assays to identify compounds that are likely to generate the AO and alternatives that are less likely to do so (Perkins et al. 2015). In settings like these, there would be relatively few incentives to challenge the results of AOPs unless the costs of safer alternatives were much higher than those of other compounds.

Another arena in which AOPs could prove to be very helpful, even in the early stages of their development, is in the prioritization of chemicals for further testing (Groh et al. 2014). There are currently about 80,000 industrial chemicals in production, and little is known about the toxicity of most of those chemicals, because they do not fall in the categories of pesticides or pharmaceuticals that have been subject to an extensive pre-market regulatory system (Cranor 2011). Therefore, as regulatory agencies strive to catch up with this massive backlog, they are faced with difficult decisions about which chemicals should be studied most intensively at present and which ones are not as important to investigate in the near future. This is another arena in which there is relatively little to lose by employing AOPs to accelerate and improve the prioritization process. For industry, there is much less to lose when chemicals are chosen for further testing than when they are threatened with actual regulation of a chemical. And for citizens and environmental groups, efforts to improve and accelerate the prioritization process are appealing, even if they are somewhat imperfect.

One might challenge our recommended focus on designing safer chemicals, assessing alternatives, and prioritizing chemicals for further testing by insisting that it prevents AOPs from meeting their full regulatory potential. From this critical perspective, the "holy grail" is for AOPs to be used in performing risk assessments that can be accepted for regulatory purposes, and the concern is that we are selling AOPs short by calling for more modest uses. Our response is that by starting with these modest uses, we "thread the needle" between two potential dangers. On one hand, if AOPs are introduced into the regulatory environment too aggressively, they risk being caught up in the political debates discussed in Sect. 19.2, which could significantly impair their future acceptance. On the other hand, the future usefulness of AOPs could be severely compromised if regulators do not play a role in their development, and it seems wasteful to avoid using AOPs until they are fully developed (Garcia-Reyero 2015). Our proposals allow regulators to begin using AOPs in ways that are relatively unlikely to generate conflict, thereby exploring the power of AOPs to play a role in more politically controversial risk assessments in the future.

As scientists and policy makers explore different uses for AOPs, it may be helpful for them to employ weight of evidence (WoE) approaches to help guide their decision making. A WoE evaluation documents the level of confidence in extrapolation beyond the empirical data. Building on the Bradford-Hill considerations (Hill 1965), WoE approaches have already been applied to AOP development, with the criteria of biological plausibility, essentiality, and empirical evidence applied to each AOP key event relationship (Becker et al. 2015). There are qualitative approaches to assessing level of confidence in a key event relationship (very strong, strong moderate, weak, very weak), but there are also quantitative weight of evidence approaches. For example, an expert panel could develop and use specific criteria to apply weights and scores to individual lines of evidence for each key event relationship within an AOP. These criteria would evaluate the mechanistic relationships between key events, using evidence of a downstream key event being impaired if an upstream key event is blocked and the consistency across a wide range of taxa and stressors for which the key event(s) occur as supporting information. Mathematical or statistical models would be developed to evaluate the weights and scores to determine the strength in the confidence of the AOP key event relationship (Becker et al. 2015). The quantitative approach is attractive because it has the potential to enhance the transparency and reproducibility of AOP WoE determinations on key event relationships.

19.4 Principles for Future Development

In addition to being strategic about the contexts in which AOPs are employed, another important step for increasing their future success is to let two principles guide their development: engagement and transparency. The STS literature highlights the importance of both principles. In the past two decades, there has been a backlash against the "deficit model" of the public's understanding of science. In the latter part of the twentieth century, many figures working in science policy acted as if most public opposition to science and technology could be attributed to a deficit or lack of scientific understanding on the part of the public. A report published by the British Royal Society (1985), *The Public Understanding of Science,* is often cited as an example of this perspective. In response to this view, many scholars have pointed out that public opposition can stem from a number of different factors, including concerns about the social impacts or risks or values associated with particular lines of research and technological innovation (see e.g., Elliott 2017; Wynne 2005).

Given that stakeholders can have legitimate concerns about the background assumptions or values or social impacts associated with a line of research, science policy makers have become increasingly concerned to incorporate broad engagement among multiple stakeholders in decision making (Elliott 2011, 2017; Guston 2008, 2014; NRC 1996). This engagement can take a variety of different forms. It can involve surveys, focus groups, or public comment periods, all of which provide opportunities for the public to inform decision makers. It can also involve various types of public meetings in which experts educate the public. Perhaps most interesting are recent efforts to promote "two-way" engagement, in which technical experts and citizens can exchange information. For example, consensus conferences, citizens' juries, and citizen advisory committees typically employ small groups of citizens that are able to hear from experts and become educated about an issue before providing feedback to policy makers.

Other approaches involve participatory collaborative modeling, which uses the structure of system dynamics to involve multiple stakeholders. System dynamics simulation models can offer significant benefits to public involvement processes. Various stakeholders can participate in virtual worlds to allow them to learn how complex systems work, by modifying different aspects of the simulation and contributing to the model building process (Sterman 1994; Stave 2002). In fact, when participatory systems modeling is involved in group decision processes (when compared to traditional group facilitation process), participants appear to have better structured discussions, stronger mental models, and reach sound decisions faster (Dwyer and Stave 2008).

A number of justifications can be given for these efforts at promoting engagement in the arena of science policy (Fiorino 1990). One justification is democratic: people have rights to be involved in decision-making processes that have the potential to impact them significantly (Elliott 2009; Shrader-Frechette 1995). Another justification is that science and technology are likely to be more effective and useful when they are guided by people with a wide range of expertise. Importantly, this expertise can involve the "local knowledge" of citizens who do not have a great deal of technical expertise but who have important on-the-ground information about how scientific and technical developments are likely to be used and received (Elliott 2009, 2017; Wynne 1989). A third justification for engagement is instrumental: new scientific and technological advances are often more likely to be accepted if a wide range of stakeholders have had a hand in developing them (Fiorino 1990).

The potential for scientific and technical advances to be more widely accepted when multiple stakeholders are involved in developing them is especially important to consider in the case of AOPs, given the tense political environment surrounding chemical regulations described in Sect. 19.2 of this chapter. There have been noteworthy cases where collaborative research efforts eased conflicts among stakeholders that were otherwise highly suspicious of one another. For example, in the wake of the Exxon Valdez oil spill, local citizens were highly suspicious of the petroleum industry. In an effort to alleviate this tension, a well-respected citizen group collaborated with the petroleum industry to perform a risk assessment of different approaches for moving barges through Prince William Sound (Busenberg 1999; Douglas 2005). Similarly, when evidence emerged that the pollen from genetically engineered Bt corn might be harming Monarch butterfly populations, a collaborative research effort involving representatives from industry, NGOs, academia, and government yielded high-quality findings that helped to resolve the controversy (Pew Initiative 2002). Of course, collaborative research projects do not always work out so well. Recent collaborative efforts to study methane emissions from hydraulic fracturing sites and to study the effects of neonicotinoid insecticides on bee hives have had limited success (Kleinman and Suryanarayanan 2015; Song and Bagley 2015). But this does not mean that multi-stakeholder engagement should be abandoned; rather, it is important to reflect further about how collaborations can be developed in ways that maintain trust and confidence in the legitimacy of the process (Elliott 2011).

So far, the development of AOPs has proceeded in a manner that captures many of the goals of engagement, not only in expertise and goals, but also in space, as AOPs are now a worldwide movement and effort. For instance, several of the workshops that have been crucial in the development of AOPs included participants from industry, academia, government, and NGOs, both US-based and international. As discussed in Garcia-Reyero (2015), a recent workshop held in 2014 (http://www.saaop.org//workshops/somma.html) that focused on the use and development of AOPs for regulatory applications not only brought together participants and sponsors from industry, NGOs, government, and academia (all both US-based and international), but also brought together human health and eco-health experts. This pioneering effort led to other workshops and efforts that maintained this set of principles on international and organizational collaboration, which spurred the development of AOPs for both human health and ecological assessment (e.g., "Adverse Outcome Pathways: From Research to Regulation," held in September 2014 in Bethesda, MD, sponsored by the National Toxicology Program; https://ntp.niehs.nih.gov/pubhealth/evalatm/3rs-meetings/past-meetings/aop-wksp-2014/index.html; SETAC Pellston Workshop: Advancing the Adverse Outcome Pathway Concept – An International Horizon Scanning Approach, April 2017).

One way in which the engagement process surrounding AOPs could potentially be improved would be to incorporate greater citizen involvement in the development process, perhaps through participatory modeling or other collaborative strategies. Ultimately, if AOPs are to be useful for risk assessments, environmental groups and other citizen organizations need to feel comfortable with their reliability. It would behoove those developing AOPs to take steps to identify the concerns that these groups might have and to explore ways to address them. A preliminary step along these lines would be to incorporate members of environmental or public-health-oriented NGOs in working groups that are developing AOPs.

Fortunately, these efforts are already underway. Several non-profit organizations, such as People for the Ethical Treatment of Animals (PETA), Physicians Committee for Responsible Medicine, the Humane Society of the United States, and ILSI Health and Environmental Science Institute, are working very closely with government agencies and regulators on alternative methods and AOP development, as well as communicating these efforts to educate and inform society. Eventually, it would also be valuable to initiate a focus group or consensus conference in which citizens could discuss their perspectives on AOPs. Some international conferences such as the Society of Environmental Chemistry and Toxicology (SETAC), the Society of Toxicology (SOT), and the American Society for Cellular and Computational Toxicology (ASCCT) annual meetings have already been heavily involved in AOP development and acceptance, with many talks, sessions, and training opportunities related to the AOP concept. Increasing international efforts in these directions and including non-scientist citizens would continue to give regulators an advance sense of major issues that need to be addressed in order to implement AOPs in a regulatory context.

In addition to engagement, transparency is a second crucial principle that should guide the development of AOPs. Efforts to promote transparency can encompass a number of different practices (Nosek et al. 2015). Some important practices include the public sharing of data, materials, and analytical techniques (e.g., Soranno et al. 2015). Other important practices that promote transparency include acknowledging crucial assumptions or judgments that underlie scientific interpretations or methodologies (Elliott 2017; Elliott and Resnik 2014). A related practice is the disclosure of relationships or conflicts of interest (COIs) that could influence one's judgment. Psychological research indicates that conflicts of interest can have significant effects on our reasoning, and many journals have recently taken steps to require disclosure of these interests in published articles (Elliott and Resnik 2015). Because of concerns about maintaining proprietary control over intellectual property, industry sometimes faces challenges with transparency (e.g., with regard to the data underlying safety studies). Despite these limitations, it is still important to pursue as much transparency as possible in all sectors.

Transparency is important for several reasons. First, it helps to facilitate effective engagement. If all stakeholders do not have access to the data and methods underlying scientific developments, they are not able to contribute effectively to subsequent deliberations about them (Soranno et al. 2015). Transparency is also crucial for promoting the quality of scientific work. Scientists and policy makers are currently very concerned that a great deal of scientific research appears not to be consistently reproducible (Alberts et al. 2015). Transparency can help alleviate this concern. By making key data, assumptions, methods, and COIs known, it allows for greater opportunities to analyze and evaluate the quality of results. Philosophers of science argue that meaningful criticism by scientists with an adequate range of perspectives is central to scientific objectivity (Douglas 2004; Elliott 2017; Longino 2002).

Fortunately, the research community appears to be taking very promising steps to promote transparency concerning research on AOPs. A clear example is the AOP-Knowledge Base (AOP-KB; http://aopkb.org/) project, a collaborative effort between the Organization for Economic Cooperation and Development (OECD), the US Environmental Protection Agency (EPA), the European Commission Joint Research Center (JRC), and the US Army Engineer Research & Development

Center (ERDC). The AOP-KB involves several modules related to AOP development, such as Effectopedia, AOP-Xplorer, and the AOP Wiki, a repository for AOPs that will be eventually approved by OECD as official AOPs when sufficient confidence and evidence is provided. The OECD launched the AOP Development programme in 2012 under the umbrella of the OECD Advisory Group on Molecular Screening and Toxicogenomics, which includes members from many OECD countries. As part of this programme, they published the Guidance Document on Developing and Assessing Adverse Outcome Pathways (2013). This programme has played a key role within AOP development, harmonizing international efforts and promoting transparency of both the data and the process. In order to maximize the future acceptability of AOPs in the regulatory context, it will be important to continue making the data and methods underlying AOPs publicly available and to be forthright about the assumptions and uncertainties associated with them.

In keeping with the principles of engagement and transparency, the development of AOPs could provide a unique opportunity to experiment with new approaches for performing chemical risk assessments. Recent initiatives along these lines include the EPA's ToxCast program that developed high throughput, in vitro, automated chemical screening technology that allows for rapid screening for changes in biological activity in thousands of assays in response to chemical exposure (Huang et al. 2016). Currently, the key safety studies that inform regulatory risk assessments are almost always performed either by the chemical industry or by contract research organizations (CROs) that are paid by industry. As discussed in Sect. 19.2, this creates financial conflicts of interest that contribute to significant public skepticism about the study results. A number of figures have recently argued that these concerns could be alleviated by creating a separate institute within a federal agency like the US Environmental Protection Agency (EPA) or an international body like the Organization for Economic Cooperation and Development (OECD) (see e.g., Krimsky 2003; Volz and Elliott 2012). This institute could be responsible for contracting out chemical safety studies to academic labs or CROs. Even though the institute would presumably be funded by taxes or fees assessed to the chemical industry, it would still lessen COIs associated with chemical safety studies, because the labs performing the studies would no longer be paid directly by industry. In some cases, the institute could also facilitate collaborative efforts in which multiple stakeholders work together to design research projects on emerging issues.

While this proposal would be politically very difficult to implement at present, the development of AOPs and the advancement of the ToxCast program could provide greater opportunities to implement something along these lines. The use of AOPs could streamline the process of risk assessment so that it would be less resource intensive and more efficient. As a result, it might be feasible to perform more of the necessary research within federal agencies rather than in external laboratories. Those developing AOPs should look for ways to facilitate this process, given its potential to lessen the polarization that has plagued recent regulatory decision making.

19.5 Conclusion

AOPs have great potential to move chemical risk assessment forward in ways that lessen costs, improve animal welfare, increase efficiency, and generate more realistic predictions of the toxic effects that are likely to be observed on endpoints that matter for regulatory purposes. Nevertheless, they also face a number of significant challenges, given the difficulties of moving from causal relationships at the molecular level to emergent toxic properties at much higher levels of biological organization. As a result, their usefulness could be seriously hampered by the extreme polarization that has characterized recent chemical regulatory policy. In order to alleviate this polarization, we recommend, at least in the near term, that AOPs be employed in "win-win" contexts such as the design of safer chemicals and the assessment of alternatives. This would enable a range of stakeholders to begin exploring and improving AOPs in settings where they do not have incentives to reject the findings out of hand. Furthermore, the usefulness and social acceptance of AOPs are likely to be increased if those developing them strive to keep promoting two principles: engagement and transparency. By doing so, they will provide opportunities for multiple stakeholders to influence their development, thereby lessening subsequent polarization.

References

Alberts B, Cicerone R, Fienberg S, Kamb A, McNutt M, Nerem R, Schekman R et al (2015) Self-correction in science at work. Science 348:1420–1422

Animal Rights and Wrongs (2011) Animal Rights and Wrongs. Nature 470(7335):435

Ankley GT, Bennett RS, Erickson RJ, Hoff DJ, Hornung MW, Johnson RD, Mount DR, Nichols JW, Russom CL, Schmieder PK et al (2010) Adverse outcome pathways: a conceptual framework to support ecotoxicology research and risk assessment. Environ Toxicol Chem 29:730–741

Becker RA, Ankley GT, Edwards SW, Kennedy SW, Linkov I, Meek B, Sachana M, Segner H, Van Der Burg B, Villeneuve DL, Watanabe H, Barton-McLaren TS (2015) Increasing scientific confidence in adverse outcome pathways: application of tailored Bradford-Hill considerations for evaluating weight of evidence. Regul Toxicol Pharmacol 72:514–537

Beder S (2000) Global spin, revised edn. Chelsea Green, White River Junction

Busenberg G (1999) Collaborative and adversarial analysis in environmental policy. Pol Stud 32:1–11

Colvin V (2003) The potential environmental impact of engineered nanoparticles. Nat Biotechnol 21:1166–1170

Cranor C (2011) Legally poisoned: how the law puts us at risk from toxicants. Oxford University Press, New York

Douglas H (2004) The irreducible complexity of objectivity. Synthese 138:453–473

Douglas H (2005) Inserting the public into science. In: Maasen S, Weingart P (eds) Democratization of expertise? Exploring novel forms of scientific advice in political decision making. Springer, Berlin, pp 153–169

Dwyer M, Stave K (2008) Group model building wins: the results of a comparative analysis. In: Proceedings of the 26th international conference of the system dynamics society, Athens, Greece

Elliott K (2009) Respect for lay perceptions of risk in the hormesis case. Hum Exp Toxicol 28:21–26

Elliott K (2011) Is a little pollution good for you? Incorporating societal values in environmental research. Oxford University Press, New York

Elliott K (2016) Environment. In: Angulo AJ (ed) Miseducation: a history of ignorance making in America and abroad. Johns Hopkins University Press, Baltimore, pp 96–119

Elliott K (2017) A tapestry of values: an introduction to values in science. Oxford University Press, New York

Elliott K, Resnik D (2014) Science, policy, and the transparency of values. Environ Health Perspect 122:647–650

Elliott K, Resnik D (2015) Scientific reproducibility, human error, and public policy. Bioscience 65:5–6

Fiorino D (1990) Citizen participation and environmental risk: a survey of institutional mechanisms. Sci Technol Hum Values 15:226–243

Forbes V, Calow P (2012) Promises and problems for the new paradigm for risk assessment and an alternative approach involving predictive systems models. Environ Toxicol Chem 31:2663–2671

Funtowicz S, Ravetz J (1992) Three types of risk assessment and the emergence of post-normal science. In: Krimsky S, Golding D (eds) Social theories of risk. Praeger, Westport, pp 251–274

Garcia-Reyero N (2015) Are adverse outcome pathways here to stay? Environ Sci Technol 49:3–9

Greek R, Menache A (2013) Systematic reviews of animal models: methodology versus epistemology. Int J Med Sci 10(3):206–221

Groh K, Carvalho R, Chipman J, Denslow N, Halder M, Murphy C, Roelofs D, Rolaki A, Schirmer K, Watanabe K (2014) Development and application of the adverse outcome pathway framework for understanding and predicting chronic toxicity: II. A focus on growth impairment in fish. Chemosphere 120:778–792

Guston D (2008) Innovation policy: not just a jumbo shrimp. Nature 454:940–941

Guston D (2014) Building the capacity for public engagement with science in the United States. Public Underst Sci 23:53–59

Hackett E, Amsterdamska O, Lynch M, Wajcman J (eds) (2007) The handbook of science and technology studies. MIT Press, Cambridge, MA

Hartung T (2013) Look back in anger – what clinical studies tell us about preclinical work. ALTEX 30(3):275–291

Hill AB (1965) The environmental and diseases: association or causation? Proc R Soc Med 58:295–300

Huang R, Xia M, Sakamuru S, Zhao J, Shahane SA, Attene-Ramos M, Zhao T, Austin CP, Simeonov A (2016) Modelling the Tox21 10K chemical profiles for in vivo toxicity prediction and mechanism characterization. Nat Commun 7:10425

Kleinman D, Suryanarayanan S (2015) Ignorance and industry: agrichemicals and honey bee deaths. In: Gross M, McGoey L (eds) Routledge international handbook of ignorance studies. Routledge, New York

Krimsky S (2003) Science in the private interest. Rowman and Littlefield, Lanham

Longino H (2002) The fate of knowledge. Princeton University Press, Princeton

Markowitz G, Rosner D (2002) Deceit and denial: the deadly politics of industrial pollution. University of California Press, Berkeley

McDonough W, Braungart M (2002) Cradle to cradle: remaking the way we make things. North Point Press, New York

Michaels D (2008) Doubt is their product: how Industry's assault on science threatens your health. Oxford University Press, New York

Myers JP, vom Saal F, Akingbemi B, Arizono K, Belcher S, Colborn T et al (2009) Why public health agencies cannot depend upon good laboratory practices as a criterion for selecting data: the case of bisphenol A. Environ Health Perspect 117:309–315

Nosek B, Alter G, Banks C, Borsboom D, Bowman S, Breckler S, Buck S et al (2015) Promoting an open research culture. Science 348:1422–1425

NRC (National Research Council) (1996) Understanding risk: informing decisions in a democratic society. National Academies Press, Washington, DC

NRC (National Research Council) (2007) Toxicity testing in the 21st century: a vision and a strategy. National Academies Press, Washington, DC

OECD Guidance document on developing and assessing adverse outcome pathways (2013) In: Series on testing and assessment, Paris ENV/JM/MONO(2013)6, Paris no. 184, p 145

Oreskes N, Conway E (2010) Merchants of doubt: how a handful of scientists obscured the truth on issues from tobacco smoke to global warming. Bloomsbury Press, New York

Perkins, E., P. Antczak, L. Burgoon, F. Falciani, N. Garcia-Reyero, S. Gutsell, G. Hodges, et al. 2015. Adverse Outcome Pathways for Regulatory Applications: Examination of Four Case Studies with Different Degrees of Completeness and Scientific Confidence. Toxicological Sciences 148:14–25

Pew Initiative on Food and Biotechnology (2002) Three years later: genetically engineered corn and the Monarch butterfly controversy. Available at: http://www.pewtrusts.org/uploadedFiles/wwwpewtrustsorg/Reports/Food_and_Biotechnology/vf_biotech_monarch.pdf. Accessed 26 July 2013

Royal Society (1985) The public understanding of science. Royal Society, London

Sarewitz D (2004) How science makes environmental controversies worse. Environ Sci Policy 7:385–403

Shrader-Frechette K (1995) Evaluating the expertise of experts. Risk: Health, Safety & Environment 6:115–126

Song L, Bagley K (2015) EDF sparks mistrust, and admiration, with its methane research. InsideClimate News (April 8): https://insideclimatenews.org/news/07042015/edf-sparks-mistrust-and-admiration-its-methane-leaks-researchnatural-gas-fracking-climate-change

Soranno P, Cheruvelil K, Elliott K, Montgomery G (2015) It's good to share: why environmental scientists' ethics are out of date. Bioscience 65:69–73

Stave KA (2002) Using system dynamics to improve public participation in environmental decisions. Syst Dyn Rev 18(2):139–167

Sterman JD (1994) Learning in and about complex systems. Syst Dyn Rev 10(2,3):291–330

Tickner J (1999) A map toward precautionary decision making. In: Raffensperger C, Tickner J (eds) Protecting public health and the environment: implementing the precautionary principle. Island Press, Washington, DC, pp 162–186

Tickner J (ed) (2003) Precaution, environmental science, and preventive public policy. Island Press, Washington, DC

Tyl R (2009) Basic exploratory research versus guideline-compliant studies used for hazard evaluation and risk assessment: bisphenol a as a case study. Environ Health Perspect 117:1644–1651

Vandenburg LN, Prins GS (2016) Clarity in the face of confusion: new studies tip the scales on Bisphenol A (BPA). Andrology. doi:10.1111/andr.12219

Volz D, Elliott K (2012) Mitigating conflicts of interest in chemical safety testing. Environ Sci Technol 46:7937–7938

Wynne B (1989) Sheep farming after Chernobyl. Environment 31:10–39

Wynne B (2005) Risk as globalizing 'democratic' discourse? Framing subjects and citizens. In: Leach M, Scoones I, Wynne B (eds) Science and citizens: globalization and the challenge of engagement. Zed Books, London, pp 66–82

Index

N. Garcia-Reyero, C.A. Murphy (eds.), *A Systems Biology Approach to Advancing Adverse Outcome Pathways for Risk Assessment*,
DOI 10.1007/978-3-319-66084-4

D

E

H

N

O

The manufacturer's authorised representative in the EU is Springer Nature Customer Service Centre GmbH, Europaplatz 3, 69115 Heidelberg, Germany. If you have any concerns regarding our products, please contact ProductSafety@springernature.com

Printed and bound by CPI Group (UK) Ltd, Croydon, CR0 4YY
30/09/2026
02225238-0013